Applied Mathematical Sciences

Founding Editors
Fritz John, Joseph Laselle and Lawrence Sirovich

Volume 185

For further volumes:
http://www.springer.com/series/34

Todd Kapitula • Keith Promislow

Spectral and Dynamical Stability of Nonlinear Waves

Springer

Todd Kapitula
Department of Mathematics
and Statistics
Calvin College
Grand Rapids, Michigan
USA

Keith Promislow
Department of Mathematics
Michigan State University
East Lansing, Michigan
USA

ISSN 0066-5452
ISBN 978-1-4939-0187-6 ISBN 978-1-4614-6995-7 (eBook)
DOI 10.1007/978-1-4614-6995-7
Springer New York Heidelberg Dordrecht London

Mathematics Subject Classification (2010): 34Lxx, 35Pxx, 35Qxx, 37Kxx, 37Lxx, 47B15, 47B25, 47Dxx, 47Exx, 47Fxx, 70Hxx, 70Kxx

Printed on acid-free paper

Springer is part of Springer Science+Business Media (www.springer.com)

Todd Kapitula: *I am grateful to my colleagues, both here at the college and elsewhere, for their support and encouragement during this endeavor. I am especially thankful to my wife, Laura, for her unwavering support during the seemingly endless rounds of revisions. Finally, I wish to thank Keith for his boundless enthusiasm for, and perseverance with, this project.*

Keith Promislow: *I thank Ronay, Sylvia, Isaac, and Levi for tolerating my absence during the writing of this book, and Todd for tolerating my presence. I am grateful to Roger Temam for teaching me the value of writing books.*

Foreword

Waves, patterns, and other permanent structures have always played a pivotal role in applied mathematics. It is the proverbial no-brainer that they should, and also that their stability to perturbations is an issue of importance. This is the first full-scale book devoted to the stability theory, developed over the past few decades, which lies at the interface of dynamical systems and functional analysis. While functional analysis provides the framework and affords the proper posing of questions, it is dynamical systems that give many of the answers.

The Evans function provides the key bridge between these two theoretical viewpoints, and much of this text concerns its development. John Evans wrote his seminal papers in the 1970s. He had trained as a medical doctor, but became enamored with mathematics just as he was about to embark on his medical career, at least at the point where his financial outlook would have improved dramatically. He gave it all up to pursue a PhD in mathematics, but his mathematical interests still reflected his medical background. Hodgkin–Huxley had formulated the equations governing the propagation of nerve impulses about 20 years earlier, and had computed, using a primitive calculator, travelling waves that represented the propagation of a nerve impulse. It was the discovery of this wave that showed they had found the mechanisms underlying the functioning of a nerve. So, it was known that the wave was present in the equations, but surely it also had to be stable. Showing its stability was the task that Evans set himself. He did not achieve it, but he did build a theory that has had applications far beyond neuroscience.

What happened next was indicative of the fortuitous circumstances that are often involved in research advances. I had heard Evans speak in 1980 while I was a postdoctoral fellow in British Columbia, and realized during his talk that his ideas provided the missing piece to a problem that had interested many of us, namely the stability of the fast traveling pulse of the FitzHugh–Nagumo equations. Although FitzHugh and Nagumo had separately formulated this system as a simplification of the Hodgkin–Huxley equations, Evans himself had resisted looking at this problem as he was after a more general result. In hindsight, had I not been lucky enough to come across this connection, and thus been able to show the power of Evans's ideas, they may never have achieved the prominence and range of applications they now enjoy. It took me most of the 1980s to convince the community of the importance of what I had by then named the Evans function. Gardner was working with me early on, and a major shift occurred when we convinced Alexander to join the effort; in particular, his topological and geometric expertise allowed us to put it in a completely new light. I should also mention Maginu in Japan who had realized early on the power of the Evans function, and his group made a number of contributions.

Simultaneously, the subject of stability was growing up for problems with Hamiltonian structure. To applied mathematicians, the division of physical problems into conservative and dissipative is fundamental, and it is not surprising that a stability theory evolved on both sides. Benjamin's stability result for water waves was perhaps the first result in this area, but much of the stability work on Hamiltonian systems was developed in Russia. Indeed, it was stimulated by the fundamental work of Arnol'd and Zakharov's work on soliton stability. This area also matured in the 1980s: Weinstein resolved the stability of standing waves of the nonlinear Schrödinger equation in his PhD thesis, and Grillakis, Strauss, and Shatah built a comprehensive theory for the stability of waves of a large class of Hamiltonian partial differential equations.

The stability story was greatly enriched by the entwining of these two threads. Of particular note was the work of Pego and Weinstein who showed that the Evans function plays a key role in determining the stability of the Korteweg–de Vries soliton. They showed that many of the calculations carried out in conservative systems could be recast in terms of the Evans function. This insight then led to the resolution of unsolved problems in the stability of wave motion for conservative partial differential equations. Zumbrun led a parallel development in his work on the stability of viscous shocks. This is an area marked by hard analytic estimates, but Zumbrun managed to push it much further through the use of insights originating in the Evans function approach.

But that is all prehistory. This book is about how all this early research grew into a full-blooded mathematical theory and how the importance of these stability methods can now be measured in a large number of applied areas. What began in nerve impulse equations has now had an impact in nonlinear optics, gas dynamics, fluid mechanics, material sciences, combustion theory, and Bose–Einstein condensates, among other areas.

For a graduate student needing to learn this material, a postdoc switching into an area that needs it, or even somebody (like me!) who claims to be expert in this area, this book will be invaluable. It covers an enormous amount of ground and does so with great insight and style. I have often been told that I should have written this book and that I should have done it long ago! I am glad I did not, as Kapitula and Promislow have done a much better job than I ever could have.

Chapel Hill, NC Christopher K.R.T. Jones

Contents

Chapter 1
Introduction

The stability of nonlinear waves has a distinguished history and an abundance of richly structured yet accessible examples, which makes it not only an important subject but also an ideal training ground for the study of linear and nonlinear partial differential equations (PDEs). While the "modern" approach to the stability of nonlinear waves can be traced back to the key papers of Joseph Boussinesq in the 1870s, the field has experienced tremendous growth over the past 30 years. Some of the growth was stimulated by T. B. Benjamin's 1972 paper, "The stability of solitary waves," which presented a treatment of the Korteweq–de Vries equation in that served to place meat on the bones of Boussinesq's ideas. A more recent avenue of growth stems from the development of dynamical systems ideas, which provide a rich complement to the functional analytic approach. In many ways these developments were stimulated by the pioneering work of Alexander et al. who recast the Evans function in a dynamical systems language. The subsequent synergy between dynamical systems and functional analysis has yielded a burst of activity and produced a unified framework for the study of stability and bifurcation in nonlinear waves.

Many graduate students in applied mathematics have been exposed to the key ideas of dynamical systems and functional analysis by the end of their first year; however, these fields are typically presented as unrelated. Within the context of nonlinear waves, the goal of this book is to show how the tools of dynamical systems provide a rich illumination of the abstract ideas of functional analysis. However, the simultaneous application of these two fields requires some sophistication. Our approach is to motivate the abstract framework by first working through detailed examples that serve to illustrate the moving parts, with the broader framework subsequently presented as a generalization of a familiar process. As much as possible, we have emphasized the structure of the framework: showing that stability and bifurcation can be understood through the dynamical systems

T. Kapitula and K. Promislow, *Spectral and Dynamical Stability of Nonlinear Waves*, Applied Mathematical Sciences 185, DOI 10.1007/978-1-4614-6995-7_1,

that characterize the underlying equilibria. We have avoided the use of exact solutions of ordinary differential equations (ODEs), the detailed calculus of which tends to obfuscate the structure behind the ideas.

Towards this goal, Chapter 2 provides a summary of background material; essentially, what is assumed of a mathematics graduate student who has completed a year of graduate PDE and functional analysis. We understand that this knowledge may be incomplete and have made an effort to integrate its application into the later chapters in a self-contained manner. In many ways Chapter 3 is the beginning of the book, applying dynamical systems ideas to illuminate a fundamental result of functional analysis: the Fredholm classification of linear differential operators on the line. The Fredholm alternative is the underpinning of many existence and bifurcation results, and its proof through dynamical systems techniques is instructive. This approach places the solvability question for a "linear system of differential equations subject to boundary conditions" inside of the bigger box of "all flows generated by the equivalent dynamical system." The task is to classify all solutions of the dynamical system, and then ask which ones satisfy the boundary conditions. This recalls the classical construction of a Green's function on a bounded domain, but the extension of the ideas to the unbounded domain, particularly the classification step, lead naturally to the idea of the Evans function, and provides a concrete formulation of the essential spectrum of important classes of linear differential operators.

After Chapter 3 the flow of the book branches, leading either to a study of nonlinear stability and bifurcation via functional analytic techniques in Chapter 4–Chapter 7 or directly to the Evans function and the key applications of eigenvalue perturbation within the essential spectrum in Chapter 8–Chapter 10. The functional analytic tract is inspired qualitatively by the work of Benjamin, and more quantitatively by the work of Grillakis et al., who provided the first comprehensive treatment of the stability of Hamiltonian systems in the presence of symmetry. Our approach introduces the unifying idea of a constrained operator and a quantitative measure of the impact of linear constraint upon eigenvalue count. The constrained operator approach naturally abuts the Hamiltonian index theory and plays a central role in the stability of critical points of Hamiltonian systems. Chapter 4 presents an overview the properties of C^0 and analytic semigroups, with attention to the role of symmetries in generating a point spectrum, and the idea that spectral stability in many situations implies orbital stability of a manifold of equilibria. In Chapter 5 we adapt these ideas to the Hamiltonian framework, using an adaptation of Benjamin's proof of the orbital stability of the solitary solution of the KdV equation to illustrate the idea of a constrained operator within a concrete framework. The framework is substantially generalized in Chapter 5.2, and is carried forward to Chapter 6 and Chapter 7 with slight modification. In Chapter 6 we present the formal perturbation structure of regular spectra of linear operators, with concrete examples that illustrate the major stability results of Chapter 4 and Chapter 5. Chapter 7 gives a detailed analysis of index theory for Hamiltonian systems, introducing the Hamiltonian–Krein index, which enumerates the potentially destabilizing point spectrum, and the Krein signature. In particular, we develop an instability criterion that is complementary to the nonlinear stability results of Chapter 5. We also discuss applications to bifurcation of point spectra with nontrivial Jordan block structure under both symmetry-breaking Hamiltonian perturbations and symmetry-preserving non-Hamiltonian perturbations.

The second part of the book develops the Evans function, an analytic function of the spectral parameter of an associated eigenvalue problem, the zeros of which coincide with the point spectrum of the operator when they are within the natural domain of the Evans function. The Evans function has analytic extensions beyond its natural domain, generically some distance into the essential spectrum. This extension plays a fundamental role in understanding bifurcations associated with point spectra that are ejected from the essential spectrum of the linearized operator under perturbation.

The machinery required to develop the Evans function is nontrivial; we first consider the simpler context of boundary-value problems on a finite domain in Chapter 8. In this context the Evans function resembles a classical Wronskian, being an entire function of the spectral parameter. The issues of branch points, branch cuts, and Riemann surfaces are deferred until Chapter 9, which addresses second-order Sturm–Liouville operators on the line. This requires a re-examination of the issue addressed in Chapter 3: the classification of all the solutions of a given dynamical system by their asymptotic behavior, which evokes the Jost functions of quantum mechanics. Even for this restricted class of problem one finds the full richness of the eigenvalue problems: branch points and branch cuts of the Evans function, and its extension beyond the natural domain and onto Riemann sheets. Within the context of second-order differential operators we address two important problems: the detection of real eigenvalues associated with instabilities, and the tracking of eigenvalues as they enter and leave the branch cuts. We also discuss the connection of the absolute spectrum to the large domain limit of the bounded domain problem for both separated and periodic boundary conditions. Chapter 10 addresses issues arising in the extension of the Evans function to higher-order linear operators, and incorporates several substantial examples that display the breadth of possible applications.

There are many relevant topics that are not touched upon in this book. All issues associated with multiple scales and (geometric) singular perturbation theory have been set aside. Eigenvalue problems can also be considered through a topological version of the Evans function, e.g., see [**38**, **93**, **142**, **245**], or by decomposing the Evans function into the product of meromorphic functions, each of which is associated with a distinct time scale, e.g., see [**71**, **72**]. The Green's function approach to nonlinear stability, pioneered by Howard and Zumbrun, is a natural outgrowth of the dynamical systems approach. Another natural extension is to weak and semi-strong interaction regimes of multipulse solutions, e.g., see [**73, 236, 248, 253, 258, 286**]. The most obvious limitation is that we have restricted ourselves to systems in one space dimension; however, these and other extensions are the domain of current research and do not seem ready for a unified treatment. The literature in the field of nonlinear waves is vast, and we have made an attempt to reference relevant papers at the end of each section; however, our effort is of necessity incomplete and we apologize in advance to the many people whose work we have not fully cited.

The content of the book has been used in a course for second-year graduate students at Michigan State University. It could also naturally be used to form two possible one-semester courses. The first course might emphasize the functional analytic approach, and include Chapters 3–7. The second course could address the Evans function and dynamical systems approach, using Chapters 3 and 8–10. Both routes would be supplemented with material from Chapter 2 as needed, and of course from the instructor's own personal repertoire of examples.

Chapter 2
Background Material and Notation

This chapter provides an overview of the background material assumed in the remainder of the book. We state major results and provide sketches of the less technical proofs, particularly where the ideas presented are instrumental in subsequent constructions. The first topic is the theory of linear systems of ordinary differential equations (ODEs). Much of this material is standard for first-year graduate courses in ODEs, such as presented in Hartman [**114**], Perko [**234**]. This is followed by a review of the basic theory of functional analysis as applied to linear partial differential equations; further details can be found in the standard references Evans [**81**], Kato [**162**]. We finish the general overview by discussing the point spectrum in the context of the Sturm–Liouville theory for second-order operators. These operators have a one-to-one relationship between the ordering of the eigenvalues and the number of zeros for the associated eigenfunctions, which is extremely useful in applications.

2.1 Linear Systems of Ordinary Differential Equations

Consider the linear system of ODEs of the form

$$\partial_x y = A(x,\lambda)y, \quad y(x_0,\lambda) = y_0(\lambda), \tag{2.1.1}$$

where $A(x,\lambda) \in \mathbb{C}^{n\times n}$ and $\lambda \in \mathbb{C}^m$ represents parameters associated with the system. We recall the following basic result concerning the existence and uniqueness of solutions to (2.1.1) (e.g., see Hartman [**114**, Chapter V]).

Lemma 2.1.1 ([**114**]). *Suppose that for fixed λ the matrix $A(x,\lambda)$ is continuous in x on the possibly infinite, open interval $I := (a,b)$. Then for each $x_0 \in I$, and $y_0 \in \mathbb{C}^n$ there exists a unique solution to (2.1.1). Furthermore, the solution is as smooth in x and λ as are the coefficient matrix $A(x,\lambda)$ and the initial data y_0.*

T. Kapitula and K. Promislow, *Spectral and Dynamical Stability of Nonlinear Waves*, Applied Mathematical Sciences 185, DOI 10.1007/978-1-4614-6995-7_2,

We fix $\boldsymbol{\lambda}$ and $x_0 \in I$, and suppress the $\boldsymbol{\lambda}$-dependence of $\boldsymbol{A}$ and $\boldsymbol{y}$. Let $S(x_0) := \{\boldsymbol{v}_1,\ldots,\boldsymbol{v}_n\} \subset \mathbb{C}^n$ be a basis. Then for each $i = 1,\ldots,n$ there exists a solution $\boldsymbol{y}_i(x)$ of (2.1.1) defined on I such that $\boldsymbol{y}_i(x_0) = \boldsymbol{v}_i$. A fundamental consequence of the uniqueness result in Lemma 2.1.1 is that for each $x \in I$, the set $S(x) := \{\boldsymbol{y}_i(x)\}_{i=1}^n$ forms a basis for $\mathbb{C}^n$. To establish this fact we first observe that for any choice of $\alpha_i \in \mathbb{C}$ for $i = 1,\ldots,n$ the function $\boldsymbol{y}(x) := \sum_{i=1}^n \alpha_i \boldsymbol{y}_i(x)$, solves (2.1.1). Fix any $x \in I$, by uniqueness of solutions, if $\boldsymbol{y}(x) = \boldsymbol{0}$, then $\boldsymbol{y} \equiv \boldsymbol{0}$ on I, hence $\boldsymbol{y}(x_0) = \boldsymbol{0}$. Since $S(x_0)$ forms a basis, we conclude that $\boldsymbol{y}(x) = \boldsymbol{0}$ if and only if each $\alpha_i = 0$, and hence $S(x)$ forms a basis. Much of the structure enjoyed by systems of linear ODEs stems from this fact: the flow induced by the system (2.1.1) preserves the rank of a set of initial data.

Concatenating the solutions constructed above, we form the matrix-valued function

$$\boldsymbol{\Psi}(x) := \left(\boldsymbol{y}_1,\ldots,\boldsymbol{y}_n\right)(x) \in \mathbb{C}^{n\times n},$$

and it is easy to verify that $\boldsymbol{\Psi}$ solves (2.1.1), that is

$$\partial_x \boldsymbol{\Psi} = \left(\partial_x \boldsymbol{y}_1,\ldots,\partial_x \boldsymbol{y}_n\right) = \left(\boldsymbol{A}\boldsymbol{y}_1,\ldots,\boldsymbol{A}\boldsymbol{y}_n\right) = \boldsymbol{A}\boldsymbol{\Psi},$$

subject to the initial condition $\boldsymbol{\Psi}(x_0) = (\boldsymbol{v}_1,\ldots,\boldsymbol{v}_n)$. A matrix-valued solution with nonzero determinant is called a fundamental matrix solution (FMS). For any constant, nonsingular matrix $\boldsymbol{B} \in \mathbb{C}^{n\times n}$, we have

$$\partial_x(\boldsymbol{\Psi}\boldsymbol{B}) = (\partial_x \boldsymbol{\Psi})\boldsymbol{B} = \boldsymbol{A}(x)(\boldsymbol{\Psi}\boldsymbol{B});$$

that is, right-multiplying by $\boldsymbol{B}$ yields a new FMS, $\boldsymbol{\Phi}(x) := \boldsymbol{\Psi}\boldsymbol{B}$. For the particular choice $\boldsymbol{B} = \boldsymbol{\Psi}(x_0)^{-1}$, then $\boldsymbol{\Phi}(x_0) = \boldsymbol{I}_n$, the $n\times n$ identity matrix. In this case we say $\boldsymbol{\Phi}(x) = \boldsymbol{\Phi}(x;x_0)$ is the principal FMS at x_0 and the solution to the initial-value problem (2.1.1) is given by $\boldsymbol{y}(x) = \boldsymbol{\Phi}(x)\boldsymbol{y}_0$.

The following property of matrix solutions generalizes Abel's identity for scalar systems. It is particularly useful in deriving properties of the Evans function in Chapter 8–Chapter 10.

Lemma 2.1.2 (Liouville's formula). *Let $\boldsymbol{\Psi}$ be a matrix solution of (2.1.1); then its determinant satisfies*

$$\det \boldsymbol{\Psi}(x) = \det \boldsymbol{\Psi}(x_0) e^{\int_{x_0}^{x} \operatorname{tr} \boldsymbol{A}(s)\,\mathrm{d}s}. \tag{2.1.2}$$

Proof. The discussion above implies that $\det \boldsymbol{\Psi}(x_0) = 0$ iff $\det \boldsymbol{\Psi}(x) = 0$ for all $x \in \mathbb{R}$, and Louiville's formula holds trivially. It suffices to establish Liouville's formula for FMS. However by uniqueness of solutions, all matrix solutions of (2.1.1) are related by a constant matrix, that is, $\tilde{\boldsymbol{\Psi}}(x) = \tilde{\boldsymbol{\Psi}}(x_0)\boldsymbol{\Psi}^{-1}(x_0)\boldsymbol{\Psi}(x)$, so establishing Liouville's formula for any FMS, $\tilde{\boldsymbol{\Psi}}$, establishes it for all FMS matrix solutions, $\boldsymbol{\Psi}$. It suffices to establish the identify

$$\partial_x \det \tilde{\boldsymbol{\Psi}}(x) = \operatorname{tr} \boldsymbol{A}(x) \det \tilde{\boldsymbol{\Psi}}(x),$$

for a particular FMS $\tilde{\Psi}$. This identity is local in x. Since A is smooth, it has the expansion $A(x) = A(x_0) + \mathcal{O}(|x - x_0|)$. Assume for the moment that $A(x_0)$ has a complete set of eigenvectors $\{v_1, \dots, v_n\}$. Form the solutions y_j of the exact problem (2.1.1) that satisfy $y_j(x_0) = v_j$, and let $\tilde{\Psi}$ denote the corresponding FMS. These solutions have the expansion $y_j(x) = e^{\lambda_j(x-x_0)}v_j + \mathcal{O}(|x - x_0|^2)$; in particular, $\partial_x y_j(x) = \lambda_j y(x) + \mathcal{O}(|x - x_0|)$. Taking ∂_x of $\tilde{\Psi}$ we calculate

$$\partial_x \det \tilde{\Psi} = \sum_{i=1}^{n} \det\Big(y_1, \dots, \partial_x y_i, \dots, y_n\Big)$$
$$= \sum_{i=1}^{n} \det\Big(y_1, \dots, \lambda_i y_i, \dots, y_n\Big) + \mathcal{O}(|x - x_0|).$$

However,

$$\det\Big(y_1, \dots, \lambda_i y_i, \dots, y_n\Big) = \lambda_i \det\Big(y_1, \dots, y_i, \dots, y_n\Big) = \lambda_i \det \tilde{\Psi}.$$

Since $\operatorname{tr} A(x_0) = \lambda_1 + \cdots + \lambda_n$, we combine these two expressions and let $x \to x_0$ to obtain

$$\partial_x \det \Psi(x_0) = \operatorname{tr} A(x_0) \det \Psi(x_0).$$

As x_0 was arbitrary we have established Liouville's formula for matrices $A(x_0)$ with a complete set of eigenvectors. If $A(x_0)$ does not have a complete set of eigenvectors, we may perturb A smoothly to obtain a family of matrices that do. This establishes Liouville's formula for the perturbed problem, and passing to the limit in Liouville's formula yields the general result. □

We may obtain solutions to the inhomogeneous linear problem

$$\partial_x y = A(x)y + f(x), \quad y(x_0) = y_0, \tag{2.1.3}$$

from the principal FMS via the variation-of-constants formula.

Lemma 2.1.3 (Variation of constants). *Consider the inhomogeneous problem (2.1.3) for $f \in L^1(I)$. Let $\Phi(x)$ be a principal FMS to (2.1.1) at $x = x_0$. The unique solution to (2.1.3) is given by*

$$y(x) = \Phi(x)y_0 + \Phi(x) \int_{x_0}^{x} \Phi^{-1}(t) f(t, y(t))\, dt.$$

Proof. It is clear that $\boldsymbol{y}(x_0) = \boldsymbol{y}_0$. Differentiating yields

$$\begin{aligned}\partial_x \boldsymbol{y}(x) &= \partial_x \boldsymbol{\Phi}(x)\boldsymbol{y}_0 + \partial_x \boldsymbol{\Phi}(x)\int_{x_0}^{x} \boldsymbol{\Phi}(t)^{-1}\boldsymbol{f}(t)\,\mathrm{d}t + \boldsymbol{\Phi}(x)\left(\boldsymbol{\Phi}(x)^{-1}\boldsymbol{f}(x)\right)\\ &= \boldsymbol{A}(x)\boldsymbol{\Phi}(x)\left(\boldsymbol{y}_0 + \int_{x_0}^{x} \boldsymbol{\Phi}(t)^{-1}\boldsymbol{f}(t)\,\mathrm{d}t\right) + \boldsymbol{f}(x)\\ &= \boldsymbol{A}(x)\boldsymbol{y}(x) + \boldsymbol{f}(x).\end{aligned}$$

Uniqueness for the inhomogeneous problem follows from the uniqueness of the homogeneous problem. □

It is important to understand the sensitivity of solutions of the homogeneous initial-value problem to changes in the initial data. Lemma 2.1.1 establishes the smoothness of solutions $\boldsymbol{y}(x, \boldsymbol{\lambda})$ with respect to $\boldsymbol{\lambda}$. Without loss of generality assume that $m = 1$, i.e., $\lambda \in \mathbb{C}$. We may Taylor expand $\boldsymbol{y}$ around $\lambda = 0$,

$$\boldsymbol{y}(x,\lambda) = \boldsymbol{y}(x,0) + \lambda\partial_\lambda \boldsymbol{y}(x,0) + \frac{1}{2}\lambda^2\partial_\lambda^2 \boldsymbol{y}(x,0) + \mathcal{O}(|\lambda|^3). \tag{2.1.4}$$

To establish a formula for $\partial_\lambda \boldsymbol{y}(x,0)$ we differentiate (2.1.1) with respect to λ and set $\lambda = 0$. This yields the linear, inhomogeneous equation for $\partial_\lambda \boldsymbol{y}$,

$$\begin{aligned}\partial_x\left(\partial_\lambda \boldsymbol{y}(x,0)\right) &= \boldsymbol{A}(x,0)\partial_\lambda \boldsymbol{y}(x,0) + \partial_\lambda \boldsymbol{A}(x,0)\boldsymbol{y}(x,0)\\ \partial_\lambda \boldsymbol{y}(x_0,0) &= \partial_\lambda \boldsymbol{y}_0(0).\end{aligned} \tag{2.1.5}$$

The variation-of-constants formula, in terms of the principal FMS $\boldsymbol{\Phi}(x,\lambda)$ at $x = x_0$, yields

$$\partial_\lambda \boldsymbol{y}(x,0) = \boldsymbol{\Phi}(x,0)\partial_\lambda \boldsymbol{y}_0(0) + \int_{x_0}^{x} \boldsymbol{\Phi}(t,0)^{-1}\partial_\lambda \boldsymbol{A}(t,0)\boldsymbol{y}(t,0)\,\mathrm{d}t. \tag{2.1.6}$$

The smoothness of solutions extends out to analyticity. Assume that the initial data $\boldsymbol{y}_0(\lambda)$ and the matrix $\boldsymbol{A}(x,\lambda)$ are analytic in λ at $\lambda = 0$ for fixed x, that is,

$$\partial_{\overline{\lambda}} \boldsymbol{A}(x,0) = \boldsymbol{0}_n, \quad \partial_{\overline{\lambda}} \boldsymbol{y}_0(0) = \boldsymbol{0},$$

where $\overline{\lambda}$ denotes the complex-conjugate of λ. Taking $\partial_{\overline{\lambda}}$ of (2.1.6), we see that $\partial_{\overline{\lambda}} \boldsymbol{y}(x,0) \equiv \boldsymbol{0}$. In other words, $\boldsymbol{y}$ is also analytic in λ for fixed x. We summarize this result in the lemma below.

Lemma 2.1.4. *If $A(x, \boldsymbol{\lambda})$ and initial data $\boldsymbol{y}_0(\boldsymbol{\lambda})$ are analytic in $\boldsymbol{\lambda}$ at $\boldsymbol{\lambda} = \boldsymbol{\lambda}_0$, then the solution $\boldsymbol{y}$ of (2.1.1) is also analytic in λ at $\boldsymbol{\lambda} = \boldsymbol{\lambda}_0$.*

2.1.1 Constant Matrices: The Matrix Exponential

For spatially constant matrices $A(x) \equiv A$ much information about the FMS can be explicitly computed. Of particular importance in the sequel are the growth rates associated with invariant subspaces of the FMS. The FMS at $x = 0$ can be constructed as a generalization of the Maclaurin series for the exponential.

Definition 2.1.5. For a constant $A \in \mathbb{C}^{n\times n}$ set

$$e^{Ax} := \sum_{n=0}^{\infty} A^n \frac{x^n}{n!}. \tag{2.1.7}$$

It should not be a surprise that e^{Ax} inherits many, but not all [see (d)] of the properties of the scalar exponential function:

Lemma 2.1.6. *For all $A, B \in \mathbb{C}^{n\times n}$, and $s, x \in \mathbb{R}$, we have*

(a) $\partial_x e^{Ax} = Ae^{Ax}$,
(b) $(e^{Ax})^{-1} = e^{-Ax}$,
(c) $e^{A(s+x)} = e^{As}e^{Ax} = e^{Ax}e^{As}$,
(d) if $AB = BA$, *then* $e^{(A+B)x} = e^{Ax}e^{Bx} = e^{Bx}e^{Ax}$,
(e) if B *is nonsingular, then* $Be^{Ax}B^{-1} = e^{BAB^{-1}x}$.

In particular, the summation given in (2.1.7) is absolutely convergent on $\mathbb{R}$ and yields the principal FMS to (2.1.1) at $x = 0$.

If A is block-diagonalizable, that is, if there exists nonsingular $P \in \mathbb{C}^{n\times n}$ such that

$$A = P\begin{pmatrix} A_1 & 0 & \cdots & 0 \\ 0 & A_2 & \cdots & 0 \\ \vdots & \vdots & \ddots & \vdots \\ 0 & 0 & \cdots & A_m \end{pmatrix}P^{-1},$$

for square submatrices $A_1, \ldots, A_m$, then it follows from Lemma 2.1.6(e) that

$$e^{Ax} = P\begin{pmatrix} e^{A_1x} & 0 & \cdots & 0 \\ 0 & e^{A_2x} & \cdots & 0 \\ \vdots & \vdots & \ddots & \vdots \\ 0 & 0 & \cdots & e^{A_mx} \end{pmatrix}P^{-1}.$$

In particular, if A has a complete set of eigenvectors $\{v_1, \ldots, v_n\}$ with associated eigenvalues μ_k, $k = 1, \ldots, n$, then

$$A = P\Lambda P^{-1}, \quad \Lambda = \mathrm{diag}(\mu_1, \ldots, \mu_n), \tag{2.1.8}$$

where

$$P = (v_1, \ldots, v_n),$$

so that

$$e^{Ax} = Pe^{\Lambda x}P^{-1} = P\,\mathrm{diag}(e^{\mu_1 x}, \dots, e^{\mu_n x})P^{-1}.$$

From this result we see that each entry of e^{Ax} is a linear combination of exponentials $e^{\mu x}$, where each μ is an eigenvalue of A. If an eigenvalue is complex-valued, i.e., $\mu = a + ib$ with $v \neq 0$, then by Euler's formula we know that $e^{\mu x} = e^{ax}\cos(bx) + ie^{ax}\sin(bx)$; consequently, if any of the eigenvalues are complex-valued, then each entry of the matrix can also be expressed as a combination of $e^{ax}\cos(bx)$ and $e^{ax}\sin(bx)$.

Unfortunately, not all matrices have a complete set of eigenvectors. Indeed, if $A = A(\boldsymbol{\lambda})$, then there may be nongeneric values of $\boldsymbol{\lambda}$, for which the diagonalization assumption breaks down. As this case is of particular interest in the study of branch points of linear operators, we discuss it in detail.

Definition 2.1.7. The spectrum of $A \in \mathbb{C}^{n\times n}$ is comprised of the eigenvalues of A,

$$\sigma(A) := \{\mu \in \mathbb{C} : \det(A - \mu I_n) = 0\}.$$

The algebraic multiplicity, $m_a(\mu)$, of $\mu \in \sigma(A)$ is the order of the zero of the characteristic polynomial $\det(A - \mu I_n) = 0$. For a given $\mu \in \sigma(A)$ let $\ker(A - \mu I_n)$ denote the kernel, i.e.,

$$\ker(A - \mu I_n) = \{v \in \mathbb{C}^n : (A - \mu I_n)v = 0\}.$$

The geometric multiplicity, $m_g(\mu)$, of $\mu \in \sigma(A)$ is the dimension of $\ker(A - \mu I_n)$. An eigenvalue $\mu \in \sigma(A)$ is simple if $m_g(\mu) = m_a(\mu) = 1$, and $\mu \in \sigma(A)$ is semi-simple (algebraically simple) if $m_g(\mu) = m_a(\mu)$.

Remark 2.1.8. If each eigenvalue of A is semi-simple, then A is diagonalizable, [**116**].

If $\mu \in \sigma(A)$ is such that $m_a(\mu) > m_g(\mu)$, then there will be at least one vector v such that $(A - \mu I_n)^2 v = 0$ with $(A - \mu I_n)v \neq 0$. It is clearly the case that $\ker(A - \mu I_n)^k \subset \ker(A - \mu I_n)^{k+1}$. A fundamental result of linear algebra states that these kernels grow until they reach full rank at $k = m_a(\mu)$, that is

$$\dim\left[\ker(A - \mu I_n)^{k-1}\right] < \dim\left[\ker(A - \mu I_n)^{k}\right], \quad k < m_a(\mu),$$

$$\dim\left[\ker(A - \mu I_n)^{k}\right] = m_a(\mu), \quad k \geq m_a(\mu).$$

Definition 2.1.9. For $\mu \in \sigma(A)$ the associated generalized eigenspace, $\mathbb{E}_\mu$, is given by

$$\mathbb{E}_\mu = \mathrm{gker}(A - \mu I_n) := \ker\left[(A - \mu I_n)^{m_a(\mu)}\right].$$

The vectors $v \in \mathrm{gker}(A - \mu I_n)\backslash\ker(A - \mu I_n)$ are called the generalized eigenvectors of A.

Example 2.1.10. If

$$\boldsymbol{A} = \begin{pmatrix} \mu & 1 \\ 0 & \mu \end{pmatrix} = \mu \boldsymbol{I}_2 + \begin{pmatrix} 0 & 1 \\ 0 & 0 \end{pmatrix},$$

then $\mu \in \sigma(\boldsymbol{A})$ satisfies $\mathrm{m_g}(\mu) = 1$ and $\mathrm{m_a}(\mu) = 2$. The eigenvector is $\boldsymbol{v} = (1,0)^{\mathrm{T}}$, and the generalized eigenvector is $\boldsymbol{w} = (0,1)^{\mathrm{T}}$.

The nilpotent matrices are the key to understanding the structure of $\mathrm{e}^{\boldsymbol{A}x}$ for nondiagonalizable matrices $\boldsymbol{A}$. For a given $j \geq 2$ define the nilpotent matrix $\boldsymbol{N}_j \in \mathbb{R}^{j \times j}$ by

$$(\boldsymbol{N}_j)_{\ell m} := \begin{cases} 1, & m = \ell + 1, \ell = 1, \dots, j-1 \\ 0, & \text{otherwise.} \end{cases} \tag{2.1.9}$$

For example,

$$\boldsymbol{N}_2 = \begin{pmatrix} 0 & 1 \\ 0 & 0 \end{pmatrix}; \quad \boldsymbol{N}_3 = \begin{pmatrix} 0 & 1 & 0 \\ 0 & 0 & 1 \\ 0 & 0 & 0 \end{pmatrix}.$$

It is easy to verify that $\boldsymbol{N}_j$ is a nilpotent matrix of order j, i.e.,

$$\boldsymbol{N}_j^j = \boldsymbol{0}; \quad \boldsymbol{N}_j^i \neq \boldsymbol{0}_j, \ i = 1, \dots, j-1;$$

hence, the matrix exponential of $\boldsymbol{N}_j$ is formed of polynomials of order no larger than $j-1$,

$$\mathrm{e}^{\boldsymbol{N}_j x} = \sum_{n=0}^{j-1} \frac{x^n}{n!} \boldsymbol{N}_j^n.$$

The Jordan canonical form states that any matrix $\boldsymbol{A}$ is block-diagonalizable with r sub-blocks of the form

$$\boldsymbol{A}_j := \mu_j \boldsymbol{I}_{\mathrm{m_a}(\mu_j)} + \boldsymbol{N}_{\mathrm{m_a}(\mu_j)}, \tag{2.1.10}$$

for the r distinct eigenvalues $\{\mu_1, \dots, \mu_r\}$ of A. This construction will not be carried out here (e.g., see [**234**, Section 1.8] and the references therein), as the procedure is quite technical. However, all that is necessary in the subsequent sections is the final result, which we state for the case of real-valued eigenvalues of a real matrix. The case of complex eigenvalues is covered in [**234**, Section 1.8].

Theorem 2.1.11 (Jordan canonical form). *Suppose that $\boldsymbol{A} \in \mathbb{R}^{n \times n}$ has distinct, real matrix eigenvalues $\mu_1, \dots, \mu_r$ for some integer $r \in [1, n]$. There then exists a basis of generalized eigenvectors $\boldsymbol{v}_1, \dots, \boldsymbol{v}_n$ such that the nonsingular matrix $\boldsymbol{P} := (\boldsymbol{v}_1, \dots, \boldsymbol{v}_n)$ block-diagonalizes $\boldsymbol{A}$ into $\boldsymbol{P}^{-1}\boldsymbol{A}\boldsymbol{P} = \mathrm{diag}(\boldsymbol{A}_1, \dots, \boldsymbol{A}_r)$, where the Jordan blocks $\boldsymbol{A}_j$ are of the form (2.1.10).*

For a Jordan block of the form $\boldsymbol{A}_j(\mu) = \mu\boldsymbol{I}_j + \boldsymbol{N}_j$ it is easy to verify that $\mu \in \sigma(\boldsymbol{A})$ with $\mathrm{m_g}(\mu) = 1$ and $\mathrm{m_a}(\mu) = j$, and from Lemma 2.1.6(d) we obtain the exponential

$$\mathrm{e}^{\boldsymbol{A}x} = \mathrm{e}^{\mu x}\mathrm{e}^{\boldsymbol{N}x} = \mathrm{e}^{\mu x}\sum_{n=0}^{j-1}\frac{x^n}{n!}\boldsymbol{N}_j^n.$$

As a consequence of Theorem 2.1.11, the formula above, and the exponential structure of block-diagonal matrices, we have the following result.

Theorem 2.1.12. *Fix* $\boldsymbol{A} \in \mathbb{C}^{n\times n}$*; then every entry of* $\mathrm{e}^{\boldsymbol{A}x}$ *is composed of linear combinations of* $p(x)\mathrm{e}^{\alpha x}\cos\beta x$ *and* $p(x)\mathrm{e}^{\alpha x}\sin\beta x$*, where* $\mu = \alpha + \mathrm{i}\beta \in \sigma(\boldsymbol{A})$ *and* $p(x)$ *is a polynomial of degree no larger than* $n-1$*.*

Example 2.1.13. Consider

$$\boldsymbol{A} = \begin{pmatrix} -1 & 1 & -2 \\ 0 & -1 & 4 \\ 0 & 0 & 1 \end{pmatrix}.$$

The eigenvalues are $\mu_1 = 1$ and $\mu_2 = -1$ with $m_\mathrm{g}(1) = m_\mathrm{a}(1) = 1$, while $m_\mathrm{g}(-1) = 1$ and $m_\mathrm{a}(-1) = 2$. For the eigenvectors $\boldsymbol{v}_1$ and $\boldsymbol{v}_2$ and generalized eigenvector $\boldsymbol{v}_3$

$$\boldsymbol{v}_1 = (0,2,1)^\mathrm{T}, \quad \boldsymbol{v}_2 = (1,0,0)^\mathrm{T}, \quad \boldsymbol{v}_3 = (0,1,0)^\mathrm{T},$$

the matrix $\boldsymbol{P} = (\boldsymbol{v}_1, \boldsymbol{v}_2, \boldsymbol{v}_3)$ block-diagonalizes $\boldsymbol{A}$,

$$\boldsymbol{P}^{-1}\boldsymbol{A}\boldsymbol{P} = \mathrm{diag}(\boldsymbol{A}_1, \boldsymbol{A}_2); \qquad \boldsymbol{A}_1 = (1), \quad \boldsymbol{A}_2 = \begin{pmatrix} -1 & 1 \\ 0 & -1 \end{pmatrix}.$$

One then has that

$$\mathrm{e}^{\boldsymbol{A}x} = \boldsymbol{P}\begin{pmatrix} \mathrm{e}^x & 0 & 0 \\ 0 & \mathrm{e}^{-x} & x\mathrm{e}^{-x} \\ 0 & 0 & \mathrm{e}^{-x} \end{pmatrix}\boldsymbol{P}^{-1},$$

so that all of the entries of $\mathrm{e}^{\boldsymbol{A}x}$ are of the form $c_1\mathrm{e}^x + c_2\mathrm{e}^{-x} + c_3x\mathrm{e}^{-x}$.

2.1.2 Constant Matrices: Invariant Subspaces and Estimates on Solutions

From our understanding of the exponential of a constant matrix we obtain sharp estimates on long-time asymptotics of solutions of the corresponding ODE. We first require the following preliminary result.

When discussing vectors, we will use $|\cdot|$ to represent the Euclidean norm on $\mathbb{C}^n$. Since all norms are equivalent on $\mathbb{C}^n$, the choice of the norm is not important. For matrices we will use the operator norm induced by the Euclidean norm. Theorem 2.1.12 and (2.1.14) yield the following estimate on solution behavior.

Lemma 2.1.14. *Set* $\sigma_{\mathrm{M}} := \max\{\mathrm{Re}\,\mu : \mu \in \sigma(\boldsymbol{A})\}$. *For each* $\epsilon > 0$ *there exists an* $M(\epsilon) \geq 1$ *such that*

$$|\mathrm{e}^{\boldsymbol{A}x}| \leq M(\epsilon)\mathrm{e}^{(\sigma_{\mathrm{M}}+\epsilon)x}, \quad x \geq 0.$$

If all $\mu \in \sigma(\boldsymbol{A})$ are semi-simple, then one has $p(x) \equiv 1$ in Theorem 2.1.12 and we have $j = 0$ in (2.1.14), in which case we may chose $\epsilon = 0$. These observations yield the following refinement of Lemma 2.1.14.

Corollary 2.1.15. *If all* $\mu \in \sigma(\boldsymbol{A})$ *are algebraically simple, then Lemma 2.1.14 holds with* $\epsilon = 0$.

The structure of the matrix exponential allows us to obtain an upper bound on its growth rate, as in Lemma 2.1.14. It also affords a lower bound, which does not depend on the multiplicity of the eigenvalue.

Lemma 2.1.16. *Let* $\sigma_{\mathrm{m}} := \min\{\mathrm{Re}\,\mu : \mu \in \sigma(\boldsymbol{A})\}$; *then for any vector* $\boldsymbol{v}$ *there is a constant* $C > 0$ *such that*

$$|\mathrm{e}^{\boldsymbol{A}x}\boldsymbol{v}| \geq C\mathrm{e}^{\sigma_{\mathrm{m}}x}|\boldsymbol{v}|, \quad x \geq 0.$$

By considering the restriction of $\boldsymbol{A}$ to its invariant subspaces we can greatly sharpen the upper and lower bounds on the decay rates of the exponential. These results stem from the following proposition.

Proposition 2.1.17. *If a subspace* $\mathbb{E} \subset \mathbb{C}^n$ *is invariant under multiplication by* $\boldsymbol{A}$, *i.e.,* $\boldsymbol{A}\mathbb{E} \subset \mathbb{E}$, *then* $\mathbb{E}$ *is invariant under multiplication by* $\mathrm{e}^{\boldsymbol{A}x}$.

Proof. Since $\boldsymbol{A}\mathbb{E} \subset \mathbb{E}$ it follows that $\boldsymbol{A}^k\mathbb{E} \subset \mathbb{E}$ for all $k \geq 0$. The result follows from the definition of the matrix exponential in Definition 2.1.5 and the closure of subspaces under addition. □

The fundamental invariant subsets of a matrix $\boldsymbol{A}$ are its generalized eigenspaces, $\{\mathbb{E}_\mu : \mu \in \sigma(\boldsymbol{A})\}$, defined in Definition 2.1.9. Their invariance arises from the simple fact that $\boldsymbol{A}$ commutes with its own polynomials,

$$(\boldsymbol{A} - \mu)^k\boldsymbol{A} = \boldsymbol{A}(\boldsymbol{A} - \mu)^k,$$

so $\boldsymbol{v} \in \mathbb{E}_\mu$ implies $\boldsymbol{A}\boldsymbol{v} \in \mathbb{E}_\mu$. Moreover, due to the Jordan canonical form, these sets are disjoint since they comprise the basis for the Jordan sub-blocks, and they collectively span $\mathbb{C}^n$. These properties make the generalized eigenspaces ideal building blocks for a decomposition.

Definition 2.1.18. A vector space A is the direct sum of two subspaces, B and C, written

$$A = B \oplus C, \tag{2.1.11}$$

if A equals the span of B and C and $B \cap C = \{0\}$.

Definition 2.1.19. To each matrix $A \in \mathbb{C}^{n\times n}$ we associate the following structures: The stable spectrum and stable eigenspace given by

$$\sigma^{\mathrm{s}}(A) := \{\mu \in \sigma(A) : \mathrm{Re}\,\mu < 0\}, \quad \mathbb{E}^{\mathrm{s}} := \oplus\{\mathbb{E}_{\mu_j} : \mu_j \in \sigma^{\mathrm{s}}(A)\},$$

the center spectrum and center subspace given by

$$\sigma^{\mathrm{c}}(A) := \{\mu \in \sigma(A) : \mathrm{Re}\,\mu = 0\}, \quad \mathbb{E}^{\mathrm{c}} := \oplus\{\mathbb{E}_{\mu_j} : \mu_j \in \sigma^{\mathrm{c}}(A)\},$$

and the unstable spectrum and unstable subspace given by

$$\sigma^{\mathrm{u}}(A) := \{\mu \in \sigma(A) : \mathrm{Re}\,\mu > 0\}, \quad \mathbb{E}^{\mathrm{u}} := \oplus\{\mathbb{E}_{\mu_j} : \mu_j \in \sigma^{\mathrm{u}}(A)\}.$$

Proposition 2.1.20. *The stable, center, and unstable eigenspaces of A enjoy the following properties*

(a) $\mathbb{C}^n = \mathbb{E}^{\mathrm{s}} \oplus \mathbb{E}^{\mathrm{c}} \oplus \mathbb{E}^{\mathrm{u}}$,
(b) $\dim[\mathbb{E}^{\mathrm{s,c,u}}]$ *equals the number of generalized eigenvectors in the basis for* $\mathbb{E}^{\mathrm{s,c,u}}$, *respectively,*
(c) $\mathrm{e}^{Ax}\mathbb{E}^{\mathrm{s,c,u}} \subset \mathbb{E}^{\mathrm{s,c,u}}$, *respectively.*

Proof. The sets $\mathbb{E}^{\mathrm{s}}$, $\mathbb{E}^{\mathrm{u}}$, and $\mathbb{E}^{\mathrm{c}}$ are mutually disjoint since their components, the generalized eigenspaces, are mutually disjoint. Since the generalized eigenspaces of A span $\mathbb{C}^n$, the direct sum decomposition of (a) follows. Part (c) follows from the invariance of the generalized eigenspaces, and part (b) is a general property of direct sums of subspaces. □

The restriction of the matrix A to its invariant subspaces $\mathbb{E}^{\mathrm{s,c,u}}$ yields a smaller matrix, whose spectrum coincides with $\sigma^{\mathrm{s,c,u}}(A)$, respectively. We introduce the following spectral quantities:

Definition 2.1.21. The upper bound of the stable and unstable spectrum is denoted,

$$\sigma_{\mathrm{M}}^{\mathrm{s,u}} := \max\{\mathrm{Re}\,\mu : \mu \in \sigma^{\mathrm{s,u}}(A)\},$$

with the corresponding lower bound

$$\sigma_{\mathrm{m}}^{\mathrm{s,u}} := \min\{\mathrm{Re}\,\mu : \mu \in \sigma^{\mathrm{s,u}}(A)\}.$$

Remark 2.1.22. We have the ordering $\sigma_{\mathrm{m}}^{\mathrm{s}} \le \sigma_{\mathrm{M}}^{\mathrm{s}} < 0 < \sigma_{\mathrm{m}}^{\mathrm{u}} \le \sigma_{\mathrm{M}}^{\mathrm{u}}$.

Applying our previous bounds Lemmas 2.1.14–2.1.16 to A restricted to $\mathbb{E}^{\mathrm{s,c,u}}$ yields the following exponential dichotomies which govern the long-term behavior of solutions of the underlying system.

Theorem 2.1.23. *For each matrix $A \in \mathbb{C}^{n\times n}$ and for all $\epsilon > 0$ there exist constants $m, M \ge 1$, dependent upon ϵ, such that for all $y_0 \in \mathbb{E}^{\mathrm{s}}$,*

$$\begin{aligned} m\mathrm{e}^{\sigma_{\mathrm{m}}^{\mathrm{s}}x}|y_0| \le |\mathrm{e}^{Ax}y_0| \le M\mathrm{e}^{(\sigma_{\mathrm{M}}^{\mathrm{s}}+\epsilon)x}|y_0| \quad x > 0 \\ m\mathrm{e}^{\sigma_{\mathrm{M}}^{\mathrm{s}}x}|y_0| \le |\mathrm{e}^{Ax}y_0| \le M\mathrm{e}^{(\sigma_{\mathrm{m}}^{\mathrm{s}}+\epsilon)x}|y_0| \quad x < 0, \end{aligned}$$

and for all $\boldsymbol{y}_0 \in \mathbb{E}^{\mathrm{u}}$,

$$m e^{\sigma_{\mathrm{m}}^{\mathrm{u}} x} |\boldsymbol{y}_0| \le |e^{\boldsymbol{A}x} \boldsymbol{y}_0| \le M e^{(\sigma_{\mathrm{M}}^{\mathrm{u}}+\epsilon)x} |\boldsymbol{y}_0| \quad x > 0$$
$$m e^{\sigma_{\mathrm{M}}^{\mathrm{u}} x} |\boldsymbol{y}_0| \le |e^{\boldsymbol{A}x} \boldsymbol{y}_0| \le M e^{(\sigma_{\mathrm{m}}^{\mathrm{u}}+\epsilon)x} |\boldsymbol{y}_0| \quad x < 0.$$

There also exists a $k \in \mathbb{N}_0$ *with* $0 \le k \le n-1$ *such that for all* $\boldsymbol{y}_0 \in \mathbb{E}^{\mathrm{c}}$,

$$m|\boldsymbol{y}_0| \le |e^{\boldsymbol{A}x} \boldsymbol{y}_0| \le M(1 + |x|^k)|\boldsymbol{y}_0|, \quad \forall x \in \mathbb{R}.$$

Remark 2.1.24. If all $\mu \in \sigma^{\mathrm{c}}(\boldsymbol{A})$ are semisimple, then $k = 0$ and all solutions of (2.1.1) residing in $\mathbb{E}^{\mathrm{c}}$ are uniformly bounded in forward and backward time.

The central implication of Theorem 2.1.23 is that the solutions residing in $\mathbb{E}^{\mathrm{s,u}}$ enjoy a sort of exponential dichotomy. Solutions in the unstable subspace $\mathbb{E}^{\mathrm{u}}$ grow exponentially in norm for $x \ge 0$, and decay for $x < 0$. Solutions in the stable subspace $\mathbb{E}^{\mathrm{s}}$ decay exponentially in norm for $x \ge 0$, and grow exponentially for $x \le 0$. The behavior of solutions in the center subspace is unclear without additional information on the Jordan structure of the center eigenspaces. All that can be said in general is that any temporal growth is polynomial in nature. The term *exponential dichotomy* has a more formal definition, given in Chapter 2.1.4, which is slightly more restrictive.

We can summarize this behavior as follows. If $\boldsymbol{A}$ is constant in x, then by Proposition 2.1.20 any initial data of (2.1.1) can be decomposed as

$$\boldsymbol{y}_0 = \boldsymbol{y}_0^{\mathrm{s}} + \boldsymbol{y}_0^{\mathrm{c}} + \boldsymbol{y}_0^{\mathrm{u}}, \quad \boldsymbol{y}_0^{\mathrm{s,c,u}} \in \mathbb{E}^{\mathrm{s,c,u}},$$

and by linearity the corresponding solution takes the form

$$e^{\boldsymbol{A}x} \boldsymbol{y}_0 = e^{\boldsymbol{A}x} \boldsymbol{y}_0^{\mathrm{s}} + e^{\boldsymbol{A}x} \boldsymbol{y}_0^{\mathrm{c}} + e^{\boldsymbol{A}x} \boldsymbol{y}_0^{\mathrm{u}}. \tag{2.1.12}$$

The long-term behavior of each component of the solution is given by Theorem 2.1.23; in particular, if $\boldsymbol{A}$ is hyperbolic, i.e., $\mathbb{E}^{\mathrm{c}} = \{\boldsymbol{0}\}$, then

$$\lim_{x \to -\infty} |e^{\boldsymbol{A}x} \boldsymbol{y}_0 - e^{\boldsymbol{A}x} \boldsymbol{y}_0^{\mathrm{s}}| = 0, \quad \lim_{x \to +\infty} |e^{\boldsymbol{A}x} \boldsymbol{y}_0 - e^{\boldsymbol{A}x} \boldsymbol{y}_0^{\mathrm{u}}| = 0.$$

2.1.3 Periodic Matrices: Floquet Theory

A second case in which the system (2.1.1) possesses extra structure is for a matrix that is periodic. Without loss of generality we assume the period to be π, that is, $\boldsymbol{A}(x + \pi, \boldsymbol{\lambda}) = \boldsymbol{A}(x, \boldsymbol{\lambda})$ for all $x \in \mathbb{R}$ and $\boldsymbol{\lambda} \in \mathbb{C}^m$. While there is a principal FMS $\boldsymbol{\Phi}(x, \boldsymbol{\lambda})$ satisfying $\boldsymbol{\Phi}(0, \boldsymbol{\lambda}) = \boldsymbol{I}_n$, the principal FMS is *not* in general the exponential of *any* matrix. Indeed, an explicit form for $\boldsymbol{\Phi}(x, \boldsymbol{\lambda})$ is unknown in general; see Exercise 2.1.3.

However, the π-periodicity of A imposes some structure on the behavior of solutions. To develop our intuition, consider the scalar problem

$$\partial_x y = a(x)y, \quad a(x+\pi) = a(x).$$

The FMS $\mathbf{\Phi}(x)$ satisfying $\mathbf{\Phi}(0) = \mathbf{I}_1$ is given by

$$\mathbf{\Phi}(x) = \mathrm{e}^{\int_0^x a(t)\,\mathrm{d}t}.$$

Define the average (mean) of $a(t)$,

$$\overline{a} := \frac{1}{\pi}\int_0^{\pi} a(t)\,\mathrm{d}t.$$

Setting

$$p(x) := \int_0^x (a(t) - \overline{a})\,\mathrm{d}t,$$

it is easy to see that $p(x)$ is π-periodic. Introducing $P(x) = \mathrm{e}^{p(x)}$, which is also π-periodic, the FMS can be factored into a periodic part and an average exponential,

$$\mathbf{\Phi}(x) = P(x)\mathrm{e}^{\overline{a}x}.$$

In particular, the asymptotic behavior of solutions is determined solely by the scalar quantity $\overline{a}$. If $\overline{a} > 0\,(< 0)$, then the solution will grow (decay) exponentially fast as $x \to +\infty$. This decomposition holds for systems, but requires a preparatory lemma.

Lemma 2.1.25. *If $\mathbf{A} \in \mathbb{C}^{n\times n}$ is nonsingular, then there exists a $\mathbf{B} \in \mathbb{C}^{n\times n}$ such that $\mathrm{e}^{\mathbf{B}} = \mathbf{A}$.*

Proof. Without loss of generality we may assume that $\mathbf{A}$ is in Jordan canonical form. Indeed, if $\mathbf{J}$ is the Jordan canonical form of $\mathbf{C}$, then $\mathbf{P}^{-1}\mathbf{A}\mathbf{P} = \mathbf{J}$ for an invertible matrix $\mathbf{P}$ and if $\mathrm{e}^{\mathbf{K}} = \mathbf{J}$, then $\mathrm{e}^{\mathbf{P}\mathbf{K}\mathbf{P}^{-1}} = \mathbf{A}$. The heart of the proof is to construct a natural logarithm of the Jordan blocks. We restrict our proof to the context of Theorem 2.1.11. Let $\mu_1,\ldots,\mu_k \in \sigma(\mathbf{A})$ have multiplicities $n_1,\ldots,n_k$. Then $\mathbf{A} = \mathrm{diag}(\mathbf{A}_1,\ldots,\mathbf{A}_k)$, where each $\mathbf{A}_j \in \mathbb{C}^{n_j\times n_j}$ with $\mathbf{A}_j = \mu_j \mathbf{I}_{n_j} + \mathbf{N}_{n_j}$, that is, $\mathbf{N}_{n_j}$ is nilpotent of order n_j. Since $\mathbf{A}$ is nonsingular, $\mu_j \neq 0$ for all j. Recalling the Taylor expansion for the natural logarithm

$$\ln(1+x) = \sum_{n=1}^{\infty}(-1)^{n+1}\frac{x^n}{n}, \quad |x| < 1,$$

we introduce

$$\mathbf{B}_j := \ln(\mu_j)\mathbf{I}_{n_j} + \mathbf{S}_j, \quad \mathbf{S}_j := \sum_{k=1}^{n_j-1}\frac{(-1)^{k+1}}{k}\left(\frac{\mathbf{N}_{n_j}}{\mu_j}\right)^k,$$

where, since each $\mu_j \in \mathbb{R}\backslash\{0\}$, we may take the branch-cut for the natural logarithm from 0 along the negative imaginary axis. The sum is finite because $\boldsymbol{N}_{n_j}$ is nilpotent. It follows from Lemma 2.1.6(d) that

$$\mathrm{e}^{\boldsymbol{B}_j} = \mu_j \mathrm{e}^{\boldsymbol{S}_j},$$

while it can be shown [**114**, pp. 61–62] that

$$\mathrm{e}^{\boldsymbol{S}_j} = \boldsymbol{I}_{n_j} + \frac{1}{\mu_j}\boldsymbol{N}_{n_j};$$

hence, $\mathrm{e}^{\boldsymbol{B}_j} = \boldsymbol{A}_j$. Defining $\boldsymbol{B} := \mathrm{diag}(\boldsymbol{B}_1,\dots,\boldsymbol{B}_k)$, we have immediately that $\mathrm{e}^{\boldsymbol{B}} = \boldsymbol{A}$. □

Remark 2.1.26. The matrix $\boldsymbol{B}$ given in Lemma 2.1.25 is not unique; indeed, one can add a multiple of $2\pi\mathrm{i}$ to each sub-block,

$$\mathrm{e}^{\boldsymbol{B}_j + 2\ell\pi\mathrm{i}\boldsymbol{I}_{n_j}} = \mathrm{e}^{\boldsymbol{B}_j}\mathrm{e}^{2\ell\pi\mathrm{i}\boldsymbol{I}_{n_j}} = \boldsymbol{A}_j\boldsymbol{I}_{n_j}, \quad \ell \in \mathbb{Z}.$$

The FMS for a periodic system of ODEs can be factored into the product of a periodic and a constant matrix exponential matrix.

Theorem 2.1.27 (Floquet's theorem). *Consider (2.1.1), where* $\boldsymbol{A}(x,\boldsymbol{\lambda}) \in \mathbb{C}^{n\times n}$ *is continuous with* $\boldsymbol{A}(x+\pi,\boldsymbol{\lambda}) = \boldsymbol{A}(x,\boldsymbol{\lambda})$. *If* $\boldsymbol{\Phi}(x,\boldsymbol{\lambda})$ *is a FMS, then there exist* $\boldsymbol{B} = \boldsymbol{B}(\boldsymbol{\lambda})$ *and* $\boldsymbol{P} = \boldsymbol{P}(x,\boldsymbol{\lambda})$, *both in* $\mathbb{C}^{n\times n}$, *such that*

$$\boldsymbol{\Phi}(x,\boldsymbol{\lambda}) = \boldsymbol{P}(x,\boldsymbol{\lambda})\mathrm{e}^{\boldsymbol{B}(\boldsymbol{\lambda})x}, \quad \boldsymbol{P}(x+\pi,\boldsymbol{\lambda}) = \boldsymbol{P}(x,\boldsymbol{\lambda}). \tag{2.1.13}$$

Proof. From the uniqueness of solutions the π periodicity of $\boldsymbol{A}(x)$ implies the factorization

$$\boldsymbol{\Phi}(x+\pi) = \boldsymbol{\Phi}(x)\boldsymbol{\Phi}(\pi).$$

Since $\Phi(\pi)$ is nonsingular, from Lemma 2.1.25 there is a $\boldsymbol{B} \in \mathbb{C}^{n\times n}$ such that $\boldsymbol{\Phi}(\pi) = \mathrm{e}^{\boldsymbol{B}\pi}$, which yields $\boldsymbol{\Phi}(x+\pi) = \boldsymbol{\Phi}(x)\mathrm{e}^{\boldsymbol{B}\pi}$. Introducing $\boldsymbol{P}(x) := \boldsymbol{\Phi}(x)\mathrm{e}^{-\boldsymbol{B}x}$, we have (2.1.13) where $\boldsymbol{P}$ is π-periodic,

$$\boldsymbol{P}(x+\pi) = \boldsymbol{\Phi}(x)\mathrm{e}^{\boldsymbol{B}\pi}\mathrm{e}^{-\boldsymbol{B}(x+\pi)} = \boldsymbol{\Phi}(x)\mathrm{e}^{-\boldsymbol{B}x} = \boldsymbol{P}(x).$$ □

Definition 2.1.28. The matrix $\mathrm{e}^{\boldsymbol{B}(\boldsymbol{\lambda})\pi}$ constructed in Theorem 2.1.27 is called the monodromy operator for $\boldsymbol{A}$ and its eigenvalues $\gamma = \gamma(\boldsymbol{\lambda}) \in \sigma\left(\mathrm{e}^{\boldsymbol{B}(\boldsymbol{\lambda})\pi}\right)$ are called the Floquet multipliers. The eigenvalues $\mu = \mu(\boldsymbol{\lambda}) \in \sigma(\boldsymbol{B}(\boldsymbol{\lambda}))$ are called the characteristic or Floquet exponents of $\boldsymbol{A}$.

The Floquet multipliers are unique; moreover, if $\boldsymbol{\Phi}(x,\boldsymbol{\lambda})$ is the principal FMS at $x = 0$, then

$$\boldsymbol{\Phi}(\pi,\boldsymbol{\lambda}) = \boldsymbol{P}(\pi,\boldsymbol{\lambda})\mathrm{e}^{\boldsymbol{B}(\boldsymbol{\lambda})\pi} = \mathrm{e}^{\boldsymbol{B}(\boldsymbol{\lambda})\pi},$$

which shows that the Floquet multipliers are precisely the eigenvalues of $\mathbf{\Phi}(\pi,\boldsymbol{\lambda})$. Conversely, the Floquet exponents are the logarithms of the eigenvalues of $\mathbf{\Phi}(\pi,\boldsymbol{\lambda})$ and hence are defined only up to a multiple of $2\pi\mathrm{i}$; that is,

$$\gamma \in \sigma(\mathbf{\Phi}(\pi,\boldsymbol{\lambda})) \quad \Leftrightarrow \quad \ln\gamma \in \sigma(\boldsymbol{B}(\boldsymbol{\lambda})\pi).$$

The Floquet multipliers play a significant role in determining the asymptotic behavior of solutions.

Lemma 2.1.29. *For each characteristic exponent $\mu(\boldsymbol{\lambda}) \in \sigma(\boldsymbol{B}(\boldsymbol{\lambda}))$ there exists a (possibly complex-valued) solution to (2.1.1) of the form $\boldsymbol{y}(x,\boldsymbol{\lambda}) = \mathrm{e}^{\mu(\boldsymbol{\lambda})x}\boldsymbol{p}(x,\boldsymbol{\lambda})$, where $\boldsymbol{p}(x,\boldsymbol{\lambda})$ is π-periodic in x for fixed $\boldsymbol{\lambda}$.*

Proof. We suppress the $\boldsymbol{\lambda}$-dependence. By Theorem 2.1.27 the principal FMS at $x = 0$ takes the form $\mathbf{\Phi}(x) = \boldsymbol{P}(x)\mathrm{e}^{\boldsymbol{B}x}$ with $\boldsymbol{P}(x) = \boldsymbol{I}_n$. Since $\mu \in \sigma(\boldsymbol{B})$, there exists a $\boldsymbol{v}$ such that $\boldsymbol{B}\boldsymbol{v} = \mu\boldsymbol{v}$. The solution to (2.1.1) emanating from $\boldsymbol{v}$ is given by $\boldsymbol{y}(x) := \mathbf{\Phi}(x)\boldsymbol{v} = \mathrm{e}^{\mu x}\boldsymbol{P}(x)\boldsymbol{v}$. Setting $\boldsymbol{p}(x) := \boldsymbol{P}(x)\boldsymbol{v}$ give the desired result. □

The results presented in Chapter 2.1.2 and Lemma 2.1.29 impose the following dichotomies on solutions of (2.1.1).

Lemma 2.1.30. *Let $\boldsymbol{A}(x,\boldsymbol{\lambda})$ be π periodic in x and let $\mathbf{\Phi}(x,\boldsymbol{\lambda})$ be the corresponding principal FMS of (2.1.1) at $x = 0$. For each $\gamma \in \sigma(\mathbf{\Phi}(\pi,\boldsymbol{\lambda}))$, then*

(a) $|\gamma| > 1$ is associated with a solution that grows with exponential rate $\ln|\gamma| > 0$ as $x \to +\infty$,
(b) $|\gamma| < 1$ is associated with a solution that decays with exponential rate $\ln|\gamma| < 0$ as $x \to +\infty$,
(c) $|\gamma| = 1$ implies that there is a solution that is nondecaying and uniformly bounded in norm.

In particular, if $\gamma = 1$, then there is a π-periodic solution.

The FMS cannot generically be determined explicitly, and one cannot directly compute the Floquet multipliers. However, partial information can be obtained via Liouville's formula; see Lemma 2.1.2.

Lemma 2.1.31. *If $\gamma_j(\boldsymbol{\lambda}) = \mathrm{e}^{\mu_j(\boldsymbol{\lambda})\pi}$ are the Floquet multipliers, then*

(a) $\displaystyle\prod_{j=1}^{n}\gamma_j(\boldsymbol{\lambda}) = \mathrm{e}^{\int_0^{\pi}\operatorname{tr}\boldsymbol{A}(t,\boldsymbol{\lambda})\,\mathrm{d}t}$,

(b) $\displaystyle\sum_{j=1}^{n}\mu_j(\boldsymbol{\lambda}) = \frac{1}{\pi}\int_0^{\pi}\operatorname{tr}\boldsymbol{A}(t,\boldsymbol{\lambda})\,\mathrm{d}t \pmod{2\mathrm{i}}$.

2.1.4 General Matrices and Exponential Dichotomies

In the general case of a matrix $A = A(x)$, much less can be said about the system (2.1.1). However, the idea of the exponential dichotomy, expressed in Theorem 2.1.23, has a natural, slightly more restrictive, extension.

Definition 2.1.32. The system (2.1.1) is said to have an exponential dichotomy if there exists a FMS, $\boldsymbol{\Phi}$, a projection $\boldsymbol{P} : \mathbb{C}^n \to \mathbb{C}^n$, and constants $M \geq 1$ and $\alpha > 0$ such that

$$|\boldsymbol{\Phi}(x)\boldsymbol{P}\boldsymbol{\Phi}^{-1}(y)| \leq M\mathrm{e}^{-\alpha(x-y)}, \quad y \leq x,$$
$$|\boldsymbol{\Phi}(x)(\boldsymbol{I}-\boldsymbol{P})\boldsymbol{\Phi}^{-1}(y)| \leq M\mathrm{e}^{-\alpha(y-x)}, \quad y \geq x.$$

If $\boldsymbol{A}$ is spatially constant, then it has an exponential dichotomy in this sense if and only if it is hyperbolic, i.e., if it has no center space. In this case we may take $\boldsymbol{\Phi} = \mathrm{e}^{\boldsymbol{A}x}$ and $\boldsymbol{P}$ to be the spectral projection onto the stable space, $\mathbb{E}^{\mathrm{s}}$, of $\boldsymbol{A}$. Then $\boldsymbol{P}$ commutes with $\boldsymbol{\Phi}$ and

$$\boldsymbol{\Phi}(x)\boldsymbol{P}\boldsymbol{\Phi}^{-1}(y) = \mathrm{e}^{\boldsymbol{A}(x-y)}\boldsymbol{P} = \boldsymbol{P}\mathrm{e}^{\boldsymbol{A}(x-y)}.$$

In particular, Theorem 2.1.23 implies that for all $\epsilon > 0$ there exists $M(\epsilon) > 0$ such that

$$|\boldsymbol{\Phi}(x)\boldsymbol{P}\boldsymbol{\Phi}^{-1}(y)| \leq M(\epsilon)\mathrm{e}^{(\sigma_M^{\mathrm{s}}+\epsilon)(x-y)}, \quad y \leq x,$$

and since $\sigma_M^{\mathrm{s}} < 0$, we may choose ϵ so small that $\alpha := -\sigma_M^{\mathrm{s}} - \epsilon > 0$. The case $y \geq x$ follows similarly, since for a hyperbolic matrix, $\boldsymbol{I} - \boldsymbol{P}$ is the spectral projection onto the unstable manifold.

In Chapter 3 we relate the exponential dichotomy generated by piecewise constant matrices

$$\boldsymbol{A}(x) = \begin{cases} \boldsymbol{A}_+, & x > 0, \\ \boldsymbol{A}_-, & x < 0, \end{cases}$$

to the properties of the stable space of $\boldsymbol{A}_+$ and the unstable space of $\boldsymbol{A}_-$. In Chapter 9 this construction is extended to matrices that converge, at an exponential rate, to constant matrices at $x \to \pm\infty$. More specifically, we consider spatially varying matrices that satisfy

$$\lim_{x\to\infty} \mathrm{e}^{rx}|\boldsymbol{A}(x) - \boldsymbol{A}_+| = 0, \quad \lim_{x\to-\infty} \mathrm{e}^{-rx}|\boldsymbol{A}(x) - \boldsymbol{A}_-| = 0,$$

for some $r > 0$, and relate the exponential dichotomy of the associated system, (2.1.1), to the properties of the stable space of $\boldsymbol{A}_+$ and the unstable space of $\boldsymbol{A}_-$.

The spectral properties of a linear operator, $\boldsymbol{L}$, defined by

$$\boldsymbol{L}\boldsymbol{y} := \partial_x \boldsymbol{y} - \boldsymbol{A}(x)\boldsymbol{y},$$

acting on its domain, the space of continuously differentiable functions in the sup-norm, is intimately tied to the exponential dichotomy of the semigroup generated by $\boldsymbol{A}$. Indeed, Palmer has shown that an operator of the form $\boldsymbol{L}$ is Fredholm, defined in Chapter 2.2.5, if and only if $\boldsymbol{A}$ generates an exponential dichotomy. Several of the theorems in Chapter 3 regarding the spectra of linear operators are special cases of Palmer's theorems, [**215**, **216**].

Exercises

Exercise 2.1.1. Recalling (2.1.6), derive a representation for $\partial_\lambda^2 \boldsymbol{y}(x,0)$ under the assumption that $\partial_\lambda^2 \boldsymbol{A}(x,0) = \boldsymbol{0}_n$, where $\boldsymbol{0}_n$ is the $n \times n$ zero matrix.

Exercise 2.1.2. Prove Lemma 2.1.6.

Exercise 2.1.3. If $\boldsymbol{A} = \boldsymbol{A}(x)$, then the exponential matrix $\boldsymbol{E}(x) := \mathrm{e}^{\int_{x_0}^{x} \boldsymbol{A}(s)\,\mathrm{d}s}$ is well-defined but *is not* generically a FMS for (2.1.1). Indeed, show that $\boldsymbol{E}$ is a FMS on an interval I if and only if $\boldsymbol{A}$ satisfies the restrictive condition

$$\boldsymbol{A}(s)\boldsymbol{A}(t) = \boldsymbol{A}(t)\boldsymbol{A}(s),$$

for all $s,t \in \boldsymbol{I}$. *Hint:* Matrices commute if and only if they are similar.

Exercise 2.1.4. For each $\epsilon > 0$, each $j \in \mathbb{N}$, and for all $x \geq 0$, show that

$$x^j \leq \frac{j!}{\epsilon^j}\mathrm{e}^{\epsilon x}. \tag{2.1.14}$$

Hint: Take j derivatives.

Exercise 2.1.5. Prove Theorem 2.1.23. Consider $\boldsymbol{A}$ restricted to its stable, center, and unstable spaces for $x > 0$. For $x < 0$ consider $-\boldsymbol{A}$.

Exercise 2.1.6. Prove Lemma 2.1.31 using Liouville's formula (see Lemma 2.1.2).

Exercise 2.1.7. Show that if the constant matrix $\boldsymbol{A}$ has a nontrivial center space $\mathbb{E}^c$, then the system (2.1.1) does not have an exponential dichotomy.

2.2 Elements of Functional Analysis

The following elements of Sobolev spaces and functional analysis are used throughout the book. For further details see, e.g., [**6**, **81**, **270**].

2.2.1 Basic Sobolev Spaces

For functions $u : \mathbb{R} \mapsto \mathbb{C}$ we define the L^p-norm for any $p \geq 1$,

$$\|u\|_p := \left(\int_{\mathbb{R}} |u(x)|^p \,\mathrm{d}x \right)^{1/p}.$$

The L^∞-norm is realized as the $p \to \infty$ limit of the L^p-norm, and is given for smooth functions by

$$\|u\|_\infty := \sup_{x \in \mathbb{R}} |u(x)|.$$

For any $p \geq 1$ the Banach space $L^p(\mathbb{R})$ is given by

$$L^p(\mathbb{R}) := \{u : \|u\|_p < \infty\}.$$

For differentiable functions we define the $W^{k,p}$-norm

$$\|u\|_{W^{k,p}} := \left(\sum_{j=0}^{k} \|\partial_x^j u\|_p^p \right)^{\frac{1}{p}},$$

and the associated space

$$W^{k,p}(\mathbb{R}) := \{u : \|u\|_{W^{k,p}} < \infty\}.$$

The Hilbert spaces $H^k := W^{k,2}$ are used frequently throughout the text; in particular, we remark that $H^0(\mathbb{R}) = L^2(\mathbb{R})$.

We introduce the inner product

$$\langle f, g \rangle := \int_{\mathbb{R}} f(x)\overline{g(x)}\,\mathrm{d}x,$$

where the overbar denotes complex conjugation. The spaces $H^k(\mathbb{R})$ are Hilbert spaces, since their norm is induced by the inner product

$$\|u\|_{H^k}^2 = \sum_{j=0}^{k} \langle \partial_x^j u, \partial_x^j u \rangle.$$

Introducing the Fourier transform of u,

$$\hat{u}(\eta) := \frac{1}{\sqrt{2\pi}} \int_{\mathbb{R}} \mathrm{e}^{-\mathrm{i}\eta x} u(x)\,\mathrm{d}x,$$

and its inverse,

$$u(x) = \frac{1}{\sqrt{2\pi}} \int_{\mathbb{R}} \mathrm{e}^{\mathrm{i}\eta x} \hat{u}(\eta)\,\mathrm{d}\eta,$$

we have Plancherel's equality

$$\|u\|_2 = \|\hat{u}\|_2. \tag{2.2.1}$$

A particularly useful property of the Fourier transform is that it exchanges differentiation for algebraic multiplication,

$$\widehat{\partial_x^\ell u} = (ik)^\ell \hat{u}. \tag{2.2.2}$$

Moreover, for each $k > 0$ the following norm is equivalent to the usual norm on $H^k(\mathbb{R})$,

$$\|u\|_{H^k}^2 = \int_{\mathbb{R}} \left(1 + |\eta|^{2k}\right) |\hat{u}(\eta)|^2 \, \mathrm{d}\eta.$$

Young's inequality states that for all $a, b \in \mathbb{R}$ and any conjugate exponents $p, q \geq 1$,

$$|ab| \leq \frac{|a|^p}{p} + \frac{|b|^q}{q}, \quad \frac{1}{p} + \frac{1}{q} = 1. \tag{2.2.3}$$

Young's inequality is used to prove Hölder's inequality, which states that for all conjugate exponents $p, q \geq 1$, then

$$|\langle f, g\rangle| \leq \|f\|_p \|g\|_q, \quad \frac{1}{p} + \frac{1}{q} = 1. \tag{2.2.4}$$

The convolution of f with g is defined by

$$f * g(x) := \int_{\mathbb{R}} f(x-y) g(y) \, \mathrm{d}y, \tag{2.2.5}$$

and for $p, q, r \geq 1$ there exists $c > 0$ for which we have the convolution inequality

$$\|f * g\|_r \leq c \|f\|_p \|g\|_q, \quad 1 + \frac{1}{r} = \frac{1}{p} + \frac{1}{q}. \tag{2.2.6}$$

There are many embeddings of Sobolev spaces. For example, from Hölder's inequality we may derive the embedding

$$\|u\|_\infty^2 = \sup_{x \in \mathbb{R}} \left| \int_x^{+\infty} \partial_y (u^2) \, \mathrm{d}y \right| \leq 2\|u\|_2 \|u_x\|_2 \leq \|u\|_2^2 + \|u_x\|_2^2 = \|u\|_{H^1}^2, \tag{2.2.7}$$

so that if $u \in H^1(\mathbb{R})$, then $u \in L^\infty(\mathbb{R})$. Moreover, $H^k(\mathbb{R})$ is a Banach algebra for any $k \geq 1$; that is, there exists a constant $C \geq 1$ such that for all $u \in H^k$,

$$\|u^\ell\|_{H^k} \leq C \|u\|_{H^k}^\ell, \quad \ell \in \mathbb{N}.$$

This property makes the map $u \mapsto u^\ell$ continuous in the $H^k(\mathbb{R})$ norm. Indeed, for any $\ell \geq 2$,

$$\|u\|_{H^k} \leq \epsilon \quad \Rightarrow \quad \|u^\ell\|_{H^k} \leq C\epsilon^{\ell-1}\|u\|_{H^k}.$$

The spaces $H^\ell(\mathbb{R}) \subset H^k(\mathbb{R})$ are dense for $\ell > k$, i.e., for each $u \in H^k(\mathbb{R})$ there is a sequence $\{u_j\} \subset H^\ell(\mathbb{R})$ such that $\|u_j - u\|_{H^k} \to 0$ as $j \to +\infty$. Moreover, if $I \subset \mathbb{R}$ is compact, then the embedding of $H^\ell(I)$ into $H^k(I)$ for $k > \ell + \frac{1}{2}$ is compact; that is, a bounded sequence in $H^k(I)$ has a convergent subsequence in $H^\ell(I)$. This result is useful for demonstrating that a linear operator is compact.

For $\boldsymbol{G} \in L^2(\mathbb{R})^{k\times l}$ with components $[\boldsymbol{G}]_{ij} = g_{ij} \in L^2(\mathbb{R})$, we denote the tensor operator $\otimes\boldsymbol{G} : L^2(\mathbb{R}) \mapsto \mathbb{R}^{k\times l}$, which acts on $h \in L^2(\mathbb{R})$, by a componentwise inner product,

$$[\otimes\boldsymbol{G}\cdot h]_{ij} = \langle g_{ij}, h\rangle. \tag{2.2.8}$$

If $\boldsymbol{F} \in L^2(\mathbb{R})^{j\times k}$, then the tensor product $\boldsymbol{F}\otimes\boldsymbol{G}$ is a finite-rank map that takes $h \in L^2(\mathbb{R})$ to $\boldsymbol{F}\otimes\boldsymbol{G}\cdot h \in L^2(\mathbb{R})^{j\times m}$ through the usual matrix multiplication of $\boldsymbol{F}$ with $\otimes\boldsymbol{G}\cdot h$.

2.2.2 Bounded and Closed Operators

We review some basic ideas associated with linear operators acting on Banach spaces: for further details, see Hărăguş and Iooss [**126**, Appendix A] and [**162**]. Let X, Y be two Banach spaces with norms $\|\cdot\|_X, \|\cdot\|_Y$ respectively, and assume that $Y \subset X$ is dense, for example $X = L^2(\mathbb{R})$ and $Y = H^k(\mathbb{R})$ for any $k \geq 1$. Consider linear operators $\mathcal{L}$, with $Y = \mathrm{D}(\mathcal{L})$, the domain of $\mathcal{L}$, dense in X and $\mathcal{L} : Y \mapsto X$. The kernel of $\mathcal{L}$ is given by

$$\ker(\mathcal{L}) := \{u \in Y : \mathcal{L}u = 0\}.$$

We say that a linear operator is closed if for any sequence $\{u_j\} \subset Y$ with

$$\lim_{j\to+\infty} \|u_j - u\|_X = 0 \quad \textit{and} \quad \lim_{j\to+\infty} \|\mathcal{L}u_j - v\|_X = 0,$$

then we have $u \in Y$ and $\mathcal{L}u = v$. This is *equivalent* to saying that the domain $\mathrm{D}(\mathcal{L})$ is complete under the graph norm of $\mathcal{L}$,

$$\|u\|_{\mathrm{D}(\mathcal{L})} := \|u\|_X + \|\mathcal{L}u\|_X.$$

We say that the operator is bounded from Y to X if

$$\sup\{\|\mathcal{L}u\|_X : u \in Y,\ \|u\|_Y = 1\} < \infty.$$

We denote the space of bounded linear operators from Y into X by $\mathcal{B}(Y,X)$ with the induced norm of $\mathcal{L}$ given by

$$\|\mathcal{L}\|_{\mathcal{B}(Y,X)} := \sup_{\|u\|_Y \neq 0} \frac{\|\mathcal{L}u\|_X}{\|u\|_Y}.$$

If $X = Y$ then the space is denoted $\mathcal{B}(X)$ and the induced norm of $\mathcal{L}$ by $\|\mathcal{L}\|_{\mathcal{B}(X)}$ or by $\|\mathcal{L}\|$ if the context is sufficiently clear. The sum of a closed operator and a bounded operator is a closed operator; however, if two closed operators have a common domain, then their sum is not necessarily a closed operator. If $\mathcal{L}$ is a closed operator with $Y = X$, then $\mathcal{L}$ is a bounded operator. If for each bounded sequence $\{u_j\} \subset Y$ the sequence $\{\mathcal{L}u_j\} \subset X$ has a convergent subsequence, then the operator $\mathcal{L}$ is said to be compact. A compact operator is bounded; furthermore, the sum of two compact operators is compact, and the composition of a compact operator and a bounded operator is compact.

2.2.3 Variational Derivatives

Consider two infinite-dimensional Hilbert spaces $Y \subset X \subset Y^*$, where Y is dense in X with respect to the norm $\|\cdot\|_X$, and $Y^* = \mathcal{B}(Y,\mathbb{R})$ is the dual of Y with respect to the X inner product. The norm on X is generated by the inner product $\langle\cdot,\cdot\rangle_X$, and that on Y by $\langle\cdot,\cdot\rangle_Y$. A Hamiltonian on X is a nonlinear functional $\mathcal{H}: Y \subset X \mapsto \mathbb{R}$.

The Hamilton $\mathcal{H}$ has a first variation, often called the first variational derivative of $\mathcal{H}$, with respect to $\langle\cdot,\cdot\rangle_X$, denoted $\delta\mathcal{H}/\delta u$, if for all $u, v \in Y$ the limit

$$\lim_{\epsilon\to 0} \frac{\mathcal{H}(u+\epsilon v) - \mathcal{H}(u)}{\epsilon},$$

exists and if the map from v to the value of the limit is bounded and linear in v. In this case, from the Riesz representation theorem there exists a functional $\delta\mathcal{H}/\delta u(u) \in Y^*$ such that

$$\lim_{\epsilon\to 0} \frac{\mathcal{H}(u+\epsilon v) - \mathcal{H}(u)}{\epsilon} = \left\langle \frac{\delta\mathcal{H}}{\delta u}(u), v \right\rangle_X.$$

Dropping the ϵ scaling on v, we observe that for each fixed $u \in Y$ the first variation induces a linear form $b_1 \in \mathcal{B}(Y,\mathbb{R})$,

$$b_1[v] := \left\langle \frac{\delta\mathcal{H}}{\delta u}(u), v \right\rangle_X.$$

If for all $M > 0$ there exists a $c > 0$ such that

$$|\mathcal{H}(u+v) - \mathcal{H}(u) - b_1[v]| \le c\|v\|_Y^2,$$

for all $\|u\|_Y, \|v\|_Y \le M$, then we say that $\mathcal{H}$ is C^1 in the Y norm. Similarly if there is a bounded, bilinear form $b_2 : Y \times Y \to \mathbb{R}$ with the property that for all $M > 0$ there exists a $c > 0$ such that

$$|\mathcal{H}(u+v) - \mathcal{H}(u) - b_1[v] - b_2[v,v]| \le c\|v\|_Y^3,$$

for all $\|u\|_Y, \|v\|_Y \le M$, then we say that $\mathcal{H}$ is C^2 in the Y norm. Such a bilinear form induces a linear map,

$$b_2[v_1, v_2] = \langle \mathcal{L} v_1, v_2 \rangle_X,$$

where

$$\mathcal{L} := \frac{\delta^2 \mathcal{H}}{\delta u^2}(u) : Y \subset X \to Y^*,$$

is called the second variational derivative of $\mathcal{H}$. In general, $\mathcal{H}$ is C^k in the Y norm if there exists bounded, multilinear forms b_j which map j copies of Y to $\mathbb{R}$ such that for all $M > 0$ there exists $c > 0$ for which

$$\left| \mathcal{H}(u+v) - \mathcal{H}(u) - \sum_{j=1}^{k} b_j[v, \dots, v] \right| \le c\|v\|_Y^{k+1},$$

for all $\|u\|_Y, \|v\|_Y \le M$. Moreover, if the first element of v_1 is frozen in b_j, then it induces an element $b_j[v_1]$ which is $j-1$ multilinear. In particular, for $j = 3$ the reduction induces a bilinear form, and hence a linear operator $\mathcal{L}[v_1] : Y \subset X \mapsto Y^*$ denoted by $\frac{\delta^3 \mathcal{H}}{\delta u^3}(u_0)[v_1]$, which satisfies

$$b_3[v_1, v, w] := \langle \mathcal{L}[v_1] v, w \rangle = \left\langle \frac{\delta^3 \mathcal{H}}{\delta u^3}(u_0)[v_1] v, w \right\rangle, \quad \text{for } v, w \in Y$$

This framework is used in Chapter 5 and Chapter 7, and in particular in Section 7.2.

2.2.4 Resolvent and Spectrum

The resolvent set of $\mathcal{L}$, $\rho(\mathcal{L})$, is the set of complex numbers $\lambda \in \mathbb{C}$ such that

(a) $\lambda \mathcal{I} - \mathcal{L}$ is invertible,
(b) $(\lambda \mathcal{I} - \mathcal{L})^{-1}$ is a bounded linear operator.

Here $\mathcal{I} : X \mapsto X$ is the identity operator, i.e., $\mathcal{I}u = u$. For $\lambda \in \rho(\mathcal{L})$ the operator $(\lambda\mathcal{I} - \mathcal{L})^{-1}$ is called the resolvent of $\mathcal{L}$. The spectrum of $\mathcal{L}$ is the complement of the resolvent set, i.e.,

$$\sigma(\mathcal{L}) = \mathbb{C}\backslash\rho(\mathcal{L}).$$

A complex number $\lambda \in \sigma(\mathcal{L})$ is called an eigenvalue if $\ker(\mathcal{L} - \lambda\mathcal{I}) \neq \{0\}$. If $\mathcal{L}$ is a closed operator, then $\sigma(\mathcal{L})$ is a closed set. If $\mathcal{L}$ is a bounded operator, then $\sigma(\mathcal{L})$ is a closed, bounded, and nonempty set.

Suppose that $\lambda \in \sigma(\mathcal{L})$ is an eigenvalue. The dimension of $\ker(\mathcal{L} - \lambda\mathcal{I})$ is called the geometric multiplicity of the eigenvalue, and is denoted by $\mathrm{m_g}(\lambda)$. If $\mathrm{m_g}(\lambda) = 1$, then the eigenvalue is called geometrically simple. If the eigenvalue is isolated, then the algebraic multiplicity of the eigenvalue, denoted $\mathrm{m_a}(\lambda)$, is the dimension of the largest subspace $Y_\lambda \subset Y$, which

(a) is invariant under the action of $\mathcal{L}$, i.e., if $u_\lambda \in Y_\lambda$, then $\mathcal{L}u_\lambda \in Y_\lambda$
(b) satisfies the property $\sigma(\mathcal{L}|_{Y_\lambda}) = \{\lambda\}$.

If $\mathrm{m_a}(\lambda) = 1$, then the eigenvalue is called algebraically simple or simple. It is true that $\mathrm{m_a}(\lambda) \geq \mathrm{m_g}(\lambda)$. An eigenvalue is called *semi-simple* if $\mathrm{m_g}(\lambda) = \mathrm{m_a}(\lambda)$. If $\mathcal{L}$ is a compact operator whose domain Y is separable, i.e. has a countably infinite dense subset, then the following hold:

(a) $0 \in \sigma(\mathcal{L})$,
(b) if $\lambda \in \sigma(\mathcal{L})$ with $\lambda \neq 0$, then λ is isolated and $\mathrm{m_a}(\lambda) < \infty$,
(c) $\sigma(\mathcal{L})$ is a countable set, and the only possible accumulation point is $\lambda = 0$.

If λ is an isolated eigenvalue, let $C \subset \mathbb{C}$ be a simple closed positively oriented curve surrounding λ that does not intersect the spectrum of $\mathcal{L}$ and whose interior contains no other points in $\sigma(\mathcal{L})$. The spectral projection $P(\lambda) : X \mapsto Y_\lambda$ is given by the Dunford integral formula

$$P(\lambda) := \frac{1}{2\pi\mathrm{i}} \oint_C (\zeta\mathcal{I} - \mathcal{L})^{-1}\,\mathrm{d}\zeta. \tag{2.2.9}$$

The operator $P(\lambda)$ commutes with $\mathcal{L}$, i.e., $P(\lambda)\mathcal{L} = \mathcal{L}P(\lambda)$; furthermore, $P(\lambda)^2 := P(\lambda)P(\lambda) = P(\lambda)$. Most significantly, the range of $P(\lambda)$ is the $\mathcal{L}$-invariant subspace Y_λ and $\mathcal{L}|_{Y_\lambda} = \mathcal{L}P(\lambda)$.

A linear operator $\mathcal{L} : Y \mapsto X$ has compact resolvent if

(a) $\rho(\mathcal{L}) \neq \emptyset$,
(b) for some $\lambda \in \rho(\mathcal{L})$ the resolvent operator $(\lambda\mathcal{I} - \mathcal{L})^{-1} : X \mapsto Y \subset X$ is compact in $\mathcal{B}(X, X)$.

If $(\lambda\mathcal{I} - \mathcal{L})^{-1}$ is a compact operator for one $\lambda \in \rho(\mathcal{L})$, then it is compact for all $\lambda \in \rho(\mathcal{L})$. If $\mathcal{L}$ is an operator with a compact resolvent, then $\sigma(\mathcal{L})$ is a countable set of isolated eigenvalues with finite algebraic multiplicities for which the only possible accumulation point is $\lambda = \infty$.

2.2.5 Adjoint and Fredholm Operators

Assume that X is a Hilbert space equipped with the inner product $\langle\cdot,\cdot\rangle$, and that $\mathcal{L}$ is a closed operator with a dense domain. The domain, $\mathrm{D}(\mathcal{L}^{\mathrm{a}})$, of the adjoint operator, $\mathcal{L}^{\mathrm{a}}$, is the set of all $v \in X$ for which the linear functional

$$u \to \langle \mathcal{L}u, v\rangle,$$

is continuous in the Hilbert norm on X. From the Riesz representation theorem we deduce that there exists $w \in X$ for which

$$\langle \mathcal{L}u, v\rangle = \langle u, w\rangle.$$

For such $v \in \mathrm{D}(\mathcal{L}^{\mathrm{a}})$ the adjoint operator is defined by the map $\mathcal{L}^{\mathrm{a}}v = w$; that is, the adjoint operator is the unique operator that satisfies

$$\langle \mathcal{L}u, v\rangle = \langle u, \mathcal{L}^{\mathrm{a}}v\rangle,$$

for all $u \in X$ and $v \in \mathrm{D}(\mathcal{L}^{\mathrm{a}})$. The adjoint operator is also closed, and its domain is also dense in X. For example, consider the second-order differential operator

$$\mathcal{L} := \partial_x^2 + a_1(x)\partial_x + a_0(x) : H^2(\mathbb{R}) \mapsto L^2(\mathbb{R}),$$

the coefficients of which are uniformly smooth and bounded. It acts upon $L^2(\mathbb{R})$, but its domain in the graph norm is $H^2(\mathbb{R})$. Integration by parts shows that the adjoint operator is given by

$$\mathcal{L}^{\mathrm{a}} = \partial_x^2 - \partial_x[a_1(x)\cdot] + a_0(x) : H^2(\mathbb{R}) \mapsto L^2(\mathbb{R}).$$

The resolvent and spectrum of an operator and its adjoint are related by

$$\rho(\mathcal{L}^{\mathrm{a}}) = \overline{\rho(\mathcal{L})}, \quad \sigma(\mathcal{L}^{\mathrm{a}}) = \overline{\sigma(\mathcal{L})},$$

and the resolvents are related through

$$(\overline{\lambda}\mathcal{I} - \mathcal{L}^{\mathrm{a}})^{-1} = \overline{(\lambda\mathcal{I} - \mathcal{L})^{-1}}.$$

In particular, this identity implies that if $\mathcal{L}$ has compact resolvent, then so does $\mathcal{L}^{\mathrm{a}}$. An operator $\mathcal{L}$ is said to be *self-adjoint* if $\mathrm{D}(\mathcal{L}) = \mathrm{D}(\mathcal{L}^{\mathrm{a}})$ and $\mathcal{L}u = \mathcal{L}^{\mathrm{a}}u$ for all $u \in \mathrm{D}(\mathcal{L})$. For self-adjoint operators it is known that $\sigma(\mathcal{L}) \subset \mathbb{R}$, and that all eigenvalues are semi-simple, i.e., $\mathrm{m_g}(\lambda) = \mathrm{m_a}(\lambda)$.

The operator $\mathcal{L}$ is a Fredholm operator if

(a) $\ker(\mathcal{L})$ is finite-dimensional,
(b) $\mathrm{R}(\mathcal{L})$ is closed with finite codimension.

Here $\mathrm{R}(\mathcal{L})$ denotes the range of $\mathcal{L}$. The Fredholm index of a Fredholm operator is defined by

$$\mathrm{ind}(\mathcal{L}) = \dim[\ker(\mathcal{L})] - \mathrm{codim}[\mathrm{R}(\mathcal{L})].$$

The operator $\mathcal{L}$ is Fredholm if and only if $\mathcal{L}^{\mathrm{a}}$ is, and the indices are related via

$$\mathrm{ind}(\mathcal{L}) = -\mathrm{ind}(\mathcal{L}^{\mathrm{a}}).$$

If $\lambda \in \sigma(\mathcal{L})$ is an isolated eigenvalue with $\mathrm{m_a}(\lambda) < \infty$, then $\lambda\mathcal{I} - \mathcal{L}$ is a Fredholm operator with index 0. It is easy to see that the range of $\mathcal{L}$ must be orthogonal to the kernel of $\mathcal{L}^{\mathrm{a}}$; indeed, if $v \in \ker(\mathcal{L}^{\mathrm{a}})$ and $\mathcal{L}u = f$ then

$$\langle f, v\rangle = \langle \mathcal{L}u, v\rangle = \langle u, \mathcal{L}^{\mathrm{a}}v\rangle = 0.$$

The sufficiency of this condition often goes by the name of the Fredholm alternative:

Theorem 2.2.1 (Fredholm alternative). *Suppose that X is a Hilbert space with inner product $\langle\cdot,\cdot\rangle$, and $\mathcal{L} : \mathcal{D}(\mathcal{L}) \subset X \mapsto X$ is a closed Fredholm operator with domain $\mathcal{D}(\mathcal{L}) \subset X$ dense in X-norm. For $f \in X$ the nonhomogeneous problem $\mathcal{L}u = f$ has a solution $u \in \mathcal{D}(\mathcal{L})$ if and only if $f \in \ker(\mathcal{L}^{\mathrm{a}})^{\perp}$; in other words,*

$$\mathrm{R}(\mathcal{L}) = \ker(\mathcal{L}^{\mathrm{a}})^{\perp}.$$

Moreover, the Fredholm index counts the dimensional mismatch between the kernels of $\mathcal{L}$ and $\mathcal{L}^{a}$,

$$\dim[\ker(\mathcal{L})] - \dim[\ker(\mathcal{L}^{\mathrm{a}})] = \mathrm{ind}(\mathcal{L}).$$

For any Fredholm operator the space X can be decomposed as

$$X = \mathrm{R}(\mathcal{L}) \oplus \ker(\mathcal{L}^{\mathrm{a}}).$$

If in addition $\mathrm{ind}(\mathcal{L}) = 0$, then $\dim[\ker(\mathcal{L})] = \dim[\ker(\mathcal{L}^{\mathrm{a}})]$ and either $\mathcal{L}$ has a kernel or the operator is one-to-one and onto—those are the alternatives in the Fredholm alternative. However, if $\mathrm{ind}(\mathcal{L}) \neq 0$, then either $\mathcal{L}$ has a kernel, in which case $\mathcal{L}$ cannot be one-to-one, or $\mathcal{L}^{\mathrm{a}}$ has a kernel, in which case $\mathcal{L}$ cannot be onto. In both of these cases $\mathcal{L}$ cannot be invertible.

Example 2.2.2. The classic example of a Fredholm operator of nonzero index is the shift operator. Let $\{\mathrm{e}_k\}_{k=0}^{\infty}$ be an orthonormal basis of a separable Hilbert space X. The right-shift operator is defined on this basis by

$$\mathcal{S}\mathrm{e}_k = \mathrm{e}_{k+1}, \quad k \geq 0,$$

and on the remainder of X by linearity. The adjoint to S is the left-shift operator

$$\mathcal{S}^{\mathrm{a}}\mathrm{e}_k = \mathrm{e}_{k-1}, \quad k \geq 1,$$

and $\mathcal{S}^{\mathrm{a}}\mathrm{e}_0 = 0$. The operator $\mathcal{S}$ has no kernel and is one-to-one, but its range has co-dimension one, spanned by e_0, so that $\mathrm{ind}(S) = 0-1 = -1$. Conversely, $\mathcal{S}^{\mathrm{a}}$ is onto but, as required by the Fredholm alternative, has a one dimensional kernel spanned by e_0. We have $\mathrm{ind}(\mathcal{S}^{\mathrm{a}}) = 1$. Indeed, $\mathcal{S}^{\mathrm{a}}$ has a right inverse while $\mathcal{S}$ has a left inverse, so that $\mathcal{S}^{\mathrm{a}}\mathcal{S} = \mathcal{I}$ while $\mathcal{S}\mathcal{S}^{\mathrm{a}}$ has a nontrivial kernel.

From this discussion we observe that if the Fredholm index is not zero, then invertibility is hopeless, while if the Fredholm index is zero, then invertibility follows if $\mathcal{L}$ has no kernel. This observation motivates the following classification of the spectral sets of operators.

Definition 2.2.3. Let X be a Banach space and let $\mathcal{L} : \mathrm{D}(\mathcal{L}) \subset X \to X$ be a closed linear operator with domain $\mathrm{D}(\mathcal{L})$ dense in X. The spectrum of $\mathcal{L}$ is decomposed into the following two sets:

(a) The essential spectrum of a Fredholm operator $\mathcal{L}$, $\sigma_{\mathrm{ess}}(\mathcal{L})$, is the set of all $\lambda \in \mathbb{C}$ such that either

- $\lambda\mathcal{I} - \mathcal{L}$ is not Fredholm, or
- $\lambda\mathcal{I} - \mathcal{L}$ is Fredholm, but $\mathrm{ind}(\lambda\mathcal{I} - \mathcal{L}) \neq 0$.

(b) The point spectrum of a Fredholm operator $\mathcal{L}$ is the set defined by

$$\sigma_{\mathrm{pt}}(\mathcal{L}) = \{\lambda \in \mathbb{C} : \mathrm{ind}(\lambda\mathcal{I} - \mathcal{L}) = 0, \text{ but } \lambda\mathcal{I} - \mathcal{L} \text{ is not invertible}\}.$$

The elements of the point spectrum are called eigenvalues of $\mathcal{L}$.

Remark 2.2.4. Different definitions of the essential spectrum can be found in the literature (e.g., the condition $\lambda\mathcal{I} - \mathcal{L}$ is not Fredholm with index 0 is sometimes replaced by $\lambda\mathcal{I} - \mathcal{L}$ is not Fredholm, or $\lambda\mathcal{I} - \mathcal{L}$ is not semi-Fredholm—see [**162**, Chapter IV, Section 5.6] and [**260**]). Definition 2.2.3 makes the essential spectrum a large set, with the advantage that the remaining spectrum, $\sigma_{\mathrm{pt}}(\mathcal{L}) = \sigma(\mathcal{L}) \setminus \sigma_{\mathrm{ess}}(\mathcal{L})$, is a discrete set consisting of isolated eigenvalues of $\mathcal{L}$.

Locating the essential spectrum requires the computation of the Fredholm index of an operator. We will employ two principal techniques to achieve this. The first technique is to perturb a known operator.

Definition 2.2.5. The operator $\mathcal{L}$ is a relatively compact perturbation of $\mathcal{L}_0$ if $(\mathcal{L}_0 - \mathcal{L})(\lambda\mathcal{I} - \mathcal{L}_0)^{-1} : X \mapsto X$ is compact for some $\lambda \in \rho(\mathcal{L}_0)$.

There are several versions of stability theorems for relatively compact perturbations of Fredholm operators [**162**, Chapter IV, Theorem 5.26] (see also [**162**, Chapter IV, Theorem 5.35]). These are frequently referred to as the *Weyl essential spectrum Theorem*.

Theorem 2.2.6 (Weyl essential spectrum theorem). *Let $\mathcal{L}$ and $\mathcal{L}_0$ be closed linear operators in a Banach space X. If $\mathcal{L}$ is a relatively compact perturbation of $\mathcal{L}_0$, then the following properties hold:*

(a) The operator $\lambda\mathcal{I} - \mathcal{L}$ is Fredholm if and only if $\lambda\mathcal{I} - \mathcal{L}_0$ is Fredholm.

(b) $\mathrm{ind}(\lambda\mathcal{I} - \mathcal{L}) = \mathrm{ind}(\lambda\mathcal{I} - \mathcal{L}_0)$.
(c) *The operators* $\mathcal{L}$ *and* $\mathcal{L}_0$ *have the same essential spectra, i.e.,* $\sigma_{\mathrm{ess}}(\mathcal{L}) = \sigma_{\mathrm{ess}}(\mathcal{L}_0)$.

The second technique to compute the Fredholm index is to show that the operator has a compact resolvent. In this second case the operator cannot have any essential spectrum.

Theorem 2.2.7. *If X is a Banach space, $Y \subset X$ is dense, and $\mathcal{L} : Y \mapsto X$ is a closed Fredholm operator with compact resolvent, then* $\mathrm{ind}(\mathcal{L}) = 0$.

2.3 The Point Spectrum: Sturm–Liouville Theory

While the essential spectrum for an operator can be determined by studying relatively compact perturbations, the point spectrum is more mutable under this class of perturbations. Indeed, for important classes of differential operators, much can be said about the location of the essential spectrum, but relatively little about the point spectrum. However, for many scalar second-order operators, there are important results.

2.3.1 *Sturm–Liouville Operators on a Bounded Domain*

A Sturm–Liouville operator $\mathcal{L}$ takes form

$$\mathcal{L}p := \partial_x^2 p + a_1(x)\partial_x p + a_0(x)p, \tag{2.3.1}$$

and will also be called a Sturmian operator. Here we consider $\mathcal{L}$ to be defined on the bounded interval $[-1,1]$, subject to boundary conditions

$$b_1^- p(-1) + b_2^- \partial_x p(-1) = 0, \quad b_1^+ p(+1) + b_2^+ \partial_x p(+1) = 0. \tag{2.3.2}$$

Such boundary conditions are called separated since they do not couple conditions at $x = 1$ with those at $x = -1$. We assume that the coefficients in (2.3.2) satisfy $(b_1^\pm)^2 + (b_2^\pm)^2 > 0$, and the coefficients $a_1(x)$ and $a_0(x)$ in $\mathcal{L}$ are C^1 and real-valued. The spectral problem is naturally posed on $H^2_{\mathrm{bc}}[-1,+1]$, where

$$H^2_{\mathrm{bc}}[-1,+1] := \{u \in H^2[-1,+1] : b_1^\pm u(\pm 1) + b_2^\pm \partial_x u(\pm 1) = 0\}.$$

The operator $\mathcal{L}$ is self-adjoint in the weighted inner product

$$\langle u, v\rangle_\rho := \int_{-1}^{1} u(x)\overline{v(x)}\rho(x)\,\mathrm{d}x, \tag{2.3.3}$$

with associated norm $\|\cdot\|_\rho$, where the weight function is

$$\rho(x) := \mathrm{e}^{\int_0^x a_1(s)\,\mathrm{d}s} > 0.$$

The associated eigenvalue problem

$$\mathcal{L}p = \lambda p, \tag{2.3.4}$$

subject to (2.3.2) satisfies the following well-known results [**114**, Chapter XI] and [**81**, Chapter 6].

Theorem 2.3.1. *Consider the Sturmian eigenvalue problem (2.3.4) with separated boundary conditions (2.3.2) on the space $H^2_{\mathrm{bc}}([-1,+1])$. All of the eigenvalues are real-valued and simple, and can be enumerated in a strictly descending order*

$$\lambda_0 > \lambda_1 > \lambda_2 > \cdots, \quad \lim_{n\to+\infty} \lambda_n = -\infty.$$

The eigenfunction $p_j(x)$ associated with the eigenvalue λ_j for $j = 0,1,2,\ldots$, can be normalized so that

(a) p_j has j simple zeros in the open interval $(-1,+1)$.
(b) The eigenfunctions are orthonormal in the ρ-weighted inner product,

$$\langle p_j, p_k\rangle_\rho = \delta_{jk},$$

where δ is the Kronecker delta.
(c) The eigenfunctions form a complete orthonormal basis of $L^2[-1,1]$ in the ρ-weighted inner product. That is, any $u \in L^2[-1,1]$ can be expressed as

$$u = \sum_{j=0}^{\infty} u_j p_j,$$

where the sum on the right-hand side converges in $\|\cdot\|_\rho$ and $u_j := \langle u, p_j\rangle_\rho$ is the jth Fourier coefficient of u with respect to $\mathcal{L}$. In particular,

$$\|u\|_\rho^2 = \sum_{j=0}^{\infty} |u_j|^2.$$

(d) The largest, or ground-state, eigenvalue can be characterized as the supremum of the bilinear form associated to $\mathcal{L}$,

$$\lambda_0 = \sup_{\|u\|_\rho=1} \langle \mathcal{L}u, u\rangle_\rho,$$

moreover the supremum is achieved at $u = p_0$, which has no zeros on $(-1,1)$.

For $\lambda \notin \sigma(\mathcal{L})$ we have the expression for the resolvent

$$(\mathcal{L}-\lambda\mathcal{I})^{-1}u=\sum_{j=0}^{\infty}\frac{\langle u,p_j\rangle}{\lambda_j-\lambda}p_j=\int_{-1}^{1}G_\lambda(x,y)u(y)\,dy, \tag{2.3.5}$$

where the Green's function for the resolvent is given by

$$G_\lambda(x,y)=\sum_{j=0}^{\infty}\frac{p_j(x)p_j(y)}{\lambda_j-\lambda}. \tag{2.3.6}$$

A similar result holds in the case that the coefficients of $\mathcal{L}$ are periodic with a common period, i.e., $a_j(x+\pi)=a_j(x)$ for $j=0,1$. Considering the eigenvalue problem on the interval $[0,\pi]$ with the periodic boundary conditions

$$p(0)=p(\pi), \quad \partial_x p(0)=\partial_x p(\pi), \tag{2.3.7}$$

the operator $\mathcal{L}$ is still self-adjoint in the weighed inner product over the space

$$H^2_{\text{per}}[0,\pi]=\{u\in H^2[0,\pi]:u(0)=u(\pi),\,\partial_x u(0)=\partial_x u(\pi)\}.$$

The boundary conditions are no longer separated, since the condition on u at $x=0$ is connected to u at $x=\pi$, and the statement of Theorem 2.3.1 must be revised; in particular, the eigenvalues need not be simple, and the correspondence between the eigenfunction and its number of zeros differs [**200**].

Theorem 2.3.2. *Consider the eigenvalue problem (2.3.4) with periodic boundary conditions (2.3.7) in the space $H^2_{\text{per}}[0,\pi]$. All of the eigenvalues are real-valued and can be enumerated in a descending order*

$$\lambda_0>\lambda_1\geq\lambda_2\geq\lambda_3\geq\lambda_4>\cdots, \quad \lim_{n\to+\infty}\lambda_n=-\infty.$$

The eigenfunction $p_j(x)$ associated with the eigenvalue λ_j for $j=0,1,2,\ldots$ can be normalized so that:

(a) for each $n\in\mathbb{N}_0$ the eigenfunctions $\{p_{2n-1},p_{2n}\}$ each have $2n$ simple zeros in the interval $[0,\pi)$;
(b) the eigenfunctions are orthonormal in the ρ-weighted inner product and form a complete basis for $L^2[0,\pi]$.
(c) The ground-state eigenvalue can be characterized as the supremum of the bilinear form associated to $\mathcal{L}$,

$$\lambda_0=\sup_{\|u\|_\rho=1}\langle\mathcal{L}u,u\rangle_\rho,$$

moreover, the supremum is achieved at $u=p_0$, which has no zeros.

2.3.2 *Sturm–Liouville Operators on the Real Line*

Consider the Sturmian operator $\mathcal{L}$ acting on $H^2(\mathbb{R})$ with smooth coefficients $a_0(x)$ and $a_1(x)$, which decay exponentially to constants at $x = \pm\infty$, i.e.,

$$\lim_{x\to\pm\infty} \mathrm{e}^{\nu|x|}|a_1(x) - a_1^{\pm}| = 0, \quad \lim_{x\to\pm\infty} \mathrm{e}^{\nu|x|}|a_0(x) - a_0^{\pm}| = 0, \tag{2.3.8}$$

for some $\nu > 0$ and constants $a_1^{\pm}, a_0^{\pm} \in \mathbb{R}$. The operator $\mathcal{L}$ is self-adjoint in the ρ-weighted inner product, where the weight has the finite asymptotic values

$$\rho_{\pm} := \lim_{x\to\pm\infty} \mathrm{e}^{-a_1^{\pm}x}\rho(x). \tag{2.3.9}$$

Moreover, the following theorem holds [272, 273].

Theorem 2.3.3. *Consider the eigenvalue problem (2.3.4) on the space $H^2(\mathbb{R})$, where the coefficients satisfy (2.3.8). The point spectrum, $\sigma_{\mathrm{pt}}(\mathcal{L})$, consists of a finite number, possibly zero, of simple eigenvalues, which can be enumerated in a strictly descending order*

$$\lambda_0 > \lambda_1 > \cdots > \lambda_N > b := \max\{a_0^-, a_0^+\}.$$

For $j = 0,\ldots,N$ the eigenfunction $p_j(x)$ associated with the eigenvalue λ_j can be normalized so that:

(a) p_j has j simple zeros.
(b) The eigenfunctions are orthonormal in the ρ-weighted inner product.
(c) The ground-state eigenvalue, if it exists, can be characterized as the supremum of the bilinear form associated to $\mathcal{L}$,

$$\lambda_0 = \sup_{\|u\|_\rho=1} \langle \mathcal{L}u, u\rangle_\rho,$$

moreover, the supremum is achieved at $u = p_0$, which has no zeros.

2.3.3 *Examples*

2.3.3.1 A Bistable Reaction–Diffusion Equation: Pulse

Consider a reaction–diffusion equation of the form

$$\partial_t u = \partial_x^2 u - W'(u), \tag{2.3.10}$$

where the smooth potential W has unequal, consecutive local minima at $u = 0$ and $u = 1$. That is, $W'(0) = W'(1) = 0$ with $W''(0), W''(1) > 0$, but $W(0) = 0$ while $W(1) < 0$. This equation possesses equilibria, $\phi(x)$, which solve the steady-state equation

$$\partial_x^2 \phi = W'(\phi). \tag{2.3.11}$$

subject to the conditions $\phi(x) \to 0$ as $x \to \pm\infty$. Viewing the steady-state equation as a dynamical system in $(\phi, \partial_x \phi)^{\mathrm{T}}$, we use a phase plane analysis to establish the existence of a homoclinic connection ϕ to the fixed point at $(0,0)^T$. Indeed, it is straightforward to see that the dynamical system has fixed points at $(0,0)^T$ and $(1,0)^T$, and that these are saddle points. We construct the homoclinic orbit by showing that the unstable manifold to $(0,0)^T$ coincides with the stable manifold. The energy

$$E(\phi, \partial_x \phi) = \frac{1}{2}(\partial_x \phi)^2 - W(\phi), \tag{2.3.12}$$

is invariant under the flow. Since the stable and unstable manifolds to $(0,0)^T$ touch $(0,0)^T$ it must be that $E(\phi, \partial_x \phi) = E(0,0) = 0$ along both manifolds, which lie along the curves

$$\partial_x \phi = \pm\sqrt{2W(\phi)},$$

in particular, both manifolds pass through the point $(\phi_m, 0)$ where ϕ_m is the zero of W on the interval $(0,1)$ (see Fig. 2.1). By translating ϕ we may make it even about $x = 0$, so that $\partial_x \phi > 0$ for $x < 0$ and $\partial_x \phi < 0$ for $x > 0$. By the mean-value theorem W' must have a zero in the interval $(0, \phi_m)$, which corresponds to a center fixed point of the dynamical system, with the remainder of the interior of the homoclinic orbit filled with closed orbits.

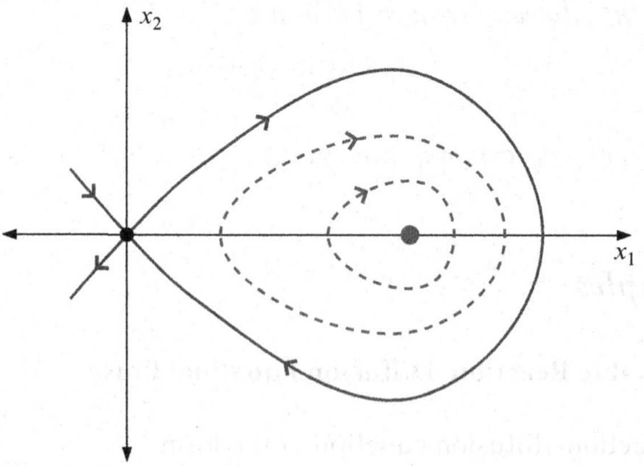

Fig. 2.1 A cartoon depiction of the phase plane for the ODE (2.3.11). The axes correspond to $(x_1, x_2)^T = (\phi, \partial_x \phi)^T$. The portion of the solid (*blue*) curve that lies in $x_1 > 0$ is the orbit homoclinic to zero. The dashed (*red*) curves inside the homoclinic are periodic orbits that correspond to spatially periodic wave solutions of the PDE (2.3.10). (Color figure online.)

For $\epsilon \ll 1$ we expand $u(x,t) = \phi(x) + \epsilon p(x)e^{\lambda t}$. Substituting the ansatz into the reaction–diffusion equation (2.3.10), and keeping terms at $\mathcal{O}(\epsilon)$ leads to the linear eigenvalue problem for p,

$$\mathcal{L}p = \lambda p, \quad \mathcal{L} = \partial_x^2 - W''(\phi).$$

Here $\mathcal{L}$ is a second-order Sturm–Liouville operator of the form (2.3.4), and since $\phi \to 0$ at an exponential rate as $x \to \pm\infty$ the coefficients $a_0(x) := -W''(\phi)$ and $a_1 := 0$ satisfy (2.3.8) with $a_0^{\pm} = -W''(0) < 0$. In particular Theorem 2.3.3 applies to $\mathcal{L}$. Let us further examine the point spectrum of $\mathcal{L}$. Differentiating the steady-sate equation (2.3.11) with respect to x yields

$$0 = \partial_x^2[\partial_x\phi] - W''(\phi)[\partial_x\phi] = \mathcal{L}(\partial_x\phi).$$

Since $\partial_x\phi(x) \to 0$ exponentially fast as $x \to \pm\infty$, $\lambda = 0$ we conclude that $\partial_x\phi \in H^2(\mathbb{R})$ is an eigenfunction of $\mathcal{L}$ with the corresponding eigenvalue $\lambda = 0$. Moreover, from its phase plane construction, $\partial_x\phi$ has precisely one zero, and by Theorem 2.3.3 we deduce that $\lambda = 0$ is the second largest eigenvalue, and there exists one positive eigenvalue $\lambda_0 > 0$, the ground state, which has an associated eigenfunction p_0 with no zeros. All other nonzero eigenvalues must be negative.

2.3.3.2 A Bistable Reaction–Diffusion Equation: Traveling Front

The bistable reaction diffusion equation (2.3.10) also possesses traveling front solutions $u(x,t) = \phi_c(x-ct)$. Introducing the traveling variable $\xi = x-ct$ and changing variables from (x,t) to (ξ,t), we obtain the evolution equation

$$\partial_t u = \partial_\xi^2 u + c\partial_\xi u - W'(u). \tag{2.3.13}$$

The front ϕ_c is an equilibrium of the traveling system, satisfying the traveling wave equation

$$\partial_\xi^2\phi + c\partial_\xi\phi_c - W'(\phi_c) = 0. \tag{2.3.14}$$

Writing the equilibrium equation as a dynamical system in $(\phi_c, \partial_\xi\phi_c)^{\mathrm{T}}$, one can verify that the energy E, introduced in (2.3.12) is monotonic in ξ, depending upon the sign of c, when evaluated at solutions of (2.3.13). It can be shown, Exercise 2.3.4, that there is a unique value of the wave speed, $c = c_*$, for which (2.3.14) possesses a heteroclinic front solution ϕ connecting the equilibrium at $(0,0)^{\mathrm{T}}$ and $(1,0)^{\mathrm{T}}$; see Fig. 2.2.

The eigenvalue problem associated to the linearization of (2.3.13) about $u = \phi_c$ takes the form

$$\mathcal{L}p = \lambda p, \quad \mathcal{L} = \partial_\xi^2 + c_*\partial_\xi - W''(\phi_{c^*}).$$

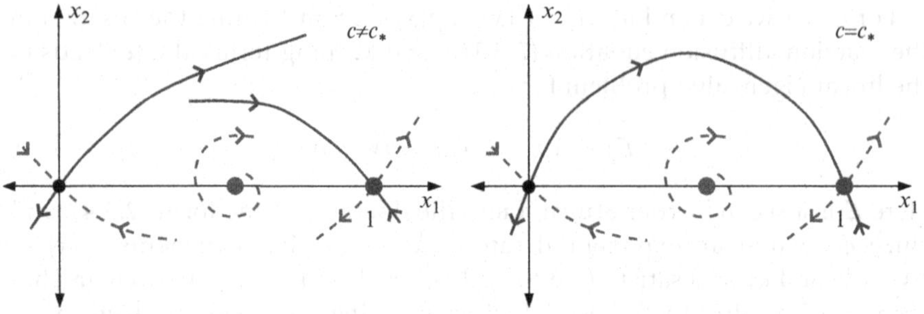

Fig. 2.2 A cartoon depiction of two different phase planes for the ODE (2.3.14). The axes correspond to $(x_1, x_2)^T = (\phi, \partial_x \phi)^T$. In the *left panel*, when $c \neq c_*$, there is no trajectory connecting the origin to $(1, 0)$: the unstable manifold of the origin [solid (*blue*) curve] does not intersect the stable manifold of $(1, 0)$ [solid (*blue*) curve]. However for $c = c_*$ the unstable manifold of $(0, 0)^T$ intersects and hence coincides with the stable manifold of $(1, 0)^T$, forming the solid (*blue*) heteroclinic orbit that lies in the first quadrant of the *right panel*. (Color figure online.)

The conditions (2.3.8) apply with $a_0^+ = -W''(1)$ and $a_0^- = -W''(0)$, while the operator $\mathcal{L}$ is self-adjoint in the inner product weighted by

$$\rho(x) = e^{c_* x}$$

(see Exercise 2.3.1). From Theorem 2.3.3 we deduce that the point spectrum is bounded from below by

$$b = \max\{-W''(0), -W''(1)\} < 0.$$

Moreover, differentiating the traveling wave equation (2.3.14) with respect to ξ shows that

$$0 = \partial_\xi^2[\partial_\xi \phi_{c_*}] + c_* \partial_\xi[\partial_\xi \phi_{c_*}] - W''(\phi_{c_*})[\partial_\xi \phi_{c_*}] = \mathcal{L}(\partial_\xi \phi_{c_*}).$$

Hence, $\lambda = 0$ is an eigenvalue with eigenfunction $\partial_\xi \phi_{c_*}$. The traveling wave is a heteroclinic connection, and is monotonically increasing in ξ, so that $p_0(x) := \partial_\xi \phi > 0$ has no zeros and $\lambda_0 = 0$ is the largest eigenvalue, and all other eigenvalues lie in the interval $(b, 0)$.

Exercises

Exercise 2.3.1. Let $\mathcal{L}$ be a Sturm–Liouville operator on $H^2(\mathbb{R})$, and define $q := p\sqrt{\rho}$ where the weight ρ is defined in (2.3.9).

(a) Show that $\langle p_1, p_2 \rangle_\rho = \langle q_1, q_2 \rangle$.
(b) Show that in the q variables the eigenvalue problem (2.3.4) becomes

$$\partial_x^2 q + \left(a_0(x) - \frac{1}{2} a_1(x)^2 - \frac{1}{2} \partial_x a_1(x)\right) q = \lambda q.$$

(c) Verify that the linear operator in part (b) is self-adjoint with respect to the unweighted $L^2(\mathbb{R})$ inner product.
(d) State the appropriate version of Theorem 2.3.3 for the eigenvalue problem of part (b) on $H^2(\mathbb{R})$. In particular, deduce that the location of the eigenvalues for $\lambda > b$ is the same for both problems.

Exercise 2.3.2. Let $\mathcal{L}$ be a Sturm–Liouville operator on $H^2(\mathbb{R})$ with piecewise constant coefficient

$$a_1(x) = \begin{cases} 0, & x < 0, \\ 1, & x > 0, \end{cases}$$

while a_0 is smooth and exponentially asymptotic. Calculate the adjoint operator $\mathcal{L}^{\mathrm{a}}$ for the ρ-weighted inner product; in particular, determine $\mathcal{D}(\mathcal{L}^{\mathrm{a}})$, showing that $\mathcal{D}(\mathcal{L}^{\mathrm{a}}) \neq \mathcal{D}(\mathcal{L}) = H^2(\mathbb{R})$.

Exercise 2.3.3. Show that the phase-plane Fig. 2.1 which shows the existence of ϕ for (2.3.11) is correct, and deduce that $\partial_x\phi$ has one zero. (*Hint:* Compute the energy for the system)

Exercise 2.3.4. Show that the traveling wave equation (2.3.14) has a heteroclinic connection, ϕ_c, for a unique value of the wave speed, $c = c_*$, by considering the intersection of the respective stable and unstable manifolds with the section $x_1 = 1/2, x_2 > 0$. Moreover, show that c_* satisfies the implicit relation

$$c_* = \frac{W(1)}{\int_{\mathbb{R}} (\partial_\xi \phi_{c_*})^2 \, \mathrm{d}\xi} < 0.$$

2.4 Additional Reading

Further details regarding the spectrum of Sturmian operators can be found in Carmona and Lacroix [**45**], Weidmann [**277, 278**].

Chapter 3
Essential and Absolute Spectra

The goal of this chapter is the characterization of the essential spectrum and Fredholm indices of two classes of linear differential operators on unbounded domains. The first class is comprised of nth-order differential operators with spatially varying coefficients that tend at an exponential rate to constant values at $\pm\infty$. We call these operators *exponentially asymptotic* and they typically arise as the linearizations of a nonlinear partial differential equation (PDE) about a heteroclinic (pulse) or homoclinic (front) solution of the equilibrium equations. The essential spectrum on unweighted function spaces, such as $L^2(\mathbb{R})$ is characterized, via the Weyl essential spectrum Theorem 2.2.6, in terms of the noninvertibility of a particular matrix. In particular, the matrix is square precisely when the Fredholm index of the operator is zero. We show that in exponentially weighted spaces the essential spectrum is shifted, and characterize the absolute spectrum as the leftmost possible shift of the boundary of the essential spectrum. The second class of operators consists of nth-order differential operators with spatially periodic coefficients with a common period. Using Floquet theory (see Chapter 2.1.3) we decompose the essential spectrum into an uncountable union of point spectrum, and establish the equivalence of the Floquet theory and the Bloch-wave theory.

3.1 The Essential Spectrum: Fronts and Pulses

The prototypical differential operator arising as the linearization of a nonlinear PDE about a heteroclinic (front) or homoclinic (pulse) equilibria takes the form

$$\mathcal{L}p := \partial_x^n p + a_{n-1}(x)\partial_x^{n-1} p + \cdots + a_1(x)\partial_x p + a_0(x)p, \tag{3.1.1}$$

where $n \geq 1$ denotes the order of the operator and $x \in \mathbb{R}$ is the spatial variable.

T. Kapitula and K. Promislow, *Spectral and Dynamical Stability of Nonlinear Waves*, Applied Mathematical Sciences 185, DOI 10.1007/978-1-4614-6995-7_3,

Definition 3.1.1. The operator $\mathcal{L}$ is said to be *exponentially asymptotic* if the coefficients $a_0,\dots,a_{n-1}$ are smooth, real-valued functions that are asymptotically constant, that is, if there exists $r>0$ such that

$$\lim_{x\to\pm\infty} e^{r|x|}|a_j(x)-a_j^{\pm}|=0.$$

The domain, $Y=H^n(\mathbb{R})$, of $\mathcal{L}$, is dense in $X=L^2(\mathbb{R})$.

Lemma 3.1.2. *If the coefficients* $\{a_j\}_{j=0}^{n-1}$ *lie in* $W^{1,\infty}(\mathbb{R})$*, then the operator* $\mathcal{L}: H^n(\mathbb{R})\subset L^2(\mathbb{R})\mapsto L^2(\mathbb{R})$ *is closed.*

Proof. Assume that $\{u_k\}_{k=1}^{\infty}\subset H^n(\mathbb{R})$ converges to u in $\|\cdot\|_{L^2}$, and that $v_k:=\mathcal{L}u_k$ converges to v in $\|\cdot\|_{L^2}$. We must show that $u\in H^n(\mathbb{R})$, and that $\mathcal{L}u=v$. From the Fourier transform we see that for all $\lambda\in\mathbb{R}_+$ the operator $\partial_x^n-i^{n-1}\lambda$ is not only invertible with $\|(\partial_x^n-i^{n-1}\lambda)^{-1}u\|_{H^{s+n-1}}\le C\|u\|_{H^s}$ for any $s\ge 0$ and C independent of u, but its resolvent also satisfies the limit

$$\lim_{\lambda\to+\infty}\|\left(\partial_x^n-i^{n-1}\lambda\right)^{-1}\|_{\mathcal{B}(H^s,H^{s+n-1})}=0.$$

By subtracting $i^{n-1}\lambda u_k$ from both sides of the definition of v_k and inverting $\partial_x^n-i^{n-1}\lambda$ we obtain

$$u_k+(\partial_x^n-i^{n-1}\lambda)^{-1}\mathcal{L}_{n-1}u_k=(\partial_x^n-i^{n-1}\lambda)^{-1}(v_k-i^{n-1}\lambda u_k),$$

where $\mathcal{L}_{n-1}:=\mathcal{L}-\partial_x^n$ is of order $n-1$, in particular, $\mathcal{L}_{n-1}\in\mathcal{B}(H^n,H^1)$. Consequently, the operator $B:=(\partial_x^n-i^{n-1}\lambda)^{-1}\mathcal{L}_{n-1}:H^n(\mathbb{R})\to H^n(\mathbb{R})$ is bounded and its induced norm tends to zero as $\lambda\to+\infty$. For fixed $\lambda\in\mathbb{R}_+$ sufficiently large, the operator $\mathcal{I}+B:H^n(\mathbb{R})\to H^n(\mathbb{R})$ is invertible. We deduce that the sequence $\{u_k\}$ is Cauchy in $H^k(\mathbb{R})$, and taking the limit as $k\to\infty$ yields the equality

$$u=(\mathcal{I}+B)^{-1}(\partial_x^n-i^{n-1}\lambda)^{-1}(v-i^{n-1}\lambda u)\in H^n(\mathbb{R}),$$

and unpacking the equality we find $\mathcal{L}u=v$. □

Remark 3.1.3. The restriction that the coefficients $\{a_j\}$ of $\mathcal{L}$ lie in $W^{1,\infty}$ can be relaxed somewhat, see Exercise 3.1.1. Moreover, for operators $\mathcal{L}$ with exponentially asymptotic and piecewise smooth coefficients, the adjoint operator is also closed, although the domain, $\mathcal{D}(\mathcal{L}^a)$, is a strict subset of $H^n(\mathbb{R})$; see Exercise 2.3.2.

In Theorem 3.1.11 we show that the exponentially asymptotic operator $\mathcal{L}$ is a relatively compact perturbation of an operator with piecewise constant coefficients, and hence from Theorem 2.2.6 we deduce that the two operators have the same essential spectrum. We call the associated piecewise constant coefficient operator the *asymptotic operator*.

Definition 3.1.4. The asymptotic operator $\mathcal{L}_\infty$ associated with the exponentially asymptotic operator $\mathcal{L}$ of (3.1.1) takes the form

$$\mathcal{L}_\infty p := \partial_x^n p + a_{n-1}^\infty \partial_x^{n-1} p + \cdots + a_1^\infty \partial_x p + a_0^\infty p, \tag{3.1.2}$$

where the coefficients a_j^∞ are piecewise constant functions obtained by replacing a_j with its limiting values on each half-line. Specifically, for $j = 0, \ldots, n-1$,

$$a_j^\infty(x) = \begin{cases} a_j^-, & x < 0 \\ a_j^+, & x \geq 0. \end{cases}$$

The operator eigenvalue $\lambda \in \mathbb{C}$ lies in the spectrum of $\mathcal{L}_\infty$ precisely when the nonhomogeneous problem

$$(\mathcal{L}_\infty - \lambda)p = f \tag{3.1.3}$$

fails to be boundedly invertible from $L^2(\mathbb{R})$ into $H^n(\mathbb{R})$. We separate the inversion of $\mathcal{L}_\infty - \lambda$ into two steps. The first step constructs solutions p, which may not lie in $H^n(\mathbb{R})$, by rewriting the spectral problem as an initial-value problem (IVP) for a first-order system of ordinary differential equations. While the existence of such solutions is guaranteed by Lemma 2.1.1, from Theorem 2.1.23 we know they will generically grow exponentially as $x \to \pm\infty$. The second step is to determine, in terms of λ and the coefficients a_j^∞, if there exists a choice of initial data for which the corresponding solution p decays exponentially as $x \to \pm\infty$, yielding $p \in H^n(\mathbb{R})$. This process replaces the question of the solvability of (3.1.3) with a question about the span of the stable and unstable spaces associated with the asymptotic problem. In many ways this is the key step to characterizing the essential spectrum of asymptotically constant differential operators.

We introduce $\boldsymbol{Y} = (p, \partial_x p, \ldots, \partial_x^{n-1} p)^{\mathrm{T}}$ and $\boldsymbol{F} = (0, 0, \ldots, f)^{\mathrm{T}}$, so that p solves (3.1.3) [without the condition $p \in H^n(\mathbb{R})$] if and only if y solves the first-order system

$$\partial_x \boldsymbol{Y} = \boldsymbol{A}_\infty(x, \lambda)\boldsymbol{Y} + \boldsymbol{F}, \tag{3.1.4}$$

where the matrix $\boldsymbol{A}_\infty(x, \lambda) \in \mathbb{C}^{n \times n}$ is piecewise constant in x, and is defined via the two asymptotic matrices

$$\boldsymbol{A}_\infty(\lambda) = \begin{cases} \boldsymbol{A}_-(\lambda), & x < 0 \\ \boldsymbol{A}_+(\lambda), & x \geq 0, \end{cases} \quad \boldsymbol{A}_\pm(\lambda) = \begin{pmatrix} 0 & 1 & \ldots & 0 & 0 \\ 0 & 0 & \ldots & 0 & 0 \\ \vdots & \vdots & \vdots\vdots\vdots & \vdots & \vdots \\ 0 & 0 & \ldots & 0 & 1 \\ \lambda - a_0^\pm & -a_1^\pm & \ldots & -a_{n-2}^\pm & -a_{n-1}^\pm \end{pmatrix}.$$

Definition 3.1.5. The eigenvalues $\{\mu_j^\pm = \mu_j^\pm(\lambda) : j = 1,\dots,n\}$ of $A_\pm(\lambda)$ are called the matrix eigenvalues. The parameter λ is called the operator eigenvalue.

Remark 3.1.6. In the literature the matrix eigenvalues are also referred to as the spatial eigenvalues, and the operator eigenvalue is also known as the temporal spectral parameter; see Remark 3.1.12 for motivation of this notation.

We first show that if the asymptotic matrices are hyperbolic, then the asymptotic operator $\mathcal{L}_\infty$ is Fredholm; in particular, we will show that $\mathrm{R}(\mathcal{L}_\infty - \lambda)$ is closed, and that both $\dim[\ker(\mathcal{L}_\infty - \lambda)]$ and $\mathrm{codim}[\mathrm{R}(\mathcal{L}_\infty - \lambda)]$ are finite. Our construction allows us to explicitly characterize the Fredholm index in terms of the dimensions of the unstable subspaces of $A_\pm(\lambda)$. For ease of presentation we suppress the λ-dependence of $A_\pm$ and the corresponding solutions.

When $A_\pm$ are hyperbolic, their stable and unstable eigenspaces (see Definition 2.1.19) yield the direct sum decomposition

$$\mathbb{C}^n = \mathbb{E}_-^{\mathrm{s}} \oplus \mathbb{E}_-^{\mathrm{u}} = \mathbb{E}_+^{\mathrm{s}} \oplus \mathbb{E}_+^{\mathrm{u}}.$$

We introduce $P_\pm^{\mathrm{s}} : \mathbb{C}^n \mapsto \mathbb{E}_\pm^{\mathrm{s}}$ and $P_\pm^{\mathrm{u}} : \mathbb{C}^n \mapsto \mathbb{E}_\pm^{\mathrm{u}}$, which are the spectral projections onto the stable and unstable subspaces of $A_\pm$ defined according to the Dunford integrals over $\sigma^{\mathrm{s,u}}(A_\pm)$, respectively (see (2.2.9)). The projections enjoy the following properties: for all $v \in \mathbb{C}^n$,

(a) $v = P_\pm^{\mathrm{s}} v + P_\pm^{\mathrm{u}} v$; furthermore, if $v \in \mathbb{E}_\pm^{\mathrm{u,s}}$, then $P_\pm^{\mathrm{u,s}} v = v$,
(b) $\mathrm{e}^{A_\pm x} P_\pm^{\mathrm{s}} v = P_\pm^{\mathrm{s}} \mathrm{e}^{A_\pm x} v$ and $\mathrm{e}^{A_\pm x} P_\pm^{\mathrm{u}} v = P_\pm^{\mathrm{u}} \mathrm{e}^{A_\pm x} v$,
(c) by Theorem 2.1.23 there are constants $C, \sigma > 0$ such that

$$\begin{aligned} |\mathrm{e}^{A_\pm x} P_\pm^{\mathrm{u}} v| &\le C\mathrm{e}^{\sigma x}|v|,\ x<0; \quad |\mathrm{e}^{A_\pm x} P_\pm^{\mathrm{s}} v| \le C\mathrm{e}^{-\sigma x}|v|,\ x>0, \\ |\mathrm{e}^{A_\pm x} P_\pm^{\mathrm{s}} v| &\ge C\mathrm{e}^{-\sigma x}|v|,\ x<0; \quad |\mathrm{e}^{A_\pm x} P_\pm^{\mathrm{u}} v| \ge C\mathrm{e}^{\sigma x}|v|,\ x>0. \end{aligned} \tag{3.1.5}$$

The bounds of (3.1.5) imply that $\mathrm{e}^{A_\pm x} v$ decays exponentially fast in norm as $x \to -\infty$ only if $v \in \mathbb{E}_\pm^{\mathrm{u}}$, and grows exponentially fast otherwise. Similarly, $\mathrm{e}^{A_\pm x} v$ decays exponentially fast in norm as $x \to +\infty$ only if $v \in \mathbb{E}_\pm^{\mathrm{s}}$, and grows exponentially fast otherwise. This *exponential dichotomy* drives the following analysis.

Given $F \in L^2(\mathbb{R})^n$, that is, each of the n components of F are in $L^2(\mathbb{R})$, then for every initial data y_0 the system (3.1.4) has a solution defined on $\mathbb{R}$. The question is, does there exist an initial data, y_0, for which the corresponding solution decays in norm as $x \to \pm\infty$; furthermore, is that initial data unique? Since the choice of the initial condition allows us n degrees of freedom, we expect that existence and uniqueness will result when the decay condition imposes n linearly independent constraints.

To characterize these constraints, we start with an arbitrary $y_0 \in \mathbb{C}^n$, and solve the inhomogeneous asymptotic system (3.1.4) for $x < 0$. By the variation of parameters formula, see Lemma 2.1.3, we have

$$y(x) = e^{A_- x} y_0 + \int_0^x e^{A_-(x-t)} F(t)\, dt, \quad x \le 0.$$

However, $v = P_-^u v + P_-^s v$, so we may decompose F and y_0 into their stable and unstable components, and rewrite the solution as

$$y(x) = e^{A_- x} P_-^u y_0 + e^{A_- x} P_-^s y_0 + \int_0^x e^{A_-(x-t)} P_-^u F(t)\, dt + \int_0^x e^{A_-(x-t)} P_-^s F(t)\, dt.$$

It is convenient to change the lower limit of integration on the inhomogeneous term with the stable projection, pushing it back to $-\infty$. In this case the solution takes the form

$$y(x) = e^{A_- x} P_-^u y_0^- + e^{A_- x} P_-^s y_0^- - \int_x^0 e^{A_-(x-t)} P_-^u F(t)\, dt + \int_{-\infty}^x e^{A_-(x-t)} P_-^s F(t)\, dt,$$

where to preserve the condition $y(0) = y_0$ we have introduced the vector

$$y_0^- := y_0 - \int_{-\infty}^0 e^{A_-(x-t)} P_-^s F(t)\, dt.$$

Introducing the Green's (matrix) function for the ODE system A_-,

$$G_-(z) = \begin{cases} -e^{A_- z} P_-^u, & z < 0 \\ +e^{A_- z} P_-^s, & z > 0, \end{cases}$$

we can rewrite the solution yet again in terms of the convolution of the matrix G_- with the vector F_-,

$$y(x) = e^{A_- x} P_-^u y_0^- + e^{A_- x} P_-^s y_0^- + (G_- * F_-)(x), \quad x \le 0, \tag{3.1.6}$$

where

$$F_-(t) = \begin{cases} F(t), & t \le 0 \\ 0, & t > 0. \end{cases}$$

By the convolution inequality (2.2.6) we see that for $F \in L^2(\mathbb{R})^n$ the inhomogeneous term is bounded in $L^q(\mathbb{R})$ for any $q \in [2, \infty]$,

$$\|G_- * F_-\|_q \le C\|G_-\|_r \|F_-\|_2 \le C\|G_-\|_r \|F\|_2, \quad r = \frac{2q}{q+2}. \tag{3.1.7}$$

Here we use the facts that $\|F_-\|_2 \le \|F\|_2$, and $\|G_-\|_r$ is bounded due to the exponential decay of $G_-(z)$ as $z \to \pm\infty$. Since the $P_-^s y_0^-$ term in (3.1.6) grows exponentially in norm as $x \to -\infty$, and the $P_-^u y_0^-$ term decays exponentially in norm as $x \to -\infty$, we see that y resides in $L^2(\mathbb{R}_-)^n$ if and only if $P_-^s y_0^- = 0$, i.e., $y_0^- \in \mathbb{E}_-^u$. Otherwise, y grows exponentially in norm as $x \to -\infty$. A similar analysis for $x \ge 0$ leads to the expression

$$y(x) = e^{A_+x} P^u_+ y_0^+ + e^{A_+x} P^s_+ y_0^+ + (G_+ * F_+)(x), \quad x \geq 0, \tag{3.1.8}$$

where G_+ and F_+ are the corresponding Green's function and inhomogeneous terms for $x > 0$. We conclude that y resides in $L^2(\mathbb{R}_+)^n$ if and only if $P^u_+ y_0^+ = 0$, i.e., $y_0^+ \in \mathbb{E}^s_+$, and y grows exponentially as $x \to \infty$ otherwise.

The composite function

$$y(x) = \begin{cases} e^{A_-x} y_0^- + (G_- * F_-)(x), & x < 0 \\ e^{A_+x} y_0^+ + (G_+ * F_+)(x), & x > 0, \end{cases}$$

solves (3.1.4) on the disjoint intervals $(-\infty, 0)$ and $(0, \infty)$. More significantly, we have established that it decays exponentially fast in norm as $x \to \pm\infty$ if and only if $y_0^- \in \mathbb{E}^u_-$ and $y_0^+ \in \mathbb{E}^s_+$. However the function y solves (3.1.4) on the whole line if and only if it is continuous at $x = 0$, that is, if and only if $y(0^+) = y(0^-)$. Recalling the tensor notation, (2.2.8), and the form of F given prior to (3.1.4), the continuity condition can be expressed as

$$y_0^- - y_0^+ = (G_+ * F_+ - G_- * F_-)\big|_{x=0} = \otimes G \cdot f, \tag{3.1.9}$$

subject to the constraints $y_0^- \in \mathbb{E}^u_-$ and $y_0^+ \in \mathbb{E}^s_+$. Here $G(x)$ is the vector given by the last column of the matrix $e^{-A_-x} P^s_-$ for $x \leq 0$, and the last column of $e^{-A_+x} P^u_+$ for $x \geq 0$. In particular, the entries of G decay exponentially in norm as both $x \to \pm\infty$. For a fixed $f \in L^2(\mathbb{R})$ the term $\otimes G \cdot f$ is a known vector in $\mathbb{C}^n$. Setting

$$n^u_- = \dim[\mathbb{E}^u_-], \quad n^s_+ = \dim[\mathbb{E}^s_+],$$

let $\{v^u_1, \ldots, v^u_{n^u_-}\}$ denote a basis for $\mathbb{E}^u_-$ and $\{v^s_1, \ldots, v^s_{n^s_+}\}$ a basis for $\mathbb{E}^s_+$. Forming the matrix

$$M = \left(v^u_1, \ldots, v^u_{n^u_-}, v^s_1, \ldots, v^s_{n^s_+}\right), \tag{3.1.10}$$

the system (3.1.9) can be written in terms of $\alpha \in \mathbb{C}^{n^u_- + n^s_+}$ as

$$M\alpha = \otimes G \cdot f. \tag{3.1.11}$$

If the matrix M is square, i.e., $n^u_- + n^s_+ = n$, and $\det M \neq 0$, then there is a unique solution α to the linear system (3.1.11). In this case

$$|\alpha| \leq C |\det M|^{-1} \|G\|_2 \|f\|_2,$$

so that we may bound each of the terms on the right-hand side of (3.1.8) in terms of $\|f\|_2$. In other words, there exists a constant $C = C(\lambda) > 0$ such that $\|y\|_2 \leq C\|f\|_2$. Returning to the original variable p, the conditions of (3.1.11) make the first $n-1$ derivatives of p continuous at $x = 0$, and the estimate on y translates to

$$\|p\|_{H^n} \leq C(\lambda)\|f\|_2, \tag{3.1.12}$$

which is precisely the desired invertability of $\mathcal{L}_\infty - \lambda$.

The invertibility of the matrix $\boldsymbol{M}$ is equivalent to the following two conditions,

$$\dim[\mathbb{E}^{\mathrm{u}}_{-}]+\dim[\mathbb{E}^{\mathrm{s}}_{+}]=n, \quad \dim[\mathbb{E}^{\mathrm{u}}_{-}\cap\mathbb{E}^{\mathrm{s}}_{+}]=0, \tag{3.1.13}$$

which can be conveniently stated as

$$\mathbb{E}^{\mathrm{u}}_{-}\oplus\mathbb{E}^{\mathrm{s}}_{+}=\mathbb{C}^n.$$

The first condition in (3.1.13) is structural, that is, it is unaffected by relatively compact perturbations. As we will see in Lemma 3.1.10, its failure leads to essential spectrum for the operator $\mathcal{L}_\infty$.

Lemma 3.1.7. *Fix $\lambda\in\mathbb{C}$. If the asymptotic matrices $\boldsymbol{A}_\pm(\lambda)$ are hyperbolic, then $\mathrm{R}(\mathcal{L}_\infty-\lambda)$ is closed; furthermore,*

$$\dim[\ker(\mathcal{L}_\infty-\lambda)]=\dim[\ker(\boldsymbol{M}(\lambda))], \quad \mathrm{codim}[\mathrm{R}(\mathcal{L}_\infty-\lambda)]=\mathrm{codim}[\mathrm{R}(\boldsymbol{M}(\lambda))],$$

where $\boldsymbol{M}(\lambda)$ is the matrix defined in (3.1.10). Moreover, $\mathcal{L}_\infty-\lambda$ is Fredholm with index

$$\mathrm{ind}(\mathcal{L}_\infty-\lambda)=n^{\mathrm{s}}_{-}(\lambda)+n^{\mathrm{s}}_{+}(\lambda)-n, \tag{3.1.14}$$

where $n^{\mathrm{s,u}}_\pm(\lambda):=\dim[\mathbb{E}^{\mathrm{s,u}}_\pm(\lambda)]$ are the dimensions of the stable and unstable subspaces of $\boldsymbol{A}_\pm(\lambda)$.

Proof. To establish the equality of the dimensions of the kernels, we observe from (3.1.11) and the discussion following it, that $f\in\mathrm{R}(\mathcal{L}_\infty-\lambda)$ if and only if $\otimes\boldsymbol{G}\cdot f\in\mathrm{R}(\boldsymbol{M}(\lambda))$. Since $\mathrm{R}(\boldsymbol{M}(\lambda))$ is closed and $f\mapsto\otimes\boldsymbol{G}\cdot f$ is continuous in $L^2(\mathbb{R})$, it follows that $\mathrm{R}(\mathcal{L}_\infty-\lambda)$ is closed. It is similarly clear that each $p\in\ker(\mathcal{L}_\infty-\lambda)$ is in one-to-one correspondence with an $\alpha\in\ker(\boldsymbol{M}(\lambda))$. Indeed, if $\dim\ker(\boldsymbol{M}(\lambda))=\dim(\mathbb{E}^{\mathrm{u}}_{-}\cap\mathbb{E}^{\mathrm{s}}_{+})=k$, then let $\{\boldsymbol{y}_{01},\ldots,\boldsymbol{y}_{0k}\}$ form a basis for $\mathbb{E}^{\mathrm{u}}_{-}\cap\mathbb{E}^{\mathrm{s}}_{+}$. For $j=1,\ldots,k$ the solution $\boldsymbol{y}_j$ of (3.1.4) with $f\equiv 0$ satisfying $\boldsymbol{y}_j(0)=\boldsymbol{y}_{j0}$ lies in $L^2(\mathbb{R})^n$. It follows from the definition of $\boldsymbol{y}_{j0}$ that its first component, p_j, lies in $H^n(\mathbb{R})$, and that p satisfies $(\mathcal{L}_\infty-\lambda)p=0$.

To demonstrate the equality of the codimensions, we consider only the case $\mathrm{codim}[\mathrm{R}(\boldsymbol{M}(\lambda))]=1$, in which case $\ker(\boldsymbol{M}^{\mathrm{a}}(\lambda))=\mathrm{span}\{\boldsymbol{v}\}$ for some vector $\boldsymbol{v}$ with $|\boldsymbol{v}|=1$. From the Fredholm theory for matrices we know that $\otimes\boldsymbol{G}\cdot f\in\mathrm{R}(\boldsymbol{M}(\lambda))$ if and only if $\boldsymbol{v}\perp\otimes\boldsymbol{G}\cdot f$; equivalently, if and only if $\boldsymbol{v}^{\mathrm{T}}\otimes\boldsymbol{G}\cdot f=0$. Let $\xi\in L^2(\mathbb{R})$ be any function that satisfies $\otimes\boldsymbol{G}\cdot\xi=\boldsymbol{v}$. The codimension-one projection onto $\mathrm{R}(\mathcal{L}_\infty-\lambda)$ is then given by

$$\pi_\infty f:=f-\xi\left(\boldsymbol{v}^{\mathrm{T}}\otimes\boldsymbol{G}\cdot f\right).$$

Indeed, $\pi_\infty f\in\mathrm{R}(\mathcal{L}_\infty-\lambda)$ since

$$\boldsymbol{v}^{\mathrm{T}}\otimes\boldsymbol{G}\cdot\pi_\infty f=\boldsymbol{v}^{\mathrm{T}}\otimes\boldsymbol{G}\cdot f-\boldsymbol{v}^{\mathrm{T}}\otimes\boldsymbol{G}\cdot\xi(\boldsymbol{v}^{\mathrm{T}}\otimes\boldsymbol{G}\cdot f)=(\boldsymbol{v}^{\mathrm{T}}\otimes\boldsymbol{G}\cdot f)(1-\boldsymbol{v}^{\mathrm{T}}\boldsymbol{v})=0.$$

Every $f \in L^2(\mathbb{R})$ can be decomposed as $f = \pi_\infty f + \beta\xi$ where $\beta := \boldsymbol{v}^{\mathrm{T}} \otimes \boldsymbol{G} \cdot f \in \mathbb{C}$. Since $\pi_\infty f \in \mathrm{R}(\mathcal{L}_\infty - \lambda)$ we have the decomposition

$$L^2(\mathbb{R}) = \mathrm{R}(\mathcal{L}_\infty - \lambda) \oplus \mathrm{span}\{\xi\},$$

which is precisely the meaning of $\mathrm{codim}[\mathrm{R}(\mathcal{L}_\infty - \lambda)] = 1 = \mathrm{codim}[\mathrm{R}(\boldsymbol{M}(\lambda))]$.

The Fredholm index of $\mathcal{L}_\infty - \lambda$ can now be expressed as the difference $\dim[\ker(\boldsymbol{M}(\lambda))] - \mathrm{codim}[\mathrm{R}(\boldsymbol{M}(\lambda))]$. However, a key result of linear algebra is that for any matrix $\boldsymbol{A} \in \mathbb{C}^{k \times \ell} : \mathbb{C}^\ell \to \mathbb{C}^k$,

$$\dim[\ker(\boldsymbol{A})] - \mathrm{codim}[\mathrm{R}(\boldsymbol{A})] = \ell - k.$$

Since the matrix $\boldsymbol{M}(\lambda) \in \mathbb{C}^{n \times (n_-^{\mathrm{u}} + n_+^{\mathrm{s}})}$, the equality (3.1.14) now follows. □

The nongeneric case of nonhyperbolic asymptotic matrices always produces an essential spectra. This is a particular case of Palmer's theorem; see Palmer [**215, 216**].

Lemma 3.1.8. *Fix $\lambda \in \mathbb{C}$. If either of the asymptotic matrices $\boldsymbol{A}_\pm(\lambda)$ is not hyperbolic, then the range of the operator $\mathcal{L}_\infty - \lambda : H^n(\mathbb{R}) \subset L^2(\mathbb{R}) \mapsto L^2(\mathbb{R})$ is not closed, and the operator is not Fredholm. In particular, $\lambda \in \sigma_{\mathrm{ess}}(\mathcal{L}_\infty)$.*

Proof. We argue by contradiction. If the range $R_\lambda := \mathrm{R}(\mathcal{L}_\infty - \lambda)$ is closed, then the restricted operator

$$\mathcal{L}_{\mathrm{r}} := (\mathcal{L}_\infty - \lambda)\big|_{\ker(\mathcal{L}_\infty - \lambda)^\perp} : \ker(\mathcal{L}_\infty - \lambda)^\perp \mapsto R_\lambda$$

will have no kernel. Since the restricted operator is closed, it is Fredholm with index zero. By the closed graph theorem we conclude that $\mathcal{L}_{\mathrm{r}}$ has a bounded inverse on R_λ. However, under the assumption that one of the asymptotic matrices is not hyperbolic we will construct, below, a Weyl sequence $\{u_k\} \subset \ker(\mathcal{L}_\infty - \lambda)^\perp$ with $\|u_k\|_{H^n} = 1$ for which $\|\mathcal{L}_{\mathrm{r}} u_k\|_{L^2} \to 0$ as $k \to \infty$. The existence of this Weyl sequence shows that the induced norm of $\mathcal{L}_{\mathrm{r}}^{-1} : L^2(\mathbb{R}) \to H^n(\mathbb{R})$ is infinite, in contradiction to the invertibility of $\mathcal{L}_{\mathrm{r}}$. Hence, the range R_λ cannot be closed.

For ease of presentation we suppress the λ dependence and assume that $\boldsymbol{A}_+ = \boldsymbol{A}_-$; the proof of the more general case is left as an exercise. Setting $\boldsymbol{A}_0 := \boldsymbol{A}_\pm$, the assumption that $\boldsymbol{A}_0$ is not hyperbolic means that it has a nontrivial center space $\mathbb{E}_0^{\mathrm{c}}$. Pick $\boldsymbol{y}_0 \in \mathbb{E}_0^{\mathrm{c}}$, and set $\boldsymbol{y}(x) = \mathrm{e}^{\boldsymbol{A}_0 x}\boldsymbol{y}_0$. The norm $|\boldsymbol{y}(x)|$ of the solution $\boldsymbol{y}(x)$ is bounded away from zero for all $x \in \mathbb{R}$; see Theorem 2.1.23. In particular, $\boldsymbol{y}$ is periodic in x with some period $T > 0$. The first component, y_1, of $\boldsymbol{y}$ satisfies $\|y_1\|_2 = \infty$; moreover, $(\mathcal{L}_\infty - \lambda)y_1 = 0$. For each $k \in \mathbb{N}$ let $\chi_k(x)$ be a smooth cut-off function that satisfies

$$\chi_k(x) = \begin{cases} 1, & x \in [-kT, kT] \\ 0, & x \in (-\infty, -kT-1] \cup [kT+1, +\infty), \end{cases}$$

and for which all derivatives are uniformly bounded, independent of k. For all $k \in \mathbb{N}$ define a sequence of functions

$$u_k(x) = \frac{1}{\|\chi_k y_1\|_{H^n}} \chi_k(x) y_1(x),$$

so that $\|u_k\|_{H^n} = 1$ for all $k \in \mathbb{N}$. However, since y_1 is a fixed smooth function and the divisor $\|\chi_k y_1\|_{H^n} \to +\infty$ as $k \to +\infty$, it follows that

$$\lim_{k \to +\infty} \|\partial_x^j u_k\|_\infty = 0, \quad j = 0, 1, \ldots, n;$$

that is, $\|u_k\|_{W^{n,\infty}} \to 0$ as $k \to \infty$. Let K_k denote the set on which the cut-off function is not constant, i.e.,

$$K_k := [-kT - 1, -kT] \cup [kT, kT + 1].$$

Since $(\mathcal{L}_\infty - \lambda)u_k = 0$ for $x \notin K_k$, it follows that

$$\|(\mathcal{L}_\infty - \lambda)u_k\|_{L^2(\mathbb{R})} = \|(\mathcal{L}_\infty - \lambda)u_k\|_{L^2(K_k)} \leq \|(\mathcal{L}_\infty - \lambda)u_k\|_{L^\infty(\mathbb{R})} \sqrt{|K_k|}.$$

However, there exists a constant $c > 0$ such that $\|(\mathcal{L}_\infty - \lambda)v\|_{L^\infty} \leq c\|v\|_{W^{n,\infty}}$, for all $v \in W^{n,\infty}(\mathbb{R})$. Since $\|u_k\|_{W^{n,\infty}} \to 0$ as $k \to +\infty$ and $|K_k|$ is uniformly bounded, we conclude that $\|(\mathcal{L}_\infty - \lambda)u_k\|_2 \to 0$ as $k \to \infty$. Moreover, in the space $L^2(\mathbb{R})$ the kernel of $\mathcal{L}_\infty - \lambda$ is trivial; since $\boldsymbol{A}_\pm = \boldsymbol{A}_0$ the operator is constant coefficient and hence the only uniformly bounded solution to $(\mathcal{L}_\infty - \lambda)p = 0$ is the spatially periodic one (e.g., $p = y_1$). Consequently, $\{u_k\} \in \ker(\mathcal{L}_\infty - \lambda)^\perp = L^2(\mathbb{R})$ and $\{u_k\}$ is a Weyl sequence. □

The essential spectrum of $\mathcal{L}_\infty$ can be completely characterized in terms of the matrix eigenvalues of its asymptotic matrices. A natural way to do so is through the Morse index of the asymptotic matrices:

Definition 3.1.9. The Morse index of a constant matrix $\boldsymbol{A}$, denoted by $\mathrm{i}(\boldsymbol{A})$, is the dimension of the unstable subspace associated to $\boldsymbol{A}$, i.e.,

$$\mathrm{i}(\boldsymbol{A}) = \dim[\mathbb{E}^u].$$

We denote the Morse indices of the asymptotic matrices for (3.1.4) by

$$\mathrm{i}_\pm(\lambda) := \mathrm{i}(\boldsymbol{A}_\pm(\lambda)), \tag{3.1.15}$$

We summarize our results in the following lemma.

Lemma 3.1.10. *For $\lambda \in \mathbb{C}$, the asymptotic operator $\mathcal{L}_\infty - \lambda$ is Fredholm if and only if the asymptotic matrices $\boldsymbol{A}_\pm(\lambda)$ are hyperbolic. The resolvent set of $\mathcal{L}_\infty$ is comprised precisely of those $\lambda \in \mathbb{C}$ for which $\mathcal{L}_\infty - \lambda$ is Fredholm and $\mathbb{C}^n = \mathbb{E}^{\mathrm{u}}_-(\lambda) \oplus \mathbb{E}^{\mathrm{s}}_+(\lambda)$, where $\mathbb{E}^{\mathrm{s,u}}_\pm(\lambda)$ are the stable and unstable eigenspaces of the asymptotic matrices. Moreover, for λ in the resolvent set there exists $C = C(\lambda) > 0$ such that*

$$\|(\mathcal{L}_\infty - \lambda)^{-1} f\|_{H^n} \leq C(\lambda)\|f\|_2.$$

For those $\lambda \in \mathbb{C}$ for which the operator is Fredholm, the Fredholm index equals the difference of the Morse indices [see (3.1.15)] of the asymptotic matrices, i.e.,

$$\operatorname{ind}(\mathcal{L}_\infty - \lambda) = \mathrm{i}_-(\lambda) - \mathrm{i}_+(\lambda). \tag{3.1.16}$$

In particular, we can characterize the essential spectrum of $\mathcal{L}_\infty$ as

$$\sigma_{\mathrm{ess}}(\mathcal{L}_\infty) = \left\{\lambda \in \mathbb{C} \,\middle|\, \mathrm{i}_-(\lambda) \neq \mathrm{i}_+(\lambda)\right\} \cup \left\{\lambda \in \mathbb{C} \,\middle|\, \dim \mathbb{E}^{\mathrm{c}}(A_\pm(\lambda)) \neq 0\right\}.$$

Proof. To characterize the Fredholm index in terms of Morse indices, we observe that if the asymptotic matrices are hyperbolic, then $n^{\mathrm{s}}_\pm(\lambda) = \dim[\mathbb{E}^{\mathrm{s}}_\pm(\lambda)] = n - \mathrm{i}_\pm(\lambda)$. This observation leads to the equality

$$n^{\mathrm{u}}_-(\lambda) + n^{\mathrm{s}}_+(\lambda) = n + \mathrm{i}_-(\lambda) - \mathrm{i}_+(\lambda).$$

In terms of the Morse indices we now see that the equality (3.1.14) implies (3.1.16). It follows that $\lambda \in \sigma_{\mathrm{ess}}(\mathcal{L}_\infty)$ if and only if $\mathrm{i}_-(\lambda) \neq \mathrm{i}_+(\lambda)$ and/or the asymptotic matrices are not hyperbolic. □

We complete this section with a characterization of the essential spectrum of the exponentially asymptotic operator $\mathcal{L}$ introduced in (3.1.1).

Theorem 3.1.11. *Assume that operator $\mathcal{L}$ given in (3.1.1) is exponentially asymptotic with $H^1(\mathbb{R})$ coefficients. Then $\mathcal{L}$ is a relatively compact perturbation of the asymptotic operator $\mathcal{L}_\infty$ given in (3.1.2). In particular,*

$$\sigma_{\mathrm{ess}}(\mathcal{L}) = \left\{\lambda \in \mathbb{C} \,\middle|\, \mathrm{i}_-(\lambda) \neq \mathrm{i}_+(\lambda)\right\} \cup \left\{\lambda \in \mathbb{C} \,\middle|\, \dim \mathbb{E}^{\mathrm{c}}(A_\pm(\lambda)) \neq 0\right\}.$$

Moreover, for each $\lambda \notin \sigma_{\mathrm{ess}}(\mathcal{L})$, either $\dim(\ker(\mathcal{L} - \lambda)) \neq 0$ or there exists $C > 0$ such that

$$\|(\mathcal{L} - \lambda)^{-1} f\|_{H^n(\mathbb{R})} \leq C \|f\|_{L^2(\mathbb{R})}. \tag{3.1.17}$$

Proof. We consider only the case that the coefficients $a_j(x)$ of $\mathcal{L}$ are constant except on a common, compact interval $I \subset \mathbb{R}$. To see that $\mathcal{L}$ is a relatively compact perturbation of $\mathcal{L}_\infty$, fix $\lambda \in \rho(\mathcal{L}_\infty)$ and observe that the nth-order derivatives in $\mathcal{L}_\infty - \mathcal{L}$ cancel, and setting aside the discontinuity at $x = 0$, we view the operator $\mathcal{L}_\infty - \mathcal{L}$ as a piecewise map from $H^n(\mathbb{R})$ into $H^1(\mathbb{R}_+)$ and into $H^1(\mathbb{R}_-)$. From Lemma 3.1.10 we know that $(\mathcal{L}_\infty - \lambda)^{-1} : L^2(\mathbb{R}) \mapsto H^n(\mathbb{R})$ is continuous, so that the composite map $(\mathcal{L}_\infty - \mathcal{L})(\mathcal{L}_\infty - \lambda)^{-1} : L^2(\mathbb{R}) \mapsto H^1(\mathbb{R}_+) \oplus H^1(\mathbb{R}_-)$ is continuous. Since the coefficients $a_j(x)$ of $\mathcal{L}$ are constant off $I \subset \mathbb{R}$, $(\mathcal{L}_\infty - \mathcal{L})(\mathcal{L}_\infty - \lambda)^{-1} : L^2(\mathbb{R}) \mapsto H^1(I_+) \oplus H^1(I_-)$ where $I_- = I \cap (-\infty, 0]$ and $I_+ = I \cap [0, \infty)$. In particular the map takes bounded sets to bounded sets. As bounded sets in $H^1(I_\pm)$ are equicontinuous and $I_\pm$ are compact, we deduce from the Arzela–Ascoli theorem that the operator $(\mathcal{L}_\infty - \mathcal{L})(\mathcal{L}_\infty - \lambda)^{-1}$ maps bounded sets of $L^2(\mathbb{R})$ into precompact sets, and hence is compact. The characterization of the essential spectrum of $\mathcal{L}$ follows from the Weyl's essential spectrum Theorem 2.2.6 and Lemma 3.1.10. The statement (3.1.17) follows from the Fredholm alternative since $\mathcal{L} - \lambda : H^n(\mathbb{R}) \to L^2(\mathbb{R})$ is Fredholm of index zero for $\lambda \notin \sigma_{\text{ess}}(\mathcal{L})$. □

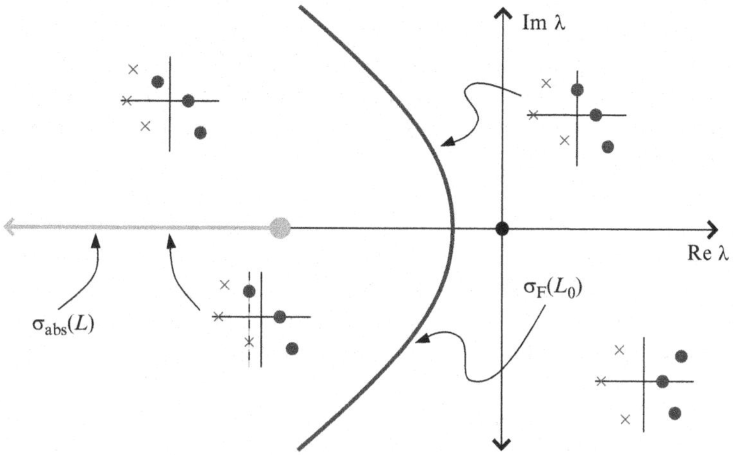

Fig. 3.1 A depiction of the matrix eigenvalues and the Fredholm border for a sixth-order problem. We consider a pulse-type operator for which $\boldsymbol{A}_- = \boldsymbol{A}_+$, so that $i_0(\lambda) := i_+(\lambda) = i_-(\lambda)$. The insets show the location of the matrix eigenvalues for various values of λ in the complex μ-plane. We assume $\mathrm{i}_0(\lambda) = 3$ for $\operatorname{Re}\lambda \gg 1$: as λ crosses the Fredholm border, one matrix eigenvalue with positive real part becomes imaginary and then attains a negative real part. The essential spectrum coincides with the Fredholm border. The absolute spectrum, $\sigma_{\text{abs}}(\mathcal{L})$, lies on the negative real axis (see Chapter 3.2). (Color figure online.)

Now that the essential spectrum has been completely characterized, we wish to study its boundary. As was discussed in the proof of Lemma 3.1.10, the boundary is determined for the values of λ for which the asymptotic matrices lose their hyperbolicity—that is, when a matrix eigenvalue becomes purely imaginary. The matrix eigenvalues, $\mu = \mu(\lambda)$, are realized as the zeros of the characteristic polynomials of $\boldsymbol{A}_\pm(\lambda)$,

$$d_\pm(\mu, \lambda) := \det(\boldsymbol{A}_\pm(\lambda) - \mu \boldsymbol{I}_n) = \mu^n + a^\pm_{n-1}\mu^{n-1} + \cdots + a^\pm_1 \mu + a^\pm_0 - \lambda. \qquad (3.1.18)$$

The equations $d_\pm(\mu, \lambda) = 0$ are referred to as the (complex) dispersion relation for the operator $\mathcal{L}$.

Remark 3.1.12. The dispersion relations relate the operator eigenvalue parameter, λ, which controls *temporal* rates of decay and growth, to the matrix eigenvalues, μ, which control *spatial* behavior, thereby assigning distinct wave speeds to different spatial modes. Indeed, if $\mathcal{L}$ is constant coefficient, then $u = \mathrm{e}^{\mu x - \lambda t}$ solves

$$\partial_t u + \mathcal{L} u = 0,$$

precisely when μ and λ solve the dispersion relation. Rewriting $u = e^{\mu(x-\lambda/\mu t)}$ it is natural to define the *phase velocity* $c_p := \lambda/\mu$ of the pure mode u. A spatially localized solution of the linear equation composed of many pure modes will spread out, or disperse, due to the differing velocities of the pure modes.

The matrix eigenvalues depend continuously upon the spectral parameter λ; hence, the Fredholm index can change only when there exists a matrix eigenvalue with zero real part. The Fredholm border(s), denoted $\sigma_F(\mathcal{L})$, are those curves in the complex λ-plane for which there exists a matrix eigenvalue with zero real part, i.e.,

$$\sigma_F(\mathcal{L}) := \{\lambda \in \mathbb{C} : d_\pm(ik, \lambda) = 0, \text{ some } k \in \mathbb{R}\} \tag{3.1.19}$$

(see Fig. 3.1). In other words, $\lambda \in \sigma_F(\mathcal{L})$ when one of the matrix eigenvalues satisfies $\mu(\lambda) = ik$ for some $k \in \mathbb{R}$. From the form of the dispersion relationship (3.1.18) we that the curves, $\lambda_\pm(k)$, of the Fredholm borders can be conveniently parameterized by the real-valued parameter k,

$$\lambda_\pm(k) := (ik)^n + a_{n-1}^\pm (ik)^{n-1} + \cdots + a_1^\pm (ik) + a_0^\pm, \quad k \in \mathbb{R}. \tag{3.1.20}$$

On these curves the asymptotic matrices lose their hyperbolicity: the Fredholm borders are the boundaries of the open regions in the complex plane on which the operator $\mathcal{L} - \lambda$ is Fredholm.

Theorem 3.1.13. *Fix an exponentially asymptotic operator $\mathcal{L}$ as in (3.1.1) and let $\sigma_F(\mathcal{L})$ denote its Fredholm borders. There exists a finite number, N, of open, disjoint, connected sets $S_j \subset \mathbb{C}$ such that*

$$\mathbb{C} \backslash \sigma_F(\mathcal{L}) = \bigcup_{j=1}^{N} S_j.$$

For each j the Fredholm index of $\mathcal{L} - \lambda$ is independent of $\lambda \in S_j$. Each set S_j is either entirely within $\sigma_{ess}(\mathcal{L})$ or is contained within $\sigma_{pt}(\mathcal{L}) \cup \rho(\mathcal{L})$. If the asymptotic matrices are equal, $\mathbf{A}_+ = \mathbf{A}_-$, then the Fredholm borders coincide and comprise the entire essential spectrum, i.e., $\sigma_{ess}(\mathcal{L}) = \sigma_F(\mathcal{L})$.

Proof. Since the Fredholm borders are realized as the zero set of an nth-order polynomial, they can have at most n crossings; thus, they can divide $\mathbb{C}$ into only a finite number of regions. Since $\mathbf{A}_\pm$ are hyperbolic and the Morse indices $i_\pm(\lambda)$ are constant on each set S_j, the Fredholm index is either zero everywhere on S_j, and the set is outside of the essential spectrum, or the index is nonzero and $S_j \subset \sigma_{ess}(\mathcal{L})$. If the asymptotic matrices are equal, then $i_+(\lambda) = i_-(\lambda)$ off of the Fredholm boundaries where the Fredholm index is zero. This implies that the essential spectrum is precisely the single Fredholm boundary $\lambda_0(k) := \lambda_\pm(k)$. □

Remark 3.1.14 (Systems). The results of Theorems 3.1.11 and 3.1.13 hold for systems in which the exponentially asymptotic operator (3.1.1) acts on

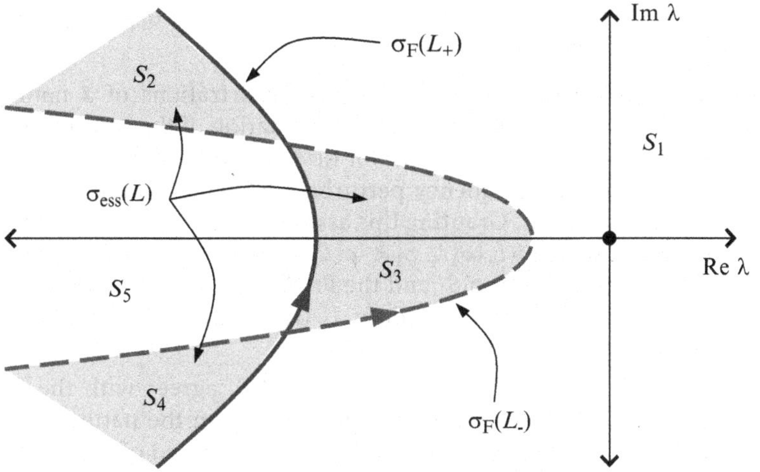

Fig. 3.2 A depiction of the sets $\{S_j\}_{j=1}^N$ of Theorem 3.1.13. The Fredholm borders are solid (*blue*) for λ_+ and dashed (*red*) for λ_-. Since $\mathrm{i}(S_1) = 0$, we have that $\mathrm{i}(S_2) = +1$, $\mathrm{i}(S_3) = -1$, $\mathrm{i}(S_4) = +1$, and $\mathrm{i}(S_5) = 0$. The essential spectrum is $\sigma_{\text{ess}}(\mathcal{L}) = \sigma_{\text{F}}(\mathcal{L}) \cup S_2 \cup S_3 \cup S_4$. (Color figure online.)

a vector-valued function $\boldsymbol{p}$ with matrix-valued coefficients $\boldsymbol{a}_0, \ldots, \boldsymbol{a}_{n-1}$. For $\boldsymbol{p} \in \mathbb{C}^m$ the dispersion relation takes the form

$$d_\pm(\mu, \lambda) := \det\left(\mu^n \boldsymbol{I}_m + \boldsymbol{a}_{n-1}^\pm \mu^{n-1} + \cdots + \boldsymbol{a}_1^\pm \mu + \boldsymbol{a}_0^\pm - \lambda \boldsymbol{I}_m\right) = 0,$$

in which $d_\pm$ is a polynomial of degree m in the operator eigenvalue λ and of degree nm in the matrix eigenvalue μ. In particular, there are now $2m$ complex roots $\lambda_\pm^\ell(\mathrm{i}k)$, $\ell = 1, \ldots, m$, that comprise the Fredholm borders of $\sigma_{\text{F}}(\mathcal{L})$

$$\sigma_{\text{F}}(\mathcal{L}) = \bigcup_{\ell=1}^{m} \Gamma_\pm^\ell, \quad \Gamma_\pm^\ell = \{\lambda_\pm^\ell(\mathrm{i}k) : k \in \mathbb{R}\}.$$

Remark 3.1.15. The Fredholm border (3.1.19) is formed by the two solution curves $\lambda_\pm(k)$, (3.1.20), of the dispersion relation (3.1.18). We denote the graphs of the solution curves by $\sigma_{\text{F}}(\mathcal{L}_\pm)$, respectively. The Fredholm borders are oriented curves in the complex plane parameterized by $k \in \mathbb{R}$, with the orientation coinciding with the direction of increasing k. As λ crosses a Fredholm border separating a domain S_j from a domain S_k, the Fredholm index of the operator $\mathcal{L} - \lambda$ may change. Indeed, it can be shown that $\text{ind}(\mathcal{L} - \lambda)$ (see Fiedler and Scheel [**82**, Proposition 2.3.1]):

(a) increases by one upon crossing the curve $\sigma_{\text{F}}(\mathcal{L}_+)$ from right to left with respect to its orientation;

(b) decreases by one upon crossing the curve $\sigma_F(\mathcal{L}_-)$ from right to left with respect to its orientation.

In particular, for operators $\mathcal{L}$ that arise as linearizations of a nonlinear parabolic PDE about an equilibria, ϕ, a manifestation of the well-posedness of the PDE is that, generically, $\lambda \in \rho(\mathcal{L})$ for $\operatorname{Re}\lambda \gg 1$. Indeed, if not, the waveform is unstable to high-frequency perturbations, which is the purview of backwards heat equations. Granting this assumption, we may order the sets S_j such that the half-line $(L,\infty] \subset S_1 \subset \rho(\mathcal{L})$ for some $L \in \mathbb{R}$. The operator $\mathcal{L}-\lambda$ has Fredholm index 0 on S_1 and the Fredholm index changes by ± 1 at the crossing of each Fredholm boundary $\lambda_\pm$ (respectively), counted according to multiplicity; see Fig. 3.2.

Remark 3.1.16. The matrix $\boldsymbol{M}$, defined in (3.1.10), agrees with the Evans function associated to the asymptotic operator $\mathcal{L}_\infty$ on the natural domain of the Evans function, the boundary of which is composed of the Fredholm borders, $\sigma_f(\mathcal{L}_\infty)$; see Chapter 9. On this natural domain, $\boldsymbol{M}$ is square, and the point spectrum of $\mathcal{L}_\infty$ within the natural domain are precisely those λ for which $\det \boldsymbol{M}(\lambda) = 0$. However, the Evans function is extended analytically beyond the Fredholm borders of $\mathcal{L}_\infty$, where the matrix $\boldsymbol{M}$ changes dimension and loses its analyticity.

3.1.1 Examples

3.1.1.1 The Generalized Korteweg–de Vries Equation

In this example we determine the essential spectrum associated with the traveling solitary-wave solution of the generalized Korteweg–de Vries equation (gKdV)

$$\partial_t u + u^p \partial_x u + \partial_x^3 u = 0,$$

where $p \geq 1$ and $(x,t) \in \mathbb{R} \times \mathbb{R}^+$. In the traveling coordinates $\xi = x - ct$ the gKdV is rewritten as

$$\partial_t u + u^p \partial_\xi u - c\partial_\xi u + \partial_\xi^3 u = 0.$$

For each $c > 0$ there exists a solitary wave $\phi_c(\xi)$ that is an equilibrium of the traveling coordinate PDE; that is, a solution of the integrated traveling wave ODE

$$\partial_\xi^2 \phi_c - c\phi_c + \frac{1}{p+1}\phi_c^{p+1} = 0. \tag{3.1.21}$$

Indeed, ϕ_c can be realized from the associated first-order dynamical system for $(\phi, \partial_\xi \phi)^{\mathrm{T}}$ by constructing the solution that is homoclinic to the origin. Since for $c > 0$ the origin is a hyperbolic fixed point of the dynamical system, we see that $\phi_c(\xi) \to 0$, together with its derivatives, at an exponential rate as $\xi \to \pm\infty$.

Linearizing the traveling coordinate PDE about ϕ_c with $u = \phi_c + v$ yields the linear PDE,

$$\partial_t v = \mathbb{L}v, \quad \mathbb{L} = \partial_\xi\left(-\partial_\xi^2 + c - \phi_c^p\right).$$

The operator $\mathbb{L}$, which is the composition of the skew-symmetric operator ∂_ξ with the self-adjoint operator $\mathcal{L} := -\partial_\xi^2 + c - \phi_c^p$, will also be denoted as the gKdV operator. Assuming separation of variables, i.e., $v = \mathrm{e}^{\lambda t}p$, leads to an eigenvalue problem of the form (3.1.1),

$$\mathbb{L}p = \lambda p. \tag{3.1.22}$$

The asymptotic operator is given by

$$\mathbb{L}_0 = -\partial_\xi^3 + c\partial_\xi.$$

The asymptotic matrices coincide, and the single Fredholm boundary is obtained from (3.1.20) as

$$\lambda_0(k) = -(\mathrm{i}k)^3 + c(\mathrm{i}k) = -\mathrm{i}k(k^2 + c),$$

so that the Fredholm border, which coincides with $\sigma_{\mathrm{ess}}(\mathcal{L})$, is the imaginary axis. The Morse index of the asymptotic matrix $A_0(\lambda)$ is given by

$$\mathrm{i}_0(\lambda) = \begin{cases} 2, & \operatorname{Re}\lambda > 0 \\ 1, & \operatorname{Re}\lambda < 0. \end{cases}$$

3.1.1.2 Exponentially Weighted Spaces

While the essential spectrum is unmoved by relatively compact perturbations, which do not effect the asymptotic operator (see Theorem 3.1.11), it is sensitive to changes at spatial infinity. In particular, the essential spectrum and Fredholm borders may move when the eigenvalue problem is recast in an exponentially weighted space. For example, consider the gKdV eigenvalue problem (3.1.22) in the exponentially weighted space $H_a^k(\mathbb{R})$ defined by the norm

$$\|u\|_{H_a^k} := \|\mathrm{e}^{ax}u\|_{H^k}.$$

Setting $q = \mathrm{e}^{ax}p$, so that $p \in L_a^2(\mathbb{R})$ if and only if $q \in L^2(\mathbb{R})$, the eigenvalue problem (3.1.22) for the gKdV operator $\mathbb{L}$ transforms to one for the conjugated operator $\mathbb{L}_a := \mathrm{e}^{ax}\mathbb{L}\mathrm{e}^{-ax}$, where

$$\mathbb{L}_a q = \lambda q, \quad \mathbb{L}_a = (\partial_\xi - a)\left(-(\partial_\xi - a)^2 + c - \phi_c^p\right). \tag{3.1.23}$$

The conjugated asymptotic operator is

$$\mathbb{L}_{a,0} = -(\partial_\xi - a)^3 + c(\partial_\xi - a),$$

and its Fredholm borders coincide with its essential spectrum, given by the graph of the curve

$$\lambda_a(k) = -(\mathrm{i}k-a)^3 + c(\mathrm{i}k-a) = a^3 - ca - 3ak^2 + \mathrm{i}k(k^2+c-3a^2).$$

For $a > 0$ the rightmost point of the essential spectrum occurs at $k = 0$ for which $\operatorname{Re}\lambda_a(0) = a^3 - ac$. If $0 < a < \sqrt{c}$, then the essential spectrum is strictly contained in the left-half of the complex plane. Furthermore, for large k the imaginary component of the essential spectrum dominates the real part, so the border is asymptotically vertical (however, it is not asymptotic to $\operatorname{Im}\lambda = -\beta$ for any $\beta > 0$). As is discussed in detail in Chapter 4, if the essential spectrum enters the right-half of the complex plane, then the traveling wave ϕ_c is generically unstable. The fact that the essential spectrum moves into the left-half complex plane when $\mathcal{L}$ is constrained to act on perturbations which decay exponentially at $+\infty$ implies that the gKdV traveling wave is more sensitive to perturbations in front of the wave than to those behind it (e.g., see [221, 252] and the references therein).

Remark 3.1.17. The exponential weight does not move the point spectrum of the operator, indeed if p is an eigenfunction of the original operator with eigenvalue λ, then $q := \mathrm{e}^{-ax}p$ is an eigenfunction of the conjugated operator with the same eigenvalue, unless the essential spectrum has moved to encompass λ, in which case q generically does not decay at one of ∞ or $-\infty$. In Chapter 9 we use an Evans function analysis to study the perturbative motion of a point spectra near the essential spectrum.

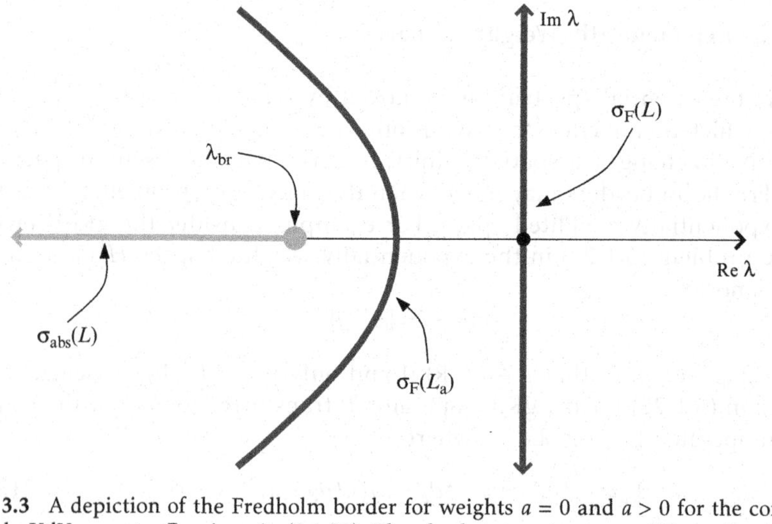

Fig. 3.3 A depiction of the Fredholm border for weights $a = 0$ and $a > 0$ for the conjugated gKdV operator $\mathbb{L}_a$ given in (3.1.23). The absolute spectrum, $\sigma_{\text{abs}}(\mathbb{L})$, as discussed in Chapter 3.2, is the leftmost limit of the Fredholm border as a is varied. (Color figure online.)

3.1.1.3 A Bistable Reaction–Diffusion Equation: Pulse

Returning to the reaction–diffusion equation of Chapter 2.3.3.1,

$$\partial_t u = \partial_x^2 u - W'(u),$$

with smooth potential W satisfying $W(0) = W'(0) = 0$, $W''(0) > 0$, and with a second transverse zero at $u = 1$, i.e., $W(1) = 0$ with $W'(1) < 0$. This system has a unique (up to translation) steady-state solution $\phi(x)$, satisfying

$$\partial_x^2 \phi = W'(\phi),$$

which is homoclinic to zero; in particular, $\partial_x^j \phi(x) \to 0$ exponentially fast as $x \to \pm\infty$ for $j = 0, 1$ (recall Fig. 2.1).

Linearizing about this equilibria yields the linear PDE,

$$\partial_t v = \mathcal{L} v, \quad \mathcal{L} = \partial_x^2 - W''(\phi), \tag{3.1.24}$$

which leads, via separation of variables, to the eigenvalue problem

$$\mathcal{L} p = \lambda p.$$

The asymptotic operator corresponding to $\mathcal{L}$ is given by

$$\mathcal{L}_0 = \partial_x^2 - W''(0),$$

and by Theorem 3.1.13 the Fredholm border $\sigma_{\mathrm{F}}(\mathcal{L})$ is given by the line

$$\sigma_{\mathrm{F}}(\mathcal{L}) = \{\lambda \in \mathbb{C} : \lambda = -k^2 - W''(0)\}$$

(see Fig. 3.4). The matrix eigenvalues are given by

$$\mu_1^0(\lambda) = \sqrt{\lambda + W''(0)}, \quad \mu_2^0(\lambda) = -\sqrt{\lambda + W''(0)},$$

so that the Morse index satisfies $\mathrm{i}_0(\lambda) = 1$ for all λ not on the Fredholm border.

3.1.1.4 A Bistable Reaction–Diffusion Equation: Front

The reaction-diffusion equation of Chapter 2.3.3.2,

$$\partial_t u = \partial_x^2 u - W'(u),$$

has a potential W with precisely two local minima at $u = 0, 1$ satisfying the conditions,

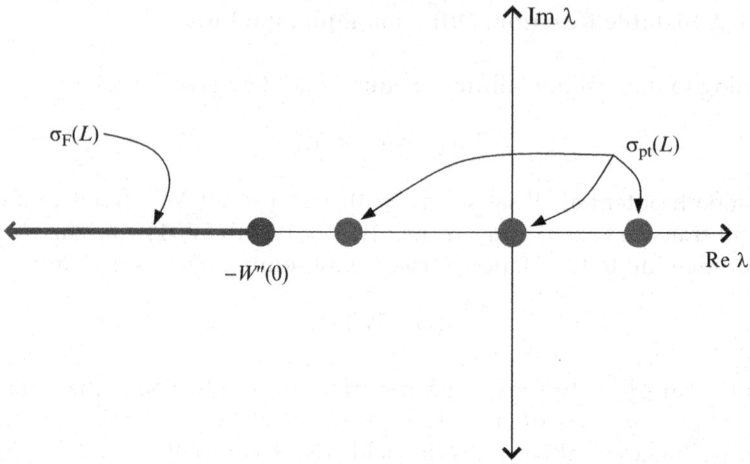

Fig. 3.4 The Fredholm border for the operator given in (3.1.24). The location of the point spectrum, $\sigma_{\mathrm{pt}}(\mathcal{L})$, is discussed in Chapter 2.3 (Color figure online)

$$W(0) = 0 > W(1), \quad W'(0) = W'(1) = 0, \quad W''(0), W''(1) > 0.$$

Introducing the traveling coordinate $\xi := x - c_* t$, there is a critical value c_* for which the system possesses a traveling front solution $u(x,t) = \phi(\xi)$, where ϕ is the unique heteroclinic solution of the ODE

$$-c_* \partial_\xi \phi = \partial_\xi^2 \phi - W'(\phi),$$

which satisfies the asymptotic relations

$$\phi(\xi) \to \begin{cases} 0, & \xi \to -\infty \\ 1, & \xi \to +\infty \end{cases}$$

(recall Fig. 2.2).

The front is a steady-state solution to the PDE in the traveling coordinates,

$$\partial_t u = \partial_\xi^2 u + c_* \partial_\xi u - W'(u).$$

Linearizing about the front yields the linear PDE,

$$\partial_t v = \mathcal{L} v, \quad \mathcal{L} = \partial_\xi^2 + c_* \partial_\xi - W''(\phi), \tag{3.1.25}$$

which for separated variables $v(x,t) = \mathrm{e}^{\lambda t} p(x)$ yields the eigenvalue problem

$$\mathcal{L} p = \lambda p.$$

The asymptotic operators are given by

$$\mathcal{L}_- = \partial_\xi^2 + c_*\partial_\xi - W''(0), \quad \mathcal{L}_+ = \partial_\xi^2 + c_*\partial_\xi - W''(1),$$

so that by Remark 3.1.15 the Fredholm borders $\sigma_F(\mathcal{L}_\pm)$ are given by the parabolic curves

$$\sigma_F(\mathcal{L}_-) = \{\lambda = -k^2 + ic_*k - W''(0) : k \in R\} = \{\lambda \in \mathbb{C} : \text{Re}\,\lambda = -W''(0) - (\text{Im}\,\lambda/c_*)^2\},$$

and

$$\sigma_F(\mathcal{L}_+) = \{\lambda = -k^2 + ic_*k - W''(1) : k \in \mathbb{R}\} = \{\lambda \in \mathbb{C} : \text{Re}\,\lambda = -W''(1) - (\text{Im}\,\lambda/c_*)^2\}.$$

If $W''(1) > W''(0) > 0$, then the curve $\sigma_F(\mathcal{L}_+)$ will be to the left of the curve $\sigma_F(\mathcal{L}_-)$, as depicted in Fig. 3.5.

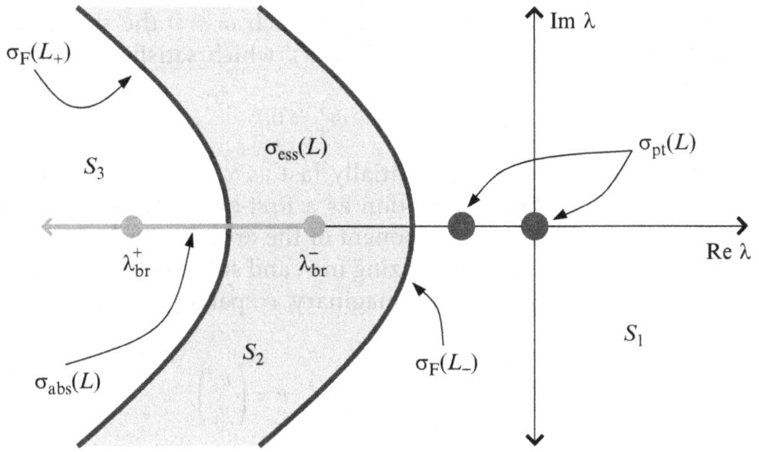

Fig. 3.5 The Fredholm borders, the essential spectrum, and the absolute spectrum for the operator given in (3.1.25) under the assumption $W''(0) < W''(1)$. The point spectrum, $\sigma_{\text{pt}}(\mathcal{L})$, and the absolute spectrum, $\sigma_{\text{abs}}(\mathcal{L})$, are discussed in Chapter 2.3 and Chapter 3.2 respectively. The two branch points of the absolute spectrum, $\lambda_{\text{br}}^\pm$, are given by $\lambda_{\text{br}}^- = -W''(0) - c_*^2/4$ and $\lambda_{\text{br}}^+ = -W''(1) - c_*^2/4$. (Color figure online.)

The matrix eigenvalues are readily calculated to be

$$\mu_1^-(\lambda) = \frac{1}{2}\left(-c_* + \sqrt{c_*^2 + 4(\lambda + W''(0))}\right), \quad \mu_2^-(\lambda) = \frac{1}{2}\left(-c_* - \sqrt{c_*^2 + 4(\lambda + W''(0))}\right)$$
$$\mu_1^+(\lambda) = \frac{1}{2}\left(-c_* + \sqrt{c_*^2 + 4(\lambda + W''(1))}\right), \quad \mu_2^+(\lambda) = \frac{1}{2}\left(-c_* - \sqrt{c_*^2 + 4(\lambda + W''(1))}\right).$$

If we denote by S_1 the region to the right of $\sigma_F(\mathcal{L}_-)$, S_2 the region between the two curves, and S_3 the region to the left $\sigma_F(\mathcal{L}_+)$, then from Remark 3.1.15 the Morse indices satisfy

$$\mathrm{i}_-(\lambda) = \begin{cases} 1, & \lambda \in S_1 \\ 0, & \lambda \in S_2 \cup S_3, \end{cases} \qquad \mathrm{i}_+(\lambda) = \begin{cases} 1, & \lambda \in S_1 \cup S_2 \\ 0, & \lambda \in S_3. \end{cases}$$

Consequently, the operator has Fredholm index 0 for $\lambda \in S_1 \cup S_3$, and Fredholm index -1 for $\lambda \in S_2$. By Theorem 3.1.13 the essential spectrum is the closure of the set S_2.

3.1.1.5 The Nonlinear Schrödinger Equation

The nonlinear Schrödinger equation (NLS), is given by

$$\mathrm{i}\partial_t u + \partial_x^2 u - \omega u + |u|^2 u = 0. \tag{3.1.26}$$

Here $u \in \mathbb{C}$, $(x,t) \in \mathbb{R} \times \mathbb{R}^+$, and $\omega > 0$. For each $\omega > 0$ the system has a real-valued time-independent solution $\phi_\omega(x)$, which satisfies the equilibrium equation

$$\partial_x^2 \phi_\omega - \omega \phi_\omega + \phi_\omega^3 = 0, \tag{3.1.27}$$

and for which $\partial_x^j \phi_\omega(x) \to 0$ exponentially fast as $x \to \pm\infty$ for $j = 0, 1$. Indeed, writing the equilibrium equation as a first-order dynamical system for $(\phi_\omega, \partial_x \phi_\omega)^{\mathrm{T}}$, ϕ_ω is the first component of the orbit that is homoclinic to the origin. Writing $u = \phi_\omega + v$, linearizing in v, and subsequently decomposing $v = v_{\mathrm{r}} + \mathrm{i} v_{\mathrm{i}}$, into its real, v_{r}, and imaginary, v_{i}, parts yields a linear PDE for $\boldsymbol{v} = (v_{\mathrm{r}}, v_{\mathrm{i}})$,

$$\partial_t \boldsymbol{v} = \underbrace{\begin{pmatrix} 0 & 1 \\ -1 & 0 \end{pmatrix} \begin{pmatrix} \mathcal{L}_+ & 0 \\ 0 & \mathcal{L}_- \end{pmatrix}}_{\mathcal{L}} \boldsymbol{v}, \quad \boldsymbol{v} = \begin{pmatrix} v_{\mathrm{r}} \\ v_{\mathrm{i}} \end{pmatrix}, \tag{3.1.28}$$

where the operators $\mathcal{L}_\pm$ are given by

$$\mathcal{L}_- v_{\mathrm{i}} = -\partial_x^2 v_{\mathrm{i}} + \omega v_{\mathrm{i}} - \phi_\omega^2 v_{\mathrm{i}}, \quad \mathcal{L}_+ v_{\mathrm{r}} = -\partial_x^2 v_{\mathrm{r}} + \omega v_{\mathrm{r}} - 3\phi_\omega^2 v_{\mathrm{r}}. \tag{3.1.29}$$

Searching for separated variables solutions, $\boldsymbol{v} = \mathrm{e}^{\lambda t} \boldsymbol{p}$, yields the eigenvalue problem

$$\mathcal{L}\boldsymbol{p} = \lambda \boldsymbol{p}.$$

Since $\phi_\omega(x) \to 0$ as $x \to \pm\infty$, the asymptotic operator is given by

$$\begin{pmatrix} 0 & 1 \\ -1 & 0 \end{pmatrix} \begin{pmatrix} \mathcal{L}_0 & 0 \\ 0 & \mathcal{L}_0 \end{pmatrix}, \quad \mathcal{L}_0 = -\partial_x^2 + \omega,$$

which is equivalent to

$$(\mathcal{L}_0^2 + \lambda^2) p_{\mathrm{r}} = 0.$$

By Theorem 3.1.13 the Fredholm border and essential spectrum coincide, and via (3.1.20) are given by the curves

$$[-(\mathrm{i}k)^2 + \omega]^2 + \lambda^2 = 0 \quad \Rightarrow \quad \lambda = \pm\mathrm{i}(\omega + k^2).$$

The essential spectrum is then a subset of the imaginary axis, i.e.,

$$\sigma_{\mathrm{ess}}(\mathcal{L}) = \{\lambda \in \mathbb{C} : |\mathrm{Im}\,\lambda| \geq \omega,\ \mathrm{Re}\,\lambda = 0\}.$$

It is not difficult to check that the Morse index is $\mathrm{i}_0(\lambda) = 2$ for $\lambda \notin \sigma_{\mathrm{ess}}(\mathcal{L}_0)$.

Exercises

Exercise 3.1.1. Verify each of the statements in the proof of Lemma 3.1.2; in particular, show that $\|\mathcal{B}\|_{\mathcal{B}(H^n)} \to 0$ as $\lambda \to +\infty$. Extend the proof to show that the operator $\mathcal{L}_\infty$ defined in (3.1.2) is also closed from $H^n(\mathbb{R}) \subset L^2(\mathbb{R})$ into $L^2(\mathbb{R})$.

Exercise 3.1.2. Prove Lemma 3.1.8 for the case that $\boldsymbol{A}_+$ is hyperbolic, while $\boldsymbol{A}_-$ has a nontrivial center space. The key issue is that $\mathcal{L}_\infty - \lambda$ can have a nontrivial kernel. Show that the sequence $\{u_k\}$ constructed in Lemma 3.1.8 can be modified so as to be orthogonal to the kernel, while still of norm one and satisfying $\|(\mathcal{L}_\infty - \lambda)u_k\|_{L^2} \to 0$ as $k \to \infty$.

Exercise 3.1.3. Prove Theorem 3.1.11 in the case that the coefficients a_j converge exponentially to their asymptotic values.

Exercise 3.1.4. Consider the 2×2 matrix $\boldsymbol{A}(x)$ given by

$$\boldsymbol{A}(x) = \begin{cases} \begin{pmatrix} 1 & 0 \\ 0 & -1 \end{pmatrix}, & x < 0 \\ \begin{pmatrix} -1 & 0 \\ \alpha & 1-\alpha \end{pmatrix}, & x \geq 0. \end{cases}$$

Determine the values of $\alpha \in \mathbb{R}$ for which the 2×2 system

$$\partial_x \boldsymbol{Y} = \boldsymbol{A}(x)\boldsymbol{Y} + \begin{pmatrix} \mathrm{e}^{-|x|} \\ 2\mathrm{e}^{-|x|} \end{pmatrix},$$

has a unique solution $\boldsymbol{Y}(x)$ that satisfies $|\boldsymbol{Y}| \in L^2(\mathbb{R})$. Explicitly construct the unique solution for those values of α.

Exercise 3.1.5. For the asymptotic linear operator $\mathcal{L}_\infty$ [see (3.1.2)] acting on it domain $H^n(\mathbb{R}) \subset L^2(\mathbb{R})$, consider the sets

$$\rho_+ = \{\lambda : \mathrm{i}_+(\lambda) = \mathrm{i}_-(\lambda) = n\}, \quad \rho_- = \{\lambda : \mathrm{i}_+(\lambda) = \mathrm{i}_-(\lambda) = 0\}.$$

For the following, suppose that $\lambda \in \rho_\pm$.

(a) Construct the matrix $\boldsymbol{M}$ of the proof of Lemma 3.1.10 to show directly that $\lambda \notin \sigma_{\mathrm{ess}}(\mathcal{L}_\infty)$ and $\lambda \notin \sigma_{\mathrm{pt}}(\mathcal{L}_\infty)$. Conclude that $\lambda \in \rho(\mathcal{L}_\infty)$, the resolvent set of $\mathcal{L}_\infty$.

(b) Construct $(\mathcal{L}_\infty - \lambda)^{-1} f$ explicitly in terms of the matrix exponentials of the asymptotic matrices.

Exercise 3.1.6. Fix $a > 0$ and let w denote a weight function that satisfies $w(x) = \mathrm{e}^{a|x|}$ for $|x| > 1$ and that is smooth and positive for $|x| < 1$. Denote by $H_w^k(\mathbb{R})$ the space of functions u for which $\|wu\|_{H^k} < \infty$. Compute $\sigma_{\mathrm{ess}}(\mathcal{L})$ for the operator $\mathcal{L} = \partial_x^2$ in the three spaces $H^2(\mathbb{R})$, $H_a^2(\mathbb{R})$, and $H_w^2(\mathbb{R})$ by studying the appropriate conjugated operators.

3.2 The Absolute Spectrum

In Chapter 3.1.1.2 the impact of an exponential weighted on the essential spectrum of the gKdV operator was studied in terms of the conjugated operator given in (3.1.23). There is a second approach to understanding exponentially weighted spaces that yields deeper insight into the nature of the spectrum, particularly into the connection between the essential spectrum on the whole line and the spectrum of the operator subject to separated boundary conditions on large but finite domains.

Consider an exponentially asymptotic operator $\mathcal{L}$ of the form (3.1.1) with smooth coefficients. Definition 3.1.4 gives the limiting operator $\mathcal{L}_\infty$, while (3.1.4) defines the asymptotic matrices $\boldsymbol{A}_\pm(\lambda)$ to the operator $\mathcal{L}_\infty - \lambda$. The existence of an exponential dichotomy [see Definition 2.1.32] for $\mathcal{L}_\infty - \lambda$ depends upon the matrix eigenvalues $\mu_j^\pm(\lambda)$ of the limiting matrices: recall that, as in Lemma 3.1.8, the operator $\mathcal{L}_\infty - \lambda$ can be Fredholm of index zero only if the limiting matrices are hyperbolic. For each limiting matrix we separated its matrix eigenvalues into groups with a positive, negative, or zero real part. A matrix eigenvalue having a positive or negative real part distinguishes whether solutions of the homogeneous version of (3.1.4), taking the associated matrix eigenvalue as initial data, decay or grow exponentially as $x \to \pm\infty$. The key result of Lemma 3.1.10 associated the boundary of the essential spectrum, i.e., the Fredholm borders, with the curves where (at least) one of the matrix eigenvalues has a zero real part.

What is the impact of an exponential weight on this framework? To this end we introduce a two-sided weight parameterized by its exponential growth rates, $a_\pm \in \mathbb{R}$; namely,

$$w_{a_\pm}(x) = \begin{cases} \mathrm{e}^{-a_- x}, & x \le -1 \\ \mathrm{e}^{-a_+ x}, & x \ge +1, \end{cases} \tag{3.2.1}$$

where w is smooth and positive for $-1 < x < +1$. Defining the spaces $H^k_{a_\pm}(\mathbb{R})$ in terms of the norm

$$\|f\|_{H^k_{a_\pm}} := \|w_{a_\pm} f\|_{H^k}, \tag{3.2.2}$$

a solution g to $(\mathcal{L}_\infty - \lambda)g = f$ lies in $H^k_{a_\pm}(\mathbb{R})$ if the solution and its derivatives decays faster then $e^{a_\pm x}$ as $x \to \pm\infty$, respectively. Thus, with respect to the weighted norm the appropriate dichotomy for the matrix eigenvalues $\mu_j^\pm(\lambda)$ of $A_\pm(\lambda)$ is not based upon the sign of their real part, but instead upon the relation of their real part to $a_\pm$. Indeed, a simple modification of the arguments leading up to Theorem 3.1.11 shows that the Fredholm borders of $\mathcal{L}_\infty$ in $H^n_{a_\pm}$ are given by those λ for which $\mathrm{Re}\,\mu_-(\lambda) = a_-$ or $\mathrm{Re}\,\mu_+(\lambda) = a_+$. A natural question now arises: what choice of exponential weight $w_{a_\pm}$ gives $\mathcal{L}$ the "largest" resolvent set? We begin with the following hypothesis.

Hypothesis 3.2.1. There exists $\alpha_0 > 0$ such that the operator $\mathcal{L} - \lambda$ acting on $H^n(\mathbb{R}) \subset L^2(\mathbb{R}) \to L^2(\mathbb{R})$ has a Fredholm index 0 for $\mathrm{Re}\,\lambda \geq \alpha_0$. Such an operator is said to be *well-posed*. For a well-posed operator, the *essential resolvent*, $\rho_{\mathrm{ess}}(\mathcal{L})$, of $\mathcal{L}$ is the connected component of the resolvent that contains $\{\mathrm{Re}\,\lambda > \alpha_0\}$.

Remark 3.2.2. Hypothesis 3.2.1 is typically a choice of orientation corresponding to the well-posedness of the underling PDE. For example, $\mathcal{L} := \partial_x^2$ satisfies the hypothesis, while $\mathcal{L} := -\partial_x^2$ does not, just as the heat equation is well-posed and the backwards heat equation is not. There exist nth-order differential operators for which neither $\mathcal{L}$ nor $-\mathcal{L}$ are well-posed.

From this hypothesis it follows from Lemma 3.1.10 that $\mathrm{i}_-(\lambda) = \mathrm{i}_+(\lambda)$ for $\mathrm{Re}\,\lambda \gg 1$: denote this common value of the Morse indices by i_∞. Order the matrix eigenvalues $\mu_\pm^j(\lambda)$, $j = 1,\dots,n$, of the asymptotic matrices $A_\pm(\lambda)$ according to the size of their real part,

$$\mathrm{Re}\,\mu_\pm^1(\lambda) \geq \cdots \geq \mathrm{Re}\,\mu_\pm^j(\lambda) \geq \cdots \geq \mathrm{Re}\,\mu_\pm^n(\lambda),$$

and introduce the stable and unstable extrema

$$\mu_\pm^{\mathrm{u}}(\lambda) := \mathrm{Re}\,\mu_\pm^{\mathrm{i}_\infty}(\lambda), \quad \mu_\pm^{\mathrm{s}}(\lambda) := \mathrm{Re}\,\mu_\pm^{\mathrm{i}_\infty+1}(\lambda).$$

That is, $\mu_\pm^{\mathrm{s}}(\lambda)$ denotes the largest real part of any of the matrix eigenvalues from $\mathbb{E}_\pm^{\mathrm{s}}(\lambda)$, and $\mu_\pm^{\mathrm{u}}(\lambda)$ denotes the smallest real part of any of the matrix eigenvalues from $\mathbb{E}_\pm^{\mathrm{u}}(\lambda)$. From our hypothesis we know that $\mu_\pm^{\mathrm{u}}(\lambda) > 0 > \mu_\pm^{\mathrm{s}}(\lambda)$ for $\mathrm{Re}\,\lambda \gg 1$. If we move the operator eigenvalue parameter λ from the far right-hand side of the complex plane towards the Fredholm border, then at least one of the matrix eigenvalues will approach the imaginary axis.

Without loss of generality, assume that a single matrix eigenvalue associated with $A_-(\lambda)$ of positive real part hits the imaginary axis as λ approaches the Fredholm boundary. For this value of λ we will have

$$\mu_-^{\mathrm{u}}(\lambda) = 0 > \mu_-^{\mathrm{s}}(\lambda), \quad \mu_+^{\mathrm{u}}(\lambda) > 0 > \mu_+^{\mathrm{s}}(\lambda).$$

(see the left two panels of Fig. 3.6). While this λ resides in $\sigma_{\mathrm{ess}}(\mathcal{L})$ as defined in an unweighted space, if we choose a weight $w_{a_\pm}$ for which $a_\pm$ satisfy

$$\mu_-^{\mathrm{u}}(\lambda) > a_- > \mu_-^{\mathrm{s}}(\lambda), \quad \mu_+^{\mathrm{u}}(\lambda) > a_+ > \mu_+^{\mathrm{s}}(\lambda), \tag{3.2.3}$$

then λ is **not** in the essential spectrum of $\mathcal{L} : H^n_{a_\pm}(\mathbb{R}) \subset L^2_{a_\pm}(\mathbb{R}) \to L^2_{a_\pm}(\mathbb{R})$ and $\mathcal{L} - \lambda$ remains Fredholm. Moreover, when either of the spectral gaps $\mu_\pm^{\mathrm{u}}(\lambda) - \mu_\pm^{\mathrm{s}}(\lambda)$ between the matrix eigenvalue extrema close (see the third panel in Fig. 3.6), it is no longer possible to choose a weight that renders the operator $\mathcal{L} - \lambda$ Fredholm with index zero on the corresponding weighted spaces. This motivates the following definition.

Definition 3.2.3. For an operator $\mathcal{L}$ satisfying Hypothesis 3.2.1 let the absolute resolvent, $\rho_{\mathrm{abs}}(\mathcal{L})$, denote the largest simply connected subset of $\mathbb{C}$ containing $\mathrm{Re}\,\lambda \gg 1$ for which

$$\mu_+^{\mathrm{u}}(\lambda) > \mu_+^{\mathrm{s}}(\lambda) \quad \text{and} \quad \mu_-^{\mathrm{u}}(\lambda) > \mu_-^{\mathrm{s}}(\lambda),$$

for all $\lambda \in \rho_{\mathrm{abs}}(\mathcal{L})$. The absolute spectrum of $\mathcal{L}$ is the boundary of $\rho_{\mathrm{abs}}(\mathcal{L})$, i.e., those λ for which

$$\mu_+^{\mathrm{u}}(\lambda) = \mu_+^{\mathrm{s}}(\lambda) \quad \text{or} \quad \mu_-^{\mathrm{u}}(\lambda) = \mu_-^{\mathrm{s}}(\lambda).$$

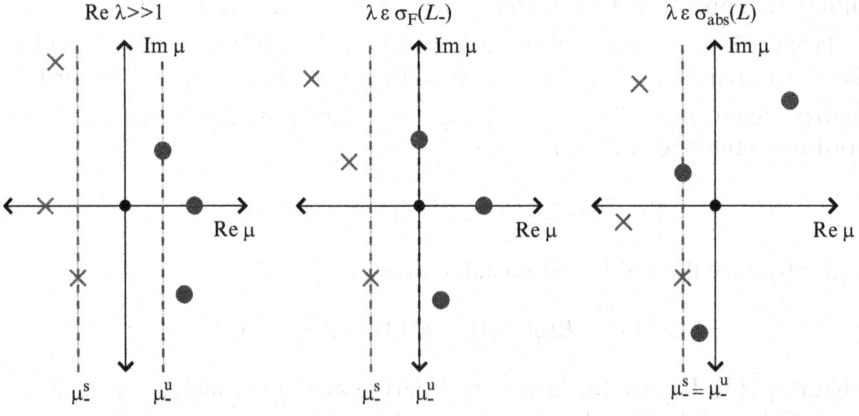

Fig. 3.6 The evolution of the matrix eigenvalues of $A_-(\lambda)$ when $n = 6$ as the spectral parameter λ is moved leftwards from $\mathrm{Re}\,\lambda \gg 1$ to cross $\sigma_{\mathrm{F}}(\mathcal{L}_-)$ and finally to touch $\sigma_{\mathrm{abs}}(\mathcal{L})$. At this last point, the gap between the analytic continuation of the stable and unstable manifolds of $A_-(\lambda)$ vanishes. In the second panel $\lambda \in \sigma_{\mathrm{ess}}(\mathcal{L})$; however, a constant $\mu_-^{\mathrm{s}} < a_- < \mathrm{u}_-^{\mathrm{u}}$ can be found so that λ is not in the essential spectrum of the conjugated operator $\mathcal{L}_{a_\pm}$. On the other hand, in the third panel no such constant a_- can be found. (Color figure online.)

The connection between the absolute spectrum of $\mathcal{L}$ and the conjugated operators

$$\mathcal{L}_{a_\pm} := w_{a_\pm}\mathcal{L}w_{a_\pm}^{-1},$$

is straightforward. The matrix eigenvalues of the asymptotic matrices $\boldsymbol{A}_+(\lambda, a_+)$ and $\boldsymbol{A}_-(\lambda, a_-)$ associated with the vectorization of the conjugated eigenvalue problem $(\mathcal{L}_{a_\pm} - \lambda)p = 0$ are simply those of $\boldsymbol{A}_+(\lambda)$ and $\boldsymbol{A}_-(\lambda)$ which have each been shifted to the left by a_+ and a_-, respectively. It is clear that the absolute spectrum is unchanged under conjugation, since the gaps between the spectral extrema do not change. Indeed, the spectral properties of $\mathcal{L}_{a_\pm}$ acting on $H^n(\mathbb{R})$ are equivalent to those of $\mathcal{L}$ acting on $H^n_{a_\pm}(\mathbb{R})$. These observations lead to the following lemma.

Lemma 3.2.4. *Suppose that $\mathcal{L}$ is a well-posed operator, satisfying Hypothesis 3.2.1. For any $a_\pm, a'_\pm \in \mathbb{R}$ the conjugated operators $\mathcal{L}_{a_\pm}$ and $\mathcal{L}_{a'_\pm}$ satisfy*

$$\sigma_{\text{abs}}\left(\mathcal{L}_{a_\pm}\right) = \sigma_{\text{abs}}(\mathcal{L}_{a'_\pm}).$$

Furthermore, the essential resolvent of each conjugated operator lies within the absolute spectrum of $\mathcal{L}$,

$$\bigcup_{a_\pm \in \mathbb{R}\times\mathbb{R}} \rho_{\text{ess}}(\mathcal{L}_{a_\pm}) \subset \rho_{\text{abs}}(\mathcal{L}).$$

Equivalently, the absolute spectrum of $\mathcal{L}$ lies outside the essential resolvent of all conjugations of $\mathcal{L}$.

The absolute spectrum also arises naturally when considering analytic continuations of the matrix eigenvalues and associated stable and unstable eigenspaces. Considering the asymptotic matrix $\boldsymbol{A}_-(\lambda)$, for λ to the right of the Fredholm border, $\sigma_{\text{F}}(\mathcal{L})$, the vectorized system

$$\partial_x \boldsymbol{Y} = \boldsymbol{A}_-(\lambda)\boldsymbol{Y}, \quad x \leq 0,$$

has naturally defined stable and unstable spaces $\mathbb{E}^{\text{u}}_-(\lambda)$ and $\mathbb{E}^{\text{s}}_-(\lambda)$. Solutions with initial data in $\mathbb{E}^{\text{u}}_-(\lambda)$ will decay exponentially fast in norm as $x \to -\infty$, while solutions with initial data in $\mathbb{E}^{\text{s}}_-(\lambda)$ will grow exponentially fast in norm as $x \to -\infty$. As λ crosses $\sigma_{\text{F}}(\mathcal{L})$ at least one matrix eigenvalue changes the sign of its real part, with a corresponding change in the dimension of the stable and unstable eigenspaces. As a consequence, the stable and unstable subspaces do not vary smoothly in λ at the Fredholm border. However, by tracking the eigenspaces associated with the matrix eigenvalues, irrespective of the sign of their real part, the associated eigenspaces generically have a unique analytic continuation that does not change dimension at or near the Fredholm border. This analytic continuation is guaranteed via an appropriate spectral projection (see Chapter 9 and Chapter 10 for the details and application to extensions of the Evans function). In other words, the stable and unstable eigenspaces of $\boldsymbol{A}_-(\lambda)$ generically have an analytical continuation beyond the Fredholm border; however the analytically continued spaces lose their exponential growth/decay properties for λ beyond the border.

The significance of this observation is that the analytic extension cannot typically be continued indefinitely. Looking at the absolute spectrum, there are nongeneric values $\lambda \in \sigma_{\mathrm{abs}}(\mathcal{L})$ for which not only does $\mu_-^{\mathrm{u}}(\lambda) = \mu_-^{\mathrm{s}}(\lambda)$, but also $\mu_-^{\mathrm{i}_\infty}(\lambda) = \mu_-^{\mathrm{i}_\infty+1}(\lambda)$ (again see Fig. 3.6, and in the third panel imagine that the (blue) filled circle and (red) cross on the line $\mu_-^{\mathrm{s}} = \mu_-^{\mathrm{u}}$ coincide). The collision of these matrix eigenvalues makes it impossible to analytically separate the eigenspaces via a spectral projection. We will see that this amounts to a branch point in the associated spectral projections, and the absolute spectrum will coincide with the associated branch cut. This discussion motivates the following definition.

Definition 3.2.5. A point $\lambda \in \sigma_{\mathrm{abs}}(\mathcal{L})$ is called a branch point if

$$\mu_-^{\mathrm{i}_\infty}(\lambda) = \mu_-^{\mathrm{i}_\infty+1}(\lambda) \quad \text{or} \quad \mu_+^{\mathrm{i}_\infty}(\lambda) = \mu_+^{\mathrm{i}_\infty+1}(\lambda).$$

3.2.1 *Examples*

3.2.1.1 The Generalized Korteweg–de Vries Equation

This is a continuation of the discussion of the linearization about the homoclinic traveling wave solution of the generalized KdV equation of Chapter 3.1.1.1. The matrix eigenvalues associated with the operator $\mathbb{L}$ of (3.1.22) satisfy the characteristic equation

$$\mu^3 - c\mu + \lambda = 0.$$

For $\mathrm{Re}\,\lambda > 0$ we have $\mathrm{i}_0(\lambda) = 2$, so that two of the matrix eigenvalues will have a positive real part, and the other one will have a negative real part. This implies that $\mathrm{i}_\infty(\mathbb{L}) = 2$. As λ moves into the left-half of the complex plane, where $\mathrm{i}_0(\lambda) = 1$, one of the matrix eigenvalues with a positive real part becomes a matrix eigenvalue with a negative real part. It is not difficult to check that for $\lambda \in \sigma_{\mathrm{abs}}(\mathcal{L})$, given by the curve

$$\sigma_{\mathrm{abs}}(\mathcal{L}) = \{\lambda \in \mathbb{C} : \mathrm{Re}\,\lambda \le -2(c/3)^{3/2},\ \mathrm{Im}\,\lambda = 0\} = \left(-\infty, -2(c/3)^{3/2}\right],$$

two of the matrix eigenvalues have the same negative real part, while the other matrix eigenvalue has a positive real part. The branch point of the absolute spectrum $b = -2(c/3)^{3/2}$ is also its rightmost point, and in particular the absolute spectrum does not intersect the essential spectrum. The interested reader should consult [252, Section 6] for more details.

3.2.1.2 A Bistable Reaction–Diffusion Equation: Front

This is a continuation of the discussion of the linearization, $\mathcal{L}$ given in (3.1.25), of the system (2.3.13) about the front solution presented in Chapter 3.1.1.4. We recall the form of the matrix eigenvalues

$$\mu_1^-(\lambda) = \frac{1}{2}\left(-c_* + \sqrt{c_*^2 + 4(\lambda + W''(0))}\right),\quad \mu_2^-(\lambda) = \frac{1}{2}\left(-c_* - \sqrt{c_*^2 + 4(\lambda + W''(0))}\right)$$
$$\mu_1^+(\lambda) = \frac{1}{2}\left(-c_* + \sqrt{c_*^2 + 4(\lambda + W''(1))}\right),\quad \mu_2^+(\lambda) = \frac{1}{2}\left(-c_* - \sqrt{c_*^2 + 4(\lambda + W''(1))}\right).$$

For $\mathrm{Re}\,\lambda > 0$ it is clear that $\mathrm{i}_\infty(\mathcal{L}) = 1$. It is straightforward to check that

$$\mathrm{Re}\,\mu_1^-(\lambda) = \mathrm{Re}\,\mu_2^-(\lambda) \quad \Leftrightarrow \quad \lambda \in \{\lambda \in \mathbb{C} : \lambda = -k^2 - W''(0) - c_*^2/4,\ k \in \mathbb{R}\},$$

and

$$\mathrm{Re}\,\mu_1^+(\lambda) = \mathrm{Re}\,\mu_2^+(\lambda) \quad \Leftrightarrow \quad \lambda \in \{\lambda \in \mathbb{C} : \lambda = -k^2 - W''(1) - c_*^2/4,\ k \in \mathbb{R}\},$$

The absolute spectrum is the union of these two sets, i.e.,

$$\sigma_{\mathrm{abs}}(\mathcal{L}) = (-\infty, -b_-] \cup (-\infty, b_+],$$

where the branch points are $b_- = -W''(0) - c_*^2/4$ and $b_+ = -W''(1) - c_*^2/4$. Unlike the gKdV example, the essential spectrum and the absolute spectrum may have nontrivial intersection; see Fig. 3.5.

3.2.2 *Absolute Spectrum and the Large Domain Limit*

It is natural, particularly when discussing numerical approximations of PDEs, to ask in what sense an eigenvalue problem posed on a large domain $[-L, L]$ corresponds to the eigenvalue problem for the same operator on an unbounded domain. It is easier to be more precise when going the other direction: given an operator $\mathcal{L}$ defined on the whole line, what relation does its spectra maintain with that spectra of the same operator restricted to a large interval, $[-L, L]$, when subject to various boundary conditions? Indeed, considering only second order operators and separated boundary conditions of the form (2.3.2), how does the choice of $b_1^\pm$ and $b_2^\pm$ impact the spectra in the large domain limit, $L \to \infty$?

To begin, compare the results of Theorems 2.3.1 and 2.3.3 for second-order linear operators. Considering the operator $\mathcal{L}$ defined in (2.3.1), when acting on the whole line, its spectra, denoted $\sigma(\mathcal{L}; \mathbb{R})$, is comprised of a finite number of point spectra (eigenvalues) and a semi-infinite interval of essential spectrum for the unbounded domain problem. The same operator, acting on a bounded domain subject to separated boundary conditions has spectra, denoted $\sigma(\mathcal{L};[-L,L])$, comprised of a countably infinite collection of point spectra (eigenvalues). It may seem reasonable to expect that the finite number, N, of eigenvalues in $\sigma_{\mathrm{pt}}(\mathcal{L};\mathbb{R})$ are well-approximated by the first N eigenvalues of $\sigma_{\mathrm{pt}}(\mathcal{L};[-L,+L])$, and in fact, this is indeed true under some broad nondegeneracy conditions (see Chapter 9.6). However, the convergence of the remaining point spectrum of $\sigma_{\mathrm{pt}}(\mathcal{L};[-L,L])$ as $L \to \infty$ is much more subtle question, with the answer depending sensitively upon the particular boundary conditions.

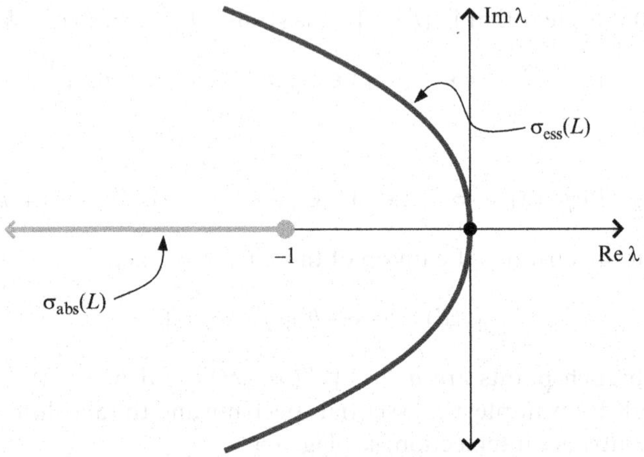

Fig. 3.7 The essential spectrum and absolute spectrum for the operator $\mathcal{L}$ of (3.2.4). The branch point is $\lambda = -1$. (Color figure online.)

We address this problem in greater detail in Chapter 9: here we content ourselves with an illustrative example. Consider the linear operator

$$\mathcal{L} := \partial_x^2 + 2\partial_x. \tag{3.2.4}$$

Acting on $L^2(\mathbb{R})$, with domain $H^2(\mathbb{R})$, the operator has no point spectrum, and the essential spectrum is given by the parabolic curve

$$\sigma_{\mathrm{ess}}(\mathcal{L};\mathbb{R}) = \{\lambda \in \mathbb{C} : \lambda = -k^2 + \mathrm{i}2k,\ k \in \mathbb{R}\}.$$

Since the matrix eigenvalues are

$$\mu_1(\lambda) = -1 + \sqrt{1+\lambda}, \quad \mu_2(\lambda) = -1 - \sqrt{1+\lambda},$$

we deduce that the absolute spectrum is given by the semi-infinite line

$$\sigma_{\mathrm{abs}}(\mathcal{L}) = \{\lambda \in \mathbb{C} : \lambda = -1 - k^2,\ k \in \mathbb{R}\},$$

with branch point at $\lambda = -1$ (see Fig. 3.7).

Now consider the eigenvalue problem for $\mathcal{L}$ of (3.2.4) on the space $H^2_{\rm bc}([-L,+L])$ with the separated Dirichlet boundary conditions

$$p(-L) = p(+L) = 0.$$

A standard calculation shows that the point spectrum is given by

$$\sigma_{\rm pt}(\mathcal{L}) = \{\lambda \in \mathbb{C} : \lambda = -1 - \left(\frac{n\pi}{L}\right)^2, \ n \in \mathbb{N}\}.$$

We see that, $\sigma_{\rm pt}(\mathcal{L}) \subset \sigma_{\rm abs}(\mathcal{L})$, and as $L \to +\infty$ the eigenvalues accumulate at the branch point $\lambda = -1$, within the absolute spectra, and not to the essential spectra associated with the unweighted space, $L^2(\mathbb{R})$.

Exercises

Exercise 3.2.1. Verify that Lemma 3.2.4 holds for the operator $\mathcal{L}$ of (3.2.4) by a directly computing $\sigma(\mathcal{L})$ in the exponentially weighted spaces $H^2_a(\mathbb{R})$ of Chapter 3.1.1.2.

Exercise 3.2.2. Consider the linear operator, $\mathcal{L}$, given in (3.2.4), acting on the space $H^2_{\rm bc}[-L,+L]$, subject to the separated boundary conditions

$$b_1^{\pm} p(\pm L) + b_2^{\pm} \partial_x p(\pm L) = 0, \quad (b_1^{\pm})^2 + (b_2^{\pm})^2 = 1.$$

(a) Show that apart from exceptional values of $b_j^{\pm}$, the point spectrum is a subset of $\sigma_{\rm abs}(\mathcal{L})$, and that for each $n \in \mathbb{N}$ the nth eigenvalue λ_n converges at the branch point, $\lambda = -1$, as $L \to +\infty$.

(b) Characterize the point spectrum for the exceptional values of $b_j^{\pm}$ and identify any distinguished accumulation points.

Exercise 3.2.3. Consider the linear operator given in (3.2.4), acting on the space $H^2_{\rm per}[-L,+L]$, subject to the nonseparated, periodic boundary conditions

$$\partial_x^j p(-L) = \partial_x^j p(+L), \quad j = 0, 1.$$

Show that as $L \to +\infty$ the point spectrum accumulates on the essential spectrum, $\sigma_{\rm ess}(\mathcal{L};\mathbb{R})$, associated with the unweighted space, $L^2(\mathbb{R})$. Furthermore, show that there are no distinguished accumulation point within $\sigma_{\rm ess}(\mathcal{L};\mathbb{R})$.

3.3 The Essential Spectrum: Periodic Coefficients

Many evolutionary PDEs possess stationary- or traveling-wave solutions which are spatially periodic. As is the case for waves that are asymptotically constant, the stability of these special solutions depends in good measure upon the spectrum of the corresponding linearization, $\mathcal{L}$, which will be of the form (3.1.1) but with spatially periodic coefficients. We begin by fixing the period, i.e.,

$$a_j(x+\pi) = a_j(x), \quad j = 0,\dots,n-1,$$

and further assume that the coefficients are smooth.

As a first result, we show that the operator $\mathcal{L} : H^n(\mathbb{R}) \subset L^2(\mathbb{R}) \to L^2(\mathbb{R})$ has no point spectrum. Introducing $\boldsymbol{Y} = (p, \partial_x p, \dots, \partial_x^{n-1} p)^{\mathrm{T}}$, the eigenvalue problem for $p \in H^n(\mathbb{R})$ becomes a first-order system,

$$\partial_x \boldsymbol{Y} = \boldsymbol{A}(x,\lambda)\boldsymbol{Y}, \quad \boldsymbol{A}(x,\lambda) = \begin{pmatrix} 0 & 1 & \dots & 0 & 0 \\ 0 & 0 & \dots & 0 & 0 \\ \vdots & \vdots & \vdots\vdots\vdots & \vdots & \vdots \\ 0 & 0 & \dots & 0 & 1 \\ \lambda - a_0(x) & -a_1(x) & \dots & -a_{n-2}(x) & -a_{n-1}(x) \end{pmatrix}, \tag{3.3.1}$$

where $\boldsymbol{A}(x+\pi,\lambda) = \boldsymbol{A}(x,\lambda)$ for all $\lambda \in \mathbb{C}$. From Theorem 2.1.27 any fundamental matrix solution $\boldsymbol{\Phi}(x,\lambda)$ is of the form

$$\boldsymbol{\Phi}(x,\lambda) = \boldsymbol{P}(x,\lambda)\mathrm{e}^{\boldsymbol{B}(\lambda)x}, \quad \boldsymbol{P}(x+\pi,\lambda) = \boldsymbol{P}(x,\lambda),$$

where $\boldsymbol{B}(\lambda) \in \mathbb{C}^{n\times n}$. The Floquet decomposition of the fundamental matrix solution allows us to form a basis for the solutions of (3.3.1) from the stable, unstable, and center eigenspaces of $\boldsymbol{B}(\lambda)$. The center eigenspace yields solutions that are uniformly bounded but nondecaying in norm for all x. The stable and unstable eigenspaces yield solutions that grow exponentially in norm as $x \to -\infty$ or $x \to +\infty$, respectively. In particular, there can be no nontrivial solutions of (3.3.1) that lie in $L^2(\mathbb{R})$. Since $\mathcal{L} - \lambda$ cannot have a kernel, the spectrum of $\mathcal{L}$ must be entirely essential spectrum.

To determine the essential spectrum we characterize those λ for which the operator $\mathcal{L} - \lambda$ has a bounded inverse. Recalling the construction of the solutions for the inhomogenous problem (3.1.4), the role of the projections onto the stable and unstable subspaces for the asymptotic matrices can be reproduced here by the stable and unstable projections P^{s} and P^{u}, respectively, associated with the Floquet matrix $\boldsymbol{B}(\lambda)$. Unfortunately, the matrix $\boldsymbol{B}(\lambda)$ is in general much more difficult to determine than are the asymptotic matrices $\boldsymbol{A}_{\pm}(\lambda)$.

A classical result, the proof of which we leave as an exercise, is that if the Floquet matrix $\boldsymbol{B}(\lambda)$ is hyperbolic, then the system (3.3.1) has an exponential dichotomy [see Definition 2.1.32] and the nonhomogeneous problem associated with (3.3.1) has a unique solution $\boldsymbol{y}(x)$ that satisfies $\|\boldsymbol{y}\|_{L^2} \le C\|\boldsymbol{F}\|_{L^2}$ for all $\boldsymbol{F} \in L^2(\mathbb{R})^n$ and some fixed $C > 0$. (e.g. see [**59**]).

As a consequence of Exercise 3.3.1 the spectrum of $\mathcal{L}$ is precisely the set of λ for which at least one Floquet exponent has a zero real part. This is analogous to determining the Fredholm borders for the spectral problem with asymptotically constant coefficients. From Floquet's Theorem 2.1.27

and Lemma 2.1.29 we know one of the Floquet exponents has a zero real part precisely when there is a solution to (3.3.1) that satisfies the *nonseparated* boundary condition

$$\boldsymbol{Y}(\pi) = \mathrm{e}^{\mathrm{i}\mu\pi}\boldsymbol{Y}(0), \quad -1 < \mu \le 1. \tag{3.3.2}$$

It is convenient to remove the μ-dependence associated with the boundary conditions. The change of variables

$$\boldsymbol{Z} := \mathrm{e}^{-\mathrm{i}\mu x}\boldsymbol{Y} \tag{3.3.3}$$

transforms the system (3.3.1) with the boundary conditions (3.3.2) to a system with periodic and μ-independent boundary conditions; namely,

$$\partial_x \boldsymbol{Z} = [\boldsymbol{A}(x,\lambda) - \mathrm{i}\mu \boldsymbol{I}_n]\boldsymbol{Z}, \quad \boldsymbol{Z}(\pi) = \boldsymbol{Z}(0). \tag{3.3.4}$$

The spectrum of $\mathcal{L}$ is typically cast in terms of the system (3.3.4). Indeed, if for some $\lambda \in \mathbb{C}$ and $\mu \in (-1,+1]$ this system possesses a nontrivial solution $\boldsymbol{Z} \in H^1_{\mathrm{per}}([0,\pi])^n$, then $\lambda \in \sigma_{\mathrm{ess}}(\mathcal{L})$. Alternatively, if $\boldsymbol{\Phi}(x,\lambda,\mu)$ denotes the fundamental matrix associated with (3.3.4), then the system possesses a π-periodic solution if and only if $1 \in \sigma(\boldsymbol{\Phi}(\pi,\lambda,\mu))$, i.e., if and only if the characteristic equation

$$E(\lambda,\mu) := \det(\boldsymbol{\Phi}(\pi,\lambda,\mu) - \boldsymbol{I}_n) = 0 \tag{3.3.5}$$

has a solution pair (λ,μ). Since the matrix $\boldsymbol{A}(\cdot,\lambda) - \mathrm{i}\mu\boldsymbol{I}$ is entire in both λ and μ, so is $\boldsymbol{\Phi}(\cdot,\lambda,\mu)$. From this analyticity and the implicit function theorem, it follows that zeros of (3.3.5) are generically not isolated; indeed, if $E(\lambda_0,\mu_0)$ and $\partial_\lambda E(\lambda_0,\mu_0) \neq 0$, then there is a simple smooth curve $\lambda = \lambda(\mu)$ with $\lambda(\mu_0) = \lambda_0$ such that $E(\lambda(\mu),\mu) = 0$ (see Chapter 8.4.1). Moreover, since $\lambda(1) = \lambda(-1)$, the curves are closed: such curves are called the spectral loops of $\mathcal{L}$.

An equivalent approach to determining the essential spectrum of $\mathcal{L}$ is the so-called *Bloch-wave decomposition* [242, Chapter 16]. This involves a change of variables that is subtly different from (3.3.3). Setting

$$p = \mathrm{e}^{\mathrm{i}\mu x}q \quad \Rightarrow \quad \partial_x^j p = \mathrm{e}^{\mathrm{i}\mu x}(\partial_x + \mathrm{i}\mu)^j q, \tag{3.3.6}$$

the operator $\mathcal{L}$ transforms to the μ-dependent operator

$$\mathcal{L}_\mu q := (\partial_x + \mathrm{i}\mu)^n q + a_{n-1}(x)(\partial_x + \mathrm{i}\mu)^{n-1}q + \cdots + a_1(x)(\partial_x + \mathrm{i}\mu)q + a_0(x)q, \tag{3.3.7}$$

and the nonseparated boundary condition (3.3.2) reduces to π-periodicity

$$\partial_x^j q(\pi) = \partial_x^j q(0), \quad j = 0,\ldots,n-1. \tag{3.3.8}$$

The equivalence between the two approaches is best understood by vectorizing the operator $\mathcal{L}_\mu$. Setting $\boldsymbol{W} = (q, \partial_x q, \ldots, \partial_x^{n-1} q)^{\mathrm{T}}$, we then calculate that

$$\boldsymbol{Y} = \mathrm{e}^{\mathrm{i}\mu x}\boldsymbol{T}(\mu)\boldsymbol{W}, \quad \boldsymbol{T}(\mu) = \begin{pmatrix} 1 & 0 & 0 & 0 & 0 & \cdots & 0 \\ \mathrm{i}\mu & 1 & 0 & 0 & 0 & \cdots & 0 \\ (\mathrm{i}\mu)^2 & 2(\mathrm{i}\mu) & 1 & 0 & 0 & \cdots & 0 \\ (\mathrm{i}\mu)^3 & 3(\mathrm{i}\mu)^2 & 3(\mathrm{i}\mu) & 1 & 0 & \cdots & 0 \\ \vdots & \vdots & \vdots & \vdots & \vdots & \vdots & \vdots \\ (\mathrm{i}\mu)^{n-1} & n(\mathrm{i}\mu)^{n-2} & \cdots & \cdots & \cdots & n(\mathrm{i}\mu) & 1 \end{pmatrix}. \tag{3.3.9}$$

Since the lower-triangular matrix $\boldsymbol{T}(\mu)$ is nonsingular, the eigenvalue problem takes the Bloch-wave form

$$\partial_x \boldsymbol{W} = \boldsymbol{T}(\mu)[\boldsymbol{A}(x,\lambda) - \mathrm{i}\mu \boldsymbol{I}_n]\boldsymbol{T}(\mu)^{-1}\boldsymbol{W}, \quad \boldsymbol{W}(\pi) = \boldsymbol{W}(0). \tag{3.3.10}$$

Upon setting

$$\boldsymbol{R} = \boldsymbol{T}(\mu)^{-1}\boldsymbol{W},$$

we see that (3.3.10) becomes

$$\partial_x \boldsymbol{R} = [\boldsymbol{A}(x,\lambda) - \mathrm{i}\mu \boldsymbol{I}_n]\boldsymbol{R}, \quad \boldsymbol{R}(\pi) = \boldsymbol{R}(0),$$

which is precisely the rescaled problem (3.3.4).

3.3.1 Example: Hill's Equation

The periodic eigenvalue problem on $H^2(\mathbb{R})$, known as Hill's equation, concerns the second-order operator

$$\mathcal{L}p := \partial_x^2 p + a_0(x)p = \lambda p, \tag{3.3.11}$$

where $a_0 : \mathbb{R} \mapsto \mathbb{R}$ is a continuous π-periodic function. This tractable simple system fully demonstrates the complexities associated with the computation of the essential spectrum (see the classical work of Magnus and Winkler [**200**], or [**112**, Chapter III.8] for further details). The Bloch-wave decomposition transforms (3.3.11) to the family of π-periodic eigenvalue problems

$$\mathcal{L}_\mu p := (\partial_x + \mathrm{i}\mu)^2 p + a_0(x)p = \lambda p. \tag{3.3.12}$$

For each $\mu \in (-1,+1]$ we search for solutions to (3.3.12) in $H^2_{\mathrm{per}}[0,\pi]$. The system (3.3.12) is self-adjoint in the standard complex inner product on $L^2[0,\pi]$,

$$\langle p, q \rangle = \int_0^\pi p(x)\overline{q}(x)\,dx,$$

so that the spectrum is purely real (see Exercise 3.3.4).

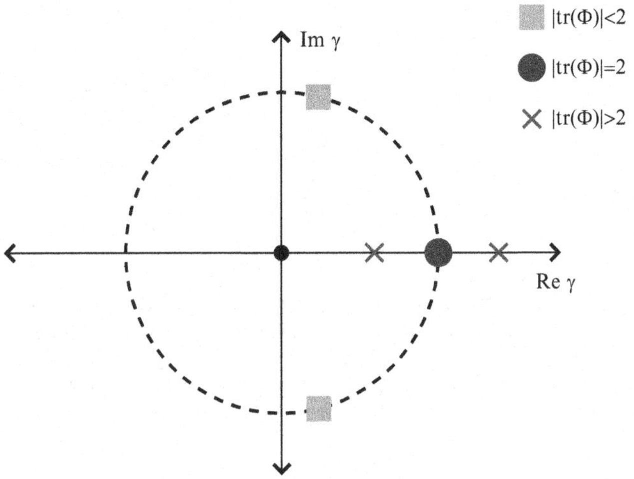

Fig. 3.8 The location of the Floquet multipliers associated with Hill's equation (3.3.11) for various values of tr($\mathbf{\Phi}(\pi, \lambda)$). (Color figure online.)

To locate the spectrum, we first rewrite (3.3.11) as the equivalent first-order system

$$\partial_x \mathbf{Y} = \begin{pmatrix} 0 & 1 \\ \lambda - a_0(x) & 0 \end{pmatrix} \mathbf{Y}, \quad \mathbf{Y}(\pi) = \gamma \mathbf{Y}(0), \tag{3.3.13}$$

where γ is a Floquet multiplier, and the nonseparated boundary condition follows from (3.3.2). We know by Exercise 3.3.2 that λ is an eigenvalue if and only if $\gamma \in \sigma(\mathbf{\Phi}(\pi, \lambda))$ satisfies $|\gamma| = 1$, where $\mathbf{\Phi}$ is a FMS for the ODE in (3.3.13). Using Lemma 2.1.31(a) the Floquet multipliers $\gamma_j = \gamma_j(\lambda)$ satisfy

$$\gamma_1 \gamma_2 = 1, \quad \gamma_1 + \gamma_2 = \mathrm{tr}\,(\mathbf{\Phi}(\pi, \lambda)), \tag{3.3.14}$$

or equivalently

$$\mathrm{tr}\,(\mathbf{\Phi}(\pi, \lambda)) = \gamma_1 + \frac{1}{\gamma_1}. \tag{3.3.15}$$

Since the trace of the fundamental matrix solution is real-valued for all $\lambda \in \mathbb{R}$, we deduce from (3.3.15) that either $|\gamma_1| = 1$ or $\gamma_1 \in \mathbb{R}$. In the first case, λ is an eigenvalue, and $|\mathrm{tr}\,(\mathbf{\Phi}(\pi, \lambda))| < 2$. In the second case either $\gamma_1 = \gamma_2 = \pm 1$ and λ is an eigenvalue, or $0 < |\gamma_1| < 1 < |\gamma_2|$ and $|\mathrm{tr}\,(\mathbf{\Phi}(\pi, \lambda))| > 2$ (see Fig. 3.8).

In summary, we have that if $|\mathrm{tr}\,(\mathbf{\Phi}(\pi, \lambda))| < 2$, then λ is an eigenvalue, whereas if $|\mathrm{tr}\,(\mathbf{\Phi}(\pi, \lambda))| > 2$, then λ is not an eigenvalue. In the former case

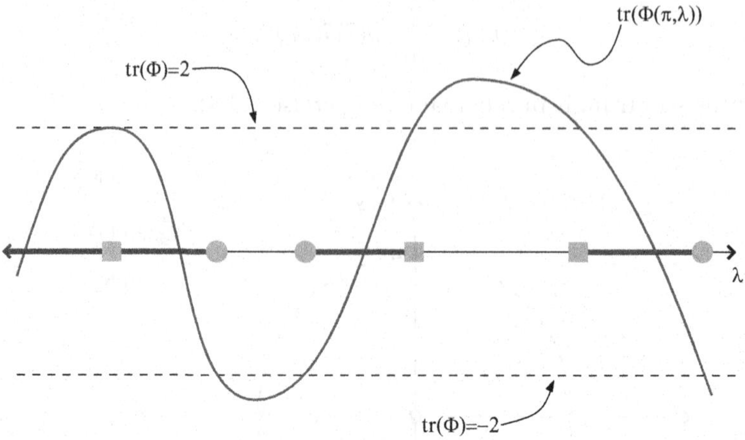

Fig. 3.9 The spectrum for Hill's equation (3.3.11). The essential spectrum is given by the thick (*blue*) line. The graph of $\mathrm{tr}(\mathbf{\Phi}(\pi,\lambda))$ is marked by a thick (*red*) curve. The periodic points where $\gamma = 1$ are marked by circles, and the antiperiodic points where $\gamma = -1$ are marked by squares. Note that the spectral gap between bands of spectrum occurs when the trace of the fundamental matrix transversely crosses the line $|\mathrm{tr}(\mathbf{\Phi})| = 2$. (Color figure online.)

$|\gamma_1| = |\gamma_2| = 1$ with $\gamma_1 = \overline{\gamma_2}$, while in the latter case the Floquet multipliers are real-valued with $0 < |\gamma_1| < 1 < |\gamma_2|$. The boundary between these two sets, i.e.,

$$|\mathrm{tr}\,(\mathbf{\Phi}(\pi,\lambda,0))| = 2,$$

occurs only when $\gamma = 1$ (a π-periodic solution) or $\gamma = -1$ (an antiperiodic solution, i.e., a solution of period 2π).

Let us examine the behavior of the spectrum near an eigenvalue λ_0 with $\gamma_1 = 1$. Since $\gamma_1 = \gamma_2$ the Jordan canonical form associated with $\mathbf{\Phi}(\pi,\lambda)$ has the form

$$\mathbf{\Phi}(\pi,\lambda) \sim \begin{pmatrix} 1 & c \\ 0 & 1 \end{pmatrix}:$$

for some $c \in \mathbb{R}$. If $c \neq 0$, then there is only one periodic solution, while if $c = 0$ there will be two linearly independent periodic solutions. Furthermore, if $c \neq 0$, then it will be the case that $\partial_\lambda \mathrm{tr}\,(\mathbf{\Phi}(\pi,\lambda_0)) \neq 0$, while $c = 0$ implies that $\partial_\lambda \mathrm{tr}\,(\mathbf{\Phi}(\pi,\lambda_0)) = 0$. We deduce that if $c \neq 0$ then $|\mathrm{tr}\,(\mathbf{\Phi}(\pi,\lambda))| > 2$ for λ either to the right or the left of λ_0, and a gap will open between the two adjacent bands of spectra. On the other hand, if $c = 0$ then, generically, there will be no gap (see Fig. 3.9).

This analysis recovers the following classical results [200]. For each $\gamma \in \{-1,1\}$ there is a countable set of eigenvalues $\lambda_j(\gamma)$ that can be ordered as $\lambda_0(\gamma) > \lambda_1(\gamma) \geq \cdots$, with $\lim_{j\to\infty} \lambda_j(\gamma) = -\infty$. There are spectral bands of the form $[\lambda_j(-1), \lambda_j(1)]$ or $[\lambda_j(1), \lambda_j(-1)]$; furthermore, these bands alternate so that the full spectrum can be decomposed as a nonoverlapping set of intervals

$$\sigma(\mathcal{L}) = \cdots \cup [\lambda_3(-1), \lambda_3(1)] \cup [\lambda_2(1), \lambda_2(-1)] \cup [\lambda_1(-1), \lambda_1(1)] \cup [\lambda_0(1), \lambda_0(-1)].$$

An eigenfunction associated with $\lambda_j(1)$ is periodic, while an eigenfunction associated with $\lambda_j(-1)$ is antiperiodic. There will be a spectral gap, e.g., $\lambda_1(1) < \lambda_0(1)$, if and only the Jordan canonical form of the fundamental matrix solution for λ at a band-edge has a Jordan block, e.g., $\partial_\lambda \operatorname{tr}(\boldsymbol{\Phi}(\pi, \lambda_0(1))) \neq 0$.

Exercises

Exercise 3.3.1. Consider the homogeneous problem

$$\partial_x \boldsymbol{Y} = \boldsymbol{A}(x)\boldsymbol{Y}, \quad \boldsymbol{A}(x+\pi) = \boldsymbol{A}(x).$$

(a) Show that if all the Floquet exponents have nonzero real part, then the fundamental matrix solution possesses an exponential dichotomy.
(b) Show that if the fundamental matrix solution possesses an exponential dichotomy, then the nonhomogeneous problem

$$\partial_x \boldsymbol{Y} = \boldsymbol{A}(x)\boldsymbol{Y} + \boldsymbol{F}(x)$$

possesses a unique solution $\boldsymbol{y}(x)$ that satisfies the estimate $\|\boldsymbol{y}\|_2 \leq C\|\boldsymbol{F}\|_2$ for some $C > 0$.
(c) If for some $\lambda \in \mathbb{C}$ one of the Floquet exponents has a zero real part, then by constructing a Weyl sequence show that the operator $\mathcal{L} - \lambda$ cannot have a bounded inverse from $L^2(\mathbb{R})$ into $H^n(\mathbb{R})$.

Exercise 3.3.2. Show that if the Floquet multiplier $1 \in \sigma(\boldsymbol{\Phi}(\pi, \lambda, \mu))$, then the Floquet multiplier $e^{i\mu\pi} \in \sigma(\boldsymbol{\Phi}(\pi, \lambda, 0))$.

Exercise 3.3.3. Consider the second-order vector system given by

$$\mathcal{L}\boldsymbol{p} := \boldsymbol{d}\partial_x^2 \boldsymbol{p} + \boldsymbol{a}_1(x)\partial_x \boldsymbol{p} + \boldsymbol{a}_0(x)\boldsymbol{p} = \lambda \boldsymbol{p},$$

where $\boldsymbol{p} : \mathbb{R} \to \mathbb{C}^n$, the matrix $\boldsymbol{d}$ is diagonal and nonsingular and the coefficient matrices $\boldsymbol{a}_i$ are π-periodic. Formulate the Bloch-wave decomposition for this problem, and show that it is equivalent to the Floquet decomposition.

Exercise 3.3.4. Verify that the Bloch-wave operator

$$\mathcal{L}_\mu = (\partial_x + i\mu)^2 + a_0(x), \quad -1 < \mu \leq 1,$$

with the boundary conditions $p(\pi) = p(0)$, $\partial_x p(\pi) = \partial_x p(0)$ is self-adjoint in the $L^2[0,\pi]$ inner product with domain $H^2_{\mathrm{per}}([0,\pi])$.

Exercise 3.3.5. Consider Hill's equation with

$$a_0(x) = A \sum_{j=-\infty}^{+\infty} \delta(x - j\pi),$$

where $\delta(x-a)$ is the dirac delta function with mass at $x = a$. Explicitly compute the bands of spectra as a function of $A > 0$.

3.4 Additional Reading

A more substantial discussion of exponential dichotomies can be found in Ben-Artzi and Gohberg [**30**], Chicone and Latushkin [**53**], Daleckii and Krein [**65**], Härterich et al. [**113**], Latushkin and Pogan [**182**].

The excellent review article by Sandstede [**250**] discusses not only the material covered in this chapter, but is also relevant to some material covered in later chapters. The connection between the Morse index, the Fredholm index, and group velocities is described in Sandstede and Scheel [**257**]. The relationship between the various types of spectra and exponential dichotomies for modulated traveling waves is given in Sandstede and Scheel [**255**]. The numerical computation of the essential and absolute spectra is discussed in Rademacher et al. [**239**].

Chapter 4
Asymptotic Stability of Waves in Dissipative Systems

A key motivation for investigating the spectrum of linear operators is to understand the stability of equilibria of nonlinear evolution equations, as well as to describe the flow in a neighborhood of manifolds of approximate equilibria. We divide this project into five steps, the first four of which have to do with the linearization about an equilibrium:

(a) identification of essential spectrum,
(b) identification of the point spectrum,
(c) resolvent estimates,
(d) semi-group estimates,
(e) nonlinear estimates.

The first step was addressed largely in Chapter 3, and the second step will be discussed in Chapter 6–Chapter 10. In this chapter we primarily address the last three of these steps, which, while occasionally technical, can be done in a "wholesale" approach. It is perhaps surprising, but the majority of the "retail" work required to prove stability of equilibria for many classes of partial differential equations (PDEs) lies in the determination of the point spectrum. This leads to a class of problems for which spectral stability, typically control of the point spectrum, implies nonlinear stability.

We start with a generic class of evolution equations of the form

$$\partial_t u = \mathcal{F}(u), \quad u(0) = u_0, \tag{4.0.1}$$

posed on a Hilbert space X with inner product $\langle\cdot,\cdot\rangle$. The nonlinear operator $\mathcal{F} : Z \subset Y \subset X \mapsto X$, where each of the subspaces is dense and continuously embedded. We assume that the flow is locally well-posed in Y; that is, for each $u_0 \in Y$ there is a time $\underline{T} = \underline{T}(\|u_0\|_Y) > 0$ for which there exists a unique solution $u(t) \in Y$ of the initial-value problem for $t \in [0, \underline{T})$. A key role is played by the "middle" space Y, its norm must be strong enough to control the nonlinearity and the generators of symmetries.

We assume that $\mathcal{F}$ has an equilibria ϕ, i.e. $\mathcal{F}(\phi) = 0$. To understand the behavior of solutions u in a neighborhood of ϕ we consider initial data of the

T. Kapitula and K. Promislow, *Spectral and Dynamical Stability of Nonlinear Waves*, Applied Mathematical Sciences 185, DOI 10.1007/978-1-4614-6995-7_4,

form $u_0 = \phi + v_0$, where $\|v_0\|_Y \ll 1$, and track the evolution of $v(t) := u(t) - \phi$. By assumption, $\mathcal{F}$ is smooth and admits an expansion of the form

$$\mathcal{F}(u) = \mathcal{F}(\phi) + \mathcal{L}v + \mathcal{N}(v).$$

Here $\mathcal{L} = \nabla_u \mathcal{F}(\phi) : Z = \mathcal{D}(\mathcal{L}) \subset X \mapsto X$ is the linearization of $\mathcal{F}$ about ϕ. We assume that the domain $\mathcal{D}(\mathcal{L})$ is dense in X. The nonlinear residual $\mathcal{N}(v) := \mathcal{F}(\phi + v) - \mathcal{F}(\phi) - \mathcal{L}v$ satisfies the quadratic estimate

$$\|\mathcal{N}(v)\|_Y \leq C\|v\|_Y^2,$$

for some constant $C > 0$ independent of $v \in Y$ with $\|v\|_Y$ sufficiently small. Since $\mathcal{F}(\phi) = 0$ and $\partial_t \phi = 0$, the evolution equation (4.0.1) can be rewritten as

$$\partial_t v = \mathcal{L}v + \mathcal{N}(v), \quad v(0) = v_0 \; (= u_0 - \phi). \tag{4.0.2}$$

If $\|v_0\|_Y$ is sufficiently small, then so long as $\|v\|_Y$ remains small it is plausible that the dynamics of the nonlinear system are primarily governed by those of the linear system,

$$\partial_t v = \mathcal{L}v, \quad v(0) = v_0. \tag{4.0.3}$$

referred to as the linearization of (4.0.1) about ϕ, which is obtained by dropping the nonlinear term. Understanding the dynamics of the linear problem (4.0.3) is the key to controlling the dynamics of the fully nonlinear flow generated by (4.0.2).

Example 4.0.1. A prototypical example is the reaction diffusion equation

$$u_t = u_{xx} - W'(u) =: \mathcal{F}(u),$$

posed on the line. The spaces are naturally chosen as $Z = H^2(\mathbb{R}) \subset Y = H^1(\mathbb{R}) \subset X = L^2(\mathbb{R})$. Assuming that $\mathcal{F}$ has an equilibrium, ϕ, which solves $\mathcal{F}(\phi) = 0$, then the linearization about ϕ is $\mathcal{L} = \partial_x^2 - W''(\phi)$ and the nonlinearity is $\mathcal{N}(v) := W'(\phi + v) - W'(\phi) - W''(\phi)v$. In particular, if $W(u) = (1-u^2)^2/4$, then one may verify that $\mathcal{N}(v) = 3\phi v^2 + v^3$, which satisfies the estimate

$$\|\mathcal{N}(v)\|_{H^1} \leq 4\|\phi\|_{H^1}\|v\|_{H^1}^2,$$

for $\|v\|_{H^1} \leq \|\phi\|_{H^1}$.

Remark 4.0.2. Unless stated otherwise, we assume that the operator $\mathcal{L}$ is exponentially asymptotic of the form (3.1.1) studied in Chapter 3. However many of the results of this chapter hold for more general families of linear operators, see [**81**] and the references therein).

4.1 Linear Dynamics

We wish to identify the relationship between the spectrum of $\mathcal{L}$ and the dynamics for the linearized system (4.0.3). It is easy to verify that if $\lambda_0 \in \sigma_{\mathrm{pt}}(\mathcal{L})$ with associated eigenfunction $\psi_0 \in Y$, then $v := \mathrm{e}^{\lambda_0 t}\psi_0$ is a solution to (4.0.3). Noting that v satisfies $\|v\|_Y = \mathrm{e}^{\mathrm{Re}\,\lambda_0 t}\|\psi_0\|_Y$, we have the following behavior of solutions of the linear equation:

(a) $\mathrm{Re}\,\lambda_0 > 0$ implies the existence of exponentially growing solutions;
(b) $\mathrm{Re}\,\lambda_0 < 0$ implies the existence of exponentially decaying solutions;
(c) $\mathrm{Re}\,\lambda_0 = 0$ implies the existence of bounded, nondecaying solutions.

Moreover, as in Theorem 2.1.23, if λ_0 is not algebraically simple, then $\mathrm{Re}\,\lambda_0 = 0$ may imply the existence of solutions that grow polynomial in time.

The impact of an essential spectrum is more subtle. To motivate the more general approach, first suppose that the operator $\mathcal{L}$ has constant coefficients, i.e.,

$$\mathcal{L} := \partial_x^n + a_{n-1}^0 \partial_x^{n-1} + \cdots + a_1^0 \partial_x + a_0^0.$$

For initial data of the form $v_0 = \mathrm{e}^{\mathrm{i}kx}$, the solution to (4.0.3) is given by

$$v(x,t) = \mathrm{e}^{\lambda(k)t + \mathrm{i}kx}, \tag{4.1.1}$$

where

$$\lambda(k) := d(\mathrm{i}k, 0) = (\mathrm{i}k)^n + a_{n-1}^0(\mathrm{i}k)^{n-1} + \cdots + a_1^0(\mathrm{i}k) + a_0^0$$

is the dispersion relation of $\mathcal{L}$ given in (3.1.18). Indeed, these λ form the Fredholm border, $\sigma_{\mathrm{F}}(\mathcal{L})$, which comprises the rightmost boundary of the essential spectrum of $\mathcal{L}$. While such solutions generically do not reside in Y (since they are not localized in space), it is clear that $\mathrm{Re}\,\lambda(k) > 0 (< 0)$ implies the existence of exponentially growing (decaying) solutions in $L^\infty(\mathbb{R})$, whereas $\mathrm{Re}\,\lambda(k) = 0$ implies the existence of solutions that are temporally bounded in $L^\infty(\mathbb{R})$ but that do not decay in time. In particular, if the Fredholm boundary lies uniformly in the left-half complex plane, then so does the essential spectrum, and all solutions of the form (4.1.1) decay exponentially in time in $L^\infty(\mathbb{R})$.

We may gauge the impact of essential spectrum on initial data in $L^2(\mathbb{R})$ via the Fourier transform. Applying the Fourier transform to (4.0.3), and recalling the Fourier-multiplier property (2.2.2), shows that the Fourier transform $\widehat{v}$ of v satisfies the ODE

$$\partial_t \widehat{v}(k,t) = \lambda(k)\widehat{v}(k,t), \quad \widehat{v}(k,0) = \widehat{v_0}(k),$$

from which we deduce

$$\widehat{v}(k,t) = \mathrm{e}^{\lambda(k)t}\widehat{v_0}(k).$$

If the essential spectrum lies uniformly in the open left-half of the complex plane, i.e., if there is a $\sigma > 0$ such that

$$\sigma_F(\mathcal{L}) \subset \{\lambda \in \mathbb{C} : \operatorname{Re}\lambda \le -\sigma\},$$

then we have the uniform exponential decay

$$|\widehat{v}(k,t)| \le e^{-\sigma t}|\widehat{v_0}(k)|, \tag{4.1.2}$$

and from the Plancheral identity (2.2.1) we can conclude from (4.1.2) that v decays exponentially in norm,

$$\|v(t)\|_2 \le e^{-\sigma t}\|v_0\|_2.$$

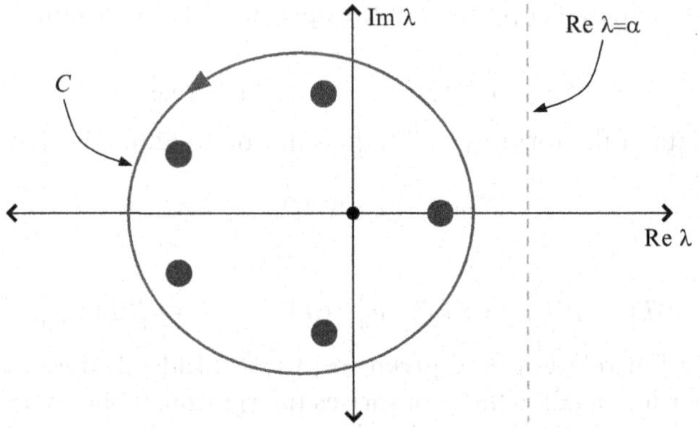

Fig. 4.1 A diagram that highlights the relationship between the solution $\boldsymbol{Y}(x)$ and its associated Laplace transform $\widetilde{\boldsymbol{Y}}(\lambda)$ for the linear ODE (4.1.3). The filled (*blue*) circles represent the eigenvalues of $\boldsymbol{A}$. The Laplace transform is well-defined for $\operatorname{Re}\lambda \ge \alpha$, i.e., for λ to the right of the (*green*) dashed line. (Color figure online.)

If the operator $\mathcal{L}$ does not have constant coefficients we may still solve the linear system via the Laplace transform. To highlight the ideas, we first consider the application of the Laplace transform to an ODE system. For a constant matrix $\boldsymbol{A} \in \mathbb{R}^{n\times n}$, the equation

$$\partial_x \boldsymbol{Y} = \boldsymbol{A}\boldsymbol{Y}, \quad \boldsymbol{Y}(0) = \boldsymbol{y}_0, \tag{4.1.3}$$

has the solution $\boldsymbol{Y}(x) = e^{\boldsymbol{A}x}\boldsymbol{y}_0$. We define the *resolvent operator* of $\boldsymbol{A}$,

$$\boldsymbol{R}(\lambda;\boldsymbol{A}) := (\lambda\mathcal{I}_n - \boldsymbol{A})^{-1},$$

which is a matrix-valued analytic function of λ for $\lambda \in \mathbb{C}\backslash\sigma(\boldsymbol{A})$. The Laplace transform of $\boldsymbol{Y}(x)$ is the function

$$\widetilde{\boldsymbol{Y}}(\lambda) = \int_0^{+\infty} \mathrm{e}^{-\lambda x}\boldsymbol{Y}(x)\,\mathrm{d}x,$$

where we assume that $\operatorname{Re}\lambda$ is sufficiently large that the integral converges (see Fig. 4.1). The key property of the Laplace transform is that it relates solutions $\boldsymbol{Y}$ of the ODE to their initial data through the resolvent operator. To establish this relationship, assume $\boldsymbol{Y}$ solves (4.1.3), and let $\widetilde{\boldsymbol{Y}}$ denote its Laplace transform. From the definition of the Laplace transform and an integration by parts we find

$$\boldsymbol{A}\widetilde{\boldsymbol{Y}}(\lambda) = \int_0^{+\infty} \mathrm{e}^{-\lambda x}\boldsymbol{A}\boldsymbol{Y}(x)\,\mathrm{d}x = \int_0^{+\infty} \mathrm{e}^{-\lambda x}\partial_x\boldsymbol{Y}(x)\,\mathrm{d}x = \lambda\widetilde{\boldsymbol{Y}}(\lambda) - \boldsymbol{y}(0),$$

and solving for $\tilde{\boldsymbol{Y}}$ we obtain the relation

$$\widetilde{\boldsymbol{Y}}(\lambda) = \boldsymbol{R}(\lambda;\boldsymbol{A})\boldsymbol{y}_0.$$

Fixing a simple, closed, positively oriented contour $C \subset \mathbb{C}$ that encircles all of $\sigma(\boldsymbol{A})$ (see Fig. 4.1), we use Cauchy's formula to invert the Laplace transform, yielding the representation,

$$\boldsymbol{Y}(x) = \frac{1}{2\pi\mathrm{i}}\oint_C \mathrm{e}^{\lambda x}\widetilde{\boldsymbol{Y}}(\lambda)\,\mathrm{d}\lambda = \left(\frac{1}{2\pi\mathrm{i}}\oint_C \mathrm{e}^{\lambda x}R(\lambda;\boldsymbol{A})\,\mathrm{d}\lambda\right)\boldsymbol{y}_0.$$

Since the solutions to the ODE are unique, the quantity in the last parenthesis must be the matrix exponential of the constant matrix $\boldsymbol{A}$, i.e.,

$$\mathrm{e}^{\boldsymbol{A}x} = \frac{1}{2\pi\mathrm{i}}\oint_C \mathrm{e}^{\lambda x}\boldsymbol{R}(\lambda;\boldsymbol{A})\,\mathrm{d}\lambda. \tag{4.1.4}$$

The solution map for the system (4.0.3) is obtained in a similar manner. Introducing the resolvent operator of $\mathcal{L}$,

$$R(\lambda;\mathcal{L}) := (\lambda\mathcal{I} - \mathcal{L})^{-1}, \tag{4.1.5}$$

we define the semi-group generated by $\mathcal{L}$ via the inverse Laplace transform of the resolvent,

$$\mathrm{e}^{\mathcal{L}t} := \frac{1}{2\pi\mathrm{i}}\int_C \mathrm{e}^{\lambda t}R(\lambda;\mathcal{L})\,\mathrm{d}\lambda. \tag{4.1.6}$$

The contour C is chosen to be a positively oriented curve that approaches ∞ at either end, and that lies to the right of $\sigma(\mathcal{L})$. The curve C must be unbounded since the essential spectrum is unbounded. The resolvent is an analytic function of λ for $\lambda \in \mathbb{C}\backslash\sigma(\mathcal{L})$, so the contour C can be continuously deformed without affecting the value of the integral, so long as the deformation does not push C across any spectrum of $\mathcal{L}$. Assuming the integral is properly convergent, the solution to the linear problem (4.0.3) is given by $v(t) = e^{\mathcal{L}t} v_0$.

It is useful to characterize linear semi-groups and their relations to the linear operators that generate them. A more detailed treatment of semi-groups can be found in [**106**, **219**].

Definition 4.1.1. Let X be a Banach space. A strongly continuous, or $\mathcal{C}^0$, semigroup is a map $S : \mathbb{R}_+ \mapsto \mathcal{B}(X)$ such that

(a) $S(0) = \mathcal{I}$,
(b) $S(t+s) = S(t)S(s)$, for all $t, s \in \mathbb{R}_+$,
(c) $\|S(t)u - u\|_X \to 0$ as $t \to 0^+$ for all $u \in X$.

The *generator*, $\mathcal{L}$, of the semi-group has domain

$$\mathcal{D}(\mathcal{L}) := \left\{ u \in X \lim_{t \to 0^+} \frac{S(t)u - u}{t}, \text{ exists in } X \right\}.$$

The action of the operator $\mathcal{L} : \mathcal{D} \subset X \to X$ on $u \in \mathcal{D}(\mathcal{L})$ is defined by the value of the limit above.

The semi-group is the solution map $u_0 \mapsto u(t)$ of the linear problem $\partial_t u = \mathcal{L}u$ subject to the initial condition $u(0) = u_0$. A linear problem has a $\mathcal{C}^0$ semigroup if the solutions recover the initial data; in many ways this is equivalent to the initial-value problem being well-posed. The Laplacian generates a $\mathcal{C}^0$ semi-group in the L^2-norm, this being the solution map for the heat equation; see Exercise 4.1.3. The negative Laplacian, which generates the backwards heat equation, does not produce a $\mathcal{C}^0$ semi-group in $L^2(\mathbb{R})$. We state the following result without proof.

Lemma 4.1.2. *Let $\mathcal{L}$ be an exponentially asymptotic nth-order differential operator of the form (3.1.1) with H^r coefficients, for $1 \le r < \infty$. If $\mathcal{L}$ is well-posed in the sense of Hypothesis 3.2.1, then $\mathcal{L}$ generates a $\mathcal{C}^0$ semigroup on $H^k(\mathbb{R})$ for all $k \le r$.*

Returning to the connections between spectra and exponential decay, we first consider the ODE example and the fundamental matrix solution e^{Ax} from (4.1.4). If there exists $\sigma > 0$ such that $\mathrm{Re}\,\lambda < -\sigma$ for all $\lambda \in \sigma(\boldsymbol{A})$, then since the associated curve C is compact and does not intersect $\sigma(\boldsymbol{A})$, the resolvent is uniformly bounded, i.e., $|\boldsymbol{R}(\lambda;\boldsymbol{A})| \le M$ for all $\lambda \in C$ and some constant $M > 0$. From these facts we readily deduce the exponential decay estimate

$$|e^{Ax}| \le \frac{M}{2\pi} \oint_C e^{(\mathrm{Re}\,\lambda)t} |d\lambda| \le \frac{M}{2\pi} |C| e^{-\sigma t},$$

where $|C|$ is the length of the curve C.

Remark 4.1.3. It is worthwhile to point out a common source of confusion. The Laplace transform is applicable to ordinary differential equations

(ODEs) in which the matrix $\boldsymbol{A}$ is constant in the evolution variable, x, and to linear operators $\mathcal{L}$ that are constant in the evolution variable t. The fact that the operator $\mathcal{L}$ may have coefficients that depend upon x is immaterial since the Laplace transform is an integral over t. The Laplace transform approach is not readily applicable to linear operators $\mathcal{L} = \mathcal{L}(t)$ that depend on the evolution variable.

Before developing decay estimates for the semi-group $e^{\mathcal{L}t}$ as defined in (4.1.6), we return to our discussion on spectral projections onto finite-rank eigenspaces (see Chapter 2.2.4). Fix $\lambda_0 \in \sigma_{\mathrm{pt}}(\mathcal{L})$, and assume the operator $\mathcal{L} - \lambda$ is Fredholm of index zero for λ in a neighborhood of λ_0. Let $C_0 \subset \mathbb{C}$ be a positively oriented, simple, closed curve that encircles λ_0, and no other spectrum, and that is small enough so that the operator $\mathcal{L} - \lambda$ is Fredholm of index zero on and in the interior of the curve. The spectral projection onto the finite-dimensional eigenspace E_{λ_0} associated with λ_0 is given by the Dunford integral formula

$$P_{\lambda_0} = \frac{1}{2\pi\mathrm{i}} \oint_{C_0} R(\lambda;\mathcal{L})\,\mathrm{d}\lambda, \tag{4.1.7}$$

and is the unique bounded operator that enjoys the following properties:

(a) $P_{\lambda_0} : X \mapsto E_{\lambda_0}$; furthermore, $P_{\lambda_0}|_{E_{\lambda_0}} = \mathcal{I}$,
(b) $P_{\lambda_0}^2 := P_{\lambda_0} P_{\lambda_0} = P_{\lambda_0}$,
(c) $P_{\lambda_0}\mathcal{L} = \mathcal{L}P_{\lambda_0}$,
(d) if $P_{\lambda_0}, P_{\lambda_1}$ are spectral projections associated with point eigenvalues $\lambda_1 \neq \lambda_0$, then $P_{\lambda_0}P_{\lambda_1} = P_{\lambda_1}P_{\lambda_0} = 0$.

Remark 4.1.4. If λ_0 is an eigenvalue of $\mathcal{L}$ with algebraic multiplicity $\mathrm{m_a}(\lambda_0)$, then the generalized kernel of the operator $\mathcal{L} - \lambda_0$ is given by

$$\mathrm{gker}(\mathcal{L} - \lambda_0) := \ker[(\mathcal{L} - \lambda_0)^{\mathrm{m_a}(\lambda_0)}],$$

and in particular $\ker(\mathcal{L} - \lambda_0) \subset \mathrm{gker}(\mathcal{L} - \lambda_0)$. The eigenspace E_{λ_0} is precisely the generalized kernel, i.e.,

$$E_{\lambda_0} = \mathrm{gker}(\mathcal{L} - \lambda_0).$$

For the adjoint operator $\mathcal{L}^{\mathrm{a}}$ we have the adjoint eigenspace $E^{\mathrm{a}}_{\overline{\lambda}_0}$, which has the same dimension as E_{λ_0}. The space has the decomposition

$$X = E_{\lambda_0} \oplus (E^{\mathrm{a}}_{\overline{\lambda}_0})^{\perp},$$

and moreover this decomposition is orthogonal.

If $S := \{\lambda_1, \ldots, \lambda_k\} \subset \sigma_{\mathrm{pt}}(\mathcal{L})$, then the spectral projection onto $E := \cup_{i=1}^k E_{\lambda_i}$ is given by the sum $P_S := P_{\lambda_1} + \cdots + P_{\lambda_k}$. This spectral projection has the properties

$$P_S\mathcal{L} = \mathcal{L}P_S, \quad \sigma\left(P_S\mathcal{L}\big|_E\right) = S.$$

We define the spectral projection complementary to $\lambda_0 \in S$ by $\Pi_{\lambda_0} := \mathcal{I} - P_{\lambda_0}$. The spectral projection complementary to the set S is given by the product $\Pi_S := \Pi_{\lambda_1} \cdots \Pi_{\lambda_k}$. The spectral projection complementary to S satisfies

$$\Pi_S : X \to (E^{\mathrm{a}})^{\perp} := \left(\bigcup_{i=1}^{k} E^{\mathrm{a}}_{\overline{\lambda}_i}\right)^{\perp}, \tag{4.1.8}$$

and enjoys the properties

$$\ker(\Pi_S) = E, \quad \Pi_S\mathcal{L} = \mathcal{L}\Pi_S, \quad \sigma\left(\Pi_S\mathcal{L}\big|_{(E^a)^{\perp}}\right) = \sigma(\mathcal{L})\backslash S.$$

In particular, the essential spectrum of $\Pi_S\mathcal{L}$ agrees with that of $\mathcal{L}$.

To establish the exponential decay properties of the linear semi-group we assume that the operator $\mathcal{L}$ is well-posed (see Hypothesis 3.2.1) with a finite collection of point spectra (see Lemma 9.3.9) and with its essential spectrum strictly in the left-half complex plane, i.e., there exists $\sigma > 0$ such that $\sigma_{\mathrm{ess}}(\mathcal{L}) \subset \{\lambda \in \mathbb{C} : \mathrm{Re}\,\lambda \leq -\sigma\}$. Let S_+ denote the finite collection of eigenvalues with nonnegative real part, and let P_+ be the spectral projection onto the associated finite-dimensional union of eigenspaces, denoted E_+, with the corresponding space of adjoint eigenfunctions denoted E^{a}_{+}. Both the complementary spectral projections, $\Pi_+ = \mathcal{I} - P_+$, and P_+, commute with $\mathcal{L}$; furthermore, the spectrum of $\Pi_+\mathcal{L}$ acting on the space $(E^{\mathrm{a}}_{+})^{\perp}$ lies strictly within the left-half complex plane. Applying these two projections to the linear problem (4.0.3), and using the relations $P_+\Pi_+ = \Pi_+P_+ = 0$, yields two uncoupled linear systems for $v^+ := P_+v$ and $v^- := \Pi_+v$,

$$\partial_t v^+ = P_+\mathcal{L}(v^+),\ v^+(0) = P_+v_0; \quad \partial_t v^- = \Pi_+\mathcal{L}(v^-),\ v^-(0) = \Pi_+v_0.$$

The v^+ equation describes the flow on the finite-dimensional invariant space E_+, whereas the v^- equation accounts for the flow on the complementary infinite-dimensional invariant space $(E^{\mathrm{a}}_{+})^{\perp}$. In particular, the finite-dimensional v^+ equation can be formulated as an ODE by projecting onto the basis vectors, with its solution behavior characterized by a fundamental matrix solution. The solution behavior for v^- is less transparent, and the resolution requires a significant tool. Since Π_+ commutes with $\mathcal{L}$ as well as its semi-group and resolvents it follows that $\Pi_+\mathrm{e}^{\mathcal{L}t} = \mathrm{e}^{\Pi_+\mathcal{L}t}$. Choosing a value of $\gamma \in \mathbb{R}$ for which the vertical contour $C = \{\lambda : \mathrm{Re}\,\lambda = \gamma\}$ avoids $\sigma(\mathcal{L})$

with S_+ on its right and $\sigma(\mathcal{L})\backslash S_+$ on its left (see Fig. 4.2), we may express the semigroup for $\mathcal{L}|_{(E_+^a)^\perp}$ as

$$e^{\Pi_+\mathcal{L}t} = e^{\mathcal{L}t}\Pi_+ = \frac{1}{2\pi i}\int_{\gamma-i\infty}^{\gamma+i\infty} e^{\lambda t}R(\lambda;\mathcal{L})\Pi_+\, d\lambda,$$

A general result of Prüss allows us to deduce exponential decay.

Theorem 4.1.5. *(Prüss [**238**, Corollary 4]) Let X be a Hilbert space and assume $\mathcal{L} : \mathcal{D}(\mathcal{L}) \subset X \to X$ generates a C^0 semi-group. Let Π be a finite codimension spectral projection associated with $\mathcal{L}$. If for some $M, \sigma > 0$ the resolvent satisfies*

$$\|R(\lambda;\mathcal{L})\Pi f\|_X \le M\|f\|_X,$$

on the set $\operatorname{Re}\lambda \ge -\sigma$, then there exists $C > 0$ such that the semi-group associated with $\Pi\mathcal{L}$ satisfies the decay estimate

$$\|e^{\mathcal{L}t}\Pi f\|_X \le Ce^{-\sigma t}\|f\|_X. \tag{4.1.9}$$

Remark 4.1.6. In the literature Theorem 4.1.5 is often referred to as the Gearhart–Prüss theorem [**18**]. In typical applications, one takes $\Pi = \Pi_+$, the spectral projection complementary to E_+.

In light of these results we make the following definition:

Definition 4.1.7. The fixed point ϕ is *spectrally stable* if its linearization $\mathcal{L}$ satisfies $\sigma(\mathcal{L}) \cap \{\lambda \in \mathbb{C} : \operatorname{Re}\lambda > 0\} = \emptyset$, i.e., there is no spectrum in the open right-half of the complex plane. Otherwise, the wave is spectrally unstable. The fixed point ϕ is *linearly exponentially stable* in X with respect to spectral projection Π if it is spectrally stable and the semi-group associated with $\Pi\mathcal{L}$ satisfies the decay estimate (4.1.9) for some $C, \sigma > 0$.

Example 4.1.8. It may be a nontrivial task to show that the resolvent is uniformly bounded (e.g., see Chapter 4.5). However, if $\mathcal{L}$ is the parabolic operator

$$\mathcal{L} = \boldsymbol{d}\partial_x^2 + \boldsymbol{a}_1(x)\partial_x + \boldsymbol{a}_0(x),$$

where $\boldsymbol{d} \in \mathbb{R}^{n\times n}$ is positive-definite, and the smooth coefficient matrices $\boldsymbol{a}_j(x) \to \boldsymbol{a}_j^{\pm}$ exponentially fast as $x \to \pm\infty$ for $j = 0, 1$, then $\mathcal{L}$ is sectorial. That is, there exists $\theta_0 \in (\pi/2, \pi)$ such that for $|\arg\lambda| > \theta_0$ and $|\lambda|$ sufficiently large,

$$\|R(\lambda;\mathcal{L})\Pi\|_{\mathcal{B}(H^1)} \le \frac{M}{|\lambda|} \tag{4.1.10}$$

see [**106, 219**]. In this case $\mathcal{L}$ generates what is known as an *analytic semigroup*.

We establish these resolvent bounds for $\mathcal{L}$ in the scalar case, with $\boldsymbol{d} = 1$. We first define the numerical range, $\mathcal{R}_X(\mathcal{L})$, of an operator $\mathcal{L} : \mathcal{D}(\mathcal{L}) \subset X \to X$, in a Hilbert space X:

$$\mathcal{R}_X(\mathcal{L}) := \{\langle \mathcal{L}u, u\rangle_X : u \in \mathcal{D}(\mathcal{L}), \|u\|_X = 1\} \tag{4.1.11}$$

(see Exercise 4.1.5 for some easily computable examples). The utility of the numerical range is demonstrated by the following lemma.

Lemma 4.1.9. *Suppose that X is a Hilbert space, $\mathcal{D}(\mathcal{L}) \subset X$ is dense, and $\mathcal{L} \in \mathcal{B}(\mathcal{D}, X)$. Then for all $\lambda \in \rho(\mathcal{L})$ the norm of the resolvent of $\mathcal{L}$ in $\mathcal{B}(X)$ is bounded by the distance of λ to the numerical range of $\mathcal{L}$ in X,*

$$\|(\lambda - \mathcal{L})^{-1}\|_{\mathcal{B}(X)} \leq (\mathrm{dist}\,(\lambda, \mathcal{R}_X(\mathcal{L})))^{-1}.$$

In particular, $\sigma(\mathcal{L}) \subset \mathcal{R}_X(\mathcal{L})$.

Proof. Take $\lambda \in \rho(\mathcal{L})$. Then for any $u \in X$ with $\|u\|_X = 1$ we form

$$v = \frac{(\lambda - \mathcal{L})^{-1} u}{\left\|(\lambda - \mathcal{L})^{-1} u\right\|_X}.$$

Since both v and u have unit norm, it follows that

$$\begin{aligned} \mathrm{dist}\,(\lambda, \mathcal{R}_X(\mathcal{L})) &\leq |\lambda - \langle \mathcal{L}v, v\rangle_X| = |\langle (\lambda - \mathcal{L})v, v\rangle_X| \\ &= \frac{|\langle u, (\lambda - \mathcal{L})^{-1} u\rangle_X|}{\left\|(\lambda - \mathcal{L})^{-1} u\right\|_X^2} \leq \frac{1}{\left\|(\lambda - \mathcal{L})^{-1} u\right\|_X}. \end{aligned}$$

The result follows by inverting the inequality and taking the supremum over $u \in X$ with unit norm. □

To bound the resolvent of $\mathcal{L}$ in $L^2(\mathbb{R})$ it suffices to examine its numerical range in $L^2(\mathbb{R})$. Let $u \in H^2(\mathbb{R})$ have unit $L^2(\mathbb{R})$ norm, then

$$\langle \mathcal{L}u, u\rangle_{L^2} = \int_{\mathbb{R}} \left(\partial_x^2 u \overline{u} + a_1(x)\partial_x u \overline{u} + a_2(x)|u|^2\right) dx.$$

In particular, the first and third terms in the integral are purely real. Since a_1 and a_2 are uniformly bounded, it is easy to establish that

$$\mathrm{Re}\langle \mathcal{L}u, u\rangle \leq -\|\partial_x u\|_{L^2}^2 + c_1, \quad |\mathrm{Im}\langle \mathcal{L}u, u\rangle| \leq c_2 \|\partial_x u\|_{L^2},$$

for some constants $c_1, c_2 > 0$. It follows that the numerical range, $\mathcal{R}_{L^2(\mathbb{R})}(\mathcal{L})$, is contained to the left of the complex-valued curve $\left\{-s^2 + c_1 + ic_2 s : s \in \mathbb{R}\right\}$, and hence lies inside a sector. The resolvent bounds (4.1.10) in $L^2(\mathbb{R})$ follow from Lemma 4.1.9. The extension to the $H^1(\mathbb{R})$-norm is accomplished in Exercise 4.1.6.

Exercises

Exercise 4.1.1. Use the Fourier transform to verify Lemma 4.1.2 for $\mathcal{L} = \partial_x^2$. Show that if $X = H^k(\mathbb{R})$, then $\mathcal{D}(\mathcal{L}) = H^{k+2}(\mathbb{R})$.

Exercise 4.1.2. Use the Cauchy residue theory for meromorphic functions to evaluate the Dunford formula (4.1.7) and construct the spectral projections for the matrices

$$\text{(a)}\begin{pmatrix}1 & 0\\ 0 & -1\end{pmatrix} \qquad \text{(b)}\begin{pmatrix}1 & 1 & 0\\ 0 & 1 & 0\\ 0 & 0 & -1\end{pmatrix} \qquad \text{(c)}\begin{pmatrix}1 & b\\ -b & -1\end{pmatrix}.$$

Exercise 4.1.3.

(a) Show that the translation operator

$$\big(T(t)u\big)(x) := u(x+t),$$

defines a C^0 semi-group on $L^2(\mathbb{R})$, with generator $\mathcal{L} = \partial_x$ on domain $H^1(\mathbb{R})$.

(b) Show that the heat equation semi-group, defined in terms of the Fourier transform,

$$\widehat{S(t)u}(k) = \mathrm{e}^{-k^2 t}\widehat{u}(k),$$

defines a C^0 semi-group on $L^2(R)$, with generator $\mathcal{L} = \partial_x^2$ on domain $H^2(\mathbb{R})$.

Exercise 4.1.4. Let $\mathcal{L}$ be an exponentially asymptotic nth-order linear operator.

(a) Suppose that $\lambda_0 \in \sigma_{\mathrm{pt}}(\mathcal{L})$ is a simple eigenvalue. Show that the spectral projection associated with λ_0 takes the form

$$P_{\lambda_0}u = \frac{\langle u, \psi_0^{\mathrm{a}}\rangle}{\langle \psi_0, \psi_0^{\mathrm{a}}\rangle}\psi_0,$$

where ψ_0 and ψ_0^{a} are the eigenfunction and adjoint eigenfunction associated with λ_0, *respectively*. It is sufficient to show that P_{λ_0} satisfies properties (a)-(d) which follow (4.1.7).

(b) Derive an explicit formula for P_{λ_0} in terms of the generalized eigenvectors of $\mathcal{L}$ in the case that it has a Jordan chain of length 2 at λ_0; that is, λ_0 satisfies $\mathrm{m_a}(\lambda_0) = 2$ and $\mathrm{m_g}(\lambda_0) = 1$.

Exercise 4.1.5. This exercise compares the numerical range and spectrum of some simple operators.

(a) Determine $\mathcal{R}_{\mathbb{C}^n}(\boldsymbol{A})$ for the matrix $\boldsymbol{A} = \begin{pmatrix}1 & 1\\ a & 1\end{pmatrix}$, where $a \in \mathbb{R}$. In particular, consider the cases $a = 0$, $a = \pm 1$, and the four complementary intervals, $(-\infty,-1), (-1,0), (0,1)$, and $(1,\infty)$. Compare the numerical range to $\sigma(\boldsymbol{A})$ in each case.

(b) Determine $\mathcal{R}_{L^2(\mathbb{R})}(\partial_x^2 + a\partial_x)$ where $a \in \mathbb{R}$ is constant in x and compare to $\sigma(\mathcal{L})$.

Exercise 4.1.6. Consider the scalar form, $\mathcal{L} = \partial_x^2 + a_1(x)\partial_x + a_0(x)$, of the exponentially asymptotic parabolic operators from Example 4.1.8. Establish the resolvent estimate (4.1.10) in the $H^1(\mathbb{R})$-norm by bounding $\mathcal{R}_{H^1(\mathbb{R})}(\mathcal{L})$.

4.2 Systems with Symmetries

The point spectrum of $\mathcal{L}$ is in general difficult to localize. However it is often the case that information on the kernel of $\mathcal{L}$ can be obtained from symmetries of the underlying system. Returning to the framework of (4.0.1), we say it possess an N-parameter symmetry operator if there exist N independent and commuting linear operators $T_j \in C^1(\mathbb{R}, \mathcal{B}(Y))$ that satisfy

$$\begin{aligned} &\text{(a)} \quad T_j(0) = \mathcal{I}, \\ &\text{(b)} \quad T_j(s+t) = T_j(s)T_j(t) = T_j(t)T_j(s), \\ &\text{(c)} \quad T_i(\gamma_i)T_j(\gamma_j) = T_j(\gamma_i)T_i(\gamma_j), \\ &\text{(d)} \quad T_j(\gamma_j)\mathcal{F}(u) = \mathcal{F}(T_j(\gamma_j)u), \end{aligned} \tag{4.2.1}$$

for $i, j = 1, \ldots, N$, and for all $s, t, \gamma_1, \ldots, \gamma_N \in \mathbb{R}$ and $u \in X$. These assumptions imply that each $T_j(\gamma_j)$ is invertible with inverse $T_j(-\gamma_j)$. We also assume that each T_j is an isometry on X, Y, and Z and that the generator T_j' of T_j, defined by the limit

$$T_j' := \lim_{\gamma_j \to 0} \frac{T_j(\gamma_j) - T_j(0)}{\gamma_j}, \tag{4.2.2}$$

exists and satisfies $T_j' \in B(Y, X)$. Moreover we require that there for each $R > 0$ there exists $M > 0$ such that

$$\|T_j(\gamma_j)u - u - \gamma_j T_j' u\|_Y \le M|\gamma_j|^2, \tag{4.2.3}$$

for all $\|u\|_Z \le R$. The group structure of conditions (4.2.1)(a)–(c) motivates the definition of the full symmetry operator

$$T(\gamma) := T_1(\gamma_1) \cdots T_N(\gamma_N)$$

for any $\gamma = (\gamma_1, \ldots, \gamma_N) \in \mathbb{R}^N$. The property (4.2.1)(d) implies that if $u = u(t)$ is a solution of (4.0.1), then so is $T(\gamma)u(t)$.

Lemma 4.2.1. *Let $\phi \in Z$ be a equilibrium solution of (4.0.1), that is, $\mathcal{F}(\phi) = 0$. If $\mathcal{F}$ has an N-parameter symmetry operator, T, then the N-dimensional manifold*

$$\mathcal{M}_T(\phi) := \left\{ T(\gamma)\phi : \gamma \in \mathbb{R}^N \right\}, \tag{4.2.4}$$

is composed of equilibrium solutions. Moreover, the spectrum of the linearization, $\mathcal{L}_\gamma$, of $\mathcal{F}$ about $T(\gamma)\phi$ is independent of γ, and the kernel of the linear operator

$\mathcal{L} := \mathcal{L}_0$ contains (at least) the N symmetry eigenfunctions

$$\ker(\mathcal{L}) \supset \{T_j'\phi : j = 1,\ldots,N\}.$$

Proof. Introducing $\mathcal{L}_\gamma = \nabla_u \mathcal{F}(T(\gamma)\phi)$, then for $\|v\|_Y \ll 1$ we have the expansion

$$\mathcal{F}(T(\gamma)(\phi + v)) = \mathcal{L}_\gamma T(\gamma)v + O(\|v\|_Y^2).$$

Since T commutes with $\mathcal{F}$, equating terms at the linear order, we find for all $v \in Y$ and γ, that $T(\gamma)\mathcal{L}v = \mathcal{L}_\gamma T(\gamma)v$. In particular, we have the relation

$$\mathcal{L}_\gamma = T(\gamma)\mathcal{L}T(-\gamma), \tag{4.2.5}$$

which implies not only that the two operators are isospectral, but also that $T(\gamma)$ maps eigenspaces of $\mathcal{L}$ to the corresponding eigenspaces of $\mathcal{L}_\gamma$. To show the existence of the symmetry modes in the kernel, for each $j = 1,\ldots,N$ we observe that $F(T_j(\gamma_j)\phi) = 0$, and differentiating with respect to γ_j yields

$$0 \equiv \frac{\partial}{\partial \gamma_j}\mathcal{F}(T_j(\gamma_j)\phi)\bigg|_{\gamma_j=0} = \mathcal{L}\left(T_j'\phi\right). \qquad \square$$

Example 4.2.2. Consider the reaction–diffusion equation

$$\partial_t u = \partial_x^2 u - W'(u),$$

that possesses the traveling front solution discussed in Chapter 2.3.3.2 and Chapter 3.1.1.4. The system has a unique traveling front solution, ϕ, with speed c_*. In the traveling variable $\xi = x - c_* t$, ϕ is the equilibrium of

$$u_t = \partial_\xi^2 u + c_* \partial_\xi u - W'(u) =: \mathcal{F}(u), \tag{4.2.6}$$

which connects the well at $u = 0$ to the well at $u = 1$. The associated linearized operator is [see (3.1.25)],

$$\mathcal{L} = \partial_\xi^2 + c_* \partial_\xi - W''(\phi).$$

Equation (4.2.6) has a translational symmetry

$$(T(\gamma)u)(\xi) := u(\xi + \gamma),$$

with generator $T' = \partial_\xi$ which maps $Y = H^1(\mathbb{R}) \to X = L^2(\mathbb{R})$. We deduce from Lemma 4.2.1 that $\lambda = 0$ is an eigenvalue with eigenfunction $\partial_\xi \phi$ generated by the translational symmetry, i.e., $\ker(\mathcal{L}) = \text{span}\{\partial_\xi \phi\}$. Moreover, since $\partial_\xi \phi$ has no nodes and $\mathcal{L}$ is Sturmian, this is the ground-state eigenfunction and $\lambda = 0$ is simple with all other eigenvalues lying strictly on the negative real axis (see Fig. 3.5); in particular, the front ϕ is spectrally stable.

Since eigenvalues are isolated, we can choose $\sigma > 0$ so that all the point spectrum, except for the simple eigenvalue at $\lambda = 0$, lies to the left of the contour $C = \{\mathrm{Re}\,\lambda = -\sigma\}$. The complementary spectral projection associated with the spectral set $S_+ = \{0\}$ is $\Pi_+ : L^2(\mathbb{R}) \mapsto \ker(\mathcal{L}^{\mathrm{a}})^\perp = \{\mathrm{e}^{c_*\xi}\partial_\xi \phi\}^\perp$. To obtain the exponential decay of the projected semi-group afforded by Theorem 4.1.5 we must bound $\|R(\lambda;\mathcal{L})\Pi_+\|_{H^1}$ for $\mathrm{Re}\,\lambda \geq -\sigma$. Since $\mathcal{L}$ is a parabolic operator, from Example 4.1.8 there exists $\theta_0 > 0$ such that the resolvent decays as $|\lambda| \to \infty$ for $|\arg\lambda| > \theta_0$. Furthermore, the projected resolvent $R(\lambda,\mathcal{L})\Pi_+$ is norm bounded for each λ satisfying $\mathrm{Re}\,\lambda > -\sigma$, as $\Pi_+\mathcal{L}$ has no point spectrum on this set. Since the norm of the resolvent is continuous in λ, the existence of a uniform bound follows from compactness and the decay of the resolvent for large $|\lambda|$. We deduce the existence of a constant $C > 0$ for which $\|\mathrm{e}^{\mathcal{L}t}\Pi_+ f\|_{H^1} \leq C\mathrm{e}^{-\sigma t}\|f\|_{H^1}$ for all $f \in H^1$, in particular ϕ is linearly exponentially stable in H^1 with respect to the projection Π_+.

Example 4.2.3. The nonlinear Schrödinger equation (3.1.26) can be written in the form (4.0.1) where

$$\mathcal{F}(u) = \mathrm{i}(\partial_x^2 u - \omega u + |u|^2 u),$$

or equivalently as a real system for the real and imaginary parts with $\boldsymbol{U} = (u_\mathrm{r}, u_\mathrm{i})^\mathrm{T}$ of $u = u_\mathrm{r} + \mathrm{i}u_\mathrm{i}$,

$$\mathcal{F}(\boldsymbol{U}) = \begin{pmatrix} -\partial_x^2 u_\mathrm{i} + \omega u_\mathrm{i} - |\boldsymbol{U}|^2 u_\mathrm{i} \\ \partial_x^2 u_\mathrm{r} - \omega u_\mathrm{r} + |\boldsymbol{U}|^2 u_\mathrm{r} \end{pmatrix}.$$

This system possesses a translational symmetry, i.e., $T_1(\gamma_1)\boldsymbol{U}(x) = \boldsymbol{U}(x+\gamma_1)$, and a rotational symmetry, i.e.,

$$T_2(\gamma_2)\boldsymbol{U}(x) = \begin{pmatrix} \cos\gamma_2 & -\sin\gamma_2 \\ \sin\gamma_2 & \cos\gamma_2 \end{pmatrix} \boldsymbol{U}(x),$$

each of which satisfy the conditions in (4.2.1). The associated full symmetry operator takes the form

$$T(\gamma_1,\gamma_2)\boldsymbol{U}(x) = \begin{pmatrix} \cos\gamma_2 & -\sin\gamma_2 \\ \sin\gamma_2 & \cos\gamma_2 \end{pmatrix} U(x+\gamma_1).$$

These symmetries have the generators

$$T_1' = \partial_x, \quad T_2' = \begin{pmatrix} 0 & -1 \\ 1 & 0 \end{pmatrix}.$$

The operator $\mathcal{F}$ has an equilibrium $\boldsymbol{U}_\omega = (\phi_\omega, 0)^{\mathrm{T}}$, where ϕ_ω solves (3.1.27), i.e.,

$$\partial_x^2\phi_\omega - \omega\phi_\omega + \phi_\omega^3 = 0. \tag{4.2.7}$$

The linearization, $\mathcal{L}$, of $\mathcal{F}$ about $\boldsymbol{U}_\omega$ is given in (3.1.28); namely,

$$\mathcal{L} = \begin{pmatrix} 0 & 1 \\ -1 & 0 \end{pmatrix}\begin{pmatrix} \mathcal{L}_+ & 0 \\ 0 & \mathcal{L}_- \end{pmatrix}, \tag{4.2.8}$$

where

$$\mathcal{L}_- = -\partial_x^2 + \omega - \phi_\omega^2, \quad \mathcal{L}_+ = -\partial_x^2 + \omega - 3\phi_\omega^2.$$

Using the defining equation (4.2.7) we can readily verify that

$$T_1'\boldsymbol{U}_\omega = \begin{pmatrix} \partial_x\phi_\omega \\ 0 \end{pmatrix}, \quad T_2'\boldsymbol{U}_\omega = \begin{pmatrix} 0 \\ \phi_\omega \end{pmatrix},$$

lie in the kernel of $\mathcal{L}$ since $\mathcal{L}_+(\partial_x\phi_\omega) = \mathcal{L}_-\phi_\omega = 0$. Moreover, the operators $\mathcal{L}_\pm$ are both Sturmian (see Theorem 2.3.3), and must therefore have simple kernels; we deduce that these functions span the kernel.

The operator $\mathcal{L}$ is not self-adjoint, and could have a generalized kernel. We look for a Jordan chain by examining the kernel of

$$\mathcal{L}^2 = \begin{pmatrix} -\mathcal{L}_-\mathcal{L}_+ & 0 \\ 0 & -\mathcal{L}_+\mathcal{L}_- \end{pmatrix}.$$

Is it possible that $\mathcal{L}_-\mathcal{L}_+v = 0$ with $v \notin \ker(\mathcal{L}_+)$? If so, then $\mathcal{L}_+v = \phi_\omega$. Since $\phi_\omega \in \ker(\mathcal{L}_+)^\perp = \mathrm{span}\{\partial_x\phi_\omega\}^\perp$, by the Fredholm alternative Theorem 2.2.1 this equation does indeed have a solution. Similarly, finding new solutions to $\mathcal{L}_+\mathcal{L}_-u = 0$ reduces to solving $\mathcal{L}_-u = \partial_x\phi_\omega$, which also has a solution as a consequence of the Fredholm alternative. In conclusion, in addition to $\ker(\mathcal{L})$ the generalized kernel, $\mathrm{gker}(\mathcal{L})$, also contains

$$\begin{pmatrix} \mathcal{L}_+^{-1}\phi_\omega \\ 0 \end{pmatrix} = \begin{pmatrix} \partial_\omega\phi_\omega \\ 0 \end{pmatrix}, \quad \text{and} \quad \begin{pmatrix} 0 \\ \mathcal{L}_-^{-1}\partial_x\phi_\omega \end{pmatrix} = \frac{1}{2}\begin{pmatrix} 0 \\ x\phi_\omega \end{pmatrix}.$$

The first of these equalities can be verified by taking ∂_ω of (4.2.7). An application of the Fredholm alternative to the equation $\mathcal{L}^3\boldsymbol{U} = 0$ yields no new solutions, so that the Jordan chain terminates with $\mathrm{m_a}(0) = \dim[\mathrm{gker}(\mathcal{L})] = 4$; see Exercise 4.2.2.

Exercises

Exercise 4.2.1. In exponentially weighted spaces, the linearization of the gKdV equation about its traveling-wave solution leads to the operator $\mathbb{L}_a$ defined in Chapter 3.1.1.2,

$$\mathbb{L}_a = (\partial_\xi - a)\left(-(\partial_\xi - a)^2 + c - \phi_c^p\right),$$

where $0 \le a < \sqrt{c}$ and $1 \le p < 4$ and ϕ_c is the solution of (3.1.21), which is homoclinic to the origin.

(a) For $a = 0$, use the fact that ϕ_c satisfies (3.1.21) to show that $\mathrm{gker}(\mathbb{L}_0) = \mathrm{span}\{\partial_x\phi_c, \partial_c\phi_c\}$. Find $\mathrm{gker}(\mathbb{L}_a)$.
(b) For $1 \le p < 4$ we will see in Chapter 5.2.1 that the wave ϕ_c is orbitally stable, and hence $\mathbb{L}_a$ has no eigenvalues with positive real part. In addition, assume that $\sigma(\mathbb{L}_a) \cap \mathrm{i}\mathbb{R} = \{0\}$, i.e., $\mathbb{L}_a$ has no spectrum on the imaginary axis except at the origin. Let $\Pi_+ : L^2(\mathbb{R}) \mapsto \mathrm{gker}(\mathbb{L}_a^{\mathrm{a}})^\perp$ be the spectral projection complementary to the spectral set $S_+ = \{0\}$. Find values of a for which there exists $M, \sigma > 0$ such that $\|R(\lambda, \mathbb{L}_{\mathrm{a}})\Pi_+\|_{\mathcal{B}(L^2)} \le M$ for all $\mathrm{Re}\,\lambda \ge -\sigma$. (*Hint:* The issue is $|\lambda| \gg 1$. Consider first the asymptotic operator with constant coefficients.)
(c) For these values of a, show that there is a $\sigma > 0$ and $C > 0$ such that the semi-group generated by $\mathbb{L}_a$ satisfies the estimate $\|\mathrm{e}^{\mathbb{L}_a t}\Pi_+\|_{\mathcal{B}(L^2)} \le C\mathrm{e}^{-\sigma t}$.

Exercise 4.2.2. Let $\mathcal{L}$ be the linearization about the pulse solution of the NLS equation given in (4.2.8). Show that the $\dim \mathrm{gker}(\mathcal{L}) = 4$.

4.3 Nonlinear Dynamics

We return to the framework of the nonlinear problem, (4.0.1), written in the expansion form (4.0.2),

$$\partial_t v = \mathcal{L}v + \mathcal{N}(v),$$

where $\mathcal{L} := \nabla_u \mathcal{F}(\phi)$ is the linearization about an equilibrium point ϕ of the original nonlinear operator $\mathcal{F}$ and

$$\mathcal{N}(v) := \mathcal{F}(\phi + v) - \mathcal{L}v. \tag{4.3.1}$$

To obtain exponential stability of ϕ to perturbations of initial data, it is natural to expect that the spectrum of $\mathcal{L}$ should lie strictly in the left-half complex plane. However, if the system has an N-fold symmetry, then Lemma 4.2.1 implies that the linearized operator $\mathcal{L}$ has an N-dimensional kernel. We can accommodate this kernel if we are willing to slightly loosen our idea of stability. In particular, we require the following assumptions.

Hypothesis 4.3.1. Regarding the linear, $\mathcal{L}$, and nonlinear, $\mathcal{N}$, operators:

(a) There is a $\sigma > 0$ such that

$$S_+ := \sigma(\mathcal{L}) \cap \{\lambda \in \mathbb{C} : \mathrm{Re}\,\lambda \ge -\sigma\} = \{0\} \tag{4.3.2}$$

(see Fig. 4.2).
(b) The generalized kernel of $\mathcal{L}$ is spanned by the symmetry eigenfunctions, with

$$\text{gker}(\mathcal{L}) = \ker(\mathcal{L}) = \text{span}\{T_1'\phi,\dots,T_N'\phi\}. \tag{4.3.3}$$

(c) There exists $M > 0$ such that spectral projection $\Pi_+ : X \mapsto \ker(\mathcal{L}^{\text{a}})^\perp$ complimentary to the set S_+ satisfies the resolvent estimate

$$\|R(\lambda;\mathcal{L})\Pi_+ f\|_Y \le M\|f\|_Y, \quad \text{Re}\,\lambda \ge -\sigma \tag{4.3.4}$$

(d) The gradient $\nabla_u\mathcal{F}$ is locally Lipschitz on bounded sets; that is, for each $R > 0$ there exists $M > 0$ such that

$$\|(\nabla_u\mathcal{F}(u) - \nabla_u\mathcal{F}(v))w\|_Y \le M\|u-v\|_Y\|w\|_Y, \tag{4.3.5}$$

as long as $\|u\|_Y, \|v\|_Y \le R$.

(e) The nonlinearity $\mathcal{N}$ is quadratic in $\|\cdot\|_Y$ near zero, i.e., there exists $R, M > 0$ such that

$$\|\mathcal{N}(v)\|_Y \le M\|v\|_Y^2, \tag{4.3.6}$$

for all $\|v\|_Y \le R$.

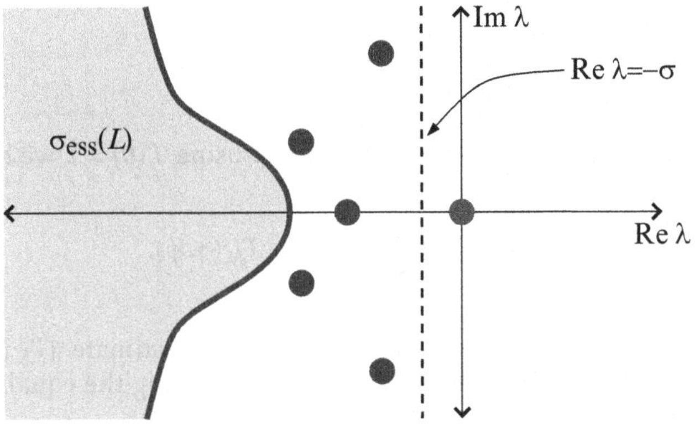

Fig. 4.2 The spectrum of $\mathcal{L}$. The only eigenvalue in the closed right-half of the complex plane is $\lambda = 0$, and the rest of the spectrum is to the left of the line $\text{Re}\,\lambda = -\sigma$. (Color figure online.)

Remark 4.3.2. Hypothesis 4.3.1(a) and (b) imply that $\mathcal{L}$ is Fredholm with index zero and $\ker(\mathcal{L})$ has algebraic and geometric multiplicity N. It follows that $\ker(\mathcal{L}^{\text{a}})$ has algebraic and geometric multiplicity N, and a basis for $\ker(\mathcal{L}^{\text{a}}) = \text{span}\{\psi_1^{\text{a}},\dots,\psi_N^{\text{a}}\}$ can be chosen out of adjoint eigenfunctions with the property that $\langle T_i'\phi, \psi_j^{\text{a}}\rangle = \delta_{ij}$, where δ_{ij} is the Kronecker delta function. The projection operators can then be written as

$$P_+u = \sum_{j=1}^{N}\langle u, \psi_j^{\text{a}}\rangle T_j'\phi, \quad \Pi_+u = u - P_+u;$$

in particular $\Pi_+ u = u$ implies that $\langle u, \psi_j^{\mathrm{a}} \rangle = 0$ for each j.

The resolvent estimate Hypothesis 4.3.1(c) implies, via Theorem 4.1.5, that the semi-group satisfies the decay estimate $\|e^{\mathcal{L}t}\Pi_+ f\|_Y \le Ce^{-\sigma t}\|f\|_Y$. However, that is not sufficient to obtain stability for the fully nonlinear problem. We must still account for the nontrivial kernel of $\mathcal{L}$. Indeed, it is not reasonable for ϕ to be asymptotically stable when it is part of a manifold of equilibria, $\mathcal{M}_T$, defined in (4.2.4). The more reasonable expectation is that the manifold itself is stable under the flow. This approach, which leads to the idea of orbital stability, requires a foliation of a neighborhood of $\mathcal{M}_T$ (see Fig. 4.3). More specifically, we show that for any $w \in Y$ sufficiently small the sum $\phi + w$ can be uniquely written as a point on $\mathcal{M}_T$ and a normal vector, i.e.,

$$\phi + w = T(\gamma)\phi + v, \quad P_+ v = 0, \tag{4.3.7}$$

where P_+ is both the spectral projection onto $\ker(\mathcal{L}^{\mathrm{a}})$ and the orthogonal projection onto the tangent space of the manifold $\mathcal{M}_T$ at ϕ.

Lemma 4.3.3. *There exists $\delta > 0$ and smooth functions $(\gamma, v) : Y \mapsto \mathbb{R}^N \times Y$ satisfying $\gamma(0) = \mathbf{0}, v(0) = 0$ such that for all $\phi \in \mathcal{M}_T$ and all $\|w\|_Y \le \delta$,*

$$\phi + w = T(\gamma(w))\phi + v(w),$$

where $v(w) \in (\ker \mathcal{L}^{\mathrm{a}})^{\perp}$.

Proof. By Taylor expanding T about $\gamma = \mathbf{0}$ and using $T(\mathbf{0}) = \mathcal{I}$ we have

$$T(\gamma)\phi = \phi + \sum_{j=1}^{N} \gamma_j T_j' \phi + T_{\mathcal{N}}(\gamma, \phi),$$

where from (4.2.3) $T_{\mathcal{N}}(\gamma, \phi)$ satisfies the quadratic estimate $\|T_{\mathcal{N}}(\gamma, \phi)\|_Y \le C|\gamma|^2$ since $\phi \in \mathcal{M}_T$ is uniformly bounded in Z. Solving the equality (4.3.7) for v we find

$$v = w - \sum_{j=1}^{N} \gamma_j T_j' \phi - T_{\mathcal{N}}(\gamma, \phi). \tag{4.3.8}$$

The constraint $P_+ v = 0$, i.e. that $v \in \mathrm{gker}(\mathcal{L}^{\mathrm{a}})^{\perp}$, is equivalent to the N equations,

$$0 = g_\ell(\gamma, w) := \langle w, \psi_\ell^{\mathrm{a}} \rangle - \sum_{j=1}^{N} \gamma_j \langle T_j' \phi, \psi_\ell^{\mathrm{a}} \rangle - \langle T_{\mathcal{N}}(\gamma, \phi), \psi_\ell^{\mathrm{a}} \rangle,$$

for $\ell = 1, \ldots, N$. Setting $\boldsymbol{g}(\gamma, w) = (g_1(\gamma, w), \ldots, g_N(\gamma, w))^{\mathrm{T}}$, we observe that $\boldsymbol{g}(\mathbf{0}, 0) = \mathbf{0}$. Moreover, by the normalization of the adjoint eigenfunctions of Remark 4.3.2 we see that $\mathrm{D}_\gamma \boldsymbol{g}(\mathbf{0}, 0) = -\boldsymbol{I}_N$ has non-zero determinant. Consequently, by the Implicit Function Theorem there exists a neighborhood of $(\mathbf{0}, 0)$ and a unique function $\gamma(w)$ such that $\boldsymbol{g}(\gamma(w), w) \equiv 0$. The dependence of v on w follows from (4.3.8). □

Definition 4.3.4. For the flow generated by (4.0.1) we say that the manifold $\mathcal{M}_T$ of equilibria is *asymptotically orbitally stable* in $\|\cdot\|_Y$ with exponential rate $\nu > 0$ if there exists $C, \delta > 0$ such that $\|u_0 - T(\gamma_0)\phi\|_Y \le \delta$ for some γ_0 implies there exists unique $\gamma_\infty = \gamma_\infty(u_0)$ such that

$$\|u(t) - T(\gamma_\infty)\phi\|_Y \le Ce^{-\nu t}\|u_0 - T(\gamma_0)\phi\|_Y. \tag{4.3.9}$$

The idea behind the proof of our main result is as follows. The foliation guaranteed by Lemma 4.3.3 allows the flexibility to eliminate the neutral nondecaying modes, spanned by $E_+ = \mathrm{gker}(\mathcal{L})$, from the system. The remaining space, spanned by $\mathrm{gker}(\mathcal{L}^a)^\perp$, is exponentially damped, and if the initial data is sufficiently small the nonlinearity can be controlled. The conclusion is that small perturbations of the underlying equilibria will decay exponentially fast to a unique member in the N-parameter manifold of equilibria generated by the symmetries.

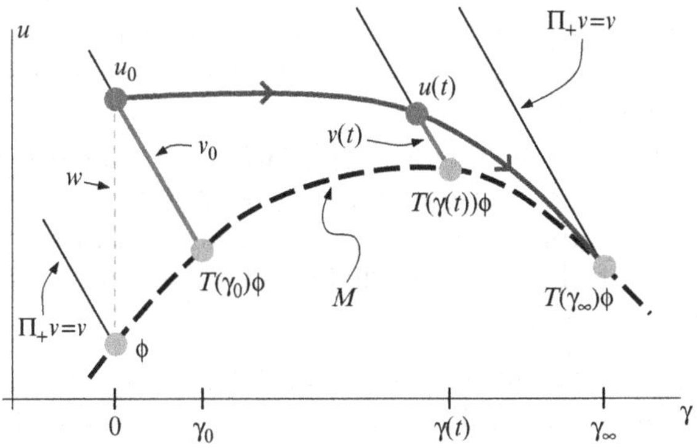

Fig. 4.3 The tubular neighborhood of $T(\gamma)\phi$ described in Lemma 4.3.3 and the asymptotic exponential stability of Theorem 4.3.5. The thick (*blue*) line represents the orbit $u = u(t)$, which emanates from the initial data u_0. The manifold $\mathcal{M} = \{T(\gamma)\phi : \gamma \in \mathbb{R}^N\}$ is denoted by a dotted (*black*) line, and is a graph above the γ-plane, depicted as the horizontal axis. The initial data $u_0 = \phi + w$ is uniquely decomposed as $u_0 = T(\gamma_0)\phi + v_0$, where $v_0 \in (\ker \mathcal{L}_0^a)^\perp$. Indeed the thin *black* lines indicate the foliation of space near $\mathcal{M}$ using offsets of the space $(\ker \mathcal{L}_0^a)^\perp$ which is complementary to $\ker \mathcal{L}_0$ which in turn equals the tangent space to $\mathcal{M}$ at ϕ. The decomposition $u(t) = T(\gamma(t))\phi + v(t)$ respects this foliation, and $\|v(t)\|_Y$ decays exponentially to 0 as $u \to T(\gamma_\infty)\phi$ in $\|\cdot\|_Y$. (Color figure online.)

Theorem 4.3.5. *Consider the nonlinear problem*

$$\partial_t u = \mathcal{F}(u),$$

where $\mathcal{F} : Z \subset X \mapsto X$ *has an* N*-fold symmetry* T*. Suppose that* $\mathcal{F}(\phi) = 0$*, and that the linearization,* $\mathcal{L} = \nabla_u \mathcal{F}(\phi)$*, and nonlinearity,* $\mathcal{N}$*, defined in (4.3.1), satisfy Hypothesis 4.3.1 for some* $\sigma > 0$*. Then for any* $\tilde{\sigma} \in (0, \sigma)$ *the manifold* $\mathcal{M}_T$ *of equilibria is asymptotically orbitally stable in* $\|\cdot\|_Y$ *with exponential rate* $\tilde{\sigma}$*.*

As we will see in the next section, it can be a nontrivial task to verify the resolvent estimate Hypothesis 4.3.1(c). However, modulo concerns about the location of the point spectrum the resolvent estimates hold for the second-order parabolic operators of Example 4.1.8, eliminating the most technical requirement of Theorem 4.3.5.

Corollary 4.3.6. *Consider the nonlinear problem*

$$\partial_t u = \mathcal{F}(u),$$

where $\mathcal{F} : Z = H^2(\mathbb{R})^n \subset Y = H^1(\mathbb{R})^n \mapsto X = L^2(\mathbb{R})^n$ *has an equilibrium* $\phi \in H^2(\mathbb{R})^n$ *and an* N*-fold symmetry* T. *Suppose that* $\mathcal{F}$ *and its nonlinearity,* $\mathcal{N}$, *about* ϕ, *defined in (4.3.1), satisfy Hypothesis 4.3.1(d) and (e). Assume in addition that* $\mathcal{L}$ *is a second-order exponentially asymptotic parabolic operator*

$$\mathcal{L} = \boldsymbol{d}\partial_x^2 + \boldsymbol{a}_1(x)\partial_x + \boldsymbol{a}_0(x), \tag{4.3.10}$$

where $\boldsymbol{d}$ *is a fixed positive-definite matrix. If* ϕ *is spectrally stable, that is, if* $\mathcal{L} := \nabla_u \mathcal{F}(\phi)$ *satisfies Hypothesis 4.3.1(a) and (b) for some* $\sigma > 0$, *then the manifold of equilibria* $\mathcal{M}_T(\phi)$ *is asymptotically orbitally stable in* $H^1(\mathbb{R})^n$ *for any exponential rate* $\tilde{\sigma} \in (0, \sigma)$.

Example 4.3.7. Returning to the reaction diffusion system of Example 4.2.2, we have already verified that the traveling front $\phi(\xi)$, $\xi = x - c_* t$, is spectrally stable with a simple eigenvalue at $\lambda = 0$ generated by the translational symmetry. From Corollary 4.3.6 we deduce that the front is asymptotically orbitally stable in $H^1(\mathbb{R})$.

Proof. (of Theorem 4.3.5) We consider an initial data u_0, which is close to $\phi \in \mathcal{M}_T$ in the Y norm; in particular, $\|u_0 - \phi\|_Y$ is sufficiently small for some ϕ. We will show that $u = u(t)$ stays close to ϕ for all $t > 0$ and in fact converges to some nearby point $T(\gamma_\infty)\phi$. Assuming that $u = u(t)$ is close to ϕ for a time $\underline{T} > 0$, then using Lemma 4.3.3 we have the decomposition

$$u(t) = T(\gamma)\phi + v, \quad v = \Pi_+ v, \tag{4.3.11}$$

where $\gamma = \gamma(t)$, and Π_+ is the complementary projection associated with the operator $\mathcal{L}_{\boldsymbol{0}} := \nabla_u \mathcal{F}(\phi)$. Without loss of generality it may be assumed that $\gamma(0) = \boldsymbol{0}$. Plugging the ansatz of (4.3.11) into the evolution equation (4.0.1), performing a Taylor expansion of $\mathcal{F}(u)$ about $T(\gamma)\phi$, and recalling that $\mathcal{F}(T(\gamma)\phi) = 0$, yields the expression

$$\partial_t v + \sum_{j=1}^{N} T_j'(T(\gamma)\phi)\partial_t \gamma_j = \mathcal{L}_\gamma v + \mathcal{N}(v;\gamma), \tag{4.3.12}$$

where the nonlinearity $\mathcal{N}(v;\gamma)$ satisfies the estimate (4.3.6).

The linear operator $\mathcal{L}_\gamma := \nabla_u \mathcal{F}(T(\gamma)\phi)$, corresponding to the linearization about $T(\gamma)\phi$, is time-dependent. To apply the semi-group estimates the linearized operator $\mathcal{L}_\gamma$ must be time-independent; in fact, the time-dependence of $\mathcal{L}$ through γ is an additional, and irksome, source of nonlinearity. We make this nonlinearity explicit through the expansion

$$\mathcal{L}_\gamma v = \mathcal{L}_0 v + [\mathcal{L}_\gamma - \mathcal{L}_0]v.$$

Going back to (4.3.12) we first observe that

$$T_j'(T(\gamma)\phi) = T_j'\phi + T_j'[T(\gamma) - \mathcal{I}]\phi.$$

From the smoothness of ϕ the quadratic estimates (4.2.3), and the semi-linearity of $\mathcal{F}$, there exists $C > 0$ such that

$$\|T_j'[T(\gamma) - \mathcal{I}]\phi\|_Y \le C|\gamma|, \quad \left\|[\mathcal{L}_\gamma - \mathcal{L}_0]v\right\|_Y \le C|\gamma|\,\|v\|_Y.$$

We can then rewrite (4.3.12) as the system

$$\partial_t v + \sum_{j=1}^{N} [T_j'\phi + t_j(\gamma)]\partial_t \gamma_j = \mathcal{L}_0 v + \mathcal{R}(\gamma, v), \tag{4.3.13}$$

where the nonlinear quantities t_j and $\mathcal{R}$ satisfy

$$\|t_j(\gamma)\|_Y = \mathcal{O}(|\gamma|), \quad \|\mathcal{R}(\gamma, v)\|_Y = \mathcal{O}(|\gamma|\,\|v\|_Y + \|v\|_Y^2). \tag{4.3.14}$$

To determine the evolution for $\gamma = \gamma(t)$ and $v = v(t)$ we act upon (4.3.13) with the spectral projection P_+ and complementary spectral projection Π_+ associated to $\mathcal{L}_0$. The role of the free parameters in γ is to account for the secular growth accumulated in the nondecaying spectrum, S_+, while the remainder v will decay exponentially. By construction, $v \in \ker(\mathcal{L}_0^{\mathrm{a}})^\perp$ for all $t \in [0, \underline{T}]$: this is equivalent to the equalities $\langle v, \psi_j^{\mathrm{a}}\rangle = 0$ for $j = 1, \ldots, N$, where $\{\psi_j^{\mathrm{a}}\}$ comprise the normalized basis for the kernel of $\mathcal{L}_0^{\mathrm{a}}$. Since this basis is independent of γ, and hence of time, it follows that $\langle \partial_t v, \psi_j^{\mathrm{a}}\rangle = 0$ for $j = 1, \ldots, N$. Taking the inner product of (4.3.13) with ψ_j for $j = 1, \ldots, N$ leads to the nonlinear ODE system for the symmetry parameters:

$$(\boldsymbol{I}_N + \boldsymbol{M}(\gamma))\partial_t \gamma = \boldsymbol{r}(\gamma, v), \tag{4.3.15}$$

where the matrix $\boldsymbol{M}$ has entries $M_{ij} = \langle t_i(\gamma), \psi_j^{\mathrm{a}}\rangle$, and the residual $\boldsymbol{r} = (r_1, \ldots, r_N)^{\mathrm{T}}$ takes the form $r_j(\gamma, v) := \langle \mathcal{R}(\gamma, v), \psi_j^{\mathrm{a}}\rangle$. Applying the complementary projection Π_+ to (4.3.13), the relation $\Pi_+ v = v$ and the commutativity of spectral projections $\Pi_+ \mathcal{L}_0 = \mathcal{L}_0 \Pi_+$ yields

$$\partial_t v + \sum_{j=1}^{N} [\Pi_+ t_j(\gamma)] \partial_t \gamma_j = \mathcal{L}_0 v + \Pi_+ \mathcal{R}(\gamma, v). \tag{4.3.16}$$

From (4.3.14) we see that $|\boldsymbol{M}| = \mathcal{O}(|\gamma|)$, and for $|\gamma|$ sufficiently small

$$(\boldsymbol{I}_N + \boldsymbol{M})^{-1} = \boldsymbol{I}_N - \boldsymbol{M} + \mathcal{O}(|\gamma|^2).$$

On the other hand, from Hölder's inequality and (4.3.14) we have

$$|r_j(\gamma, v)| = \mathcal{O}(|\gamma| \, \|v\|_Y + \|v\|_Y^2).$$

Inverting $\boldsymbol{I}_N + \boldsymbol{M}$ in (4.3.15) and applying the estimate above we obtain

$$\partial_t \gamma = \mathcal{O}(|\gamma| \, \|v\|_Y + \|v\|_Y^2). \tag{4.3.17}$$

Combining (4.3.17) with the estimate (4.3.14) on $\mathcal{R}$, we rewrite (4.3.16) as

$$\partial_t v = \mathcal{L}_0 v + \mathcal{R}_{\mathrm{f}}(\gamma, v), \tag{4.3.18}$$

where the nonlinearity $\mathcal{R}_{\mathrm{f}}$ satisfies

$$\|\mathcal{R}_{\mathrm{f}}(\gamma, v)\|_Y = \mathcal{O}(|\gamma| \, \|v\|_Y + \|v\|_Y^2). \tag{4.3.19}$$

With these estimates we may obtain uniform bounds on the solutions of the coupled system (4.3.17)–(4.3.18). We fix $\widetilde{\sigma} \in (0, \sigma)$ and introduce

$$M_v(t) := \sup_{0 \le s \le t} \left(\mathrm{e}^{\widetilde{\sigma} s} \|v(s)\|_Y \right), \quad M_\gamma(t) := \sup_{0 \le s \le t} |\gamma(s)|.$$

The quantity M_v affords the estimate

$$\|v(s)\|_Y \le \mathrm{e}^{-\widetilde{\sigma} s} M_v(t), \quad 0 < s \le t, \tag{4.3.20}$$

so if M_v is uniformly bounded, then $\|v\|_Y$ decays with exponential rate $\widetilde{\sigma}$ as $t \to +\infty$. Since $\mathcal{L}_0$ is independent of t, the solution to the linear problem

$$\partial_t v = \mathcal{L}_0 v, \quad v(0) = v_0$$

is given by $v(t) = \mathrm{e}^{\mathcal{L}_0 t} v_0$. Consequently, we may apply the variation of constants formula (recall the ODE formulation in Lemma 2.1.3) to the nonlinear problem (4.3.18) to obtain the solution

$$v(t) = \mathrm{e}^{\mathcal{L}_0 t} v_0 + \int_0^t \mathrm{e}^{\mathcal{L}_0 (t-s)} \mathcal{R}_{\mathrm{f}}(\gamma(s), v(s)) \, \mathrm{d}s. \tag{4.3.21}$$

From Hypothesis 4.3.1(c) and Theorem 4.1.5 we have the semigroup exponential decay estimate (4.1.9). Applying this to (4.3.21) yields

$$\|v(t)\|_Y \leq C\mathrm{e}^{-\sigma t}\|v_0\|_Y + C\int_0^t \mathrm{e}^{-\sigma(t-s)}\|\mathcal{R}_{\mathrm{f}}(\gamma(s), v(s))\|_Y \,\mathrm{d}s,$$

which in light of the estimate (4.3.19) reduces to

$$\|v(t)\|_Y \leq C\mathrm{e}^{-\sigma t}\|v_0\|_Y + C\int_0^t \mathrm{e}^{-\sigma(t-s)}\left(|\gamma(s)|\,\|v(s)\|_Y + \|v(s)\|_Y^2\right)\mathrm{d}s.$$

Applying (4.3.20) to the integral terms yields the estimate

$$\|v(t)\|_Y \leq C\mathrm{e}^{-\sigma t}\|v_0\|_Y + C\mathrm{e}^{-\sigma t}\int_0^t \left(\mathrm{e}^{(\sigma-\widetilde{\sigma})s}M_\gamma(t)M_v(t) + \mathrm{e}^{(\sigma-2\widetilde{\sigma})s}M_v^2(t)\right)\mathrm{d}s.$$

The functions M_v and M_γ are independent of s: fixing $\widetilde{\sigma} \in (\sigma/2, \sigma)$ we evaluate the integrals to obtain

$$\|v(t)\|_Y \leq C\left(\mathrm{e}^{-\widetilde{\sigma}t}\|v_0\|_Y + \mathrm{e}^{-\widetilde{\sigma}t}M_\gamma(t)M_v(t) + \mathrm{e}^{-2\widetilde{\sigma}t}M_v^2(t)\right),$$

where $C = C(\widetilde{\sigma})$. Since $t \in [0, \underline{T}]$ is arbitrary, we can fix $0 < t' < t$, replace t with t' and multiply by $\mathrm{e}^{\widetilde{\sigma}t'}$, obtaining

$$\mathrm{e}^{\widetilde{\sigma}t'}\|v(t')\|_Y \leq C\left(\|v_0\|_Y + M_\gamma(t')M_v(t') + M_v^2(t')\right)$$

Since $M_v(t') \leq M_v(t)$ and $M_\gamma(t') \leq M_\gamma(t)$, taking the supremum over $t' \in [0, t]$ yields the bound

$$M_v(t) \leq C\left(\|v_0\|_Y + M_\gamma(t)M_v(t) + M_v^2(t)\right). \tag{4.3.22}$$

To bound M_γ we integrate (4.3.17) from 0 to t, and use (4.3.20) to obtain

$$M_\gamma(t) \leq C_1\left(M_\gamma(t)M_v(t) + M_v^2(t)\right), \tag{4.3.23}$$

for some $C_1 > 0$.

We now show that the inequalities of (4.3.22) and (4.3.23) imply the desired decay. Assume that $\|v_0\|_Y$ and $\underline{T}$ are such that $M_v(t) \leq 1/(2C_1)$ for all $t \in [0, \underline{T}]$: we will show that $\underline{T} = +\infty$. With this restriction on M_v we have from (4.3.23) that

$$M_\gamma(t) \leq \frac{1}{2}M_\gamma(t) + C_1 M_v^2(t),$$

and collecting the two terms in M_γ we obtain

$$M_\gamma(t) \leq 2C_1 M_v^2(t), \quad t \in [0, \underline{T}]. \tag{4.3.24}$$

With this bound we rewrite the inequality (4.3.22) as

$$M_v(t) \le C_2\left(\|v_0\|_Y + M_v^2(t) + M_v^3(t)\right), \tag{4.3.25}$$

where $C_2 > 1$. Now consider the polynomial

$$p(r) = C_2\|v_0\|_Y - r + C_2 r^2 + C_2 r^3,$$

and note that the inequality (4.3.25) corresponds to the set of all r such that $p(r) \ge 0$. For $\|v_0\|_Y$ sufficiently small $p(r)$ has two successive positive zeros r_1 and r_2 satisfying

$$0 < r_1 = C_2\|v_0\|_Y + \mathcal{O}(\|v_0\|_Y^2) \ll r_2;$$

furthermore, $p(r) > 0$ for $0 < r < r_1$ and $r_2 < r$. Taking $C_2 > 1$ we have $r_1 = C_2\|v_0\|_Y + O(\|v_0\|_Y^2)$ and $M_v(0) = \|v_0\|_Y < r_1$, then by continuity of $M_v(t)$ in t it follows that

$$M_v(t) \le r_1 = \mathcal{O}(\|v_0\|_Y), \quad t \in [0, \underline{T}]. \tag{4.3.26}$$

If $\|v_0\|_Y$ is sufficiently small that $r_1 < 1/(2C_1)$, we may extend this process until $\underline{T} = \infty$, so that by using (4.3.20),

$$\|v(t)\|_Y \le C_3\|v_0\|_Y e^{-\tilde{\sigma}t}, \quad t \ge 0. \tag{4.3.27}$$

Now, from (4.3.24) and (4.3.26) we see that $|\gamma(t)| = \mathcal{O}(\|v_0\|_Y^2)$. Returning this estimate and (4.3.27) to (4.3.17) yields

$$\partial_t \gamma = \mathcal{O}(e^{-\tilde{\sigma}t}\|v_0\|_Y^3 + e^{-2\tilde{\sigma}t}\|v_0\|_Y^2),$$

which upon integrating from $t = t_1$ to $t = t_2 > t_1$ implies the existence of a $C_4 > 0$ such that

$$|\gamma(t_1) - \gamma(t_2)| \le C_4 e^{-\tilde{\sigma}t_1}\|v_0\|_Y^2.$$

The sequence $\{\gamma(t)\}_{t>0}$ is Cauchy, and moreover it converges at an exponential rate to a limit γ_∞. This result, together with (4.3.27), yields the asymptotic orbital stability of $\mathcal{M}_T$. □

Remark 4.3.8. If the evolution equation (4.0.1) is perturbed so that one or more of the symmetries are broken, then the manifold $\mathcal{M}_T$ will no longer be composed of equilibria. However, it may retain its stability in the sense that initial data that starts sufficiently close to $\mathcal{M}_T$ may converge into a thin neighborhood about it. On the other hand, the flow in the tangent plane—that is, the evolution of γ—may not necessarily settle down to a limit, or at least not one close to the initial data. In this case one will be required to update the "base point"—the value $\gamma_0 = 0$, which was frozen in the linearized operator and the projection—episodically as $\gamma(t)$ migrates away from γ_0. For examples of the application of this technique the interested reader should see [73, 236].

4.4 Example: Scalar Viscous Conservation Law

The scalar viscous conservation law is a PDE of the form,

$$\partial_t u = \partial_x^2 u - \partial_x(f(u)), \tag{4.4.1}$$

where $(x,t) \in \mathbb{R} \times \mathbb{R}^+$, and the flux function $f : \mathbb{R} \to \mathbb{R}$ is smooth. In the absence of viscosity—when the $\partial_x^2 u$ term is dropped—the equation supports "shock solutions." These are discontinuous traveling waves of the form $u(x,t) = S(x-ct)$, where S is piecewise constant, taking the values ϕ_+ for $\xi := x - ct > 0$ and ϕ_- for $\xi < 0$ (e.g., see [**265**, Part III]). The wave-speed c satisfies the Rankine–Hugoniot relation

$$c = \frac{f(\phi_+) - f(\phi_-)}{\phi_+ - \phi_-}. \tag{4.4.2}$$

Such a shock solution is an "entropy" solution, and hence a weak solution of the PDE, only if (a) the Oleĭnik entropy condition, and (b) the Lax entropy condition, which relate f and the end-states, are satisfied:

(a) $f(u) - f(\phi_-) < c(u - \phi_-)$ for $\phi_+ < u < \phi_-$,
(b) $f'(\phi_+) < c < f'(\phi_-)$.

Reinstating the viscosity, one can search for a smooth traveling wave (often called a viscous shock) which connects the two end-states, $\phi_\pm$. In the moving frame, $\xi = x - ct$, the conservation law (4.4.1) becomes

$$\partial_t u = \partial_\xi^2 u + c\partial_\xi u - \partial_\xi f(u) =: \mathcal{F}(u). \tag{4.4.3}$$

There exists an equilibrium solution to (4.4.3) that satisfies $\phi(\xi) \to \phi_\pm$ exponentially fast as $\xi \to \pm\infty$ if and only if the speed c satisfies the Rankine–Hugoniot relation and the two entropy conditions hold (see Fig. 4.4).

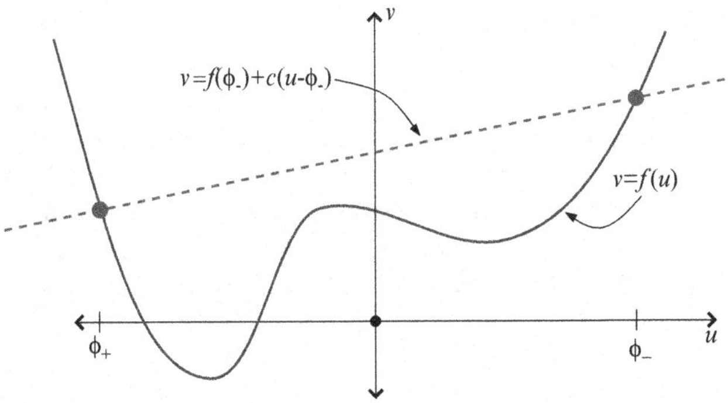

Fig. 4.4 The Oleinik entropy condition, and the Lax entropy condition are satisfied if the graph of f (solid *blue*) lies below the secant line (dashed *red*) connecting the points $(\phi_+, f(\phi_+))$ and $(\phi_-, f(\phi_-))$. The wave speed c is the slope of the secant line. (Color figure online.)

There are several complications to studying the spectral and asymptotic stability of the viscous shock. Indeed, the x-derivative acting on the flux function of (4.4.1) raises two chief obstacles to a direct application of the framework developed in Chapter 4.3. First, the nonlinearity is too strong; neither the semi-linearity condition (4.3.5) nor the nonlinear estimate (4.3.6) hold in the usual Sobolev spaces. As we will see, this obstacle can be overcome by studying an "integrated" form of (4.4.1) that "factors" an x-derivative out of the nonlinearity. The second obstacle is that the essential spectrum of the linearization about the viscous shock will generically touch the origin, so that the spectral hypothesis (4.3.2) fails. This issue is resolved by using an exponential weight to push the essential spectrum into the open left-half of the complex plane.

To begin, we fix values of $\phi_\pm$ for which the Oleinik and Lax entropy conditions hold, denote by ϕ the corresponding viscous shock, and determine c from the Rankine–Hugoniot condition, (4.4.2). In light of the conservation of mass property outlined in Exercise 4.4.2, we consider initial data u_0 for (4.4.3) which converges exponentially to $\phi_\pm$ as $\xi \to \pm\infty$. The conservation law system possesses translational symmetry, $T(\gamma)u(\xi,t) = u(\xi-\gamma,t)$, so we anticipate that the solution u will approach some translation $T(\gamma_\infty)\phi$ as $t \to \infty$. Our goal is to show that the remainder, $u - T(\gamma_\infty)\phi$ decays in an appropriate sense; however, due to the obstacles above we study this perturbation in an "integrated" framework, introducing

$$v(\xi,t) := \int_{-\infty}^{\xi} u(s,t) - \phi(s-\gamma_\infty)\,\mathrm{d}s.$$

Motivated by Exercise 4.4.2(b) we choose the limiting value γ_∞ so that $u_0 - T(\gamma_\infty)\phi$ has zero total mass. This choice forces the integrated perturbation $v(x,t)$ to decay to zero at both $\xi = \pm\infty$. If $v(\cdot,t) \to 0$ in $H^s(\mathbb{R})$, then $u(\cdot,t)$ will converge to $T(\gamma_\infty)\phi$ in $H^{s-1}(\mathbb{R})$.

Analyzing the integrated perturbation has two advantages. First, it will allow us to "factor" the ∂_ξ out of the nonlinearity and commute it with the semi-group, yielding a nonlinearity that satisfies (4.3.6). The second advantage is that the limiting value of the translational invariant is fixed a priori, so that we can linearize about the final translation of the viscous pulse, which avoids the need for the Lipschitz estimate (4.3.5). Indeed, the translational symmetry is broken in the integrated version of the conservation law. Without loss of generality, we assume $\gamma_\infty = 0$.

We substitute of $u = \partial_\xi v + \phi$ into (4.4.3), and factor out ∂_ξ, obtaining the evolution equation

$$v_t = \mathcal{L}v + \mathcal{N}(\partial_\xi v), \quad v(\xi,0) = v_0, \tag{4.4.4}$$

where the linearized operator and nonlinearity have the form

$$\mathcal{L} = \partial_\xi^2 + (c - f'(\phi))\partial_\xi, \quad \mathcal{N}(w) = -[f(\phi + w) - f(\phi) - f'(\phi)w].$$

Applying variation of parameters to (4.4.4) yields the integral formulation

$$v(t) = \mathrm{e}^{\mathcal{L}t} v_0 + \int_0^t \mathrm{e}^{\mathcal{L}(t-s)} \mathcal{N}(\partial_\xi v(s))\,\mathrm{d}s. \tag{4.4.5}$$

While the nonlinearity $\mathcal{N}$ is smooth, the composition $\mathcal{N}(\partial_\xi \cdot)$ does not satisfy (4.3.6) in typical Sobolev spaces. Indeed the optimal estimate is of the form

$$\|\mathcal{N}(\partial_\xi v)\|_{H_1} \leq C \|\partial_\xi v\|_{H_1}^2 \leq C \|v\|_{H_2}^2.$$

This obstacle is surmounted with a subtle step: differentiate (4.4.5) with respect to ξ, and introduce $z := \partial_\xi v = u - T(\gamma_\infty)\phi$. While this may seem like a return to the u variable, in fact the two changes of variable have commuted ∂_ξ and the semi-group, yielding the integral equation

$$z(t) = \partial_\xi \mathrm{e}^{\mathcal{L}t} v_0 + \int_0^t \partial_\xi \mathrm{e}^{\mathcal{L}(t-s)} \mathcal{N}(z(s))\,\mathrm{d}s. \tag{4.4.6}$$

Since f is smooth, the nonlinearity is now properly quadratic in $H^s(\mathbb{R})$ for $s \geq 1$. However, we need decay estimates on $\partial_\xi \mathrm{e}^{\mathcal{L}t}$. Since the linear evolution equation $\partial_t y = \mathcal{L}y$ is essentially the heat equation, it is not surprising that the $L^2(\mathbb{R})$-norm of the derivatives of y would decay at rates similar to those of $\|y\|_{L^2}$, particularly since we are interested in initial data y_0 with zero total mass. We do not prove these results here; rather, [**219**] demonstrates that for Sturm–Liouville operators such as $\mathcal{L}$, decay of y in $H^1(\mathbb{R})$ implies decay of $\partial_\xi^2 y$ in $L^2(\mathbb{R})$, i.e.,

$$\|\mathrm{e}^{\mathcal{L}t} y\|_{H^1} \leq C\mathrm{e}^{\delta t}\|y\|_{H^1} \quad \Rightarrow \quad \|\partial_\xi \mathrm{e}^{\mathcal{L}t} y\|_{H^1} \leq C \begin{cases} \mathrm{e}^{\delta t}/\sqrt{t}\,\|y\|_{H^1}, & t > 0 \\ \mathrm{e}^{\delta t}\|y\|_{H^2}, & t \geq 0 \end{cases} \tag{4.4.7}$$

(e.g., see Exercise 4.4.4). Since $\mathcal{L}$ is a second-order parabolic operator of the class discussed in Example 4.1.8, the resolvent estimates in $H^1(\mathbb{R})$ are guaranteed for any δ that bounds the spectrum of $\mathcal{L}$ to its left. However, as we will see shortly, in an unweighted norm δ cannot be taken to be negative.

To locate the essential spectrum of $\mathcal{L}$ we define

$$a_1(\xi) := c - f'(\phi(\xi)),$$

and note that the Lax entropy condition implies,

$$\lim_{\xi \to \pm\infty} a_1(\xi) = a_1^{\pm}, \quad a_1^- < 0 < a_1^+.$$

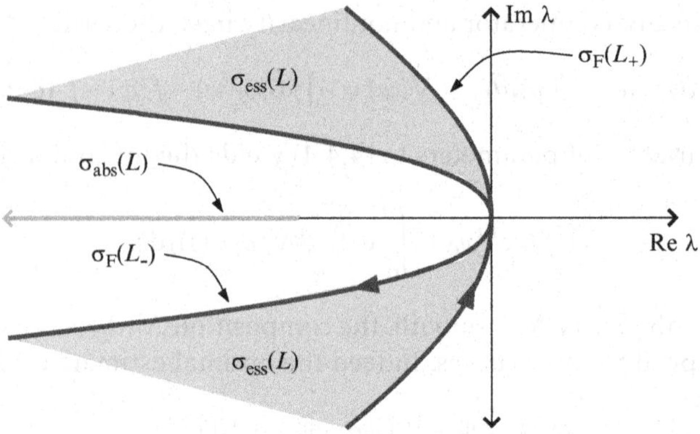

Fig. 4.5 The spectrum of the operator $\mathcal{L}$ given in (4.4.4) under the assumption that $a_1^+ < -a_1^-$. The arrows on each Fredholm border give the orientation of the curve. (Color figure online.)

The asymptotic operators take the form $\mathcal{L}_\pm = \partial_\xi^2 + a_1^\pm \partial_\xi$, and the Fredholm borders are given by the parabolic curves

$$\sigma_F(\mathcal{L}_\pm) = \{\lambda \in \mathbb{C} : \lambda = -k^2 + \mathrm{i}a_1^\pm k,\ k \in \mathbb{R}\}.$$

The two curves have opposite orientation; hence, the essential spectrum is between the Fredholm borders (see Fig. 4.5). Moreover these curves intersect at the origin, so that $\lambda = 0$ is embedded in the essential spectrum.

We must move the essential spectrum, both to better understand the point spectrum of $\mathcal{L}$, and to satisfy the spectral hypothesis (4.3.2). Fortunately, from Exercise 4.4.3 the absolute spectrum does not coincide with the essential spectrum. From the discussion of exponentially weighted spaces in Chapter 3.1.1.2 this suggests the possibility of shifting the essential spectrum into the left-half of the complex plane. However, care must be taken to verify semi-group decay and the quadratic estimate (4.3.6) in the new norm.

Pick $\eta \in (0, \min\{-a_1^-/2, a_1^+/2\})$, and consider the weight function $w(\xi) := \cosh(\eta x)$, which satisfies $w(\xi) \geq 1$ and grows like $\mathrm{e}^{\eta|\xi|}/2$ for large $|\xi|$. Multiplying both sides of the eigenvalue problem for $\mathcal{L}$ by the weight w yields

$$\mathcal{L}p = \lambda p \quad \Leftrightarrow \quad \mathcal{L}_w[wp] = \lambda[wp], \quad \mathcal{L}_w q := w\mathcal{L}(q/w). \tag{4.4.8}$$

The asymptotic operators $\mathcal{L}_{w,\pm}$ associated to $\mathcal{L}_w$ are given by

$$\mathcal{L}_{w,\pm} = (\partial_\xi \mp \eta)^2 + a_1^\pm(\partial_\xi \mp \eta)$$

and for $\mathcal{L}_w$ the Fredholm borders are given by the parabolic curves

$$\sigma_F(\mathcal{L}_{w,\pm}) = \{\lambda \in \mathbb{C} : \lambda = \eta(\eta \mp a_1^\pm) - k^2 + \mathrm{i}(a_1^\pm \mp 2\eta)k,\ k \in \mathbb{R}\}.$$

The two curves still have opposite orientation, and the essential spectrum again lies between the Fredholm borders. More importantly, the restrictions on η guarantee that $\sigma_0 := -\max\{\eta(\eta \mp a_1^{\pm})\} > 0$ and that the essential spectrum is contained in $\{\lambda \in \mathbb{C} : \mathrm{Re}\,\lambda < -\sigma_0\}$ (see Fig. 4.6).

We wish to estimate the solution z in the weighted space $H_w^2(\mathbb{R})$ with norm $\|f\|_{H_w^2}^2 = \|wf\|_{H^2}^2$. Returning to (4.4.6), we multiply both sides by the weight w to find,

$$wz(t) = w\partial_\xi \mathrm{e}^{\mathcal{L}t} v_0 + \int_0^t w\partial_\xi \mathrm{e}^{\mathcal{L}(t-s)} \mathcal{N}(z(s))\,\mathrm{d}s. \tag{4.4.9}$$

Since w is independent of time, the uniqueness of solutions of the associated linear PDEs implies the relation,

$$w\mathcal{L}f = \mathcal{L}_w[wf] \quad \Rightarrow \quad w\mathrm{e}^{\mathcal{L}t} f = \mathrm{e}^{\mathcal{L}_w t}[wf].$$

With this identity, we may rewrite the derivative of the semi-group as

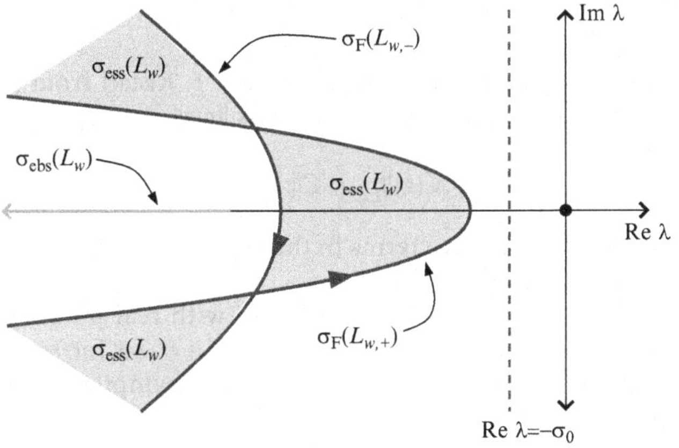

Fig. 4.6 The spectrum for the operator $\mathcal{L}_w$ given in (4.4.8) under the assumption that $a_1^+ < -a_1^-$. The arrows on each Fredholm border give the orientation of the curve. The vertical dashed (*red*) line is the curve $\mathrm{Re}\,\lambda = -\sigma_0$. Compare with Fig. 4.5. (Color figure online.)

$$w\partial_\xi \mathrm{e}^{\mathcal{L}t} f = \partial_\xi(w\mathrm{e}^{\mathcal{L}t} f) - (\partial_\xi w)\mathrm{e}^{\mathcal{L}t} f = \partial_\xi(\mathrm{e}^{\mathcal{L}_w t}[wf]) - (\partial_\xi w)\mathrm{e}^{\mathcal{L}t} f.$$

The function $\partial_\xi w$ is also hyperbolic, with the same growth rates at infinity as w. To simplify the presentation we equate the two functions, i.e.,

$$(\partial_\xi w)\mathrm{e}^{\mathcal{L}t} f \sim w\mathrm{e}^{\mathcal{L}t} f = \mathrm{e}^{\mathcal{L}_w t}[wf].$$

Modulo this approximation, we obtain the bounds

$$\|w\partial_\xi e^{\mathcal{L}t} f\|_{H^1} \le C\left(\|\partial_\xi(e^{\mathcal{L}_w t}[wf])\|_{H^1} + \|e^{\mathcal{L}_w t}[wf]\|_{H^1}\right) \le C\left\|e^{\mathcal{L}_w t}[wf]\right\|_{H^2}. \tag{4.4.10}$$

The operator $\mathcal{L}_w$ is second-order and parabolic-type with no essential spectrum in the set $\{\mathrm{Re}\,\lambda > -\sigma_0\}$. We will verify below that $\mathcal{L}_w$ also has no point spectrum on this set. Consequently, from Example 4.1.8 and (4.1.9) we obtain decay estimates on $e^{\mathcal{L}_w t}$, which from (4.4.7) and (4.4.10) imply

$$\|w\partial_\xi e^{\mathcal{L}t} f\|_{H^1} \le C \begin{cases} e^{-\sigma t}/\sqrt{t}\|wf\|_{H^1}, & t > 0 \\ e^{-\sigma t}\|wf\|_{H^2}, & t \ge 0. \end{cases}$$

These are the exponential semi-group bounds we require.

We now verify the quadratic estimate (4.3.6) on the nonlinearity $\mathcal{N}$ in the weighted norm $\|\cdot\|_{H^1_w}$. Fortunately, the weighted norm behaves well with respect to nonlinearity. Indeed, for z small we have

$$\|w\mathcal{N}(z)\|_{L^2}^2 = \int_{\mathbb{R}} (w(\xi)\mathcal{N}(z(\xi)))^2\, d\xi \le C \int_{\mathbb{R}} (wz)^2 z^2\, d\xi \le C\|wz\|_{L^2}^2 \|wz\|_{L^\infty}^2.$$

The last equality follows from the fact that $w \ge 1$. Recall from (2.2.7) that $\|z\|_\infty \le C\|z\|_{H_1}$. Following that argument yields $\|wz\|_{L^\infty} \le C\|z\|_{H^1_w}$, so that we can conclude

$$\|\mathcal{N}(z)\|_{L^2_w} \le C\|z\|_{H^1_w}^2.$$

The treatment of the derivative terms in the $H^1_w(\mathbb{R})$-norm is similar, and the quadratic estimate follows.

The operators $\mathcal{L}$ and $\mathcal{L}_w$ are both Sturmian with real point spectra. We first show that $\sigma_{\mathrm{pt}}(\mathcal{L}) \subset \sigma_{\mathrm{pt}}(\mathcal{L}_w)$. Suppose that $\mathcal{L}u = \lambda u$ for some λ to the right of $\sigma_{\mathrm{abs}}(\mathcal{L})$. The eigenfunctions of $\mathcal{L}$ have the asymptotic behavior (see Chapter 9 for the details),

$$u(\xi) \sim \begin{cases} \mathcal{O}(e^{r_- \xi}), & \xi \to -\infty \\ \mathcal{O}(e^{r_+ \xi}), & \xi \to +\infty \end{cases}, \quad r_\pm := \frac{1}{2}\left(-a_1^\pm \mp \sqrt{(a_1^\pm)^2 + 4\lambda}\right).$$

It is significant that $r_- > -a_1^-/2$, and $r_+ < -a_1^+/2$ for all real λ to the right of the edge of $\sigma_{\mathrm{abs}}(\mathcal{L})$. It is this fact that motivated the choice of the parameter η in the weight, w. It follows that $\|u\|_{L^2_w} < \infty$, and in particular $wu \in L^2(\mathbb{R})$ is an eigenfunction of $\mathcal{L}_w$ corresponding to eigenvalue λ, which proves the claim.

The remaining task is to show that $\mathcal{L}_w$ has no non-negative real point spectrum. In principle, this is equivalent to showing that $\mathcal{L}$ has no non-negative real point spectrum, however the essential spectrum of $\mathcal{L}$ touches the origin. It is instructive to investigate the impact this has on the kernel of $\mathcal{L}$. We first observe that ϕ is an equilibrium of (4.4.3) from which we deduce that $\mathcal{L}\phi = 0$. Since $\phi_+ < \phi_-$, it is clearly the case that $\phi \notin L^2(\mathbb{R})$. This is contrary to the expectation that $T'\phi = \partial_\xi \phi \in \ker(\mathcal{L})$, but is consistent with the fact that $\mathcal{L}$ arises from the integrated version of the conservation law, for which the translational symmetry is broken. We also have that $\mathcal{L}(1) = 0$, and since $\mathcal{L}$ is a second-order differential operator, the solution space of $\mathcal{L} = 0$ is spanned by $\{\phi, 1\}$. In particular, since $\phi_- \neq \phi_+$ no linear combination of the solutions to $\mathcal{L}f = 0$ decay at both $\xi = \pm\infty$. Thus, the operator $\mathcal{L}$ has no kernel in $H^2(\mathbb{R})$, and we cannot apply the Sturmian theory to the solutions u of $\mathcal{L}u = 0$.

The most direct method to obtain information on the point spectrum of $\mathcal{L}_w$ is to return to the non-integrated operator,

$$\mathcal{L}_0\psi := \partial_\xi^2 \psi + c\partial_\xi \psi - \partial_\xi \left(f'(\phi)\psi\right),$$

obtained by linearizing (4.4.3) about the viscous shock ϕ. It is straightforward to verify that $\lambda \in \sigma_{\rm pt}(\mathcal{L})$ with associated eigenfunction $v \in H^2(\mathbb{R})$ implies that $\lambda \in \sigma_{\rm pt}(\mathcal{L}_0)$ with associated eigenfunction $\psi = \partial_\xi v$; hence, $\sigma_{\rm pt}(\mathcal{L}) \subset \sigma_{\rm pt}(\mathcal{L}_0)$. At $\lambda = 0$ we see that ϕ, which solves $\mathcal{L}\phi = 0$, maps onto $\partial_\xi \phi$, i.e., $\mathcal{L}_0(\partial_\xi \phi) = 0$. While $\partial_\xi \phi \in L^2(\mathbb{R})$, $\lambda = 0$ is also in the essential spectrum of $\mathcal{L}_0$, which touches the origin. However, unlike $\mathcal{L}$, the operator $\mathcal{L}_0$ has an $H^2(\mathbb{R})$ eigenvalue embedded in the essential spectrum at $\lambda = 0$. To shift the essential spectrum of $\mathcal{L}_0$ we reintroduce the weight w, and the associated weighted operator

$$\mathcal{L}_{0,w} q := w\mathcal{L}_0(q/w),$$

(see Fig. 4.6). Arguing as before, we have that $\sigma_{\rm pt}(\mathcal{L}_0) \subset \sigma_{\rm pt}(\mathcal{L}_{0,w})$. Thus, if $\mathcal{L}_{0,w}$ has no real positive eigenvalues, then neither will $\mathcal{L}_0$, which in turn implies that $\mathcal{L}$, and finally $\mathcal{L}_w$, have no positive eigenvalues. Since the essential spectrum of $\mathcal{L}_{0,w}$ coincides with that of $\mathcal{L}_w$, we see that $0 \notin \sigma_{\rm ess}(\mathcal{L}_{0,w})$. Moreover,

$$\mathcal{L}_0(\partial_\xi \phi) = 0 \quad \Rightarrow \quad \mathcal{L}_{0,w}(w\partial_\xi \phi) = 0,$$

so that $\{0\} \subset \sigma_{\rm pt}(\mathcal{L}_{0,w})$ is an isolated eigenvalue. Since $w\partial_\xi \phi$ has no zeros, we deduce that $\lambda = 0$ is the ground-state eigenvalue of the Sturmian operator $\mathcal{L}_{w,0}$, and any other point spectrum of $\mathcal{L}_{w,0}$ would be negative. We conclude that the only possible non-negative eigenvalue of $\mathcal{L}w$ is $\lambda = 0$, but if $\mathcal{L}wu = 0$, then $\mathcal{L}(u/w) = 0$, which implies that $u \in \operatorname{span}\{w\phi, w\}$ which does not reside in $L^2(\mathbb{R})$. Hence, the point spectrum of $\mathcal{L}_w$ is strictly negative.

We are now ready to complete the asymptotic stability proof. We invoke the spirit, if not the language, of Theorem 4.3.5 (also see Exercise 4.4.5 with $Y = H_w^1(\mathbb{R})$ and $Z = H_w^2(\mathbb{R})$). The system (4.4.6) has no symmetries and the conjugated operator $\mathcal{L}_w$ associated with the system (4.4.9) has no kernel; moreover, it satisfies the spectral assumption (4.3.2). The nonlinearity is quadratic in $\|\cdot\|_{H_w^2}$. As there are no symmetries, we do not require the semilinearity assumption (4.3.5). We deduce that if $\|wv_0\|_{H^2}$ is sufficiently small, then for some $\sigma > 0$ we have the fully nonlinear estimate on the solution z of (4.4.9),

$$\|wz(t)\|_{H^1} \leq Ce^{-\sigma t}\|wv_0\|_{H^2}, \quad t \geq 0.$$

From the definition of z and v we have

$$wz(t) = w[u(t) - T(\gamma_\infty)\phi],$$

and we can rewrite the asymptotic stability result as

$$\|w[u(t) - T(\gamma_\infty)\phi]\|_{H^1} \leq Ce^{-\sigma t}\left\| w\int_{-\infty}^{\xi} [u_0(s) - T(\gamma_\infty)\phi(s)]\,ds \right\|_{H^2}.$$

Remark 4.4.1. This asymptotic orbital stability result of an exponentially weighted $L^\infty(\mathbb{R})$-space was originally obtained by Sattinger [**259**]. The result was extended to spaces with polynomial weights in Kawashima and Matsumura [**168**]. Unlike exponential weights, polynomial weights do not shift the essential spectrum into the open left-half plane and the relaxation rate to equilibria is only polynomial in time. The authors used an energy method to prove the result, which required a convexity assumption on the nonlinearity. In Jones et al. [**139**] the convexity assumption was removed, replacing the energy methods with semi-group estimates based upon a careful analysis of the resolvent near the origin. A novel reformulation of the problem in terms of pointwise estimates on the Green's function associated with the resolvent led to a new proof in Howard [**117**]. The Green's function approach has since been substantially generalized by Howard and Zumbrun [**118**], Zumbrun [**290**, **292**], Zumbrun and Howard [**295**].

Exercises

Exercise 4.4.1. Use a phase plane analysis to show that the Oleinik and Lax entropy conditions guarantee the existence of a monotonically decreasing viscous profile $\phi(\xi)$ with limit states $\phi_\pm$ which is a stationary solution of (4.4.3) when the speed c is given by the Rankine–Hugoniot condition (4.4.2).

Exercise 4.4.2. Consider a solution u of (4.4.3) with initial data u_0 satisfying

$$\lim_{\xi\to\pm\infty} u_0(\xi) = \phi_\pm, \quad \lim_{\xi\to\pm\infty} \partial_\xi u_0(\xi) = 0,$$

with the approach being exponentially fast.

(a) Let $\phi_\pm$ satisfy the Oleinik and Lax entropy conditions with ϕ denoting the associated viscous shock. Assume that the solution $u(\cdot, t)$ of (4.4.3) enjoys the same limiting behavior as u_0, and the speed c is given by the Rankine–Hugoniot condition. Show that the mass of u relative to ϕ is conserved, i.e.,

$$\int_{\mathbb{R}} (u(s,t) - \phi(s))\,\mathrm{d}s = \int_{\mathbb{R}} (u_0(s) - \phi(s))\,\mathrm{d}s, \quad t > 0.$$

(b) Introduce the function

$$g(\gamma) := \int_{\mathbb{R}} u_0(s) - \phi(s-\gamma)\,\mathrm{d}s.$$

Show that there is a unique γ_∞ such that $g(\gamma_\infty) = 0$.

Exercise 4.4.3. Show that the absolute spectrum for the operator $\mathcal{L}$ of (4.4.4) is given by the line segment

$$\sigma_{\text{abs}}(\mathcal{L}) = \{\lambda \in \mathbb{C} : \lambda = -\min\left\{(a_1^-)^2/4, (a_1^+)^2/4\right\} - k^2,\ k \in \mathbb{R}\}.$$

Exercise 4.4.4. Consider the operator $\mathcal{L}$ given by

$$\mathcal{L} := \partial_x^2 - \sigma, \quad \sigma > 0.$$

Using the Fourier transform, show that the semi-group generated by $\mathcal{L}$ satisfies the estimates

$$\|\mathrm{e}^{\mathcal{L}t}u\|_{H_1} \le C\mathrm{e}^{-\sigma t}\|u\|_{H^1}, \quad \|\partial_x \mathrm{e}^{\mathcal{L}t}u\|_{H^1} \le C\begin{cases} \mathrm{e}^{-\sigma t}/\sqrt{t}\,\|u\|_{H^1}, & t > 0 \\ \mathrm{e}^{-\sigma t}\|u\|_{H^2}, & t \ge 0. \end{cases}$$

Exercise 4.4.5. Consider the integral equation

$$v(t) = \mathrm{e}^{\mathcal{L}t}v_0 + \int_0^t \mathrm{e}^{\mathcal{L}(t-s)}\mathcal{N}(v(s))\,\mathrm{d}s.$$

Suppose that for some $\sigma > 0$ the semi-group satisfies the estimates

$$\|\mathrm{e}^{\mathcal{L}t}f\|_Y \le C\begin{cases} \mathrm{e}^{-\sigma t}/\sqrt{t}\,\|f\|_Y, & t > 0 \\ \mathrm{e}^{-\sigma t}\,\|f\|_Z, & t \ge 0, \end{cases}$$

where Y is continuously embedded in Z and the constant C is independent of time. Further suppose that $\|\mathcal{N}(v)\|_Y \le C\|v\|_Y^2$. Show that if $\|v_0\|_Z$ is sufficiently small, then $\|v(t)\|_Y \le C\mathrm{e}^{-\sigma t/2}\|v_0\|_Z$.

4.5 Example: Nonlinear Schrödinger-Type Equations

The parametrically forced nonlinear Schrödinger equation (PNLS) takes the form,

$$\mathrm{i}\partial_t u + \partial_x^2 u - u + |u|^2 u + \mathrm{i}\left(\mu u - \mathrm{e}^{-\mathrm{i}\theta}\overline{u}\right) = 0, \tag{4.5.1}$$

where $u(x,t) \in \mathbb{C}$, $(x,t) \in \mathbb{R} \times \mathbb{R}^+$, $0 < \mu < 1$, and $-\pi/2 < \theta < \pi/2$. For this problem the linearized operator is not second-order parabolic, and the central obstacle to applying Theorem 4.3.5 is to verify the resolvent estimate Hypothesis 4.3.1(c). Our approach follows [152]. Writing $u = u_\mathrm{r} + \mathrm{i}u_\mathrm{i}$, and forming the vector $\boldsymbol{U} = (u_\mathrm{r}, u_\mathrm{i})^\mathrm{T}$ we rewrite the PNLS as a vector system

$$\partial_t \boldsymbol{U} = \begin{pmatrix} 0 & -1 \\ 1 & 0 \end{pmatrix} \left(\partial_x^2 \boldsymbol{U} + |\boldsymbol{U}|^2 \boldsymbol{U} + \begin{pmatrix} -(1+\sin\theta) & -(\mu+\cos\theta) \\ \mu - \cos\theta & -(1-\sin\theta) \end{pmatrix} \boldsymbol{U} \right). \tag{4.5.2}$$

Here we will study the particular case $\mu = \cos\theta$. Note that for each value of μ there are two possible values of θ.

First consider the existence problem. The eigenvalues of the matrix multiplying $\boldsymbol{U}$ take the form $\lambda_\pm = -(1 \pm \sin\theta)$, with associated eigenvectors $\boldsymbol{V}_+ = (1,0)^\mathrm{T}$ and $\boldsymbol{V}_- = (\cos\theta, -\sin\theta)^\mathrm{T}$. The eigenvectors correspond to a complex phase $\gamma_+ = 0$ or $\gamma_- = -\theta$. The system supports standing solutions of constant complex phase,

$$\boldsymbol{U}_{\theta,\pm}(x) = \phi_{\theta,\pm}(x) \boldsymbol{V}_\pm,$$

where $\phi_{\theta,\pm}$ solves

$$\partial_x^2 \phi_{\theta,\pm} - A_\pm^2 \phi_{\theta,\pm} + \phi_{\theta,\pm}^3 = 0, \quad A_\pm = \sqrt{1 \pm \sin\theta}. \tag{4.5.3}$$

For fixed values of θ and $\mu < 1$ there are two distinct solutions: a narrower one with larger amplitude, $\boldsymbol{U}_+$, and a broader one with smaller amplitude, $\boldsymbol{U}_-$. The two pulses coincide when $\mu = 1$ $(\theta = 0)$.

We consider the spectral stability of the zero-phase solution, $\boldsymbol{U}_+$. To simplify notation, we drop the $\pm$ subscript, denoting the solution as

$$\boldsymbol{U}_\theta(x) = \phi_\theta(x) \begin{pmatrix} 1 \\ 0 \end{pmatrix}, \quad \phi_\theta(x) = \sqrt{2} A_+ \operatorname{sech}(A_+ x).$$

Setting $\boldsymbol{U} = \boldsymbol{U}_\theta + \boldsymbol{V}$, and keeping only those terms linear in $\boldsymbol{V}$, we arrive at the linearized problem

$$\partial_t \boldsymbol{V} = \mathcal{L} \boldsymbol{V}, \quad \mathcal{L} = \begin{pmatrix} 0 & \mathcal{L}_2 \\ -\mathcal{L}_1 & -2\mu \end{pmatrix}, \tag{4.5.4}$$

where

$$\mathcal{L}_1 = -\partial_x^2 + A_+^2 - 3\phi_\theta^2, \quad \mathcal{L}_2 = -\partial_x^2 + A_-^2 - \phi_\theta^2.$$

To determine the spectrum of $\mathcal{L}$, we first find the essential spectrum by considering the asymptotic operator

$$\mathcal{L}_\infty = \begin{pmatrix} 0 & -\partial_x^2 + A_-^2 \\ \partial_x^2 - A_+^2 & -2\mu \end{pmatrix}.$$

This operator has the dispersion relation

$$\lambda^2 + 2\mu\lambda + (k^2 + A_+^2)(k^2 + A_-^2) = 0,$$

from which we see that the essential spectrum (Fredholm border) is the set

$$\sigma_{\text{ess}}(\mathcal{L}) = \left\{\lambda \in \mathbb{C} : \lambda = -\mu + \mathrm{i}k\sqrt{2+k^2},\ k \in \mathbb{R}\right\},$$

i.e., the vertical line $\operatorname{Re}\lambda = -\mu$ (see Fig. 4.7).

Now consider the point spectrum. Solutions to (4.5.2) are invariant under the action of spatial translation, i.e., $T_1(\gamma_1)\boldsymbol{U}(x,t) = \boldsymbol{U}(x+\gamma_1,t)$, so that $\lambda = 0$ is an eigenvalue and moreover

$$\mathcal{L}(\partial_x \boldsymbol{U}_\theta) = \boldsymbol{0} \quad \Rightarrow \quad \mathcal{L}_1(\partial_x \phi_\theta) = 0.$$

Since $\mathcal{L}_1$ is the negative of a Sturmian operator we know that $\ker(\mathcal{L}_1) = \operatorname{span}\{\partial_x\phi_\theta\}$. Since $\partial_x\phi_\theta$ has a single zero, we deduce that $\mathcal{L}_1$ has a single negative eigenvalue, its ground state, while its essential spectrum is given by

$$\sigma_{\text{ess}}(\mathcal{L}_1) = \{\lambda \in \mathbb{C} : \lambda = A_+^2 + k^2,\ k \in \mathbb{R}\}.$$

There can be an additional element in the kernel of $\mathcal{L}$ only if there is a nontrivial solution to $\mathcal{L}_2 v_2 = 0$. However, the operator $\mathcal{L}_2$ is amenable to analysis by special functions; in particular, it can be shown that

$$\sigma_{\text{pt}}(\mathcal{L}_2) = \{-2\sin\theta\}, \qquad \sigma_{\text{ess}}(\mathcal{L}_2) = \{\lambda \in \mathbb{C} : \lambda = A_-^2 + k^2,\ k \in \mathbb{R}\},$$

so that if $\theta \neq 0$ the operator $\mathcal{L}_2$ is invertible (see Chapter 9.3.2 for details). Thus, we can conclude that

$$\ker(\mathcal{L}) = \operatorname{span}\{\partial_x \boldsymbol{U}_\theta\}.$$

Regarding $\operatorname{gker}(\mathcal{L})$, the operator $\mathcal{L}$ can have a Jordan chain only if there is a $\boldsymbol{V}$ that solves $\mathcal{L}\boldsymbol{V} = \partial_x \boldsymbol{U}_\theta$. Such a $\boldsymbol{V} = (v_1, v_2)^{\mathrm{T}}$ would take the form

$$\mathcal{L}_1 v_1 = -2\mu v_2, \quad v_2 = \mathcal{L}_2^{-1}(\partial_x\phi_\theta).$$

By the Fredholm alternative v_1 exists only if $\langle v_2, \partial_x\phi_\theta\rangle = 0$, which substituting for v_2 yields $\langle \mathcal{L}_2^{-1}(\partial_x\phi_\theta), \partial_x\phi_\theta\rangle = 0$. However, the operator $\mathcal{L}_2$ preserves parity, mapping odd functions to odd functions and even functions to even functions. Furthermore, the spectrum of $\mathcal{L}_2$ is positive, except possibly for its ground-state eigenvalue, which has an even eigenfunction; thus, $\mathcal{L}_2$, and hence $\mathcal{L}_2^{-1}$, is positive on odd functions. Since $\partial_x\phi_\theta$ is odd about $x = 0$, we have $\langle \mathcal{L}_2^{-1}(\partial_x\phi_\theta), \partial_x\phi_\theta\rangle > 0$. We deduce that $\lambda = 0$ is a simple eigenvalue for $0 < \mu < 1$.

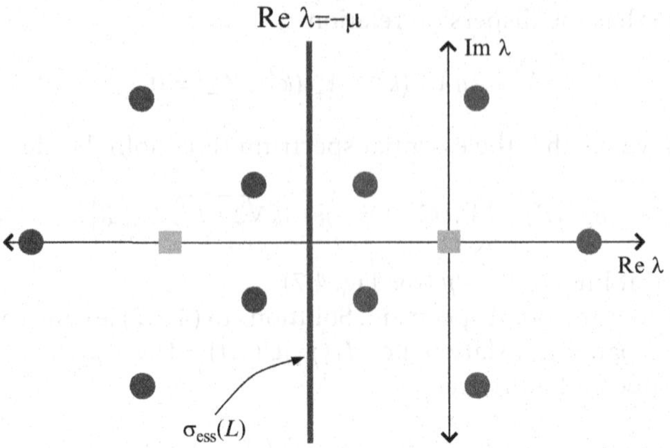

Fig. 4.7 The spectrum for the operator $\mathcal{L}$ given in (4.5.4). The essential spectrum is definitive, as are the eigenvalues at $\lambda = 0, -2\mu$ (squares). The other eigenvalues (circles) are shown only to demonstrate the symmetries of the point spectrum with respect to the essential spectrum and the real axis. (Color figure online.)

The spectrum of $\mathcal{L}$ has symmetries. Indeed, rewriting the eigenvalue problem as

$$\mathcal{L}_1\mathcal{L}_2 v_2 = -z v_2, \quad z := \lambda^2 + 2\mu\lambda, \tag{4.5.5}$$

we see that z is an eigenvalue for (4.5.5) if and only if there are a pair of eigenvalues

$$\lambda = \lambda_\pm := -\mu \pm \sqrt{\mu^2 + z} \tag{4.5.6}$$

of (4.5.4). This pair of eigenvalues is symmetric with respect to the essential spectrum. In particular the simple eigenvalue of $\mathcal{L}$ at $\lambda = 0$, has a mate who is a simple eigenvalue at $\lambda = -2\mu$. In addition, $\lambda \in \sigma_{pt}(\mathcal{L})$ if and only if $\overline{\lambda} \in \sigma_{pt}(\mathcal{L})$, so that the point spectrum of $\mathcal{L}$ is symmetric with respect to the real axis. Figure 4.7 depicts a possible realization of $\sigma(\mathcal{L})$.

A complete investigation of $\sigma_{pt}(\mathcal{L})$ is undertaken in Exercises 6.1.5 and 7.1.4. In particular it is determined that if $\sin\theta < 0$, then $\mathcal{L}$ will have a positive real eigenvalue, while if $\sin\theta > 0$ and θ is not too large, then $\sigma_{pt}(\mathcal{L})\backslash\{0\}$ lies strictly in the left-half complex plane. In particular, $\boldsymbol{U}_\theta$ is spectrally unstable if $\sin\theta < 0$ and spectrally stable if $\sin\theta > 0$ and $\theta > 0$ is not too large.

The major thrust of this example is to verify points (c) and (d) of Hypothesis 4.3.1. Letting $\mathcal{F}(\boldsymbol{U})$ denote the right-hand side of (4.5.2), the Lipschitz estimate (4.3.5) of $\nabla_U\mathcal{F}$ over the space $H^1(\mathbb{R})$ is apparent, since the only nonzero terms in the difference $\nabla_U\mathcal{F}(\boldsymbol{U}) - \nabla_U\mathcal{F}(\boldsymbol{V})$ are multiplier operators arising from the cubic term in $\mathcal{F}$. Using the algebraic nature of the $H^1(\mathbb{R})$ norm, it is easy to obtain the bound

$$\begin{aligned}\|(\nabla_U\mathcal{F}(\boldsymbol{U}) - \nabla_U\mathcal{F}(\boldsymbol{V}))\boldsymbol{W}\|_{H^1} &\le C\|\nabla_U\mathcal{F}(\boldsymbol{U}) - \nabla_U\mathcal{F}(\boldsymbol{V})\|_{H^1}\|\boldsymbol{W}\|_{H^1}\\ &\le C\|\boldsymbol{U} - \boldsymbol{V}\|_{H^1}\|\boldsymbol{W}\|_{H^1},\end{aligned}$$

where $C > 0$ may depend upon $\|U\|_{H^1}$ and $\|V\|_{H^1}$. Motivated by the PNLS equation, we establish the following resolvent estimates on a slight generalization of $\mathcal{L}$.

Theorem 4.5.1 (NLS-type resolvent estimates). *Consider a linear operator of the form*

$$\mathcal{L} = \begin{pmatrix} 0 & \mathcal{L}_2 \\ -\mathcal{L}_1 & -b \end{pmatrix}, \tag{4.5.7}$$

where $b \in \mathbb{R}^+$, and for $k = 1, 2$ each of the suboperators is of the form

$$\mathcal{L}_k = -\partial_x^2 + V_k(x),$$

with the smooth potential $V_k(x)$ tending exponentially to constant limits $V_k^{\pm}$ as $x \to \pm\infty$. If $\sigma_{\text{ess}}(\mathcal{L}) \subset \{\lambda \in \mathbb{C} : \operatorname{Re}\lambda \leq -\sigma\}$ for some $\sigma > 0$, then the resolvent estimates (4.3.4) hold in $Y = H^1(\mathbb{R}) \times H^1(\mathbb{R})$ for $\mathcal{L}$, where Π_+ is the spectral projection complementary to the spectral set $S_+ := \{\lambda \in \sigma(\mathcal{L}) : \operatorname{Re}\lambda > -\sigma\}$; see (4.1.8). Moreover, the semi-group estimates (4.1.9) hold.

It is straightforward to check that Hypothesis 4.3.1 holds on the space Y. Consequently, we invoke Theorem 4.3.5 to establish the following.

Corollary 4.5.2. *If the zero-phase wave for (4.5.2) is spectrally stable, then it is asymptotically stable.*

Proof. (of Theorem 4.5.1) Denote by E_+ the spectral subspace associated to S_+, and its orthogonal complement by $(E_+^{\text{a}})^\perp$. Since the spectrum of $\mathcal{L}$ is related to the spectrum of the fourth-order differential operator $\mathcal{L}_1\mathcal{L}_2$ [see (4.5.5)], the Evans function theory of Chapter 10 implies that E_+ is finite-dimensional. Moreover, both E_+ and E_+^{a} are invariant under $\mathcal{L}$. In particular, $\mathcal{L}\Pi_+$ has a kernel given by E_+, while $\mathcal{L}|_{(E_+^{\text{a}})^\perp} : (E_+^{\text{a}})^\perp \mapsto (E_+^{\text{a}})^\perp$ has no kernel. As a consequence, the equation $\mathcal{L}\boldsymbol{f} = \boldsymbol{g}$ is solvable for all $\boldsymbol{g} \in (E_+^{\text{a}})^\perp$ and the induced norm of the resolvent $\|(\mathcal{L} - \lambda \boldsymbol{I}_2)^{-1}\Pi_+\|_{\mathcal{B}(H^1)}$ is finite for $\operatorname{Re}\lambda > -\sigma$. Indeed, the resolvent is an analytic function of λ on that set. The goal is to show a uniform bound as $|\lambda| \to \infty$ within S_+. We sketch the main ideas of the proof: details may be found in [**152**, Section 7.1].

We assume that $\boldsymbol{g} = (g_1, g_2)^{\mathrm{T}} \in (E_+^{\text{a}})^\perp$ and drop Π_+ from the resolvent expression. We first invert the matrix element of $R(\lambda, \mathcal{L}) = (\mathcal{L} - \lambda \boldsymbol{I}_2)^{-1}$. If $(\mathcal{L} - \lambda \boldsymbol{I}_2)\boldsymbol{f} = \boldsymbol{g}$, i.e.,

$$\begin{pmatrix} -\lambda & \mathcal{L}_2 \\ -\mathcal{L}_1 & -(b+\lambda) \end{pmatrix} \begin{pmatrix} f_1 \\ f_2 \end{pmatrix} = \begin{pmatrix} g_1 \\ g_2 \end{pmatrix},$$

then $\boldsymbol{f} = R(\lambda, \mathcal{L})\boldsymbol{g}$, and solving for $\boldsymbol{f}$ via row reduction we obtain an expression for the resolvent,

$$R(\lambda; \mathcal{L}) = \begin{pmatrix} -(b+\lambda)(\mathcal{L}_2\mathcal{L}_1 + z)^{-1} & -\mathcal{L}_2(\mathcal{L}_2\mathcal{L}_1 + z)^{-1} \\ \mathcal{L}_1(\mathcal{L}_1\mathcal{L}_2 + z)^{-1} & -\lambda(\mathcal{L}_1\mathcal{L}_2 + z)^{-1} \end{pmatrix}, \quad z := \lambda^2 + b\lambda.$$

Note that for $|\lambda|$ sufficiently large $|z| = \mathcal{O}(|\lambda|^2)$. We will have an L^2-bound on the resolvent if we can show

$$\|(\mathcal{L}_1\mathcal{L}_2 + z)^{-1}\|_{B(L^2)} \le \frac{C}{|\lambda|}, \quad \|\mathcal{L}_1(\mathcal{L}_1\mathcal{L}_2 + z)^{-1}\|_{B(L^2)} \le C, \tag{4.5.8}$$

for $\mathrm{Re}\,\lambda \ge 0$ and $|\lambda| \gg 1$: the estimates for the other terms are similar.

For self-adjoint operators, estimates of the form (4.5.8) are relatively straightforward to obtain since the resolvent decays at a rate proportional to the distance to the spectrum. This argument can be extended to non–self-adjoint operators which can be written as a sum of a self-adjoint part and a sufficiently lower-order perturbation. In the case at hand, the key observation is to write $\mathcal{L}_1\mathcal{L}_2$ in the form

$$\mathcal{L}_1\mathcal{L}_2 = \mathcal{S} + \mathcal{T} \tag{4.5.9}$$

where

$$\mathcal{S}f = \partial_x^4 f - V_2\partial_x^2 f - \partial_x^2(V_2 f) + (V_1V_2 + \eta^2)f, \quad \mathcal{T}f = 2(\partial_x V_2 - \partial_x V_1)\cdot\partial_x f - \eta^2 f.$$

The operator $\mathcal{S}$ is fourth-order and self-adjoint in $L^2(R)$. The important point is that the non–self-adjoint component, $\mathcal{T}$, is merely first-order. The spectrum of $\mathcal{S}$ is real, and the parameter $\eta > 0$ is chosen sufficiently large that the spectrum is positive, i.e.,

$$\sigma(\mathcal{S}) \subset \{\lambda \in \mathbb{C} : \mathrm{Re}\,\lambda > \sigma\}.$$

Before proceeding to the full problem, we illustrate the ideas for the constant-coefficient (and self-adjoint) operator $\partial_x^\ell(-\partial_x^4 - z)^{-1}$ for $\ell \in [0,2]$. We estimate the resolvent using the Fourier transform, which, via the multiplier formula (2.2.2) and the Plancherel identity (2.2.1), yields

$$\|\partial_x^\ell(-\partial_x^4 - z)^{-1} f\|_{H^s}^2 = \int_{\mathbb{R}} |k|^{2\ell}\frac{(1+|k|^{2s})|\widehat{f}(k)|^2}{|k^4+z|^2}\,dk \le \left(\sup_{k\in\mathbb{R}}\frac{|k|^{2\ell}}{|k^4+z|^2}\right)\|f\|_{H^s}^2.$$

Recalling that $z = \lambda^2 + b\lambda$, we particularly wish to bound the resolvent for λ of the form $\lambda = -\sigma + it$, for $|t| \gg 1$. Breaking $z = z_1 + iz_2$ into real and complex parts, we see that for $|\lambda| \gg 1$ we have $z_1 \sim -t^2 \sim -|\lambda|^2$ and $|z_2| \sim |\lambda|$. In this case the supremum scales like

$$\left(\sup_{k\in\mathbb{R}}\frac{|k|^{2\ell}}{|k^4+z|^2}\right) \sim |z_1|^{\ell/2}|z_2|^{-2} \sim |\lambda|^{\ell-2},$$

from which we deduce the large $|\lambda|$ resolvent estimate

$$\|\partial_x^\ell(-\partial_x^4 - z)^{-1} f\|_{H^s} \le \frac{C}{|\lambda|^{1-\ell/2}}\|f\|_{H^s}.$$

In our analogy with the spatially varying operators we equate $\mathcal{L}_1\mathcal{L}_2 \sim \mathcal{S} \sim \partial_x^4$ and $\mathcal{L}_1 \sim \mathcal{L}_2 \sim \partial_x^2$, and obtain estimates equivalent to (4.5.8) by taking $\ell = 0$ and $\ell = 2$.

The operator $\mathcal{S}$ is not a constant coefficient, but self-adjoint operators of the form $\mathcal{S}$ give rise to integral transforms similar to the Fourier transform (see [**274**] for examples). For each $\rho \in \sigma(\mathcal{S})$ we denote by P_ρ a properly normalized $\mathcal{S}$-spectral projection onto the associated eigenspace (e.g., see [**244**, Chapter VII.3]). The positive operator $\mathcal{S}^\ell$ has strictly positive spectrum for any $\ell \geq 0$, which engenders the following representation

$$\mathcal{S}^\ell f = \int_{\rho\in\sigma(\mathcal{S})} \rho^\ell \,\mathrm{d}P_\rho(f).$$

In particular, for $\ell = 0$, since the projections are mutually $L^2(\mathbb{R})$ orthogonal we recover Plancherel,

$$\|f\|_{L^2}^2 = \int_{\mathbb{R}} \int_{\rho_1\in\sigma(\mathcal{S})} \int_{\rho_2\in\sigma(\mathcal{S})} \mathrm{d}P_{\rho_1}(f)\,\mathrm{d}P_{\rho_2}(f)\,\mathrm{d}x = \int_{\rho\in\sigma(\mathcal{S})} |\mathrm{d}P_\rho(f)|^2.$$

For $\ell > 0$ the fractional operator $\mathcal{S}^\ell$ controls $\partial_x^{4\ell}$, and generates a norm equivalent to that of $H^{4\ell}(\mathbb{R})$; that is, there exists $0 < m < M$ such that for all $f \in H^{4\ell}(\mathbb{R})$,

$$m\|f\|_{H^{4\ell}} \leq \|\mathcal{S}^\ell f\|_{L^2} \leq M\|f\|_{H^{4\ell}}. \tag{4.5.10}$$

Moreover, we can represent the action of $\mathcal{S}^\ell(\mathcal{S}+z)^{-1}$ as

$$\mathcal{S}^\ell(S+z)^{-1} f = \int_{\sigma(\mathcal{S})} \frac{\rho^\ell}{\rho+z}\mathrm{d}P_\rho(f).$$

It is most difficult to bound the resolvent over the set $\lambda = -\sigma + it$ with $|t| \gg 1$ for which $z = z_1 + iz_2$ satisfies $z_1 \sim -|\lambda|^2$ and $|z_2| \sim |\lambda|$. On this set we have the bound

$$\sup_{\rho>0}\left|\frac{\rho^\ell}{\rho+z}\right| \leq \frac{C|z_1|^\ell}{|z_2|} \leq \frac{C}{|\lambda|^{1-2\ell}},$$

and for any $0 \leq \ell \leq \frac{1}{2}$ we have the resolvent estimate,

$$\|\mathcal{S}^\ell(\mathcal{S}+z)^{-1} f\|_{L^2} \leq \frac{C}{|\lambda|^{1-2\ell}}\left(\int_{\sigma(\mathcal{S})} |\mathrm{d}P_\rho(f)|^2\right)^{1/2} = \frac{C}{|\lambda|^{1-2\ell}}\|f\|_{L^2}. \tag{4.5.11}$$

To establish the first estimate in (4.5.8) we observe that if

$$(\mathcal{L}_1\mathcal{L}_2 + z)f = g \quad \Rightarrow \quad (\mathcal{S}+\mathcal{T}+z)f = g,$$

then

$$\left(\mathcal{I} + \mathcal{T}(\mathcal{S}+z)^{-1}\right)(\mathcal{S}+z)f = g,$$

so that upon inverting

$$(\mathcal{L}_1\mathcal{L}_2+z)^{-1}=(\mathcal{S}+z)^{-1}\left(\mathcal{I}+\mathcal{T}(\mathcal{S}+z)^{-1}\right)^{-1}. \tag{4.5.12}$$

We first bound the rightmost term through the estimate

$$\|\mathcal{T}(\mathcal{S}+z)^{-1}\|_{\mathcal{B}(L^2)}=\|\mathcal{T}\mathcal{S}^{-\ell}\mathcal{S}^{\ell}(\mathcal{S}+z)^{-1}\|_{\mathcal{B}(L^2)}\le\|\mathcal{T}\mathcal{S}^{-\ell}\|_2\|\mathcal{S}^{\ell}(\mathcal{S}+z)^{-1}\|_{\mathcal{B}(L^2)}.$$

Since $\mathcal{T}$ is only a first-order differential operator, we may take $\ell=1/4$ and from (4.5.10) we see that $\mathcal{T}\mathcal{S}^{-\frac{1}{4}}:L^2(\mathbb{R})\mapsto L^2(\mathbb{R})$ is a bounded operator. On the other hand, from (4.5.11) we have the bound

$$\|\mathcal{S}^{1/4}(\mathcal{S}+z)^{-1}\|_{\mathcal{B}(L^2)}\le\frac{C}{|\lambda|^{1/2}}.$$

Thus, for $|\lambda|$ sufficiently large $\|\mathcal{T}(\mathcal{S}+z)^{-1}\|_{\mathcal{B}(L^2)}\ll 1$, so that $\mathcal{I}+\mathcal{T}(\mathcal{S}+z)^{-1}:L^2(\mathbb{R})\mapsto L^2(\mathbb{R})$ has a bounded inverse. It is precisely at this point that we fully exploit the decomposition (4.5.9). If we had separated $\mathcal{L}_1\mathcal{L}_2$ into a fourth self-adjoint and a second-order non–self-adjoint term, we would have been forced to take $\ell=\frac{1}{2}$, and $\|\mathcal{T}(\mathcal{S}+z)^{-1}\|_{\mathcal{B}(L^2)}$ would not have been small for $|\lambda|\gg 1$. Using (4.5.11) with $\ell=0$ on the first term of (4.5.12), we conclude that

$$\left\|(\mathcal{L}_1\mathcal{L}_2+z)^{-1}\right\|_{\mathcal{B}(L^2)}\le\left\|(\mathcal{S}+z)^{-1}\right\|_{\mathcal{B}(L^2)}\left\|\left(\mathcal{I}+\mathcal{T}(\mathcal{S}+z)^{-1}\right)^{-1}\right\|_{\mathcal{B}(L^2)}\le\frac{C}{|\lambda|},$$

which establishes the first estimate of (4.5.8).

To establish the second estimate of (4.5.8) we use the identity

$$\mathcal{L}_1(\mathcal{L}_1\mathcal{L}_2+z)^{-1}=\mathcal{L}_1\mathcal{S}^{-1/2}\mathcal{S}^{1/2}(\mathcal{S}+z)^{-1}\left(\mathcal{I}+\mathcal{T}(\mathcal{S}+z)^{-1}\right)^{-1}.$$

Since $\mathcal{L}_1$ is a second-order operator, we see from (4.5.10) that $\mathcal{L}_1\mathcal{S}^{-1/2}:L^2(\mathbb{R})\mapsto L^2(\mathbb{R})$ is a bounded operator. Furthermore, (4.5.11) implies that $\|\mathcal{S}^{1/2}(\mathcal{S}+z)^{-1}\|_{\mathcal{B}(L^2)}$ is uniformly bounded for large $|\lambda|$ with $\mathrm{Re}\,\lambda>-\sigma$. This completes the proof of (4.5.8).

To extend the resolvent estimates to $H^1(\mathbb{R})$ requires an additional technical argument, in which the H^1 norm is replaced by the equivalent norm generated by $\mathcal{S}^{1/4}$ and using the commutativity of functions of $\mathcal{S}$. The details may be found in [237, Section 4.1]. □

4.6 Additional Reading

An extension of Prüss's Theorem 4.1.5 to Banach spaces is given in Chicone and Latushkin [53] (also see Engel and Nagel [**76**], Helffer and Sjöstrand [**115**], Latushkin and Yurov [**186**]). A spectral mapping theorem for NLS-type problems, for which the essential spectrum resides on the imaginary axis and exponential decay estimates are not available for the semi-group, can be found in Gesztesy et al. [**99**].

Resolvent and semi-group estimates for the case that the essential spectrum is not uniformly bounded away from the origin, based upon an analysis in polynomially and exponentially weighted spaces, can be found in, e.g., Ghazaryan et al. [**102, 103, 104**], Kapitula [**141**]. There is an extensive literature on the stability of traveling-wave solutions of systems of nonlinear conservation laws. The Green's function approach, pioneered by Gardner and Zumbrun [**96**], Zumbrun and Howard [**295**], is strongly connected to the Evans function. In particular, see the lecture notes by Zumbrun [**293**].

A geometric formulation of the dynamical theory presented in Chapter 4.3 can be found in Bates and Jones [**25**]. Stability analysis of manifolds composed of quasi-steady structures, for which the natural linearized operators are weakly time-dependent, can be found in Bellsky et al. [**29**], Doelman and Kaper [**70**], Doelman et al. [**73**], Promislow [**236**], Sandstede [**249**].

The stability theory presented in this chapter can be extended to discrete systems or systems with delay, see, Ablowitz et al. [**5**], Beck et al. [**28**], Hupkes and Sandstede [**123**], Kevrekidis and Weinstein [**171**], Pelinovsky et al. [**232, 233**], Samaey and Sandstede [**247**] for examples.

Examples of the methods presented in this chapter can be found in a great number of applications arising in a wide variety of fields. For example, biological applications have been considered by Bressloff and Folias [**40**], Cytrynbaum and Keener [**64**], Folias and Bressloff [**84, 85**], Gardner and Jones [**95**], Jones and Romeo [**138**], Sandstede [**251**]. Applications in optical communications, lasers, and Bose–Einstein condensation have been considered by Kapitula et al. [**160**], Kevrekidis et al. [**172**], Moore and Promislow [**209**], Moore et al. [**210**], Promislow and Kutz [**237**], and in the books [**1**] and Kevrekidis [**169**], Pelinovsky [**223**], Yang [**284**].

Chapter 5
Orbital Stability of Waves in Hamiltonian Systems

Hamiltonian systems arise in a myriad of applications where damping can be neglected, from the motion of celestial bodies, to the spinning of rigid tops, to interactions of particles in molecular systems. They are also imbued with a rich structure that arises from the conservation of the underlying energy, the Hamiltonian, as well as other quantities such as mass and momentum. In this chapter we present a theory for the nonlinear stability of generalized traveling-wave solutions of Hamiltonian systems. This field has a long history, starting with a conjecture of Boussinesq, dating to 1872 [**39**], in which he suggested the constraint structure could be used to understand the stability of the critical points of the Hamiltonian.

The theory presented in Chapter 4 does not directly apply to Hamiltonian systems as the mechanism of their stability is fundamentally different. Indeed, we will see that the spectrum of a stable wave in a Hamiltonian system must be purely imaginary—there can be no expectation of exponential decay in unweighted norms—and in many cases the absolute spectra is also purely imaginary, so a shift to an exponentially weighted space may be fruitless. Instead of asymptotic stability, the goal of this chapter is to demonstrate the role of constraint in generating simple stability. Following Boussinesq's intuition, the orbital stability within a Hamiltonian system is characterized in terms of the spectrum of the second variation of a Lagrangian functional constructed out of the conserved quantities of the system. We build the theory in stages, showing in a finite-dimensional context that unconstrained minimizers of the Hamiltonian are nonlinearly stable [**105, 202**]. We extend this approach to an infinite-dimensional context, demonstrating the orbital stability of the traveling-wave solutions of the generalized Korteweg–de Vries equation by characterizing them as constrained minimizers of the Hamiltonian. We conclude with a general framework, motivated by the approach of Grillakis–Shatah-Strauss [**109, 110**], which characterizes the critical points of systems with symmetry and conserved quantities via the analysis of a constrained operator.

T. Kapitula and K. Promislow, *Spectral and Dynamical Stability of Nonlinear Waves*, Applied Mathematical Sciences 185, DOI 10.1007/978-1-4614-6995-7_5,

5.1 Finite-Dimensional Systems

The classical finite-dimensional Hamiltonian system consists of a state vector $\boldsymbol{u} \in \mathbb{R}^{2n}$ for some $n \geq 1$, and a Hamiltonian $H : \mathbb{R}^{2n} \mapsto \mathbb{R}$, which depends smoothly upon $\boldsymbol{u}$, that describes the conserved energy of the system in terms of its degrees of freedom, $\boldsymbol{u}$. The corresponding Hamiltonian ODE for the evolution of the state vector takes the form

$$\partial_t \boldsymbol{u} = J \nabla_{\boldsymbol{u}} H(\boldsymbol{u}), \tag{5.1.1}$$

where J is a $2n \times 2n$ nonsingular matrix that is skew-symmetric with respect to the usual Euclidean inner product, i.e., $J^{\mathrm{T}} = -J$. Skew-symmetric matrices map a vector into its perpendicular subspace, i.e.,

$$J\boldsymbol{x} \cdot \boldsymbol{x} = \boldsymbol{x} \cdot J^{\mathrm{T}} \boldsymbol{x} = -\boldsymbol{x} \cdot J\boldsymbol{x} \quad \Rightarrow \quad J\boldsymbol{x} \cdot \boldsymbol{x} = 0. \tag{5.1.2}$$

In this light, the Hamiltonian ODE (5.1.1) can be read as "the flow of $\boldsymbol{u}$ is orthogonal to the gradient of the energy," which suggests that the Hamiltonian is conserved under the flow.

Proposition 5.1.1. *Let $\boldsymbol{u}$ be the solution of (5.1.1) corresponding to initial data $\boldsymbol{u}(0) = \boldsymbol{u}_0$. Then $H(\boldsymbol{u}(t)) = H(\boldsymbol{u}_0)$ for all $t \geq 0$.*

Proof. Consider the evolution of $H(\boldsymbol{u}(t))$, taking the time derivative we have

$$\partial_t H(\boldsymbol{u}) = \nabla_{\boldsymbol{u}} H(\boldsymbol{u}) \cdot \partial_t \boldsymbol{u} = \nabla_{\boldsymbol{u}} H(\boldsymbol{u}) \cdot J \nabla_{\boldsymbol{u}} H(\boldsymbol{u}) = 0,$$

where the last equality follows from (5.1.2) with $\boldsymbol{x} = \nabla_{\boldsymbol{u}} H(\boldsymbol{u})$. Since the time derivative is zero, the functional is constant. □

The canonical Hamiltonian system is derived from Newton's second law. The skew matrix J takes the form

$$J = \begin{pmatrix} \boldsymbol{0}_n & \boldsymbol{I}_n \\ -\boldsymbol{I}_n & \boldsymbol{0}_n \end{pmatrix},$$

where $\boldsymbol{I}_n \in \mathbb{R}^{n \times n}$ is the identity matrix and $\boldsymbol{0}_n \in \mathbb{R}^{n \times n}$ is the zero matrix. The state vector is written as $\boldsymbol{u} = (\boldsymbol{p}, \boldsymbol{q})^{\mathrm{T}}$ for $\boldsymbol{p}, \boldsymbol{q} \in \mathbb{R}^n$, and the Hamiltonian system takes the form

$$\partial_t p_i = \frac{\partial H}{\partial q_i}, \quad \partial_t q_i = -\frac{\partial H}{\partial p_i},$$

for $i = 1, \ldots, n$. This formulation is, e.g., applied to molecular systems, where p_i denotes the momentum of the ith molecule and q_i the associated position. The Hamiltonian describes the total system energy as a combination of kinetic and potential energy arising from interactions between molecules.

In many situations there are additional conserved quantities beyond the Hamiltonian. These take the form of functionals $Q : \mathbb{R}^{2n} \mapsto \mathbb{R}$ for which

$\partial_t Q(\boldsymbol{u}(t)) = 0$ for any solution $\boldsymbol{u}$ of (5.1.1). The conserved quantities restrict the degrees of freedom present in the system. In the most extreme case, when there are n independent conserved quantities, the system (5.1.1) can generically be transforming into the so-called action–angle variables to give the system

$$\partial_t \boldsymbol{I} = \boldsymbol{0}, \quad \partial_t \boldsymbol{\theta} = \boldsymbol{\omega}(\boldsymbol{I}),$$

where $(\boldsymbol{I}, \boldsymbol{\theta}) \in \mathbb{R}^n \times [0, 2\pi]^n$, and $\boldsymbol{\omega}$ is a nonlinear vector-valued function [**20**, **105**, **281**]. The action variable $\boldsymbol{I}$ encodes the values of the conserved quantities, while $\boldsymbol{\theta}$ denotes the angles, the n remaining degrees of freedom in the system which precess at constant rates. For example, the action–angle decomposition can be applied to the motion of an undamped, spinning top. In this case the conservation of linear and angular momentum constrain the evolution of the system to the point that an explicit solution is possible, including such apparently complex behavior as rotation, precession, and nutation of a spinning object.

Turning to stability issues, consider a critical point, $\boldsymbol{\phi}$, of the Hamiltonian energy, i.e., a solution of $\nabla_{\boldsymbol{u}}(H(\boldsymbol{\phi})) = \boldsymbol{0}$. Clearly, $\boldsymbol{\phi}$ is also an equilibrium of the Hamiltonian system (5.1.1). We are interested in studying the dynamics of solutions with initial data $\boldsymbol{u}_0$ that lies close to $\boldsymbol{\phi}$. If $|\boldsymbol{u}_0 - \boldsymbol{\phi}| \ll 1$, what can be said about $\boldsymbol{u}(t)$? Asymptotic stability is generically ruled out for finite-dimensional Hamiltonians. Indeed if $H(\boldsymbol{u}_0) \neq H(\boldsymbol{\phi})$, then $\boldsymbol{u}(t)$ cannot converge to $\boldsymbol{\phi}$, for if it did $H(\boldsymbol{u}_0) = H(\boldsymbol{u}(t)) \to H(\boldsymbol{\phi})$ as $t \to +\infty$, which is a contradiction. The most that can be hoped for is that $\boldsymbol{u}(t)$ remains close to $\boldsymbol{\phi}$.

To address the stability we study the structure of the Hamiltonian about $\boldsymbol{\phi}$. Introducing the perturbation $\boldsymbol{v} := \boldsymbol{u} - \boldsymbol{\phi}$, a Taylor expansion of H about $\boldsymbol{\phi}$ yields

$$H(\boldsymbol{u}) = H(\boldsymbol{\phi}) + \nabla_{\boldsymbol{u}} H(\boldsymbol{\phi}) \cdot \boldsymbol{v} + \boldsymbol{v} \cdot \boldsymbol{L}\boldsymbol{v} + \mathcal{O}(|\boldsymbol{v}|^3),$$

where the Hessian matrix $\boldsymbol{L} := \nabla^2_{\boldsymbol{u}} H(\boldsymbol{\phi}) \in \mathbb{R}^{2n \times 2n}$ has entries

$$\boldsymbol{L}_{ij} = \frac{\partial^2 H}{\partial u_i \partial u_j}(\boldsymbol{\phi}).$$

It is significant in the sequel that the operator $\boldsymbol{L}$ is symmetric, i.e., $\boldsymbol{L}^{\mathrm{T}} = \boldsymbol{L}$. Since $\boldsymbol{\phi}$ is a critical point of H the first derivative is zero, $\nabla_{\boldsymbol{u}} H(\boldsymbol{\phi}) = \boldsymbol{0}$, and the Hamiltonian has the expansion

$$H(\boldsymbol{u}) - H(\boldsymbol{\phi}) = \frac{1}{2}\boldsymbol{v} \cdot \boldsymbol{L}\boldsymbol{v} + \mathcal{O}(|\boldsymbol{v}|^3). \tag{5.1.3}$$

Taking the $\nabla_{\boldsymbol{v}}$ of (5.1.3) and substituting for $\nabla_{\boldsymbol{v}} H$, we see that the Hamiltonian system (5.1.1) can be rewritten in a form evocative of (4.0.2),

$$\partial_t \boldsymbol{v} = \boldsymbol{J}\boldsymbol{L}\boldsymbol{v} + \boldsymbol{N}(\boldsymbol{v}), \quad |\boldsymbol{N}(\boldsymbol{v})| = \mathcal{O}(|\boldsymbol{v}|^2),$$

where JL is the linearization of the flow about ϕ. The following proposition hints at the structure typical of such linearizations.

Proposition 5.1.2. *The point spectrum $\sigma_{\mathrm{pt}}(JL)$ is symmetric with respect to the real and imaginary axes of the complex plane, so that the eigenvalues of JL come in quartets of the form $\{\pm\lambda, \pm\overline{\lambda}\}$. In particular, either $\sigma_{\mathrm{pt}}(JL) \subset \mathrm{i}\mathbb{R}$, or the critical point ϕ is linearly exponentially unstable.*

Proof. Suppose that $\lambda \in \sigma_{\mathrm{pt}}(JL)$ with the associated eigenvector x. Since JL has real-valued entries,

$$JLx = \lambda x \quad \Rightarrow \quad JL\overline{x} = \overline{\lambda}\overline{x},$$

i.e., $\overline{\lambda} \in \sigma_{\mathrm{pt}}(JL)$, where the overline denotes complex conjugation. In addition,

$$JLx = \lambda x \quad \Rightarrow \quad -LJ(J^{-1}x) = (-\lambda)J^{-1}x,$$

and since $(JL)^{\mathrm{T}} = -LJ$ we deduce that $-\lambda \in \sigma_{\mathrm{pt}}((JL)^{\mathrm{T}})$. However, every matrix satisfies $\sigma_{\mathrm{pt}}(A) = \overline{\sigma_{\mathrm{pt}}(A^{\mathrm{T}})}$, and we deduce that $-\overline{\lambda} \in \sigma_{\mathrm{pt}}(JL)$. The statement on spectral stability follows from the spectral symmetry, since an eigenvalue of a negative real part implies the existence of an eigenvalue of a positive real part. □

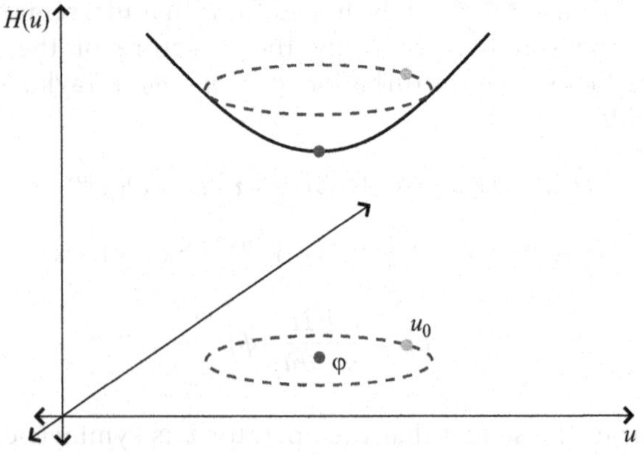

Fig. 5.1 A bounded level set $H(u) = H(u_0)$ near a local minima ϕ is depicted by a (*blue*) dashed curve. (Color figure online.)

Proposition 5.1.2 lends weight to the statement that asymptotic stability of equilibria is unlikely in finite-dimensional Hamiltonian systems. However, it is possible for a critical point ϕ of the Hamiltonian to be stable under the flow. Indeed, if ϕ is a nondegenerate minima of H, then as depicted in Fig. 5.1 the nearby level sets will be bounded. Lemma 5.1.3 shows that this sufficient for nonlinear stability.

Lemma 5.1.3. *Suppose that ϕ is a critical point for the Hamiltonian system (5.1.1). If ϕ is a strict local minimum, i.e., $\boldsymbol{L}$ is a positive-definite matrix, then ϕ is stable. Specifically, there exists $C, \delta > 0$ such that for $|\boldsymbol{u}_0 - \phi| \leq \delta$, the solution u of (5.1.1) satisfies*

$$|\boldsymbol{u}(t) - \phi| \leq C|\boldsymbol{u}_0 - \phi|, \quad t \geq 0. \tag{5.1.4}$$

Proof. Set $\boldsymbol{v} = \boldsymbol{u} - \phi$, and recall that the Taylor expansion of H about ϕ leads to the expression,

$$H(\boldsymbol{u}) - H(\phi) = \frac{1}{2}\boldsymbol{v} \cdot \boldsymbol{L}\boldsymbol{v} + \mathcal{O}(|\boldsymbol{v}|^3).$$

Since $\boldsymbol{L}$ is symmetric, all of its eigenvalues, μ_j for $j = 1, \ldots, 2n$, are real-valued. By assumption, $\boldsymbol{L}$ is positive-definite and its eigenvalues are positive, indeed $\alpha_- := \min\{\mu_j : j = 1, \ldots, 2n\} > 0$. Moreover the quantity $\alpha_+ := \max\{j = 1, \ldots, 2n\} \geq \alpha_-$, these constants afford the bounds

$$\alpha_-|\boldsymbol{v}|^2 \leq \boldsymbol{v} \cdot \boldsymbol{L}\boldsymbol{v} \leq \alpha_+|\boldsymbol{v}|^2 \tag{5.1.5}$$

with the equality attained at the corresponding eigenvectors. Bounding the remainder term in the Taylor expansion of the Hamiltonian implies the existence of a $\delta > 0$ such that for $|\boldsymbol{v}| \leq \delta$ we have the estimates,

$$\frac{1}{2}\alpha_-|\boldsymbol{v}|^2 \leq H(\boldsymbol{u}) - H(\phi) \leq 2\alpha_+|\boldsymbol{v}|^2. \tag{5.1.6}$$

The lower bound implies that the initial data controls the norm of the perturbation through the invariance of the Hamiltonian,

$$|\boldsymbol{v}(t)|^2 \leq 2\frac{H(\boldsymbol{u}(t)) - H(\phi)}{\alpha_-} = 2\frac{H(\boldsymbol{u}_0) - H(\phi)}{\alpha_-}. \tag{5.1.7}$$

The upper bound in (5.1.6) is a manifestation of the level sets within δ of ϕ being closed. It permits us to rewrite the proximity of the values of the Hamiltonian in terms of proximity in Euclidean space, that inserting the upper bound into (5.1.7) yields

$$|\boldsymbol{v}(t)|^2 \leq 2\frac{H(\boldsymbol{u}_0) - H(\phi)}{\alpha_-} \leq 4\frac{\alpha_+}{\alpha_-}|\boldsymbol{v}_0||^2.$$

The conclusion (5.1.4) is achieved with $C = 2\sqrt{\alpha_+/\alpha_-}$. Indeed, near ϕ the level sets of $\mathcal{H}$ are approximately ellipsoids and the quantity α_+/α_- is the ratio of their major and minor axes. □

The key point of Lemma 5.1.3 is that stability of a critical point ϕ of the Hamiltonian H under *any* Hamiltonian flow associated with H can be deduced independent of the dynamics of the system near ϕ: the nonlinear stability is independent of the choice of skew matrix $\boldsymbol{J}$. In the next section we extend this approach to infinite-dimensional Hamiltonian systems, which requires that we overcome two obstacles. First, if the system (5.1.1) possesses symmetries $\{T_j\}_{j=1}^N$, then as in Lemma 4.2.1 the image of the critical point under these symmetries will generate a manifold of critical points, and the set $\{T_j'\phi\}$ will lie in the kernel of the linearization $\boldsymbol{J}\boldsymbol{L}$ about ϕ. This suggests that $\boldsymbol{L}$ will have a null space and can be at best positive semi-definite, i.e. $\alpha_- = 0$ in Lemma 5.1.3. As in Chapter 4.3 this obstacle will be overcome through the notion of orbital stability. The second obstacle is that in many instances $\boldsymbol{L}$ has eigenvalues of negative real part, which suggests that the critical point is not a minimizer, and hence cannot be stable. However, we will see that each symmetry T_j generates a conserved quantity Q_j associated with the flow. As mentioned in the discussion of action–angle variables, these conserved quantities constrain the flow. We introduce the idea of a constrained operator and derive conditions under which it is positive semi-definite. Indeed, the characterization of the spectrum of constrained operators is a main thrust of this chapter and the framework developed, which culminates in Theorem 5.3.2, has significant applications.

5.2 Infinite-Dimensional Hamiltonian Systems with Symmetry

Consider two infinite-dimensional Hilbert spaces $Y \subset X \subset Y^*$, where Y is dense in X in $\|\cdot\|_X$ and Y^* is the dual of Y in the X inner product. The norm on X is generated by the inner product $\langle\cdot,\cdot\rangle_X$, and that on Y by $\langle\cdot,\cdot\rangle_Y$. A Hamiltonian on X is a nonlinear functional $\mathcal{H} : Y \subset X \mapsto \mathbb{R}$. The associated Hamiltonian system takes the form

$$\partial_t u = \mathcal{J}\frac{\delta\mathcal{H}}{\delta u}(u), \tag{5.2.1}$$

where $\mathcal{J} : Y \subset X \mapsto X$ is a densely defined, skew-symmetric operator with respect to the X-inner product, i.e., $\langle\mathcal{J}u, v\rangle_X = -\langle u, \mathcal{J}v\rangle_X$ for all $u, v \in Y$, and in particular $\langle\mathcal{J}u, u\rangle_X = 0$ for all $u \in Y$. For a sufficiently smooth Hamiltonian $\mathcal{H}$ the first variation with respect to $\langle\cdot,\cdot\rangle_X$, denoted $\delta\mathcal{H}/\delta u$, is defined as the element of Y^* that satisfies the limit

$$\lim_{\epsilon\to 0}\frac{\mathcal{H}(u+\epsilon v)-\mathcal{H}(u)}{\epsilon} = \left\langle\frac{\delta\mathcal{H}}{\delta u}(u), v\right\rangle_X,$$

for all $v \in Y$. From the chain rule we see that smooth solutions of (5.2.1) conserve the Hamiltonian, i.e.,

$$\partial_t\mathcal{H}(u(t)) = \left\langle\frac{\delta\mathcal{H}}{\delta u}(u), \partial_t u\right\rangle_X = \left\langle\frac{\delta\mathcal{H}}{\delta u}, \mathcal{J}\frac{\delta\mathcal{H}}{\delta u}\right\rangle_X = 0.$$

Pursuing a generalization of the finite-dimensional expansion given in (5.1.3), we fix $\phi \in Y$ and consider a smooth Hamiltonian $\mathcal{H}$. For $u = \phi + \epsilon v$ with $\|v\|_Y = \mathcal{O}(1)$ the Hamiltonian admits a *formal* Taylor expansion

$$\mathcal{H}(\phi + \epsilon v) - \mathcal{H}(\phi) = \epsilon \left\langle \frac{\delta \mathcal{H}}{\delta u}(\phi), v \right\rangle_X + \epsilon^2 \frac{1}{2} \langle \mathcal{L} v, v \rangle_X + \mathcal{O}\left(\epsilon^3\right),$$

where the self-adjoint linear operator $\mathcal{L}$ is called the second variation of $\mathcal{H}$, see Chapter 2.2.3,

$$\mathcal{L} := \frac{\delta^2 \mathcal{H}}{\delta u^2}(\phi) : \mathcal{D}(\mathcal{L}) \subset Y \mapsto X.$$

If ϕ is a critical point of $\mathcal{H}$, i.e.,

$$\frac{\delta \mathcal{H}}{\delta u}(\phi) = 0,$$

then the Taylor expansion reduces to an infinite-dimensional version of (5.1.3),

$$\mathcal{H}(u) - \mathcal{H}(\phi) = \epsilon^2 \frac{1}{2} \langle \mathcal{L} v, v \rangle + \mathcal{O}(\epsilon^3). \tag{5.2.2}$$

However, there are differences between the symmetric matrix $\boldsymbol{L}$ and the self-adjoint operator $\mathcal{L}$; in particular, the operator $\mathcal{L}$ is generically unbounded and has a nontrivial kernel, which in the finite-dimensional case would correspond to $\alpha_- = 0$ and $\alpha_+ = \infty$. Before proceeding to a general framework for the orbital stability problem, we work out a detailed example in Chapter 5.2.1.

5.2.1 The Generalized Korteweg–de Vries Equation

Consider the initial-value problem for the generalized Korteweg–de Vries equation (gKdV),

$$\partial_t u + u^p \partial_x u + \partial_x^3 u = 0, \quad u(x,0) = u_0(x), \tag{5.2.3}$$

where $p \geq 1$ and $(x,t) \in \mathbb{R} \times \mathbb{R}^+$. The gKdV appears as a model equation for a surprising number of physical systems, including the description of long wavelength motion of the free surface of a fluid, e.g., see [**3**, Chapter 1] and [**211**, Chapter 2]. We take $Y = H^1(\mathbb{R}) \subset L^2(\mathbb{R}) = X$ with the standard inner products. Unless stated otherwise, the inner product used throughout this example will be that associated with X. The gKdV equation can be written in the Hamiltonian form (5.2.1) with L^2-skew operator $\mathcal{J} = \partial_x : H^1(\mathbb{R}) \subset L^2(\mathbb{R}) \mapsto L^2(\mathbb{R})$ and Hamiltonian

$$\mathcal{H}(u) = \int_{\mathbb{R}} \left(\frac{1}{2} (\partial_x u)^2 - \frac{1}{(p+1)(p+2)} u^{p+2} \right) \mathrm{d}x. \tag{5.2.4}$$

More specifically, the gKdV equation can be written as

$$\partial_t u = \mathcal{J}\frac{\delta\mathcal{H}}{\delta u}(u) = \partial_x\left(-\partial_x^2 u - \frac{1}{p+1}u^{p+1}\right),$$

subject to the initial data $u(x,0) = u_0(x)$ (compare with Exercise 5.2.1). In addition to the Hamiltonian, the form of the skew operator $\mathcal{J}$ gives a second conserved quantity: the charge

$$Q(u) := \frac{1}{2}\langle u, u\rangle = \frac{1}{2}\|u\|_2^2.$$

The conservation of charge arises from the spatial translation symmetry of the gKdV, i.e., if $u(x,t)$ is a solution, then so is $T(\gamma)u(x,t) = u(x+\gamma,t)$. Introducing the Hamiltonian density $H(u)$, the integrand associated to the Hamiltonian, we see that the invariance of Q with time requires only that $H(u)$ vanish at $x = \pm\infty$; indeed,

$$\partial_t Q(u) = \langle \partial_t u, u\rangle = \left\langle \mathcal{J}\frac{\delta\mathcal{H}}{\delta u}(u), u\right\rangle = -\left\langle \frac{\delta\mathcal{H}}{\delta u}(u), \partial_x u\right\rangle = -\int_{\mathbb{R}} \partial_x H(u)\,\mathrm{d}x = 0.$$

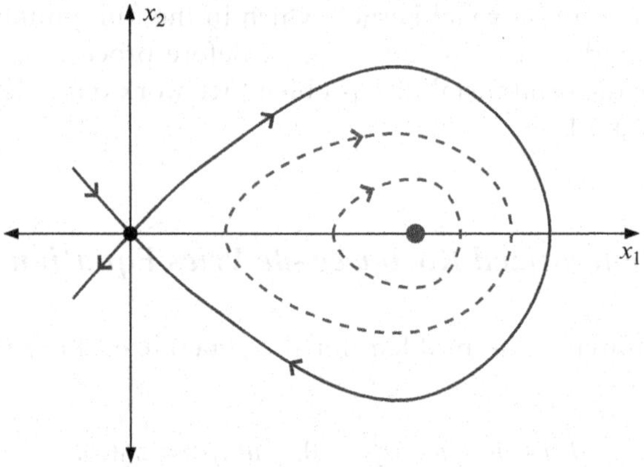

Fig. 5.2 The phase plane for the ODE system (5.2.6). The axes correspond to $(x_1, x_2) = (\phi_c, \partial_\xi\phi_c)$. The portion of the solid (*blue*) curve that lies in $x_1 > 0$ is the orbit homoclinic to zero, and corresponds to the traveling-wave solution we study. The dashed (*red*) curves inside the homoclinic are periodic orbits that correspond to spatially periodic cnoidal wave solutions of the gKdV equation. (Color figure online.)

The gKdV equation possesses a family of traveling-waves $u_c(x,t) = \phi_c(\xi)$ that are parameterized by their speed $c > 0$. In traveling coordinates $\xi := x - ct$ the gKdV can be rewritten as

$$\partial_t u = \partial_\xi \left(-\partial_\xi^2 u + cu - \frac{1}{p+1} u^{p+1} \right), \quad u(\xi,0) = u_0(\xi). \tag{5.2.5}$$

The wave form ϕ_c satisfies the traveling-wave ODE

$$\partial_\xi \left(-\partial_\xi^2 \phi_c + c\phi_c - \frac{1}{p+1} \phi_c^{p+1} \right) = 0.$$

Assuming that ϕ_c approaches zero as $\xi \to \pm\infty$, this equation can be integrated once, reducing it to the second-order ODE,

$$-\partial_\xi^2 \phi_c + c\phi_c - \frac{1}{p+1} \phi_c^{p+1} = 0, \tag{5.2.6}$$

which can be put into Hamiltonian form

$$\partial_\xi \begin{pmatrix} \phi_c \\ \partial_\xi \phi_c \end{pmatrix} = \begin{pmatrix} 0 & 1 \\ -1 & 0 \end{pmatrix} \nabla_{\boldsymbol{u}} E(\phi_c, \partial_\xi \phi_c),$$

where

$$E(u_1, u_2) = \frac{1}{(p+1)(p+2)} u_1^{p+2} - \frac{c}{2} u_1^2 + \frac{1}{2} u_2^2.$$

In the $(\phi_c, \partial_\xi \phi_c)$ phase plane this system has a saddle point at $(0,0)$, and since E is conserved along orbits, the unstable manifold at $(0,0)$ must intersect the stable manifold, yielding a homoclinic orbit (see Fig. 5.2). It can be shown that the homoclinic orbit has the explicit formula

$$\phi_c(\xi) = \left[\frac{(p+1)(p+2)}{2} c \right]^{1/p} \operatorname{sech}^{2/p} \left(\frac{\sqrt{c}\, p}{2} \xi \right) \tag{5.2.7}$$

(e.g., see[**220**]); however, we emphasize that none of the calculations addressing the stability require knowledge of this explicit form.

The first step of the stability analysis is a consideration of the linearization of (5.2.5) about ϕ_c,

$$\partial_t v = \mathbb{L} v, \tag{5.2.8}$$

where $\mathbb{L} := \mathcal{J}\mathcal{L}$ is the gKdV operator, and the second variation is given by

$$\mathcal{L} := -\partial_\xi^2 + c - \phi_c^p. \tag{5.2.9}$$

In Chapter 3.1.1.1 we computed that $\sigma_{\mathrm{ess}}(\mathbb{L}) = \mathrm{i}\mathbb{R}$. Since the system (5.2.5) has a translational symmetry, we deduce from Lemma 4.2.1 that $T'\phi_c = \partial_\xi \phi_c \in \ker(\mathbb{L})$. Since $\mathcal{J} = \partial_\xi$ has no kernel on $L^2(\mathbb{R})$, we suspect that $\partial_\xi \phi_c \in \ker(\mathcal{L})$. This is readily verified by acting ∂_ξ on (5.2.6),

$$0 = \partial_\xi \left(-\partial_\xi^2 \phi_c + c\phi_c - \frac{1}{p+1} \phi_c^{p+1} \right) = \mathcal{L}(\partial_\xi \phi_c).$$

Indeed, since $\mathcal{L}$ is Sturm–Liouville operator, $\ker(\mathcal{L})$ is simple and we deduce that $\ker(\mathbb{L}) = \operatorname{span}(\partial_\xi \phi_c)$. However, $\mathbb{L}$ is not self-adjoint, and differentiating the traveling-wave ODE with respect to c yields

$$\mathcal{L}(\partial_c \phi_c) = -\phi_c \quad \Rightarrow \quad \partial_\xi \mathcal{L}(\partial_c \phi_c) = \mathbb{L}(\partial_c \phi_c) = -\partial_\xi \phi_c. \tag{5.2.10}$$

In particular, $\operatorname{span}\{\partial_\xi \phi_c, \partial_c \phi_c\} \subset \operatorname{gker}(\mathbb{L})$, and $\mathbb{L}$ has a Jordan chain of algebraic multiplicity of at least two. We will see that Jordan chains are generic for the linearizations of Hamiltonian PDEs.

For Hamiltonian systems the Jordan chain associated with the kernel of the linearization generically terminates at length two. For the gKdV we look for a generalized eigenvalue of algebraic multiplicity three by solving

$$\mathbb{L} v = \partial_c \phi_c \quad \Rightarrow \quad \mathcal{L} v = \int_{-\infty}^{\xi} \partial_c \phi_c(s)\, \mathrm{d}s.$$

From the Fredholm alternative we can solve for v only if the right-hand side is orthogonal to $\ker(\mathcal{L})$, i.e.,

$$0 = \left\langle \int_{-\infty}^{\xi} \partial_c \phi_c(s)\, \mathrm{d}s, \partial_\xi \phi_c \right\rangle = -\langle \partial_c \phi_c, \phi_c \rangle = -\frac{1}{2} \partial_c \langle \phi_c, \phi_c \rangle.$$

The solvability condition is equivalent to $\partial_c Q(\phi_c) = 0$. This connection between the symmetry T, the kernel of $\mathbb{L}$, and the conserved quantity Q is generic. To evaluate this expression we scale c out of (5.2.6) by writing $\phi_c(\xi) = c^{1/p} \phi_0(\sqrt{c}\,\xi)$, where ϕ_0 satisfies,

$$-\partial_\xi^2 \phi_0 + \phi_0 - \frac{1}{p+1} \phi_0^{p+1} = 0. \tag{5.2.11}$$

It follows that

$$Q(\phi_c) = c^{2/p} \int_{\mathbb{R}} \phi_0^2(\sqrt{c}\,\xi)\, \mathrm{d}\xi = c^{(4-p)/(2p)} \|\phi_0\|_2^2, \tag{5.2.12}$$

and hence $\partial_c Q(\phi_c) \neq 0$ for $p \neq 4$. We deduce that for these values of p, $\dim[\operatorname{gker}(\mathbb{L})] = 2$.

The spectrum of $\sigma(\mathbb{L})$ violates the stability Hypothesis 4.3.1 in two ways: the essential spectrum is not bounded away from the imaginary axis, and the eigenvalue at zero is not simple (see Fig. 5.3). Furthermore, by following the arguments leading to Proposition 5.1.2 it can be shown that $\sigma_{\mathrm{pt}}(\mathbb{L})$ has the four-fold symmetry

$$\lambda \in \sigma_{\mathrm{pt}}(\mathbb{L}) \implies \{\pm\lambda, \pm\overline{\lambda}\} \in \sigma_{\mathrm{pt}}(\mathbb{L}).$$

To account for the kernel of $\mathbb{L}$, which arises from the translational symmetry, we introduce the following sense of orbital stability.

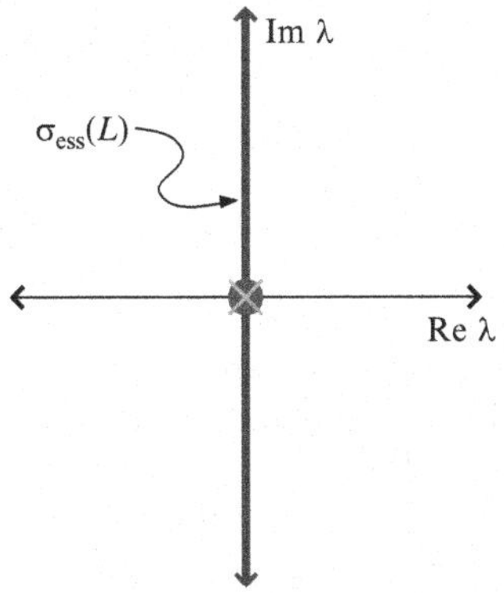

Fig. 5.3 The information on spectrum of $\sigma(L)$ determined by the analysis. The Jordan chain at $\lambda = 0$ is denoted by the (*green*) cross overlaying the (*red*) circle. We have not yet ruled out other point spectrum; however, the nonzero eigenvalues satisfy the Hamiltonian eigenvalues symmetry of $\{\pm\lambda, \pm\overline{\lambda}\} \subset \sigma_{pt}(L)$, and one can infer, retroactively, from the nonlinear stability result Theorem 5.2.2 that $\sigma(L) \subset i\mathbb{R}$. (Color figure online.)

Definition 5.2.1 (Orbital stability). Consider a critical point ϕ of a Hamiltonian PDE (5.2.1), which has an N-fold symmetry group $T = T(\gamma)$, satisfying (4.2.1) for $\mathcal{F} = \mathcal{J}\frac{\delta\mathcal{H}}{\delta u}$. The N-dimensional manifold $\mathcal{M}_T$ introduced in (4.2.4) is orbitally stable in a norm $\|\cdot\|$ if for all $\epsilon > 0$ there exists $\delta > 0$ such that

$$\inf_{\gamma} \|u_0 - T(\gamma)\phi\| \leq \delta \quad \Rightarrow \quad \inf_{\gamma} \|u(t) - T(\gamma)\phi\| < \epsilon.$$

The semi-norm $\inf_\gamma \|u(t) - T(\gamma)\phi\|$ is called the *orbital norm* over $\mathcal{M}_T$.

For the gKdV equation the manifold $\mathcal{M}_T = \{T(\gamma)\phi_c : \gamma \in \mathbb{R}\}$ consists of all the translations of the wave ϕ_c. The notion of orbital stability permits us to slide ϕ_c around to get a "best fit" before taking the norm of the difference. This is a subtle way of imposing a hidden constraint on the difference between the solution u and the shifted pulse profile $T(\gamma)\phi_c$. In particular, it yields a local coordinate system with the properties described in Lemma 4.3.3.

To see this in the context of the gKdV, we fix $u \in H^1(\mathbb{R})$ sufficiently close to $\mathcal{M}_T$ and introduce the smooth function

$$f(\gamma) = \langle u - T(\gamma)\phi_c, u - T(\gamma)\phi_c \rangle,$$

where T is the translational symmetry. Since $\partial_\gamma T(\gamma)\phi_c = T(\gamma)T'\phi_c = T(\gamma)\partial_\xi\phi_c$, the minima is attained at an interior point $\gamma \in \mathbb{R}$ which solves,

$$0 = f'(\gamma) = -2\langle u - T(\gamma)\phi_c, T(\gamma)\partial_\xi\phi_c \rangle = -2\langle u, T(\gamma)\partial_\xi\phi_c \rangle.$$

From the translational symmetry of the traveling-wave ODE (5.2.6) we know that $\ker(\mathcal{L}_\gamma) = \text{span}\{T(\gamma)\partial_\xi\phi_c\}$, where $\mathcal{L}_\gamma = -\partial_\xi^2 + c - T(\gamma)\phi_c^p$ is the linearization about the translated wave. Since $\mathcal{L}_\gamma$ is self-adjoint, we can invoke Lemma 4.3.3 to find a neighborhood B of $\mathcal{M}_T$ for which $u = u(t) \in B$ may be decomposed as

$$u(t) = T(\gamma(t))\phi_c + h(t), \quad \langle h(t), T(\gamma(t))\partial_\xi \phi_c \rangle = 0. \tag{5.2.13}$$

We would like to exploit the structure of the Hamiltonian near ϕ_c, as illuminated in (5.2.2), to bound the distance of a solution $u(t)$ to the manifold $\mathcal{M}_T$. However, the traveling wave is not a critical point of the Hamiltonian given in (5.2.4), but, as we will see more generally, it is a "symmetry" solution and can be cast as a critical point of the constrained problem associated to a linear combination of the Hamiltonian and the invariant Q. To this effect we define the Lagrangian for the gKdV

$$\Lambda(u) = \mathcal{H}(u) + \lambda Q(u), \tag{5.2.14}$$

where λ can be viewed as the Lagrange multiplier associated with the constraint $Q(u) = Q(u_0)$. The variational derivative of Λ takes the form

$$\frac{\delta \Lambda}{\delta u} = -\partial_\xi^2 u + \lambda u - \frac{1}{p+1} u^{p+1},$$

and we see from the traveling-wave ODE (5.2.6) that ϕ_c is a critical point of Λ for the choice $\lambda = c$.

The Lagrangian is invariant under both the gKdV flow and the translational invariant T. For any solution $u = u(t)$ of gKdV we have the equality

$$\Lambda(u(t)) - \Lambda(T(\gamma(t))\phi_c) = \Lambda(u_0) - \Lambda(T(\gamma(0))\phi_c).$$

If u_0 and $u(t)$ reside within the foliation neighborhood, B, of $\mathcal{M}_T$, then $\gamma(t)$ may be chosen to enforce the decomposition (5.2.13). The next task is to show that for $1 \leq p < 4$, the Lagrangian is uniformly continuous on bounded sets in the $H^1(\mathbb{R})$-norm. This is accomplished in Exercise 5.2.2.

For a solution $u = u(t)$ of the gKdV, we desire the bound M, and hence δ, in Exercise 5.2.2 to depend only upon the initial data through $\|u_0\|_{H^1}$. This result follows if we can control the $H^1(\mathbb{R})$ norm of u through a linear combination of the invariants, H and Q. First observe that

$$\begin{aligned} \|u\|_{H^1}^2 &= 2(\mathcal{H}(u) + Q(u)) + \frac{1}{(p+1)(p+2)} \int_{\mathbb{R}} u(x)^{p+2}\, dx \\ &\leq 2(\mathcal{H}(u) + Q(u)) + \frac{1}{(p+1)(p+2)} \|u\|_{p+2}^{p+2}. \end{aligned} \tag{5.2.15}$$

On the other hand, the $L^{p+2}(\mathbb{R})$ norm is controlled by $L^\infty(\mathbb{R})$ and $L^2(\mathbb{R})$ through

$$\|u\|_{p+2}^{p+2} = \int_{-\infty}^{+\infty} |u(x)|^{p+2}\,\mathrm{d}x \leq \|u\|_\infty^p \|u\|_2^2 \leq 2\|u\|_\infty^p Q(u),$$

and the Sobolev embedding (2.2.7), yields the $L^\infty(\mathbb{R})$ norm bound

$$\|u\|_\infty^2 \leq 2\|u\|_2 \|\partial_x u\|_2 = 2^{3/2}\|\partial_x u\|_2 Q(u)^{1/2}. \tag{5.2.16}$$

Returning these estimates to (5.2.15) and simplifying yields

$$\|u\|_{H^1}^2 \leq 2(\mathcal{H}(u) + Q(u)) + \frac{2^{(3p+4)/4}}{(p+1)(p+2)}\|\partial_x u\|_2^{p/2} Q(u)^{(p+4)/4}.$$

For $p < 4$ we may apply Young's inequality (2.2.3) with conjugate exponents $p_1 = 4/p$ and $p_2 = 4/(4-p)$ to the last term on the right-hand side, obtaining

$$\|u\|_{H^1}^2 \leq 2(\mathcal{H}(u) + Q(u)) + \frac{1}{2}\|u_x\|_2^2 + \nu Q(u)^{(p+4)/(4-p)},$$

for an appropriately chosen constant $\nu > 0$. Finally, since $\|\partial_x u\|_2 \leq \|u\|_{H^1}$, combining the $H^1(\mathbb{R})$-norm terms and using the time-invariance of $\mathcal{H}, Q$ yields the global-in-time bound

$$\|u\|_{H^1}^2 \leq 4(\mathcal{H}(u_0) + Q(u_0)) + 2\nu Q(u_0)^{(p+4)/(4-p)}. \tag{5.2.17}$$

The remaining task is to show that the Lagrangian Λ for the gKdV controls the $H^1(\mathbb{R})$ norm; that is, for u in the foliated neighborhood B of $\mathcal{M}_T$ and γ given in (5.2.13) there is an $\alpha > 0$ such that

$$\|u - T(\gamma)\phi_c\|_{H_1}^2 \leq \alpha(\Lambda(u) - \Lambda(T(\gamma)\phi_c)).$$

In the main result of this section, the existence of such an $\alpha > 0$ is characterized solely by the derivative of the charge of the wave, $Q(\phi_c)$, with respect to the wave speed c.

Theorem 5.2.2. *Fix $c > 0$. The manifold $\mathcal{M}_T$ generated by the translational symmetry T acting on the traveling-wave solution ϕ_c of (5.2.7) of the gKdV equation (5.2.3) is $H^1(\mathbb{R})$-orbitally stable so long as*

$$\partial_c Q(\phi_c) > 0. \tag{5.2.18}$$

Moreover, this condition is satisfied for all $p \in [1,4)$.

Proof. Given the initial data u_0, let $u = u(t)$ be the corresponding solution of gKdV. We choose the speed c so that $Q(u_0) = Q(\phi_c)$. Since the charge is invariant it follows that $Q(u(t)) = Q(\phi_c)$ for all $t \geq 0$. Let $\epsilon > 0$ be given. We must choose $\delta = \delta(\epsilon) > 0$ such that for all $t \geq 0$,

$$\inf_{\gamma\in\mathbb{R}} \|u_0 - T(\gamma)\phi_c\|_{H^1} \le \delta \quad \Rightarrow \quad \inf_{\gamma\in\mathbb{R}} \|u(t) - T(\gamma)\phi_c\|_{H^1} \le \epsilon. \tag{5.2.19}$$

We know that for δ sufficiently small there exists a continuous function $\gamma(t)$ for which the decomposition (5.2.13) holds for $t \in (0, t_m)$ for some $t_m > 0$. To prove the orbital stability, we show that the $H^1(\mathbb{R})$-norm of the perturbation

$$h(\xi, t) := u(\xi, t) - T(\gamma(t))\phi_c$$

remains uniformly small for all $t \in (0, t_m)$, and hence we can extend t_m iteratively to $+\infty$.

Define the time-invariant quantity

$$\overline{\Lambda} := \Lambda(u) - \Lambda(T(\gamma(t))\phi_c),$$

and note by Exercise 5.2.2 that the size of $\overline{\Lambda}$ is controlled by the orbital norm, so that by the appropriate choice of δ we may make $|\overline{\Lambda}|$ as small as required. For notational convenience we assume that $\gamma(t) \equiv 0$, so that $T(\gamma(t)) = \mathcal{I}$. The expansion of $\overline{\Lambda}$ in powers of h yields the expansion

$$\overline{\Lambda} = \Lambda(\phi_c + h) - \Lambda(\phi_c) = \left\langle \frac{\delta\Lambda}{\delta u}(\phi_c), h \right\rangle + \frac{1}{2}\left\langle \frac{\delta^2\Lambda}{\delta u^2}(\phi_c)h, h \right\rangle + \mathcal{O}\left(\|h\|_{H^1}^3\right),$$

where the higher-order terms are polynomial in h, so that the $\|\cdot\|_{H^1}$ bound follows from the inequality (2.2.7). We have already seen that ϕ_c is a critical point of the Lagrangian. The second variation of the Lagrangian takes precisely the form given in (5.2.9), i.e.,

$$\mathcal{L} := \frac{\delta^2\Lambda}{\delta u^2}(\phi_c) = -\partial_\xi^2 + c - \phi_c^p, \tag{5.2.20}$$

and we may then express $\overline{\Lambda}$ as

$$\overline{\Lambda} = \langle \mathcal{L}h, h \rangle + \mathcal{O}\left(\|h\|_{H^1}^3\right). \tag{5.2.21}$$

The constraint $Q(u) = Q(\phi_c)$ and the decomposition (5.2.13) imply that the perturbation h lies in the nonlinear admissible set

$$\mathcal{A}_c = \{h \in H^1(\mathbb{R}) : Q(\phi_c + h) = Q(\phi_c),\ \langle h, \phi_c' \rangle = 0\}. \tag{5.2.22}$$

In Lemma 5.2.3 we show that subject to (5.2.18) there exist positive constants α, β such that

$$\langle \mathcal{L}h, h \rangle \ge \alpha\|h\|_{H^1}^2 - \beta\|h\|_{H^1}^3, \quad h \in \mathcal{A}_c. \tag{5.2.23}$$

Assuming this result, the work is essentially done, since (5.2.21) implies that

$$|\overline{\Lambda}| \geq \overline{\Lambda} \geq \alpha\|h\|_{H^1}^2 - \tilde{\beta}\|h\|_{H^1}^3, \tag{5.2.24}$$

for some $\tilde{\beta} > 0$. Recalling the discussion from (4.3.25) and (4.3.26), the inequality (5.2.24) yields an *excluded region* defined by the two positive roots $0 < r_1 < r_2$ of the cubic

$$f(r) = |\overline{\Lambda}| - \alpha r^2 + \tilde{\beta} r^3,$$

(see Fig. 5.4). The condition $f(r) > 0$ implies either $r < r_1 = \mathcal{O}(|\overline{\Lambda}|^{1/2})$ or $r > r_2 = \mathcal{O}(1)$. Recalling that $\|h_0\|_{H^1} \leq \delta$, we take $\delta \leq r_1(|\overline{\Lambda}|)$, and taking δ smaller yet if necessary, we can force $r_1(|\overline{\Lambda}|) \leq \epsilon$. Moreover, since $\|h(t)\|_{H^1}$ is continuous for $t \in [0, t_m)$, the existence of the excluded region implies that $\|h\|_{H^1}$ cannot escape the connected set $(0, r_1) \subset (0, \epsilon)$. In particular, for all $t \in [0, t_m]$,

$$\inf_{\gamma \in \mathbb{R}} \|u(t) - T(\gamma(t))\phi_c\|_{H^1} \leq \|h(t)\|_{H^1} \leq \epsilon.$$

Since $\|h(t_m)\|_{H^1}$ is uniformly bounded, independent of t_m, we can choose ϵ so small that $u(t) \in B$, the ball about $\mathcal{M}_T$, and iterate the process to extend t_m by a uniform amount. Arguing inductively we can extend t_m to $+\infty$, which proves the orbital stability modulo (5.2.18). The positivity of $\partial_c Q(\phi_c)$ for $p \in [1, 4)$ and $c \in \mathbb{R}$ follows from (5.2.12). □

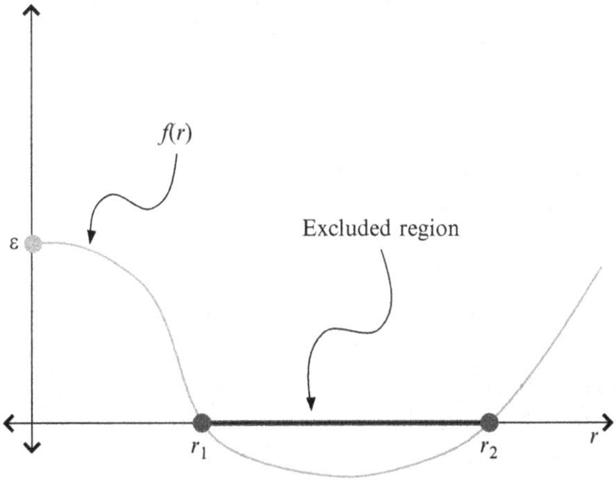

Fig. 5.4 The graph of the function $f(r)$ given in (5.2.24) is denoted by a solid (*green*) curve. The two zeros of the function are shown as (*red*) circles, and the excluded region is shown as a solid (*blue*) line. (Color figure online.)

It remains to establish the central estimate (5.2.23). The first obstacle is the fact that space $\mathcal{A}_c$ is nonlinear. Since we consider only small h, it is sufficient to develop the inequality (5.2.23) for h in the tangent plane to the admissible space at $h = 0$. Taylor expanding Q yields

$$Q(\phi_c + h) - Q(\phi_c) = \left\langle \frac{\delta Q}{\delta u}(\phi_c), h \right\rangle + \frac{1}{2}\|h\|_{L^2}^2 = \langle \phi_c, h\rangle + \frac{1}{2}\|h\|_{L^2}^2.$$

We deduce that the linearized version of the nonlinear constraint in $\mathcal{A}_c$ is the condition $\langle \phi_c, h\rangle = 0$. Accordingly, we define the linear admissible space

$$\mathcal{A}_c' := \left\{ h \in H^1(\mathbb{R}) : \langle h, \phi_c\rangle = \langle h, \partial_\xi \phi_c\rangle = 0 \right\}. \tag{5.2.25}$$

We remark that $\mathcal{A}_c' = \text{gker}(\partial_\xi \mathcal{L})^\perp$ is the tangent plane to $\mathcal{A}_c$ at ϕ_c. Exercise 5.2.3 shows that the linear and the nonlinear spaces differ by a quadratic amount for h small; in particular it is sufficient to establish (5.2.60) for sufficiently small h.

The constraint $h \in \mathcal{A}_c'$ in the estimate (5.2.60) is significant. The essential spectrum of $\mathcal{L}$, $\sigma_{\text{ess}}(\mathcal{L}) = \{\lambda \in \mathbb{C} : \lambda = c + k^2,\, k \in \mathbb{R}\}$, is contained in the open right-half of the complex plane. However, the operator $-\mathcal{L}$ is Sturmian with a simple kernel spanned by ϕ_c', which has a single internal zero, the rightmost point of the blue orbit of Fig. 5.2. From Chapter 2.3 we deduce that $\mathcal{L}$ has one negative eigenvalue $\lambda_0 < 0$, with associated ground-state eigenfunction ψ_0 that has no zeros. Not only is $\mathcal{L}$ not uniformly positive, but also $\langle \mathcal{L}\psi_0, \psi_0\rangle = \lambda_0 \|\psi_0\|_2^2 < 0$. The operator $\mathcal{L}$ cannot satisfy the inequality (5.2.60) for all $h \in H^1(\mathbb{R})$.

To obtain the coercivity of $\mathcal{L}$ we must take into account the fact that it acts on a constrained space. To unravel the role of the constraint it is natural to define the bilinear form

$$b[u,v] = \langle \mathcal{L}u, v\rangle, \quad u, v \in \mathcal{A}_c'. \tag{5.2.26}$$

If $h \in H^1(\mathbb{R})$, i.e., if the bilinear form is not constrained, then b induces the operator $\mathcal{L}$ [244, Chapter VIII]. The question is:

What operator does the constrained bilinear form induce?

There are two layers to the problem: the first is the kernel of $\mathcal{L}$, and the second is the negative eigenvalue. We attack each layer by using an element of the generalized kernel of L.

We first consider the kernel of $\mathcal{L}$, which is also the kernel of $L = \mathcal{J}\mathcal{L}$. We introduce $P_0 : L^2(\mathbb{R}) \mapsto \ker(\mathcal{L})$, the spectral projection onto the kernel of $\mathcal{L}$,

$$P_0 u = \frac{\langle u, \partial_\xi \phi_c\rangle}{\langle \partial_\xi \phi_c, \partial_\xi \phi_c\rangle} \partial_\xi \phi_c,$$

and its complement $\Pi_0 = \mathcal{I} - P_0$ which has the range,

$$X_0 = \left\{ h \in L^2(\mathbb{R}) : \langle h, \partial_\xi \phi_c\rangle = 0 \right\} = \ker(\mathcal{L})^\perp.$$

Both of these projections are self-adjoint. Recalling that $Y = H^1(\mathbb{R})$ we observe

$$\mathcal{A}_c' = \text{span}\{\partial_\xi \phi_c\}^\perp \cap \text{span}\{\phi_c\}^\perp \cap Y,$$

and decompose the new base space $Y_0 := X_0 \cap Y$ into the space we wish to study, and the remainder of the Jordan block of $\mathbb{L}$, i.e.,

$$Y_0 = \mathcal{A}_c' \oplus \text{span}\{\phi_c\}.$$

Subject to the constraint $u, v \in Y_0$ we have the relationships

$$b[u,v] = \langle \mathcal{L}u, v\rangle = \langle \mathcal{L}u, \Pi_0 v\rangle = \langle \Pi_0 \mathcal{L}u, v\rangle.$$

This motivates the definition $\mathcal{L}_0 := \Pi_0 \mathcal{L}$, which maps $\mathcal{L}_0 : \mathcal{D}(\mathcal{L}) \cap Y_0 \subset X_0 \mapsto X_0$. The operator $\mathcal{L}_0$ is the operator induced by the bilinear form b constrained to act on Y_0. Since P_0 is a spectral projection of $\mathcal{L}$, $\mathcal{L}_0$ is very similar to $\mathcal{L}$. In particular, $\sigma(\mathcal{L}_0) = \sigma(\mathcal{L})\backslash\{0\}$, so that $\mathcal{L}_0$ is boundedly invertible, has a single negative eigenvalue λ_0 with associated eigenfunction ψ_0, with the rest of $\sigma(\mathcal{L}_0)$ strictly positive.

We further constrain the bilinear form b to act on $\mathcal{A}_c'$. We introduce the self-adjoint projection

$$\Pi u = u - \frac{\langle u, \phi_c\rangle}{\langle \phi_c, \phi_c\rangle}\phi_c. \tag{5.2.27}$$

This is *not* a spectral projection for $\mathcal{L}$, nor for $\mathcal{L}_0$. However, the projection operator does commute with Π_0, i.e., $\Pi_0 \Pi = \Pi \Pi_0$ and hence $\Pi : Y_0 \mapsto \mathcal{A}_c'$. The bilinear form b constrained to $\mathcal{A}_c'$ induces the *constrained operator*

$$\mathcal{L}_\Pi := \Pi \mathcal{L}_0 : \mathcal{D}(\mathcal{L}) \cap \mathcal{A}_c' \subset X_0 \to \Pi X_0.$$

The constrained operator $\mathcal{L}_\Pi$ is self-adjoint, so its spectrum is real-valued. Moreover, the difference $\mathcal{L}_0 - \Pi \mathcal{L}_0$, acting on $\mathcal{A}_c'$ is rank-one, and hence compact, so the Weyl essential spectrum theorem implies that $\sigma_{\text{ess}}(\Pi \mathcal{L}_0) = \sigma_{\text{ess}}(\mathcal{L}_0)$. If the point spectra of $\mathcal{L}_\Pi$ is strictly positive, then we can obtain the bound (5.2.60) we seek. The lemma below, motivated by [**279**, Appendix A], gives a sharp characterization of the lower bound of $\sigma(\mathcal{L}_\Pi)$. This result is generalized considerably in Chapter 5.3.

Lemma 5.2.3. *The ground-state eigenvalue α_0 of the constrained operator $\mathcal{L}_\Pi$ acting on $\mathcal{A}_c'$ is strictly positive if and only if*

$$\partial_c Q(\phi_c) > 0. \tag{5.2.28}$$

Moreover, under this condition there exists $\alpha > 0$ such that

$$b[h,h] = \langle \mathcal{L}h, h\rangle \geq \alpha \|h\|_{H^1}^2, \quad h \in \mathcal{A}_c'. \tag{5.2.29}$$

Proof. The ground-state eigenvalue, α_0, of $\mathcal{L}_\Pi$ has the variational characterization,

$$\alpha_0 = \inf_{\psi \in \mathcal{A}'_c} \frac{\langle \mathcal{L}\psi, \psi \rangle}{\langle \psi, \psi \rangle} \tag{5.2.30}$$

[**60**]. The ground state eigenvalue $\lambda_0 < 0$ of $\mathcal{L}$ has a similar formulation, with the infimum over $\psi \in H^1(\mathbb{R})$. We immediately deduce that $\lambda_0 \leq \alpha_0$, since λ_0 is the infimum of the same functional over a larger space. Moreover, since both ψ_0 and ϕ_c are of one sign over $\mathbb{R}$, $\langle \psi_0, \phi_c \rangle \neq 0$, so $\psi_0 \notin \mathcal{A}'_c$, and hence $\lambda_0 < \alpha_0$.

The infimum of (5.2.30) is achieved at $\psi \in \mathcal{A}'_c$, [**81**, Chapter 8], which solves the eigenvalue problem

$$\mathcal{L}_\Pi \psi = \alpha_0 \psi \quad \Rightarrow \quad \mathcal{L}_0 \psi = \alpha_0 \psi + \mu \phi_c, \tag{5.2.31}$$

for some value of the Lagrange multipliers $\alpha_0, \mu \in \mathbb{R}$. As $\mathcal{L}_0$ is invertible on X_0, we may solve for $\psi \in \ker(\mathcal{L})^\perp$ to obtain

$$\psi(\alpha_0) = \mu (\mathcal{L}_0 - \alpha_0)^{-1} \phi_c, \tag{5.2.32}$$

where (without loss of generality) the Lagrange multiplier $\mu = \mu(\alpha_0) > 0$ is chosen to satisfy the constraint $\langle \psi, \psi \rangle = 1$. To enforce the condition $\psi \in \mathcal{A}'_c$ we impose the constraint $\langle \psi, \phi_c \rangle = 0$: it is this condition that determines the unknown α_0.

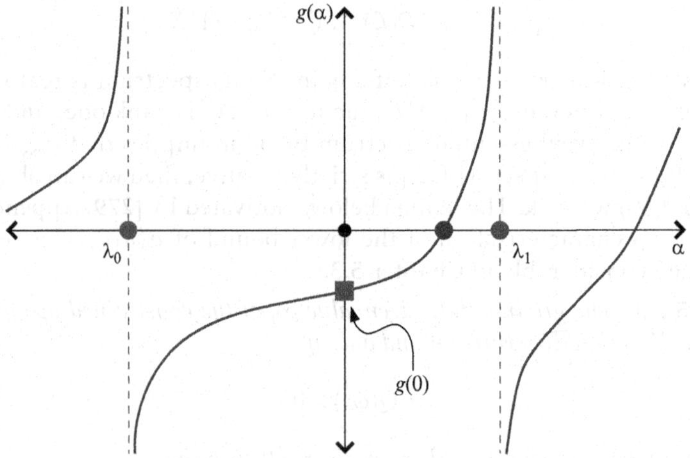

Fig. 5.5 A typical plot of $g(\alpha_0)$ is given by the solid (*blue*) curve. The vertical asymptotes associated with the first two eigenvalues $\lambda_0 < 0 < \lambda_1$ are denoted by dashed (*red*) lines. In this portrayal, $g(0) < 0$, so that the ground-state eigenvalue, α_0, of $\mathcal{L}_\Pi$, which is denoted by a (*blue*) circle, is positive. (Color figure online.)

We use (5.2.32) to rewrite the constraint as a function of α_0 alone,

$$g(\alpha_0) := \frac{1}{\mu}\langle\psi,\phi_c\rangle = \left\langle(\mathcal{L}_0-\alpha_0)^{-1}\phi_c,\phi_c\right\rangle, \tag{5.2.33}$$

so that $g(\alpha_0) = 0$ precisely when α_0 is an eigenvalue of $\mathcal{L}_\Pi$. In particular, if $g(\alpha_0)$ is nonzero for all $\alpha_0 \le 0$, then the ground-state eigenvalue must be positive. The function g is (real) analytic for $\alpha_0 \notin \sigma(\mathcal{L}_0)$, with pole singularities possible at the eigenvalues of $\mathcal{L}_0$, and is strictly increasing where it is smooth, since

$$g'(\alpha_0) = \langle(\mathcal{L}_0-\alpha_0)^{-2}\phi_c,\phi_c\rangle = \left\|(\mathcal{L}_0-\alpha_0)^{-1}\phi_c\right\|_2^2 > 0 \tag{5.2.34}$$

(see Exercise 5.2.4). The operator $\mathcal{L}_0$ has no spectrum on (λ_0,λ_1), where $0 < \lambda_1$ is the smallest element of $\sigma(\mathcal{L}_0)\backslash\{\lambda_0\}$. Consequently, the function g is analytic and monotonically increasing on this interval. Moreover, the condition $\langle\phi_c,\psi_0\rangle \neq 0$ implies

$$\lim_{\alpha_0\to\lambda_0^+} g(\alpha_0) = -\infty \tag{5.2.35}$$

(see Exercise 5.2.4). It follows that the first zero, α_0, of g is negative if $g(0) > 0$, positive if $g(0) < 0$ and equals zero if $g(0) = 0$ (see Fig. 5.5). This zero is the ground-state eigenvalue of $\mathcal{L}_\Pi$.

To relate the value of $g(0)$ to the derivative of the charge $Q(\phi_c)$ with respect to wave speed we take the derivative of the traveling-wave ODE (5.2.6) with respect to speed c, and as in (5.2.10) we find that $\mathcal{L}\partial_c\phi_c = -\phi_c$. From parity considerations $P_0\phi_c = 0$, so that $\phi_c \in (\ker(\mathcal{L}))^\perp$ and we may invert $\mathcal{L}_0$

$$\mathcal{L}_0\partial_c\phi_c = -\phi_c \quad\Rightarrow\quad \partial_c\phi_c = -\mathcal{L}_0^{-1}\phi_c. \tag{5.2.36}$$

This permits us to rewrite $g(0)$ as

$$g(0) = \langle\mathcal{L}_0^{-1}\phi_c,\phi_c\rangle = -\langle\partial_c\phi_c,\phi_c\rangle = -\partial_c Q(\phi_c).$$

We see that the condition (5.2.28) is intimately related to the sign of α_0, and in particular $\partial_c Q(\phi_c) > 0$ implies that $\alpha_0 > 0$.

Now we consider the estimate (5.2.29). In the case that $\alpha_0 > 0$ the variational characterization (5.2.30) gives the $L^2(\mathbb{R})$ coercivity

$$\langle\mathcal{L}v,v\rangle \ge \alpha_0\|v\|_2^2, \quad v\in\mathcal{A}_c'.$$

To obtain $H^1(\mathbb{R})$ coercivity we argue by contradiction. If the bilinear form is not $H^1(\mathbb{R})$ coercive over $\mathcal{A}_c'$, then there is a sequence of functions $\{v_n\}_{n=1}^\infty \subset H^1(\mathbb{R})\cap\mathcal{A}_c'$ with $\|v_n\|_{H^1} = 1$ for which $\langle\mathcal{L}v_n,v_n\rangle \to 0$ as $n\to\infty$. The $L^2(\mathbb{R})$ coercivity implies that $\lim_{n\to\infty}\|v_n\|_2 = 0$, and since their $H^1(\mathbb{R})$ norms are unity we have

$$\lim_{n\to\infty}\langle\partial_\xi v_n,\partial_\xi v_n\rangle = 1.$$

For the potential $V(\xi) := c - \phi_c^p(\xi)$ associated with $\mathcal{L}$ we have the limit

$$\lim_{n\to\infty} |\langle V(\xi)v_n, v_n\rangle| \leq \|V\|_\infty \lim_{n\to\infty} \langle v_n, v_n\rangle = 0.$$

These limits bring us to the contradiction; namely,

$$0 = \lim_{n\to\infty} \langle \mathcal{L}v_n, v_n\rangle = \lim_{n\to\infty} \Big(\langle \partial_\xi v_n, \partial_\xi v_n\rangle + \langle V(\xi)v_n, v_n\rangle\Big) = 1,$$

which establishes the $H^1(\mathbb{R})$ coercivity (5.2.29) for some $\alpha > 0$. □

Remark 5.2.4. For $p > 4$ not only do the uniform $H^1(\mathbb{R})$ bounds (5.2.17) fail, but the solutions of gKdV are not globally well-posed in time; in fact, they exhibit finite-time blow up for initial data of sufficiently localized mass. The traveling-wave solutions are also spectrally unstable for these values of p (see Chapter 10.3.1).

5.2.2 *General Orbital Stability Result*

The ideas leading to the nonlinear stability of the traveling-wave solution of the generalized KdV equation can be placed within a general framework that in many ways is the culmination of Boussinesq's remarks from 1872. Our presentation is a slight extension of the 1990 work of Grillakis et al. who, in their seminal paper **[110]**, first developed a comprehensive treatment for stability in Hamiltonian systems in the presence of symmetries. We return to the Hamiltonian systems of Chapter 5.2, i.e.,

$$\partial_t u = \mathcal{J}\frac{\delta\mathcal{H}}{\delta u}(u), \tag{5.2.37}$$

and impose the following conditions on $\mathcal{J}$ and the Hamiltonian $\mathcal{H}$:

Hypothesis 5.2.5. Hamiltonian Framework

(a) $\mathcal{H} : Y \mapsto \mathbb{R}$ has two continuous derivatives, $\delta\mathcal{H}/\delta u : Y \mapsto Y^*$, and $\delta^2\mathcal{H}/\delta u^2 : \mathcal{D} \subset X \mapsto X$ generates a continuous bilinear form on Y. In particular for each critical point ϕ of $\mathcal{H}$ there exists $c > 0$ such that

$$\left|\left\langle \frac{\delta^2\mathcal{H}}{\delta u^2}(\phi)v, v\right\rangle\right| \leq c\|v\|_Y^2,$$

and moreover

$$\left|\mathcal{H}(\phi+v) - \mathcal{H}(\phi) - \left\langle \frac{\delta^2\mathcal{H}}{\delta u^2}(\phi)v, v\right\rangle\right| \leq c\|v\|_Y^3, \tag{5.2.38}$$

for all $\|v\|_Y$.

(b) $\mathcal{J} : Y \mapsto X$ is skew-symmetric with respect to the X inner product and has a finite-dimensional kernel, i.e., there exists $M > 0$ such that

$$\ker(\mathcal{J}) = \text{span}\{\psi_1, \cdots, \psi_M\}, \tag{5.2.39}$$

with the ψ_j's orthonormal in X.

(c) Both $\mathcal{J}$ and the Hamiltonian $\mathcal{H}$ possess an N-dimensional symmetry group $T(\gamma) : Y \mapsto Y$ satisfying (4.2.1) and (4.2.2); in particular

$$\mathcal{H}(T(\gamma)u) = \mathcal{H}(u), \quad \mathcal{J}T(\gamma) = T(\gamma)\mathcal{J}, \tag{5.2.40}$$

for all $u \in Y$ and $\gamma \in \mathbb{R}^N$.

(d) The symmetry T is an isometry on X; that is,

$$\langle T(\gamma)u, T(\gamma)v\rangle_X = \langle u, v\rangle_X \tag{5.2.41}$$

for all $u, v \in X$ and all $\gamma \in \mathbb{R}^N$.

(e) For each $j = 1, \ldots, N$, the symmetry generator $T_j' : Y \subset X \to \ker(\mathcal{J})^\perp \subset X$. Moreover for each j the operator $\mathcal{J}^{-1}T_j : X \mapsto Y^*$ is bounded.

The first step in the development of the theoretical framework is to identify the quantities conserved by the system. These have two principal sources, the symmetries and the kernel of $\mathcal{J}$.

Lemma 5.2.6. *For each symmetry T_j the quadratic functional*

$$Q_j(u) := \frac{1}{2}\langle \mathcal{J}^{-1}T_j'u, u\rangle, \quad j = 1, \ldots, N. \tag{5.2.42}$$

is continuous in $\|\cdot\|_Y$ and invariant under the action of T and is conserved by the Hamiltonian flow. Furthermore, $\mathcal{J}^{-1}T_j'$ is self-adjoint in X and

$$\frac{\delta Q}{\delta u} = \mathcal{J}^{-1}T_j'. \tag{5.2.43}$$

Proof. From Hypothesis 5.2.5 (e) it follows that $\mathcal{J}^{-1}T_j' : Y \mapsto Y^*$ is bounded, while from Hypothesis 5.2.5(c) and (4.2.1)(b) it follows that T commutes with $\mathcal{J}^{-1}T_j'$. The assumption Hypothesis 5.2.5(d) thus implies that Q_j is invariant under the action of T. Moreover, since T_j is an isometry on X it follows that

$$0 = \partial_\gamma \langle T_j(\gamma)u, T_j(\gamma)\rangle\Big|_{\gamma=0} = \langle T_j'u, u\rangle + \langle u, T_j'u\rangle,$$

and hence T_j' is skew-symmetric on Y in the X inner product. By assumption, $\mathcal{J}$ commutes with T_j and hence with T_j'. Finally $\mathcal{J}^{-1}$ is skew-symmetric on its domain, $\ker(\mathcal{J})^\perp$. Let $\Pi_{\mathcal{J}}$ denote the orthogonal projection onto

$\ker(\mathcal{J})^\perp$. Then we find that

$$\langle \mathcal{J}^{-1}T_j' u, v\rangle = \langle \mathcal{J}^{-1}T_j' u, \Pi_{\mathcal{J}} v\rangle = \langle u, T_j' \mathcal{J}^{-1}\Pi_{\mathcal{J}} v\rangle.$$

However, T_j' commutes with $\mathcal{J}^{-1}$ and hence with $\Pi_{\mathcal{J}}$, while from Hypothesis 5.2.5(e) we see that $\Pi_{\mathcal{J}} T_j' = T_j'$; combining these two observations with the line above we deduce

$$\langle \mathcal{J}^{-1}T_j' u, v\rangle = \langle u, \mathcal{J}^{-1}T_j' v\rangle,$$

and hence $\mathcal{J}^{-1}T_j'$ is self-adjoint in the X inner product. The statement (5.2.43) follows immediately. Finally, the string of equalities

$$\begin{aligned}\partial_t Q_j(u) &= \left\langle \frac{\delta Q_j}{\delta u}(u), \partial_t u\right\rangle = -\left\langle T_j' u, \frac{\delta \mathcal{H}}{\delta u}(u)\right\rangle \\ &= -\left\langle T_j(\gamma)T_j' u, \frac{\delta \mathcal{H}}{\delta u}(T(\gamma)u)\right\rangle = -\partial_{\gamma_j}\mathcal{H}(T(\gamma)u) = 0,\end{aligned}$$

show that Q_j is conserved under the Hamiltonian flow. □

Remark 5.2.7. While each symmetry generates a quadratic conserved quantity, each element $\psi_j \in \ker(\mathcal{J})$ generates a linear functional $\langle u, \psi_j\rangle$ that is conserved under the flow

$$\partial_t\langle u, \psi_j\rangle = \langle \partial_t u, \psi_j\rangle = \left\langle \mathcal{J}\frac{\delta \mathcal{H}}{\delta u}, \psi_j\right\rangle = -\left\langle \frac{\delta \mathcal{H}}{\delta u}, \mathcal{J}\psi_j\right\rangle = 0.$$

Remark 5.2.8. The gKdV equation does not directly fit into the framework Hypothesis 5.2.5. Although the skew-symmetric operator $\mathcal{J} = \partial_x$ has no kernel on $Y = H^1(\mathbb{R})$, it is not boundedly invertible on any finite-codimension subset of $L^2(\mathbb{R})$. However the gKdV is special in that the skew operator is the generator of the translational symmetry: $T' = \partial_x = \mathcal{J}$. In particular $\mathcal{J}^{-1}T' = \mathcal{I}$ on $Y = H^1(\mathbb{R})$. In this nongeneric situation we can replace the requirement that $\mathcal{J}^{-1} : \ker(\mathcal{J})^\perp \subset X \to Y$ is bounded with $\mathcal{J}^{-1}T_j' : Y \to Y$ is bounded and self-adjoint. This permits the definition of Q_j in (5.2.42) satisfying (5.2.43).

The goal of this section is to establish criteria that determine the stability of generalized traveling-waves of (5.2.37). For each collection $\boldsymbol{c} = (c_1, \ldots, c_N)^{\mathrm{T}}$ of speeds these are solutions to (5.2.37) of the form

$$u_c(x,t) := T(-\boldsymbol{c}t)\phi_c(x) = T_1(-c_1 t)\cdots T_N(-c_N t)\phi_c(x).$$

The speeds c_j correspond to rates of evolution of the symmetry parameters, e.g., for the nonlinear Schrödinger equation (NLS) equation the speeds correspond to rates of translation and complex rotation. Inserting this ansatz into (5.2.37), and using the invariance of T, we find that ϕ_c must satisfy the generalized traveling-wave ODE,

$$-\sum_{j=1}^{N} c_j T_j' \phi_c = \mathcal{J} \frac{\delta \mathcal{H}}{\delta u}(\phi_c). \tag{5.2.44}$$

The traveling-wave ODE is the variational derivative of the Lagrangian,

$$\Lambda(u;\boldsymbol{c}) := \mathcal{H}(u) + \sum_{i=1}^{N} c_i Q_i(u). \tag{5.2.45}$$

Most significantly, if $u = u(t)$ is a solution of (5.2.37), then the boosted function $\tilde{u}(t) := T(\boldsymbol{c}t)u(t)$ solves the "boosted" Hamiltonian system

$$\partial_t \tilde{u} = \mathcal{J} \frac{\delta \Lambda}{\delta u}(\tilde{u}), \tag{5.2.46}$$

so that ϕ_c is a critical point of Λ.

This motivates us to postulate the existence of critical points of Λ:

(f) There is an open, connected set $\Omega \subset \mathbb{R}^N$ such that for all $\boldsymbol{c} \in \Omega$ there is a ϕ_c that solves

$$\frac{\delta \Lambda}{\delta u}(\phi_c) = 0; \tag{5.2.47}$$

moreover, ϕ_c varies smoothly with $\boldsymbol{c}$ in $\|\cdot\|_Y$.

Since T is an isometry on X, the stability of the traveling wave u_c under the system generated by $\mathcal{H}$ is the same as that of the critical point ϕ_c under the boosted system. From the invariance of Λ with respect to T, each $\boldsymbol{c} \in \Omega$ yields a manifold

$$\mathcal{M}_T(\boldsymbol{c}) := \left\{ T(\gamma)\phi_c : \gamma \in \mathbb{R}^N \right\}, \tag{5.2.48}$$

of critical points of Λ. The orbital stability of $\mathcal{M}_T(\boldsymbol{c})$ hinges upon the spectrum of the second variation of the Lagrangian,

$$\mathcal{L} := \frac{\delta^2 \Lambda}{\delta^2 u}(\phi_c).$$

For each γ, $T(\gamma)\phi_c$ is also a critical point of Λ, and taking the derivative ∂_{γ_j} of the critical point equation we see that

$$\partial_{\gamma_j} \frac{\delta \Lambda}{\delta u}(T(\gamma)\phi_c) = 0 \quad \Rightarrow \quad \mathcal{L}\left(T_j' \phi_c\right) = 0,\ j = 1, \ldots, N.$$

We introduce the following notation

Definition 5.2.9. Consider a self-adjoint linear operator $\mathcal{L} : \mathcal{D}(\mathcal{L}) \subset X \mapsto X$. Suppose that $\mathcal{L}$ generates a continuous bilinear form on $Y \subset X$ in the X inner product. We introduce the negative index $\mathrm{n}_Y(\mathcal{L})$ that denotes the dimension of the largest subspace $K \subset Y$ over which the bilinear form b associated with $\mathcal{L}$ is negative,

$$b[v,v] := \langle \mathcal{L}v, v\rangle < 0, \quad v \in K\backslash\{0\}. \tag{5.2.49}$$

While the subspace is not unique, the dimension is well-defined, (e.g., see [**188**, Theorem 8.1]). We call the corresponding linear space the *negative cone* and denote it $\mathrm{N}_Y(\mathcal{L})$. We define the positive index $\mathrm{p}_Y(\mathcal{L})$ and the corresponding positive cone $\mathrm{P}_Y(\mathcal{L})$, while $\mathrm{z}_Y(\mathcal{L})$ denotes the dimension of $\ker(\mathcal{L})$, and $\mathrm{Z}_Y(\mathcal{L}) = \ker(\mathcal{L})$. If the space $Y \times Y$ is the natural domain of the bilinear form, then the Y subscript will be omitted.

Regarding the operator $\mathcal{L}$ we add the following assumptions:

(g) $\mathrm{z}(\mathcal{L}) = N$ and $\mathrm{Z}(\mathcal{L}) = \mathrm{span}\{T_1'\phi_c, \dots, T_N'\phi_c\}$.
(h) The dimension of the negative space of $\mathcal{L}$ is finite; i.e., $\mathrm{n}(\mathcal{L}) < \infty$.
(i) There exists $\delta > 0$ such that

$$b[v,v] \ge \delta \|v\|_Y^2, \quad v \in \mathrm{P}(\mathcal{L}).$$

Remark 5.2.10. For exponentially asymptotic self-adjoint operators of the class (3.1.1), hypothesis (i) holds if the essential spectrum is uniformly bounded away from the origin, i.e., $\sigma_{\mathrm{ess}} \subset [\delta_0, \infty)$ for some $\delta_0 > 0$.

The proof of orbital stability begins with the foliation of a neighborhood B of the manifold $\mathcal{M}_T(c)$ traced out by critical point ϕ_c under the action of the symmetries. If the solution $u(t)$ of the boosted Hamiltonian system, (5.2.46), is sufficiently close to $\mathcal{M}_T(c)$, then from Lemma 4.3.3 and Hypothesis 5.2.5(g) we may decompose it as

$$u(t) = T(\gamma(t))\phi_c + h(t),$$

where $h(t) \in \ker(\mathcal{L}_\gamma)^\perp$ and $\mathcal{L}_\gamma$ is the second variational derivative of Λ at $T(\gamma)\phi_c$. If $\mathcal{L}_\gamma$ has negative eigenvalues, then the critical point $T(\gamma)\phi_c$ cannot be a minimizer of Λ; however, it can be a constrained minimizer. Since the spectrum of $\mathcal{L}_\gamma$ and the conserved quantities are all independent of γ, and since we focus only on showing the coercivity of $\mathcal{L}$ subject to the constraints, we suppress the γ dependence. The perturbation h and symmetry parameters γ satisfy

$$\partial_t h - \sum_{j=1}^{N} T_j'\phi_c \partial_t \gamma_j = \mathcal{J}\frac{\delta \Lambda}{\delta u}(\phi_c + h), \quad h(0) = u_0 - \phi_c =: h_0 \tag{5.2.50}$$

[compare with (4.3.12)]. As for the gKdV example, we will determine the orbital stability of $\mathcal{M}_T(c)$ not from the nonlinear flow, but rather from the properties of $\mathcal{L}$.

Since the values of the conserved quantities are set by the initial data, the perturbation h evolves within the nonlinear admissible set

$$\mathcal{A}_c = \{h \in Y : Q_j(\phi_c + h) = Q_j(\phi_c), j = 1,\dots,N; h \in \ker(\mathcal{J})^\perp; h \in \ker(\mathcal{L})^\perp\},$$

where the first set of constraints arises from the symmetries, the second from the kernel of $\mathcal{J}$, and the third from the foliation decomposition. In particular, the speeds $\boldsymbol{c}$ are set from the initial data through the conditions $Q_j(\phi_c) = Q_j(u_0)$ for $j = 1,\dots,N$. The tangent space to $\mathcal{A}_c$ at ϕ_c forms the admissible space

$$\mathcal{A}'_c = \{h \in Y : \langle h, \mathcal{J}^{-1} T'_j \phi_c\rangle = 0,\ j = 1,\dots,N; h \in \ker(\mathcal{J})^\perp \cap \ker(\mathcal{L})^\perp\}. \quad (5.2.51)$$

The heart of the orbital stability proof is to show the coercivity of the bilinear form b of (5.2.49) when constrained to act upon the admissible space $\mathcal{A}'_c$.

Before addressing the constrained bilinear form, we first address the kernel of $\mathcal{L}$ by forming the spaces $X_0 := \ker(\mathcal{L})^\perp$ and $Y_0 = Y \cap X_0$, and defining the operator $\mathcal{L}_0 := \Pi_0 \mathcal{L} : \mathcal{D}(\mathcal{L}) \cap Y_0 \subset X_0 \to X_0$, where Π_0 is the orthogonal (spectral) projection onto X_0. Since Π_0 is a spectral projection and commutes with $\mathcal{L}$, the operator $\mathcal{L}_0 = \mathcal{L}\big|_{\mathcal{D}(\mathcal{L})\cap Y_0}$ and $\sigma(\mathcal{L}_0) = \sigma(\mathcal{L})\backslash\{0\}$. That is, $\mathcal{L}_0$ merely loses the kernel of $\mathcal{L}$, and in particular $\mathrm{n}(\mathcal{L}_0) = \mathrm{n}_{Y_0}(\mathcal{L}) = \mathrm{n}(\mathcal{L})$. Moreover, by Hypothesis 5.2.5(g–i) we see that the inverse operator $\mathcal{L}_0^{-1} : X_0 \to Y_0$ is bounded. This restriction represents the orbital element of the stability via the orthogonal foliation of a neighborhood of $\mathcal{M}_T(\boldsymbol{c})$.

The essential step is to gauge the impact of the constraints upon the invertible operator $\mathcal{L}_0$. From the form of (5.2.51) we observe that the constraints arising from the conserved quantities form the space $\mathcal{J}^{-1}\ker(\mathcal{L})$. This motivates the definition of the *constraint space*, $\mathcal{S}$, as the orthogonal complement of the admissible space $\mathcal{A}'_c$ inside of X_0,

$$S := (\mathcal{A}'_c)^\perp \cap X_0 = \ker(\mathcal{J}) \oplus \mathcal{J}^{-1}\ker(\mathcal{L}) = \mathrm{span}\{s_j : j = 1,\dots,M+N\}, \quad (5.2.52)$$

where we have introduced the basis elements,

$$s_j = \begin{cases} \mathcal{J}^{-1} T'_j \phi_c, & j = 1,\dots,N \\ \psi_{j-N}, & j = N+1,\dots,M+N, \end{cases}$$

where the ψ_j's span the kernel of $\mathcal{J}$, see (5.2.39). Constraining the bilinear form b of (5.2.49) to act on $\mathcal{A}'_c \subset Y_0$ induces the constrained operator

$$\mathcal{L}_\Pi := \Pi\mathcal{L}_0 : \mathcal{D}(\mathcal{L}) \cap \mathcal{A}'_c \subset \Pi X_0 \mapsto \Pi X_0,$$

where Π is the X-orthogonal projection $\Pi : Y_0 \mapsto \mathcal{A}'_c$. It is significant that Π is not a spectral projection for $\mathcal{L}_0$ and $\ker(\Pi) = \mathcal{S}$ is not an eigenspace of $\mathcal{L}_0$. From Hypothesis 5.2.5(i) the constrained operator is Y-coercive if

$\mathrm{n}(\mathcal{L}_\Pi) = \mathrm{n}_{\mathcal{A}'_c}(\mathcal{L}) = 0$. Determining this index requires us to gauge the impact of the conserved quantities on the curvatures of the energy surface of Λ near ϕ_c. The central result, presented in Chapter 5.3, quantifies the impact of the constraint space S on $\mathcal{L}_0$ through the negative index of the symmetric *constraint matrix* $\boldsymbol{D} \in \mathbb{R}^{(M+N)\times(M+N)}$ with entries

$$\boldsymbol{D}_{ij} := \left\langle \mathcal{L}^{-1} s_i, s_j \right\rangle. \tag{5.2.53}$$

Specifically, Theorem 5.3.2 shows that the negative index $\mathrm{n}(\boldsymbol{D})$ equals the number of negative eigenvalues that are removed from $\sigma_{\mathrm{pt}}(\mathcal{L}_0)$ when $\mathcal{L}_0$ is constrained to act on $\mathcal{A}'_c$.

To highlight the connection between the constraint matrix $\boldsymbol{D}$ and the conserved quantities, we reformulate the matrix. Differentiating the existence equation (5.2.44) with respect to c_i for $i = 1,\dots,N$ yields

$$\mathcal{L}\partial_{c_i}\phi_c = -\mathcal{J}^{-1}T_i'\phi_c = -s_i = -\frac{\delta Q_i}{\delta u}(\phi_c). \tag{5.2.54}$$

In particular, for $i,j \le N$ we have the expression

$$\boldsymbol{D}_{ij} = -\left\langle \frac{\delta Q_i}{\delta u}(\phi_c), \partial_{c_j}\phi_c \right\rangle = -\frac{\partial Q_i}{\partial c_j}(\phi_c). \tag{5.2.55}$$

To extend this reformulation to the remainder of $\boldsymbol{D}$ we assume that ϕ_c can be smoothly extended to a larger family of equilibria of (5.2.37) which are not critical points of $\mathcal{H}$, but which arise from the impact of the kernel of $\mathcal{J}$.

(f') For some $\delta > 0$ there is an open connected set $\Omega' \subset \mathbb{R}^N \times [-\delta,\delta]^M$ such that for all $(\boldsymbol{c},\tilde{\boldsymbol{c}}) = (\boldsymbol{c}, c_{N+1},\dots,c_{N+M}) \in \Omega'$ there is a $\phi_{c,\tilde{c}}$ that is smooth in $\|\cdot\|_Y$ on Ω' and solves

$$\frac{\delta\Lambda}{\delta u}(\phi) = -\sum_{j=1}^{M} c_{N+j}\psi_j \in \ker(\mathcal{J}). \tag{5.2.56}$$

Taking ∂_{c_j} of (5.2.56), for $j \ge N$ we see that the equilibrium $\phi_{c,\tilde{c}}$ has the property

$$\mathcal{L}\partial_{c_j}\phi_{c,\tilde{c}}\big|_{\tilde{c}=0} = -\psi_{j-N} = -s_j.$$

Renaming the conserved quantities induced by $\ker(\mathcal{J})$ as

$$Q_{N+j}(u) := \left\langle \psi_j, u \right\rangle = \left\langle s_{N+j}, u \right\rangle, \tag{5.2.57}$$

we may rewrite the constraint matrix $\boldsymbol{D}$ in the compact form

$$\boldsymbol{D}_{ij} = -\frac{\partial Q_i}{\partial c_j}(\phi_{c,\tilde{c}}), \quad i,j = 1,\dots,N+M.$$

The constrained operator $\mathcal{L}_\Pi$ is Y-coercive if $\mathrm{n}(\mathcal{L}_\Pi) = 0$. This is guaranteed by Proposition 5.3.1 if $\boldsymbol{D}$ has no kernel and (5.2.58) holds. In this case, we may apply the proof of Theorem 5.2.2 to the solution h of (5.2.50) to deduce the orbital stability of $\mathcal{M}_T(\boldsymbol{c})$. The remaining details are omitted. The orbital stability theorem was first established within this framework, subject to the condition that $\mathcal{J}$ is boundedly invertible, in Grillakis et al. [**110**].

Theorem 5.2.11. *Suppose that Hypothesis 5.2.5 (a)–(i) are satisfied, and in particular for each $\boldsymbol{c} \in \Omega$ let ϕ_c denote the critical point of Λ afforded by Hypothesis (f). For each $\boldsymbol{c} \in \Omega$ for which the constraint matrix $\boldsymbol{D} = \boldsymbol{D}(\boldsymbol{c})$ defined in (5.2.53) is nonsingular and satisfies*

$$\mathrm{n}(\mathcal{L}) = \mathrm{n}(\boldsymbol{D}), \tag{5.2.58}$$

then the corresponding manifolds $\mathcal{M}_T(\boldsymbol{c})$ are $\|\cdot\|_Y$-orbitally stable under the flow (5.2.37). In particular, for each $\epsilon > 0$ there is a $\delta > 0$ such that for any initial data, u_0, which satisfies

$$Q_j(u_0) = Q_j(\phi_c), \quad j = 1, \ldots, M+N, \tag{5.2.59}$$

then the corresponding solution u of (5.2.37) satisfies

$$\inf_{\gamma \in \mathbb{R}^N} \|u_0 - T(\gamma)\phi_c\|_Y \le \delta \quad \Rightarrow \quad \inf_{\gamma \in \mathbb{R}^N} \|u(t) - T(\gamma)\phi_c\|_Y \le \epsilon.$$

Remark 5.2.12. We emphasize that the constraints

$$Q_j(u_0) = Q_j(\phi_c) \quad j = N+1, \ldots, N+M,$$

enforce the condition $h_0 \in \mathcal{A}_c$ on the initial data of (5.2.50). The stability of ϕ_c is only for perturbations h_0 that do not change the values of the conserved quantities. If the perturbations change the conserved quantities, one must identify a new value of $\boldsymbol{c}$ that satisfies (5.2.59) and verify that (5.2.58) holds. The Hypothesis 5.2.5(f') is only required to reformulate $\boldsymbol{D}$ as in (5.2.58): the stability result holds in the absence of this strengthened assumption. An application to a Hamiltonian system for which $\mathcal{J}$ has a non-trivial kernel is presented in Chapter 6.1.2.

Remark 5.2.13. The generalized NLS with power nonlinearity

$$\mathrm{i}\partial_t u + \partial_x^2 u + |u|^p u = 0, \quad p \ge 1,$$

possesses a solitary wave $\phi_\omega(x)$ for each value of $p \ge 1$. Moreover,

$$\partial_\omega Q_2(\phi_\omega) \begin{cases} > 0, & 1 \le p < 4 \\ < 0, & 4 < p, \end{cases}$$

so that, as for gKdV, the solitary wave is stable for $p \in (1,4)$. The sign of $\partial_\omega Q_2(\phi_\omega)$ is known as the Vakhitov–Kolokolov stability criterion for solitary waves.

Exercises

Exercise 5.2.1. Consider $Y = H^1(\mathbb{R}) \subset L^2(\mathbb{R}) = X \subset Y^* = H^{-1}(\mathbb{R})$, where Y has the usual $H^1(\mathbb{R})$ inner product

$$\langle u,v\rangle_{H^1} = \int_{\mathbb{R}} (uv + \partial_x u \partial_x v)\,\mathrm{d}x.$$

Fix $p \geq 0$ and consider the Hamiltonian

$$\mathcal{H}(u) = \int_{\mathbb{R}} \left(\frac{1}{2}(\partial_x u)^2 + \frac{1}{(p+1)(p+2)}u^{p+2}\right)\mathrm{d}x.$$

Verify that the first variation of $\mathcal{H}$ with respect to the standard $L^2(\mathbb{R})$ inner product is given by

$$\frac{\delta\mathcal{H}}{\delta u}(u) = -\partial_x^2 u + \frac{1}{p+1}u^{p+1} \in H^{-1}(\mathbb{R})$$

where the $\partial_x^2 u$ term arises from an integration-by-parts. Show that for $p \geq 1$ the second variation of $\mathcal{H}$ with respect to the standard $L^2(\mathbb{R})$ inner product yields the linear operator

$$\mathcal{L} := \frac{\delta^2\mathcal{H}}{\delta u^2} = -\partial_x^2 + u^p : H^2(\mathbb{R}) \subset L^2(\mathbb{R}) \mapsto L^2(\mathbb{R}).$$

Exercise 5.2.2. Show that the gKdV Lagrangian defined in (5.2.14) is continuous in $H^1(\mathbb{R})$ on bounded sets; that is, for all $M, \epsilon > 0$ there exists $\delta > 0$ such that for all $\|u\|_{H^1}, \|v\|_{H^1} \leq M$,

$$\|u - v\|_{H^1} \leq \delta \quad \Rightarrow \quad |\Lambda(u) - \Lambda(v)| \leq \epsilon.$$

Exercise 5.2.3. Show that any $h \in \mathcal{A}_c$ [see (5.2.22)] with $\|h\|_{H^1}$ sufficiently small can be written as

$$h = h_1 + \nu\phi_c,$$

where $h_1 \in \mathcal{A}_c'$ [see (5.2.25)] and $|\nu| = \mathcal{O}\left(\|h\|_{H^1}^2\right)$. Use this result to show that if for $h \in \mathcal{A}_c'$ there is an $\alpha > 0$ such that

$$\langle \mathcal{L}h, h\rangle \geq \alpha\|h\|_{H^1}^2, \tag{5.2.60}$$

then the estimate (5.2.23) holds over $\mathcal{A}_c$.

Exercise 5.2.4. Let $\mathcal{L}$ be a Sturmian operator of the form (2.3.4) on a bounded domain $[-1,1]$, subject to separated boundary conditions of the form (2.3.2), so that Theorem 2.3.1 applies. Moreover, assume that the first-order coefficient of $\mathcal{L}$ satisfies $a_1 \equiv 0$, so that the eigenfunctions $\{\psi_j\}_{j=0}^{\infty}$ of $\mathcal{L}$ are orthonormal with respect to the usual $L^2(\mathbb{R})$ inner product, and assume the associated eigenvalues satisfy $\lambda_0 < 0 < \lambda_1 < \cdots$. Let $\phi_c > 0$ be a given function in $L^2[-1,1]$.

(a) Use the Green's function for the resolvent, (2.3.5), to obtain an explicit representation for the function

$$g(\lambda) := \langle (\mathcal{L} - \lambda)^{-1}\phi_c, \phi_c\rangle.$$

and prove (5.2.34) for this function g.

(b) Show that the function g is positive for $\lambda < \lambda_0$. Derive the limit of g as $\lambda \to -\infty$ and as $\lambda \to \lambda_0^+$.

Exercise 5.2.5. Suppose that the operator $\mathcal{L}_0$ in Lemma 5.2.3 has $M+1$ negative eigenvalues and no kernel; that is, $\lambda_0 < \lambda_1 < \cdots < \lambda_M < 0 < \lambda_{M+1} < \cdots$. Let $\phi \in H^1(\mathbb{R})$ be such that $\langle \phi, \psi_j\rangle \neq 0$ for $j = 0, \ldots, M$, where ψ_j is the eigenfunction associated with λ_j. Let Π be the projection off of ϕ, as in (5.2.27), and let $X_0 := \{\phi\}^\perp \subset L^2(\mathbb{R})$. Show that the operator $\mathcal{L}_\Pi := \Pi\mathcal{L} : \mathcal{D}(\mathcal{L}) \cap X_0 \subset X_0 s \to X_0$ has $M+1-\mathrm{n}(\langle \mathcal{L}_0^{-1}\phi, \phi\rangle)$ negative eigenvalues, where

$$\mathrm{n}(\langle \mathcal{L}_0^{-1}\phi, \phi\rangle) = \begin{cases} 0, & \langle \mathcal{L}_0^{-1}\phi, \phi\rangle > 0 \\ 1, & \langle \mathcal{L}_0^{-1}\phi, \phi\rangle < 0. \end{cases}$$

Exercise 5.2.6. Consider the operator $\mathbb{L} := \mathcal{J}\mathcal{L}$, obtained by linearizing (5.2.46) about ϕ_c.

(a) Show that $\mathbb{L}$ has an N-dimensional kernel, and at least a $2N$-dimensional generalized kernel. In particular, show that

$$\ker(\mathcal{L}) \oplus \mathcal{J}^{-1}\ker(\mathcal{L}) \subset \mathrm{gker}(\mathbb{L}), \tag{5.2.61}$$

where this is an orthogonal direct sum. *Hint:* Take the derivative of (5.2.47) with respect to c_j to establish the Jordan chain structure and the X orthogonality of the two spaces.

(b) Show that the two sets in (5.2.61) are equal, and moreover

$$(\mathcal{A}_c')^\perp = \mathrm{gker}(\mathbb{L}) \oplus \ker(\mathcal{J}) = \ker(\mathcal{L}) \oplus \mathcal{J}^{-1}\ker(\mathcal{L}) \oplus \ker(\mathcal{J}), \tag{5.2.62}$$

if the matrix $\boldsymbol{D}$ defined in (5.2.53) is nonsingular. Here $\mathcal{A}_c'$ is defined in (5.2.51). *Hint:* Use the Fredholm alternative.

Exercise 5.2.7. Show that if there exists an $\alpha > 0$ such that

$$b[h,h] \geq \alpha \|h\|_Y^2,$$

for all $h \in \mathcal{A}_c'$, then there exists $\beta > 0$ such that

$$b[h,h] \geq \alpha\|h\|_Y^2 - \beta\|h\|_Y^3,$$

for all $h \in \mathcal{A}_c$. The bilinear form is defined in Definition 5.2.9.

Exercise 5.2.8. Show that $\mathrm{n}(\boldsymbol{D}) \leq \mathrm{n}(\mathcal{L})$, where $\boldsymbol{D}$ is defined in (5.2.53).

Exercise 5.2.9. If $\tilde{c} \neq 0$, then the equilibrium $\phi_{c,\tilde{c}}$ is not a critical point of Λ.

(a) Show that for $j = 1,\dots,N$,

$$\mathcal{L}T_j'\phi_{c,\tilde{c}} = -\sum_{k=1}^{M} c_{N+k}T_j'\psi_k, \tag{5.2.63}$$

so that the symmetries need not generate elements of the kernel of $\mathcal{L}$.

(b) Find an augmented Lagrangian of which $\phi_{c,\tilde{c}}$ is a critical point, and show that the Hamiltonian flow can be written in terms of the new Lagrangian. Moreover, show that if $T_j'\psi_k = 0$ for all $j = 1,\dots,N$ and $k = 1,\dots,M$, then the augmented Lagrangian has the same symmetry group as the original. If the symmetry group of the augmented Lagrangian is smaller, then for small nonzero values of $\tilde{c}$, the perturbation methods of Chapter 7.2.1 can be used to check for instabilities arising from perturbations to the kernel of $\mathcal{JL}$.

Exercise 5.2.10. This exercise applies Theorem 5.2.11 to establish the orbital stability of the NLS (see [**46**, **279**, **280**] and the references therein for the original stability results on this problem). The NLS equation takes the form

$$\mathrm{i}\partial_t u + \partial_x^2 u + |u|^2 u = 0,$$

where $u \in \mathbb{C}$ and $(x,t) \in \mathbb{R} \times \mathbb{R}^+$ with $X = L^2(\mathbb{R})$ and $Y = H^1(\mathbb{R}) \subset X$. The NLS has translational and rotational symmetries and supports a two-parameter family of complex-valued generalized traveling waves, $u(x,t) = \mathrm{e}^{\mathrm{i}c_2 t}\phi(x - c_1 t)$, where ϕ solves

$$\partial_\xi^2\phi - \mathrm{i}c_1\partial_\xi\phi - c_2\phi + |\phi|^2\phi = 0, \tag{5.2.64}$$

in terms of the traveling variable $\xi = x - c_1 t$.

(a) Verify that for any $c \in \mathbb{R}$ the Gauge transformation,

$$u(x,t) \mapsto \mathrm{e}^{\mathrm{i}(cx/2 - c^2 t/4)}u(x - ct, t),$$

maps solutions of NLS to solutions. That is, if u solves NLS, then so does its Gauge transform. Show that for stability considerations it is sufficient to study standing waves $u(x,t) = \mathrm{e}^{\mathrm{i}\omega t}\phi_\omega(x)$, where ϕ_ω is the real solution of

$$\partial_x^2\phi - \omega\phi + \phi^3 = 0, \tag{5.2.65}$$

with $\omega = c_2 - c_1^2/4$.

(b) Show that the NLS equation can be written as a real system for $\boldsymbol{U} = (\operatorname{Re} u, \operatorname{Im} u)^{\mathrm{T}}$, which satisfies

$$\partial_t \boldsymbol{U} = -\mathcal{J}\left(\partial_x^2 \boldsymbol{U} + |\boldsymbol{U}|^2 \boldsymbol{U}\right) = \mathcal{J}\frac{\delta \mathcal{H}}{\delta \boldsymbol{U}}(\boldsymbol{U}),$$

where the skew-symmetric matrix and Hamiltonian are given, respectively, by

$$\mathcal{J} = \begin{pmatrix} 0 & -1 \\ 1 & 0 \end{pmatrix}, \quad \mathcal{H}(\boldsymbol{U}) = \frac{1}{2}\int_{\mathbb{R}} |\partial_x \boldsymbol{U}|^2 - \frac{1}{2}|\boldsymbol{U}|^4 \,\mathrm{d}x.$$

(c) The vector form of the NLS is invariant under the symmetries

$$T_1(\gamma_1)\boldsymbol{U}(x) = \boldsymbol{U}(x+\gamma), \quad T_2(\gamma_2)\boldsymbol{U} = \begin{pmatrix} \cos(\gamma_2) & \sin(\gamma_2) \\ -\sin(\gamma_2) & \cos(\gamma_2) \end{pmatrix}\boldsymbol{U},$$

for $\gamma_1 \in \mathbb{R}$ and $\gamma_2 \in [0, 2\pi)$. Verify that T_1 and T_2 satisfy (4.2.1) and hypotheses (c) and (d) of this section. Determine the operators T_1' and T_2', and find the conserved quantities $Q_1(u)$ and $Q_2(u)$ associated with these symmetries.

(d) Consider the Lagrangian given by

$$\Lambda(\boldsymbol{U}) = \mathcal{H}(\boldsymbol{U}) + \omega Q_2(\boldsymbol{U}), \quad \omega > 0.$$

Verify that $\Phi_\omega(x) = (\phi_\omega(x), 0)^{\mathrm{T}}$ is a critical point of Λ where ϕ_ω is the real solution of (5.2.65) which is homoclinic to zero. Because of the Gauge transform, one invariant is enough.

(e) Let $\mathcal{L}$ denote the second variation of Λ at ϕ_ω. Show that

$$\ker(\mathcal{L}) = \operatorname{span}\{T_1'\Phi_\omega, T_2'\Phi_\omega\},$$

and that

$$\operatorname{gker}(\mathcal{L}) = \operatorname{span}\left(\{T_1'\Phi_\omega, T_2'\Phi_\omega\} \cup \{\mathcal{L}^{-1}\mathcal{J}^{-1}T_1'\Phi_\omega, \mathcal{L}^{-1}\mathcal{J}^{-1}T_2'\Phi_\omega\}\right).$$

Use the Guage transform to relate Φ_c, the vectorized solution of (5.2.64), to ϕ_ω and show that

$$\partial_{c_1}\Phi_c = \begin{pmatrix} 0 \\ x\phi_\omega/2 \end{pmatrix}, \quad \partial_{c_2}\Phi_c = \begin{pmatrix} -\partial_\omega \phi_\omega \\ 0 \end{pmatrix}, \tag{5.2.66}$$

when $\boldsymbol{c} = (0, \omega)$. Use these expressions to obtain explicit formulas for $\operatorname{gker}(\mathcal{L})$.

(f) Determine $\mathrm{n}(\mathcal{L})$.

(g) Form the matrix $\boldsymbol{D}$ of (5.2.53), and show that

$$\mathrm{n}(\boldsymbol{D}) = \mathrm{n}(-\partial_\omega Q_2(\phi_\omega)).$$

(h) Verify the remainder of Hypothesis 5.2.5 apply and use Theorem 5.2.11 to deduce the orbital stability of the manifold $\mathcal{M}_T(\omega)$ in $H^1(\mathbb{R})$.

5.3 Eigenvalues of Constrained Self-Adjoint Operators

In this section we develop an index theorem that determines the negative eigenvalue count of self-adjoint operators when constrained to operate on subspaces of finite codimension. The principal result, Theorem 5.3.2, has several applications, one of which is to complete the proof of Theorem 5.2.11. Following the notation of Chapter 5.2.2, we consider a self-adjoint operator $\mathcal{L} : D(\mathcal{L}) \subset X \to X$, with dense domain $D(\mathcal{L})$ and a space $Y \subset X \subset Y^*$ such that $\mathcal{L} \in \mathcal{B}(Y, Y^*)$. We define the bilinear form

$$b[u,v] := \langle \mathcal{L}u, v \rangle_X,$$

which is continuous in $\|\cdot\|_Y$, and fix a finite codimensional subspace $\mathcal{A} \subset Y$, called the *admissible space*, with X orthogonal projection, $\Pi : X \to \mathcal{R}(\Pi) \subset X$ which satisfies $\Pi Y = \mathcal{A}$. Restricting the bilinear form b to act upon $\mathcal{A}$ induces the operator $\mathcal{L}_\Pi := \Pi\mathcal{L} : \mathcal{D}(\mathcal{L}) \cap \mathcal{A} \subset \Pi X \to \Pi X$. Recalling Definition 5.2.9, we assume that $\mathrm{n}(\mathcal{L}) + \mathrm{z}(\mathcal{L}) < \infty$. In this section we focus upon:

(a) the relation between $\mathrm{n}(\mathcal{L})$ and $\mathrm{n}_\mathcal{A}(\mathcal{L}) = \mathrm{n}(\mathcal{L}_\Pi)$.
(b) the location of the eigenvalues of $\mathcal{L}_\Pi$ relative to those of $\mathcal{L}$.

We assume that the finite-dimensional space $S := \mathcal{A}^\perp \subset X$, called the *constraint space*, is X-orthogonal to $\ker(\mathcal{L})$, and write $Y = S \oplus \mathcal{A}$ where

$$S = \mathrm{span}\{s_1, \ldots, s_m\} \subset \ker(\mathcal{L})^\perp, \quad \mathcal{A} = \{h \in Y : \langle h, s_j \rangle = 0,\ j = 1, \ldots, m\}.$$

The first step is to characterize $\ker(\mathcal{L}_\Pi)$. We define the inverse of $\mathcal{L}$ on its range by the map $\mathcal{L}^{-1} : [\ker(\mathcal{L})]^\perp \mapsto [\ker(\mathcal{L})]^\perp$, which was denoted by $\mathcal{L}_0^{-1}$ in Lemma 5.2.3. There are two opportunities for an element $s^\perp \in \mathcal{A}$ to lie in $\ker(\mathcal{L}_\Pi)$: either (a) $s^\perp \in \ker(\mathcal{L})$, or (b) $\mathcal{L}s^\perp \in S$. For case (a), since $\ker(\mathcal{L}) \subset \mathcal{A}$, it follows that $\ker(\mathcal{L}) \subset \ker(\mathcal{L}_\Pi)$. For case (b), since $\mathcal{L}$ is invertible on S, we obtain an element of the kernel precisely when there exists an $s \in S$ with $\mathcal{L}^{-1}s \in \mathcal{A}$; indeed, for such s

$$\mathcal{L}_\Pi \mathcal{L}^{-1} s = \Pi \mathcal{L} \mathcal{L}^{-1} s = \Pi s = 0.$$

Initially, we make the (generic) assumption that the two spaces $\mathcal{L}^{-1}(S)$,

$$\mathcal{L}^{-1}(S) := \mathrm{span}\{\mathcal{L}^{-1}s_1, \ldots, \mathcal{L}^{-1}s_m\},$$

and $\mathcal{A}$ have trivial intersection; that is, $\mathcal{L}^{-1}(S)\cap\mathcal{A}=\{0\}$. The $m\times m$ Hermitian constraint matrix takes the form

$$D_{ij}=\langle s_i,\mathcal{L}^{-1}s_j\rangle, \tag{5.3.1}$$

this assumption is equivalent to the nonsingularity of the constraint matrix $\boldsymbol{D}$. Indeed, if $\boldsymbol{D}$ is singular then we may use any element of the kernel of $\boldsymbol{D}$ to construct an nonzero $s\in S$ for which $\mathcal{L}^{-1}s\in\mathcal{A}$. The following result relates $\mathrm{n}(\mathcal{L}_\Pi)$ to $\mathrm{n}(\mathcal{L})$:

Proposition 5.3.1. *Suppose that $S\subset\ker(\mathcal{L})^\perp$ is an m-dimensional subspace and the Hermitian constraint matrix $\boldsymbol{D}\in\mathbb{C}^{m\times m}$ defined in (5.3.1) is nonsingular. The difference in the negative eigenvalue count of $\mathcal{L}$ and $\mathcal{L}_\Pi$ equals the negative count of $\boldsymbol{D}$, i.e.,*

$$\mathrm{n}(\mathcal{L}_\Pi)=\mathrm{n}(\mathcal{L})-\mathrm{n}(\boldsymbol{D}). \tag{5.3.2}$$

Proof. The first step of the proof is to show that any $h\in X$ can be written as

$$h=\mathcal{L}^{-1}s+s^\perp, \tag{5.3.3}$$

where $s\in S$ and $s^\perp\in\mathcal{A}$. To determine s we define $P=\mathcal{I}-\Pi$, the orthogonal projection onto S, and write $Ph=\sum a_is_i$ and $s=\sum b_is_i$. Projecting (5.3.3) with P, we obtain

$$\sum a_is_i=P\sum b_i\mathcal{L}^{-1}s_i.$$

Taking the inner product with s_j for $j=1,\dots,m$ yields a system for the unknown $\boldsymbol{b}$ with unique solution

$$\boldsymbol{b}=\boldsymbol{D}^{-1}\boldsymbol{a}.$$

This determines s, and we define $s^\perp:=\Pi h-\Pi\mathcal{L}^{-1}s$.

Each $h\in X$ can be decomposed in two ways: (a) via the eigenspaces of $\mathcal{L}$, and (b) as in (5.3.3). Using this latter decomposition, and exploiting the orthogonality of S and $\mathcal{A}$, we expand the bilinear form as

$$\langle\mathcal{L}h,h\rangle=\langle s+\mathcal{L}s^\perp,\mathcal{L}^{-1}s+s^\perp\rangle=\langle s,\mathcal{L}^{-1}s\rangle+\langle s^\perp,\mathcal{L}s^\perp\rangle. \tag{5.3.4}$$

Denoting $d_1:=\mathrm{n}_S(\mathcal{L}^{-1})$ and $d_2:=\mathrm{n}_\mathcal{A}(\mathcal{L})$, with the corresponding maximal subspaces $S_{d_1}:=\mathrm{N}_S(\mathcal{L}^{-1})$ and $\mathcal{A}_{d_2}:=\mathrm{N}_\mathcal{A}(\mathcal{L})$ we enumerate the basis elements

$$S_{d_1}=\mathrm{span}\{t_1,\dots,t_{d_1}\},\quad \mathcal{A}_{d_2}=\mathrm{span}\{t_1^\perp,\dots,t_{d_2}^\perp\}.$$

From these basis elements we form the subspace

$$X_{d_1+d_2}:=\mathrm{span}\{\mathcal{L}^{-1}t_1,\dots,\mathcal{L}^{-1}t_{d_1},t_1^\perp,\dots,t_{d_2}^\perp\}.$$

From (5.3.4) we see for each $h \in X_{d_1+d_2}$ there is an $s \in S_{d_1}$ and an $s^\perp \in \mathcal{A}_{d_2}$ such that

$$\langle \mathcal{L}h, h\rangle = \langle s, \mathcal{L}^{-1}s\rangle + \langle s^\perp, \mathcal{L}s^\perp\rangle < 0.$$

Moreover, $\dim(X_{d_1+d_2}) = d_1 + d_2$, since if not, there exists $s \in S_{d_1}$ such that $\mathcal{L}^{-1}s \in \mathcal{A}$, which contradicts $\mathrm{z}(\boldsymbol{D}) = 0$. We deduce the lower bound,

$$\mathrm{n}_X(\mathcal{L}) \geq \dim(X_{d_1+d_2}) = \mathrm{n}_S(\mathcal{L}^{-1}) + \mathrm{n}_{\mathcal{A}}(\mathcal{L}).$$

On the other hand, denoting $d = \mathrm{n}_X(\mathcal{L})$ with the corresponding maximal space given by $X_d = \mathrm{span}\{h_1, \dots, h_d\}$, we again decompose each basis element as $h_i = \mathcal{L}^{-1}t_i + t_i^\perp$, and form the spaces

$$S_d := \mathrm{span}\{t_1, \dots t_d\}, \quad \mathcal{A}_d := \mathrm{span}\{t_1^\perp, \dots, t_d^\perp\}.$$

For $h \in X_d$ we may write $h = \sum a_j h_j$, and use (5.3.4) to write

$$\langle \mathcal{L}h, h\rangle = \sum_{i,j=1}^{d} a_i a_j \left(\langle t_i, \mathcal{L}^{-1}t_j\rangle + \langle t_i^\perp, \mathcal{L}t_j^\perp\rangle\right) = \boldsymbol{a}\cdot(\boldsymbol{A}+\boldsymbol{B})\boldsymbol{a}.$$

Here $\boldsymbol{A}, \boldsymbol{B}$ are symmetric $d \times d$ matrices with entries defined by

$$\boldsymbol{A}_{ij} := \langle t_i, \mathcal{L}^{-1}t_j\rangle, \quad \boldsymbol{B}_{ij} := \langle t_i^\perp, \mathcal{L}t_j^\perp\rangle.$$

From their construction it follows that $\mathrm{n}(\boldsymbol{A}) \leq \mathrm{n}_S(\mathcal{L}^{-1})$ and $\mathrm{n}(\boldsymbol{B}) \leq \mathrm{n}_{\mathcal{A}}(\mathcal{L})$. Since the negative index is subadditive on symmetric matrices we have the upper bound,

$$\mathrm{n}_X(\mathcal{L}) = \mathrm{n}(\boldsymbol{A}+\boldsymbol{B}) \leq \mathrm{n}(\boldsymbol{A}) + \mathrm{n}(\boldsymbol{B}) \leq \mathrm{n}_S(\mathcal{L}^{-1}) + \mathrm{n}_{\mathcal{A}}(\mathcal{L}),$$

and with the lower bound we conclude the equality,

$$\mathrm{n}_X(\mathcal{L}) = \mathrm{n}_S(\mathcal{L}^{-1}) + \mathrm{n}_{\mathcal{A}}(\mathcal{L}). \tag{5.3.5}$$

To finish the proof, we expand $s \in S$ as $s = \sum a_j s_j$, and from (5.3.1) find that

$$\langle s, \mathcal{L}^{-1}s\rangle = \boldsymbol{a}\cdot\boldsymbol{D}\boldsymbol{a}, \quad \boldsymbol{a} = (a_1, \dots, a_m)^{\mathrm{T}};$$

from which we deduce $\mathrm{n}_S(\mathcal{L}^{-1}) = \mathrm{n}(\boldsymbol{D})$. Substituting this equality and $\mathrm{n}_{\mathcal{A}}(\mathcal{L}) = \mathrm{n}(\mathcal{L}_\Pi)$ into (5.3.5) establishes (5.3.2). □

In the theorem below we remove the assumption that $\mathcal{L}^{-1}(S) \cap \mathcal{A} = \{0\}$, so that the theorem encompasses the case that the matrix $\boldsymbol{D}$ is singular. This technical difficulty is surmounted by analytically perturbing the constraint space S so as to push the zero eigenvalues of $\boldsymbol{D}$ onto the negative real axis. We apply Proposition 5.3.1 to the perturbed problem, and obtain the result for the unperturbed problem as a limit.

Theorem 5.3.2 (Index theorem). *Let $\mathcal{L}$ be a self-adjoint operator on a Hilbert space X. Fix a finite codimension admissible space $\mathcal{A} \subset X$ with constraint space $S := \mathcal{A}^\perp \subset \ker(\mathcal{L})^\perp$, and let $\mathcal{L}_\Pi$ be the constrained operator associated with $\mathcal{A}$. The constraint matrix $\boldsymbol{D} \in \mathbb{C}^{m\times m}$, defined in (5.3.1), characterizes the difference in negative indices of the constrained and the unconstrained operators,*

$$\mathrm{n}(\mathcal{L}_\Pi) = \mathrm{n}(\mathcal{L}) - \mathrm{n}(\boldsymbol{D}) - \mathrm{z}(\boldsymbol{D}); \tag{5.3.6}$$

moreover, the dimension of the kernel of the constrained operator satisfies

$$\mathrm{z}(\mathcal{L}_\Pi) = \mathrm{z}(\mathcal{L}) + \mathrm{z}(\boldsymbol{D}). \tag{5.3.7}$$

Remark 5.3.3. The index theorem was first proved at this level of generality in [**63**, Lemma 3.4]; however, the proofs of this theorem and Proposition 5.3.1, while motivated by [**110**, Section 3], first appeared in [**148**]. The interested reader should also consult [**198**].

Proof. To deal with the singularity of $\boldsymbol{D}$ we introduce the subspaces

$$S_1 := \{s \in S : \mathcal{L}^{-1}s \in \mathcal{A}\}, \quad T_1 := \{\mathcal{L}^{-1}s : s \in S_1\} \subset \mathcal{A}, \tag{5.3.8}$$

as well as the orthogonal complement S_1^c of S_1 relative to S, which yields the decomposition $S = S_1 \oplus S_1^c$. Since $\mathcal{L}$ is one-to-one on S_1, we may define the common value $k := \dim(T_1) = \dim(S_1)$. Recall that $\ker(\mathcal{L}) \subset \ker(\mathcal{L}_\Pi)$, the first goal is to establish the decomposition

$$\ker(\mathcal{L}_\Pi) = \ker(\mathcal{L}) \oplus T_1. \tag{5.3.9}$$

To this end we choose orthonormal bases $\{\phi_1,\ldots,\phi_k\}$ for S_1, and $\{\phi_{k+1},\ldots,\phi_m\}$ for S_1^c, and form the $m\times m$ Hermitian matrix $\boldsymbol{D}$ as in (5.3.1). From the definition of S_1, and the orthogonality of $\mathcal{A}$ and S, we see that $\langle \mathcal{L}^{-1}\phi_i, \phi_j\rangle = 0$ for $i = 1,\ldots,k$ and $j = 1,\ldots,m$. Since $\boldsymbol{D}$ is symmetric, it follows that $\boldsymbol{D}$ is block-diagonal,

$$\boldsymbol{D} = \begin{pmatrix} \boldsymbol{0} & \boldsymbol{0} \\ \boldsymbol{0} & \boldsymbol{D}_1 \end{pmatrix}, \tag{5.3.10}$$

where $\boldsymbol{D}_1 \in \mathbb{C}^{(m-k)\times(m-k)}$. Moreover, $\boldsymbol{D}_1$ is nonsingular; indeed, if it were not, then we could construct a nontrivial element of $S_1^c \cap S_1$. By construction, $T_1 \perp \ker(\mathcal{L})$, and has the maximal dimension, k, of any such subspace; the decomposition (5.3.9) follows.

The proof of Theorem 5.3.2 in the case $k = 0$ follows from Proposition 5.3.1. The general case $k = \mathrm{z}(\boldsymbol{D}) \geq 1$ follows by analytically perturbing the constraint space S to push the zero eigenvalues of the constraint matrix $\boldsymbol{D}$ onto the negative real axis. We perturb only the space, S_1, writing its basis elements as $\phi_j(\epsilon) = \phi_j + \epsilon\phi_j^1$, for $j = 1,\ldots,k$ and $0 \leq \epsilon \ll 1$, where the perturbations $\{\phi_j^1\}$ are taken from $\ker(\mathcal{L})^\perp$. By assumption, the essential spectrum of $\mathcal{L}$ is bounded away from the origin; it follows from classical results, e.g., Kato [162], that the kernels of $\mathcal{L}$ and $\boldsymbol{D}$ perturb analytically. The matrix $\boldsymbol{D} = \boldsymbol{D}(\epsilon)$ has entries $\boldsymbol{D}(\epsilon)_{ij} := \langle \phi_i(\epsilon), \mathcal{L}^{-1}\phi_j(\epsilon)\rangle$, and the k zero eigenvalues of $\boldsymbol{D}(0)$ satisfy the regular perturbation expansion $\lambda_j = \epsilon\lambda_j^1 + \mathcal{O}(\epsilon^2)$. The leading order term λ_j^1 is an eigenvalue of the Hermitian matrix $\boldsymbol{M} \in \mathbb{C}^{k\times k}$, which has the entries

$$\boldsymbol{M}_{ij} := -\langle \phi_i, [P'(0)\mathcal{L}P(0) + P(0)\mathcal{L}P'(0)]\phi_j\rangle = -\langle \phi_i, [P'(0)\mathcal{L} + \mathcal{L}P'(0)]\phi_j\rangle$$

(e.g., see Chapter 6.1). Furthermore, $P(\epsilon)$, the orthogonal projection onto S, has the expansion

$$P(\epsilon) = P(0) + \epsilon P'(0) + \mathcal{O}(\epsilon^2), \quad P'(0)f := \sum_{j=1}^{m} \left(\langle f, \phi_j\rangle\phi_j^1 + \langle f, \phi_j^1\rangle\phi_j\right).$$

The task is to choose the perturbations $\{\phi_j^1\}$ such that $\{\lambda_j^1\}$ are strictly negative. The perturbed matrix $\boldsymbol{D}(\epsilon)$ will then be nonsingular, so that we can apply the result of Proposition 5.3.1. We start by requiring $\{\phi_j^1\} \subset \mathcal{A}$, which implies that the perturbation to S_1 is orthogonal to S. With this constraint and the orthonormality of the basis $\{\phi_j\}_{j=1}^m$ we see that

$$P'(0)\phi_i = \sum_{j=1}^{m} \left(\langle \phi_i, \phi_j\rangle\phi_j^1 + \langle \phi_i, \phi_j^1\rangle\phi_j\right) = \phi_i^1.$$

Since $P'(0)$ is self-adjoint, the matrix $\boldsymbol{M}$ reduces to

$$\boldsymbol{M}_{ij} = -\langle \phi_i, (P'(0)\mathcal{L} + \mathcal{L}P'(0))\phi_j\rangle = -2\langle \phi_i^1, \mathcal{L}\phi_j\rangle.$$

We thus choose $\phi_i^1 = \mathcal{L}^{-1}\phi_i/2$ for $i = 1,\ldots,k$, so that $\{\phi_i^1\} \subset T_1 \subset \mathcal{A}$. Again using the orthonormality of $\{\phi_i\}_{i=1}^k$ we have for $i, j = 1,\ldots,k$,

$$\boldsymbol{M}_{ij} = -2\langle \phi_i^1, \mathcal{L}\phi_j\rangle = -2\left\langle \frac{1}{2}\mathcal{L}^{-1}\phi_i, \mathcal{L}\phi_j\right\rangle = -\delta_{ij}, \tag{5.3.11}$$

where δ_{ij} is the Kronecker delta. That is, $\boldsymbol{M} = -\boldsymbol{I}_{k\times k}$, and the k zero eigenvalues of $\boldsymbol{D}$ perturb as $\lambda_j^1 = -\epsilon + O(\epsilon^2)$ for $j = 1,\ldots,k$.

To complete the proof, for $\epsilon > 0$ the matrix $\boldsymbol{D}(\epsilon)$ is invertible, and from (5.3.2) we conclude that

$$\mathrm{n}(\mathcal{L}_{\Pi(\epsilon)}) = \mathrm{n}(\mathcal{L}) - \mathrm{n}(\boldsymbol{D}(\epsilon)); \qquad \Pi(\epsilon) := \mathcal{I} - P(\epsilon).$$

However, for $\epsilon > 0$ sufficiently small, there are k eigenvalues of $\boldsymbol{D}(\epsilon)$ that are $\mathcal{O}(\epsilon)$, and each of these eigenvalues resides in the left-half complex plane. From the continuity of the eigenvalues in ϵ, we deduce that

$$\mathrm{n}(\boldsymbol{D}(\epsilon)) = \mathrm{n}(\boldsymbol{D}(0)) + \mathrm{z}(\boldsymbol{D}(0)),$$

which gives an ϵ independent expression for $\mathrm{n}(\mathcal{L}_{\Pi(\epsilon)})$. Taking the $\epsilon \to 0^+$ limit yields (5.3.6). □

In the remainder of this section we assume $\sigma_{\mathrm{ess}}(\mathcal{L}) \subset [\mu_{\mathrm{ess}}, \infty)$ for some $\mu_{\mathrm{ess}} > 0$, and apply the index Theorem 5.3.2 to locate the point spectrum of $\mathcal{L}_\Pi$ relative to the point spectrum of $\mathcal{L}$ on $(-\infty, \mu_{\mathrm{ess}})$. We will use the fact that for any $\lambda_1 < \lambda_2 < \mu_{\mathrm{ess}}$ the difference $\mathrm{n}(\mathcal{L} - \lambda_2 \mathcal{I}) - \mathrm{n}(\mathcal{L} - \lambda_1 \mathcal{I})$ counts the number of eigenvalues of $\mathcal{L}$ in $[\lambda_1, \lambda_2)$.

To begin, we enumerate the distinct point spectra of $\mathcal{L}$

$$\sigma_{\mathrm{pt}}(\mathcal{L}) = \{\lambda_0 < \lambda_1 < \cdots < \mu_{\mathrm{ess}}\}. \tag{5.3.12}$$

Recalling the constraint space S, we define the meromorphic constraint matrix $\boldsymbol{D}(\lambda) \in \mathbb{R}^{m \times m}$ via

$$\boldsymbol{D}_{ij}(\lambda) = \langle s_i, (\mathcal{L} - \lambda \mathcal{I})^{-1} s_j \rangle. \tag{5.3.13}$$

Since $\{s_j\} \subset \ker(\mathcal{L})^\perp$, $\boldsymbol{D}(\lambda)$ is regular (analytic or a removable singularity) at $\lambda = 0$. For real $\lambda \in (-\infty, \mu_{\mathrm{ess}})$, $\boldsymbol{D}(\lambda)$ is meromorphic and self-adjoint, with a generically nonzero (but not necessarily full-rank) residue at each $\lambda_j \in \sigma_{\mathrm{pt}}(\mathcal{L})$ (see [**279**, Lemma E.1] or [**63**, Lemma 3.4]). This implies that some, but not necessarily all, of the eigenvalues of $\boldsymbol{D}(\lambda)$ may have poles at each λ_j. For each $\lambda_j \in \sigma_{\mathrm{pt}}(\mathcal{L})$ define

$$\boldsymbol{D}_j = \lim_{\lambda \to \lambda_j} (\lambda_j - \lambda) \boldsymbol{D}(\lambda), \tag{5.3.14}$$

i.e., $\boldsymbol{D}_j$ is the negative of the residue of $\boldsymbol{D}(\lambda)$ at $\lambda = \lambda_j$.

Since $\boldsymbol{D}(\lambda)$ is self-adjoint and real-meromorphic, its eigenvalues, $\{d_j(\lambda)\}_{j=1}^m$, and eigenvectors, $\{\boldsymbol{v}_j(\lambda)\}_{j=1}^m$, are also real meromorphic (e.g., see [**162**]). Now,

$$\boldsymbol{D}'_{ij}(\lambda) = \langle (\mathcal{L} - \lambda \mathcal{I})^{-1} s_i, (\mathcal{L} - \lambda \mathcal{I})^{-1} s_j \rangle,$$

so for any $\boldsymbol{c} \in \mathbb{R}^m$ and $\lambda \notin \sigma_{\mathrm{pt}}(\mathcal{L})$ we may introduce $s = \sum c_i s_i$ for which

$$\boldsymbol{c} \cdot \boldsymbol{D}'(\lambda) \boldsymbol{c} = \langle (\mathcal{L} - \lambda \mathcal{I})^{-1} s, (\mathcal{L} - \lambda \mathcal{I})^{-1} s \rangle \geq 0.$$

When $\boldsymbol{D}(\lambda)$ is regular the matrix $\boldsymbol{D}'(\lambda)$ is positive-definite, which implies that its eigenvalues are strictly increasing functions of λ, since

$$d'_i(\lambda) = \frac{\boldsymbol{v}_i(\lambda) \cdot \boldsymbol{D}'(\lambda) \boldsymbol{v}_i(\lambda)}{|\boldsymbol{v}_i(\lambda)|^2} > 0. \tag{5.3.15}$$

Definition 5.3.4. We denote by n_j the number of eigenvalues of $\boldsymbol{D}(\lambda)$ that are regular at $\lambda = \lambda_j$ and are strictly negative, and by z_j the number of eigenvalues of $\boldsymbol{D}(\lambda)$ that are regular at $\lambda = \lambda_j$ and take the value 0.

Theorem 5.3.5 (Eigenvalue Interlacing Theorem). *Let $\mathcal{L}$ be a self-adjoint operator with $\sigma_{\text{ess}}(\mathcal{L}) \subset [\mu_{\text{ess}}, \infty)$ for some $\mu_{\text{ess}} > 0$, and distinct point spectrum enumerated as in (5.3.12). Let $\mathcal{A}$ be a finite co-dimensional admissible space with orthogonal projection Π and associated constrained operator $\mathcal{L}_\Pi$. For each $\lambda_j \in \sigma_{\text{pt}}(\mathcal{L})$ the dimension of the kernel of $\mathcal{L}_\Pi - \lambda_j \mathcal{I}$ is given by*

$$\mathrm{z}\big(\mathcal{L}_\Pi - \lambda_j \mathcal{I}\big) = \mathrm{z}\big(\mathcal{L} - \lambda_j \mathcal{I}\big) + \mathrm{z}_j - \mathrm{rank}(\boldsymbol{D}_j). \tag{5.3.16}$$

The number of eigenvalues of $\mathcal{L}_\Pi$ in the open interval $(\lambda_j, \lambda_{j+1})$ is given by

$$\mathrm{n}(\mathcal{L}_\Pi - \lambda \mathcal{I})\Big|_{\lambda_j^+}^{\lambda_{j+1}^-} = \mathrm{rank}(\boldsymbol{D}_j) + \mathrm{n}_j - (\mathrm{n}_{j+1} + \mathrm{z}_{j+1}). \tag{5.3.17}$$

Finally, the number of eigenvalues of $\mathcal{L}_\Pi$ in the interval $[\lambda_j, \lambda_{j+1})$ is given by

$$\mathrm{n}(\mathcal{L}_\Pi - \lambda \mathcal{I})\Big|_{\lambda_j}^{\lambda_{j+1}^-} = (\mathrm{z}_j + \mathrm{n}_j) - (\mathrm{n}_{j+1} + \mathrm{z}_{j+1}). \tag{5.3.18}$$

Proof. The index $\mathrm{n}(\mathcal{L}_\Pi - \lambda \mathcal{I})$ counts the point spectra, by multiplicity, of $\mathcal{L}_\Pi$ to the left of λ. Since the point spectra are isolated, for $\epsilon > 0$ sufficiently small, applying (5.3.6) to $\mathcal{L} - (\lambda_j \pm \epsilon)\mathcal{I}$ yields

$$\mathrm{n}(\mathcal{L}_\Pi - \lambda \mathcal{I})\Big|_{\lambda_j - \epsilon}^{\lambda_j + \epsilon} = \mathrm{z}(\mathcal{L} - \lambda_j \mathcal{I}) - (\mathrm{n} + \mathrm{z})(\boldsymbol{D}(\lambda))\Big|_{\lambda_j - \epsilon}^{\lambda_j + \epsilon}. \tag{5.3.19}$$

On the other hand, for $\epsilon > 0$ sufficiently small we have

$$\mathrm{n}(\mathcal{L}_\Pi - \lambda \mathcal{I})\Big|_{\lambda_j - \epsilon}^{\lambda_j + \epsilon} = \mathrm{z}\big(\mathcal{L}_\Pi - \lambda_j \mathcal{I}\big). \tag{5.3.20}$$

Since the eigenvalues of $\boldsymbol{D}(\lambda)$ are strictly increasing in λ at regular points, [see (5.3.15)], and since $m - \mathrm{rank}(\boldsymbol{D}_j)$ of the eigenvalues will have a removable singularity at $\lambda = \lambda_j$ (e.g., see the argument in either of [**63**, Lemma 3.4] or [**146**, p. 13], in which it is shown that $\boldsymbol{D}_j$ is positive semi-definite, but is not necessarily full rank), we find the equalities

$$(\mathrm{n} + \mathrm{z})(\boldsymbol{D}(\lambda_j - \epsilon)) = \mathrm{z}_j + \mathrm{n}_j, \quad \text{and} \quad (\mathrm{n} + \mathrm{z})(\boldsymbol{D}(\lambda_j + \epsilon)) = \mathrm{n}_j + \mathrm{rank}(\boldsymbol{D}_j),$$

(see Fig. 5.6 for an illustrative example). Subtracting these two equalities yields

$$(\mathrm{n} + \mathrm{z})(\boldsymbol{D}(\lambda))\Big|_{\lambda_j - \epsilon}^{\lambda_j + \epsilon} = \mathrm{rank}(\boldsymbol{D}_j) - \mathrm{z}_j. \tag{5.3.21}$$

Substitution of (5.3.20) and (5.3.21) into (5.3.19) yields the result of (5.3.16).

To address (5.3.17) and (5.3.18) we assume that $\lambda_j < \lambda_{j+1}$ are successive, distinct eigenvalues of $\mathcal{L}$. Since $\mathcal{L}$ has no spectra in $(\lambda_j, \lambda_{j+1})$ we have

$$\mathrm{n}\big(\mathcal{L} - (\lambda_{j+1} - \epsilon)\mathcal{I}\big) = \mathrm{n}\big(\mathcal{L} - (\lambda_j + \epsilon)\mathcal{I}\big),$$

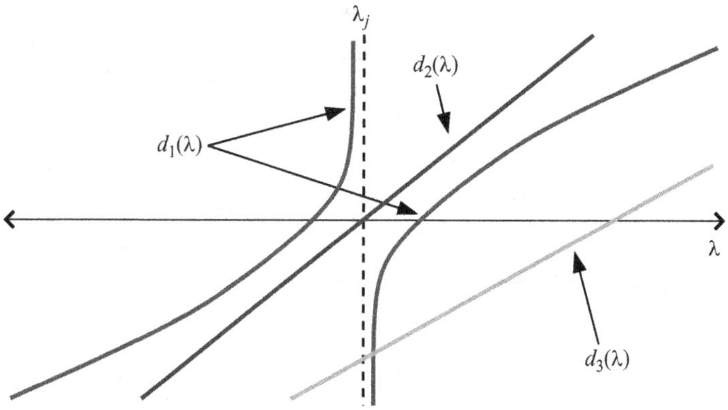

Fig. 5.6 The eigenvalues $d_i(\lambda)$ of $\boldsymbol{D}(\lambda)$ near $\lambda = \lambda_j$ in the case that $m = 3$ with $\mathrm{z}_j = \mathrm{n}_j = \mathrm{rank}(\boldsymbol{D}_j) = 1$. Since $\mathrm{rank}(\boldsymbol{D}_j) - \mathrm{z}_j = 0$ we have $(\mathrm{n}+\mathrm{z})(\boldsymbol{D}(\lambda_j - \epsilon)) = 2 = (\mathrm{n}+\mathrm{z})(\boldsymbol{D}(\lambda_j + \epsilon))$, which is in agreement with (5.3.21) (Color figure online)

and applying the index (5.3.6) yields

$$\mathrm{n}(\mathcal{L}_\Pi - \lambda\mathcal{I})\Big|_{\lambda_j+\epsilon}^{\lambda_{j+1}-\epsilon} = -(\mathrm{n}+\mathrm{z})(\boldsymbol{D}(\lambda))\Big|_{\lambda_j+\epsilon}^{\lambda_{j+1}-\epsilon}.$$

However,

$$(\mathrm{n}+\mathrm{z})(\boldsymbol{D}(\lambda_j+\epsilon)) = \mathrm{n}_j + \mathrm{rank}(\boldsymbol{D}_j), \quad (\mathrm{n}+\mathrm{z})(\boldsymbol{D}(\lambda_{j+1}-\epsilon)) = \mathrm{z}_{j+1} + \mathrm{n}_{j+1},$$

and taking the limit $\epsilon \to 0^+$ we obtain (5.3.17).

Finally, combining (5.3.16) and (5.3.17) yields (5.3.18) since

$$\mathrm{z}\left(\mathcal{L}_\Pi - \lambda_j\mathcal{I}\right) + \mathrm{n}(\mathcal{L}_\Pi - \lambda\mathcal{I})\Big|_{\lambda_j^+}^{\lambda_{j+1}^-} = \mathrm{n}\left(\mathcal{L}_\Pi - \lambda_{j+1}\mathcal{I}\right) - \mathrm{n}\left(\mathcal{L}_\Pi - \lambda_j\mathcal{I}\right). \qquad \square$$

Remark 5.3.6. For the case $\lambda = 0$ the result of (5.3.16) of the eigenvalue interlacing theorem is a generalization of (5.3.7), which was derived under the assumption that $\mathrm{rank}(\boldsymbol{D}(0)) = 0$.

Remark 5.3.7. In [**146**, p. 13] it was observed that $\mathrm{rank}(\boldsymbol{D}_j) \le \mathrm{z}(\mathcal{L} - \lambda_j\mathcal{I})$; consequently, when $m = \dim(S) \ge 2$ the residue $\boldsymbol{D}_j$ will generically not have full rank.

In the case that the constraint space, S, has dimension $m = 1$, the eigenvalue interlacing theorem can be further refined. If the residue $\boldsymbol{D}_j \ne 0$, then it will necessarily be true that $\mathrm{z}_j = \mathrm{n}_j = 0$. This observation yields the following corollary, which could by used to provide an alternate proof of Lemma 5.2.3.

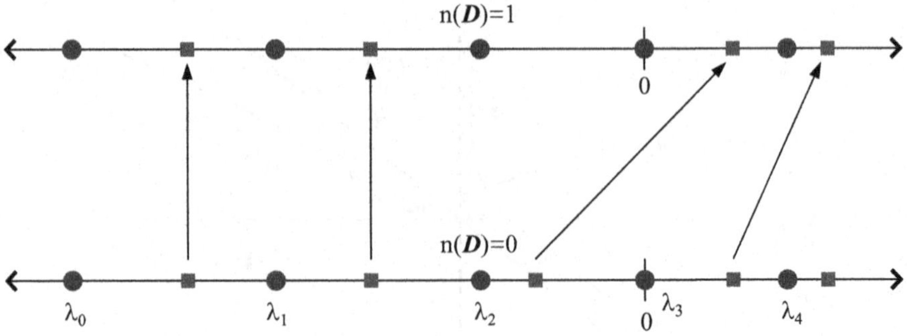

Fig. 5.7 The spectrum of $\mathcal{L}$ (*blue* circles) and $\mathcal{L}_\Pi$ (*red* squares) for two choices, S_0 and S_1, of the constraint space. In both cases the dimension of the constraint space is one, so that Corollary 5.3.8 of the eigenvalue interlacing theorem applies. In both cases $\mathrm{n}(\mathcal{L}) = 3$, $z(\mathcal{L}) = 1$ with $\ker(\mathcal{L}) \subset S_j^\perp$ for $j = 0, 1$, and all of the eigenvalues of $\sigma(\mathcal{L})$ are geometrically simple with $\boldsymbol{D}_j \neq 0$ for all j except for $\lambda = 0$. In the *top figure* $\mathrm{n}(\boldsymbol{D}) = 1$, and in the *bottom figure* $\mathrm{n}(\boldsymbol{D}) = 0$, so that by the index theorem $\mathrm{n}(\mathcal{L}_\Pi) = \mathrm{n}(\mathcal{L}) - \mathrm{n}(\boldsymbol{D}) = 2$ (*bottom figure*) or 3 (*top figure*). Moreover, by the Corollary 5.3.8 to the eigenvalue Interlacing Theorem there is one eigenvalue of $\mathcal{L}_\Pi$ in each interval $(\lambda_j, \lambda_{j+1})$, except in the interval (λ_2, λ_3) in the case $\mathrm{n}(\boldsymbol{D}) = 1$ (*top figure*). The arrows indicate possible motion of the eigenvalues of $\mathcal{L}_\Pi$ as S_0 homotopies to S_1. (Color figure online.)

Corollary 5.3.8. *Suppose that* $m := \dim(S) = 1$, $\lambda_j, \lambda_{j+1} \in \sigma_{\mathrm{pt}}(\mathcal{L})$, *and that the residues* $\boldsymbol{D}_j, \boldsymbol{D}_{j+1} \neq 0$. *If the geometric multiplicity of* λ_j *is* ℓ, *then it is an eigenvalue of* $\mathcal{L}_\Pi$ *of multiplicity* $\ell - 1$; *that is,*

$$\mathrm{z}\left(\mathcal{L}_\Pi - \lambda_j \mathcal{I}\right) = \mathrm{z}(\mathcal{L} - \lambda_j \mathcal{I}) - 1.$$

Moreover, the missing eigenvalue moves to the right and lies in the interval $(\lambda_j, \lambda_{j+1})$, *i.e.,*

$$\mathrm{n}\left(\mathcal{L}_\Pi - \lambda \mathcal{I}\right)\Big|_{\lambda_j^+}^{\lambda_{j+1}^-} = 1.$$

In particular, if all the eigenvalues of $\mathcal{L}$ *are simple, and if* $\lambda = 0$ *is an eigenvalue that corresponds to a removable singularity (i.e.,* $\ker(\mathcal{L}) \subset S^\perp$*), and if* $\boldsymbol{D}_j \neq 0$ *for all* j *such that* $\lambda_j \neq 0$, *then the eigenvalues of the reduced operator interlace those of* $\mathcal{L}$ *(see Fig. 5.7).*

5.4 Additional Reading

The approach to orbital stability of equilibria of Hamiltonian systems that culminates in Theorem 5.2.11 is based upon the approach of Grillakis, et al. developed in [**109**, **110**]. This formulation generalizes a collection of earlier work; see Albert et al. [7], Bona and Sachs [**35**] and references therein.

Stability results for which the matrix D is singular can be found in Comech and Pelinovsky [**57**], Comech et al. [**58**].

Hamiltonian systems can also generate asymptotic stability. These results require substantially more delicate analysis of the linear semi-group. Such analysis can be found in Bona et al. [37], Kevrekidis et al. [**173**], Marzuola and Weinstein [**204**], Miller and Weinstein [**206**], Mizumachi and Pelinovsky [**208**], Pelinovsky and Stefanov [**229, 230**], Soffer and Weinstein [**266, 267**], Tao [**269**].

When the Hamiltonian system is integrable, or near-integrable, then a whole host of new tools comes into play, which may be used to determine stability. There is a tremendous literature on this topic. The books by Abdullaev [2], Ablowitz and Clarkson [**3**], Ablowitz et al. [**5**], Newell [**211**] are a good starting place. The seminal papers by Ablowitz et al. [**4**], Gardner et al. [**90**] are the beginning of it all. Some applications and extensions of the theory can be found in Beals and Coifman [**26**], Beals et al. [27], Kapitula [**145**], Kaup [**163, 164**], Kaup and Lakoba [**165**], Kaup and Newell [**166**], Kaup and Yang [**167**], Lakoba and Kaup [**181**], Maddocks and Sachs [**199**], Yang [**282, 283**].

Hamiltonian systems can also support invariant tori. Discussions of the existence and stability of invariant tori can be found in Bambusi [21, 22], Bambusi et al. [23], Craig and Wayne [**61, 62**], Wayne [**276**].

Chapter 6
Point Spectrum: Reduction to Finite-Rank Eigenvalue Problems

The word *bifurcation* refers to changes in the number and stability of equilibria supported by a governing system as its parameters are varied. The classical bifurcation problem begins with an analysis of the point spectrum of the linearized operator associated with the equilibria under investigation. In this chapter we investigate finite-rank bifurcations for which a finite number of point eigenvalues cross the imaginary axis, either transversely or more degenerately, as the system parameters are varied. In particular, we derive the perturbative motion of such point spectra. This analysis is most informative in those cases for which the associated linearized operator initially has purely imaginary eigenvalues, and a small change in parameters moves the eigenvalues decisively off the imaginary axis. Typically, the unperturbed system possesses extra structure, and the interesting question is often: What happens to the critical spectra when the special structure is broken?

We consider the case of a single bifurcation parameter, which we denote ϵ, and without loss of generality assume the bifurcation occurs at the value $\epsilon = 0$. We assume the operator $\mathcal{L} = \mathcal{L}(\epsilon)$ has domain $\mathcal{D}$, which is independent of ϵ, and moreover $\mathcal{L} \in \mathcal{B}(\mathcal{D}, X)$ admits an expansion of the form

$$\mathcal{L} = \mathcal{L}_0 + \epsilon\mathcal{L}_1 + \mathcal{O}(\epsilon^2) : \mathcal{D}(\mathcal{L}) \subset X \mapsto X, \tag{6.0.1}$$

and is a smooth function ϵ in the graph norm induced by $\mathcal{L}_0$. The associated eigenvalue problem is

$$\mathcal{L}u = \lambda u, \tag{6.0.2}$$

where u lives in a Hilbert space X with inner product $\langle\cdot,\cdot\rangle$. We consider eigenvalues λ for which the operator $\mathcal{L} - \lambda\mathcal{I}$ is Fredholm with index zero, i.e., $\lambda \in \sigma_{\mathrm{pt}}(\mathcal{L})$. We take the spectrum of $\mathcal{L}_0$, including eigenfunctions, to be known. In this context the smooth dependence of the eigenvalues and eigenfunctions upon λ is classical (e.g., see [**162**]). In order to track the eigenvalues as ϵ is varied we construct an appropriate projection that reduces the (generically) infinite-dimensional eigenvalue problem for $\mathcal{L}$ to an

T. Kapitula and K. Promislow, *Spectral and Dynamical Stability of Nonlinear Waves*, Applied Mathematical Sciences 185, DOI 10.1007/978-1-4614-6995-7_6,

eigenvalue problem for a finite-dimensional matrix, the size of which depends on the multiplicity of the unperturbed eigenvalue.

We study two scenarios. In the first it is assumed that for the unperturbed problem $\lambda = \lambda_0$ is an algebraically simple eigenvalue, i.e., $\mathrm{m_g}(\lambda_0) = \mathrm{m_a}(\lambda_0)$, with a trivial Jordan-block structure. In this case all of the eigenvalues perturb analytically in ϵ so that each of the $\mathrm{m_g}(\lambda_0)$ eigenvalues has an expansion $\lambda(\epsilon) = \lambda_0 + \mathcal{O}(\epsilon)$. The second case involves an unperturbed problem with an eigenvalue at $\lambda = \lambda_0$ that has $\mathrm{m_g}(\lambda_0) = 1$ with $\mathrm{m_a}(\lambda) = p \geq 2$; in other words, the eigenvalue is geometrically simple, but the eigenspace has a nontrivial Jordan-block structure. In this case the perturbed eigenvalues will not depend analytically upon ϵ; in fact, they will generically satisfy $\lambda = \lambda_0 + \mathcal{O}(\epsilon^{1/p})$.

6.1 Perturbation of an Algebraically Simple Eigenvalue

Suppose the unperturbed member $\mathcal{L}_0$ of the operator $\mathcal{L}$ of (6.0.1) has a semi-simple eigenvalue $\lambda_0 \in \sigma_{\mathrm{pt}}(\mathcal{L}_0)$ with $\mathrm{m_g}(\lambda_0) = \mathrm{m_a}(\lambda_0) = m \geq 1$, in particular $\mathcal{L} - \lambda_0$ is Fredholm of index zero. Then there is a basis $\{\psi_1, \ldots, \psi_m\}$ of the eigenspace

$$\mathrm{gker}(\mathcal{L}_0 - \lambda_0 \mathcal{I}) = \ker(\mathcal{L}_0 - \lambda_0 \mathcal{I}) = \mathrm{span}\{\psi_1, \ldots, \psi_m\}.$$

We search for the eigenvalues and associated eigenfunctions of $\mathcal{L}$ via a regular perturbation expansion of the form

$$\lambda = \lambda_0 + \epsilon\lambda_1 + \mathcal{O}(\epsilon^2), \quad u = \sum_{j=1}^{m} v_j \psi_j + \epsilon u_1 + \mathcal{O}(\epsilon^2). \tag{6.1.1}$$

Plugging the expansions into (6.0.1) and collecting orders of ϵ yields the equalities,

$$\begin{aligned} \mathcal{O}(1): &\quad (\mathcal{L}_0 - \lambda_0 \mathcal{I})\left(\sum_{j=1}^{m} v_j \psi_j\right) = 0 \\ \mathcal{O}(\epsilon): &\quad (\mathcal{L}_0 - \lambda_0 \mathcal{I}) u_1 = \sum_{j=1}^{m} v_j (\lambda_1 \mathcal{I} - \mathcal{L}_1)\psi_j. \end{aligned} \tag{6.1.2}$$

The $\mathcal{O}(1)$ equation is trivially satisfied. From the Fredholm alternative, Theorem 2.2.1, applied to $\mathcal{L} - \lambda_0$, there is a solution to (6.1.2) to the $\mathcal{O}(\epsilon)$ equation if and only if

$$\sum_{j=1}^{m} v_j (\lambda_1 \mathcal{I} - \mathcal{L}_1)\psi_j \in \ker(\mathcal{L}_0^{\mathrm{a}} - \overline{\lambda_0} \mathcal{I})^{\perp}, \tag{6.1.3}$$

where $\mathcal{L}_0^{\mathrm{a}}$ is the adjoint operator of $\mathcal{L}_0$. Since $\mathcal{L}_0 - \lambda_0 \mathcal{I}$ is Fredholm with index zero, one can choose basis vectors $\{\psi_1^{\mathrm{a}}, \dots, \psi_m^{\mathrm{a}}\}$ for $\ker(\mathcal{L}_0^{\mathrm{a}} - \overline{\lambda_0}\mathcal{I})$ with the property that $\langle \psi_i, \psi_j^{\mathrm{a}} \rangle = \delta_{ij}$, where δ_{ij} is the Kronecker delta function (see Exercise 6.1.1).

Taking the X-inner product of the $\mathcal{O}(\epsilon)$ equation with ψ_i^{a} for $i = 1, \dots, m$, and using the fact that the right-hand side of the $\mathcal{O}(\epsilon)$ equation in (6.1.2) must satisfy the condition (6.1.3), we obtain the linear system for the coefficients v_j,

$$\sum_{j=1}^{m} \left(\lambda_1 \delta_{ij} - \langle \mathcal{L}_1 \psi_j, \psi_i^{\mathrm{a}} \rangle \right) v_j = 0, \quad i = 1, \dots, m. \tag{6.1.4}$$

This system has a matrix formulation in terms of $\boldsymbol{L}_1 \in \mathbb{C}^{m \times m}$ with entries

$$(\boldsymbol{L}_1)_{ij} := \langle \mathcal{L}_1 \psi_j, \psi_i^{\mathrm{a}} \rangle, \tag{6.1.5}$$

for which the system (6.1.4) is equivalent to the matrix eigenvalue problem

$$\boldsymbol{L}_1 \boldsymbol{v} = \lambda_1 \boldsymbol{v}, \quad \boldsymbol{v} := (v_1, \dots, v_m)^{\mathrm{T}}. \tag{6.1.6}$$

The matrix $\boldsymbol{L}_1$ will generically have m simple eigenvalues, denoted λ_1^j for $j = 1, \dots, m$. For the operator $\mathcal{L}$ each of the $\mathcal{O}(\epsilon)$-corrections to λ_0 will be of the form $\lambda \sim \lambda_0 + \epsilon \lambda_1^j$; furthermore, for each j the associated eigenfunction for each perturbed eigenvalue is to leading order a linear combination of the unperturbed eigenfunctions, where the coefficients of the linear combination are given by the associated eigenvector $\boldsymbol{v}$. We conclude that, the m-dimensional eigenspace $\ker(\mathcal{L}_0 - \lambda_0 \mathcal{I})$ generically breaks in m one-dimensional spaces, each of which lie inside $\ker(\mathcal{L}_0 - \lambda_0 \mathcal{I})$ at leading order (see Fig. 6.1).

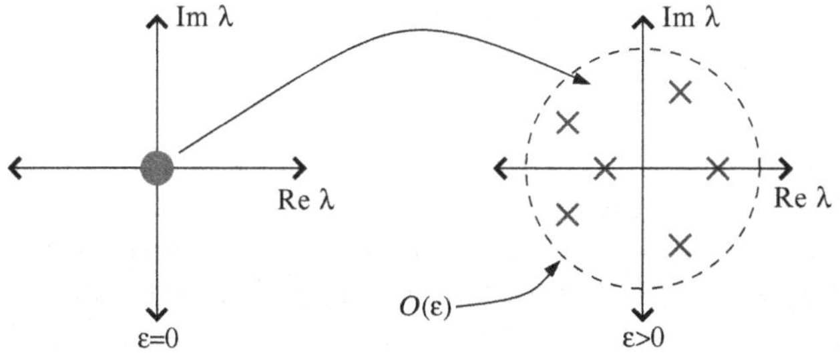

Fig. 6.1 The spectrum of $\mathcal{L}_0$ and of $\mathcal{L}$ for $\epsilon > 0$. When $\epsilon = 0$, $\lambda = 0$ is an algebraically simple eigenvalue of $\mathcal{L}_0$ with $\mathrm{m_g}(0) = 6$. For $\epsilon > 0$ there will be six eigenvalues which are $\mathcal{O}(\epsilon)$. The leading order term in the asymptotic expansion for each eigenvalue is found by solving the 6×6 eigenvalue problem (6.1.6). (Color figure online.)

6.1.1 Example: Parametrically Forced Ginzburg–Landau Equation

The parametrically forced Ginzburg–Landau equation (pfGL) is given by

$$\partial_t u = \partial_x^2 u - u + |u|^2 u + \mathrm{i}\epsilon\left(\mu u - \mathrm{e}^{-\mathrm{i}\theta}\overline{u}\right), \tag{6.1.7}$$

where $(x,t) \in \mathbb{R}\times\mathbb{R}^+$, $u(x,t) \in \mathbb{C}$, $0 < \mu < 1$, $\epsilon \geq 0$, and $-\pi/2 < \theta < \pi/2$ satisfies $\cos\theta = \mu$ (compare with the PNLS in Chapter 4.5). The review article of Aranson and Kramer [**17**] presents some of the physical and mathematical significance of the Ginzburg-Landau equation. Writing $u = u_\mathrm{r} + \mathrm{i}u_\mathrm{i}$, where u_r is the real part and u_i is the imaginary part, permits (6.1.7) to be rewritten as a system

$$\partial_t \boldsymbol{U} = \partial_x^2 \boldsymbol{U} - \begin{pmatrix} A_+^2 & 2\epsilon\mu \\ 0 & A_-^2 \end{pmatrix}\boldsymbol{U} + |\boldsymbol{U}|^2\boldsymbol{U}, \quad \boldsymbol{U} = \begin{pmatrix} u_\mathrm{r} \\ u_\mathrm{i} \end{pmatrix}, \tag{6.1.8}$$

where

$$A_\pm^2 := 1 \pm \epsilon\sin\theta.$$

The 2×2 matrix in (6.1.8) has eigenvalues $A_\pm^2$ with associated eigenvectors, $\boldsymbol{v}_+ := (1,0)^\mathrm{T}$ and $\boldsymbol{v}_- := (\cos\theta, -\sin\theta)^\mathrm{T}$. This observation yields the steady-state solutions of the form

$$\boldsymbol{U}_\pm(x) = \phi_\pm(x)\boldsymbol{v}_\pm, \tag{6.1.9}$$

where the waves $\phi_\pm$ are solutions of the scalar equation

$$\partial_x^2\phi_\pm - A_\pm^2\phi_\pm + \phi_\pm^3 = 0 \tag{6.1.10}$$

that are homoclinic to zero. Rescaling this equation, we see that $\phi_\pm(x) = A_\pm\phi_0(A_\pm x)$, where ϕ_0 is the unique homoclinic solution to

$$\partial_x^2\phi_0 - \phi_0 + \phi_0^3 = 0. \tag{6.1.11}$$

Because the system (6.1.11) is invariant under $x \mapsto -x$ and has translational symmetry, we may assume that $\phi_0(x)$ is even.

When $\epsilon = 0$ the two profiles $\phi_\pm$ coincide with ϕ_0. Indeed, when $\epsilon = 0$ the system (6.1.8) possesses the translational and rotational symmetries associated with the NLS

$$T_1(\gamma_1)\boldsymbol{U}(x,t) = \boldsymbol{U}(x+\gamma_1, t), \quad T_2(\gamma_2)\boldsymbol{U}(x,t) = \begin{pmatrix} \cos\gamma_2 & -\sin\gamma_2 \\ \sin\gamma_2 & \cos\gamma_2 \end{pmatrix}\boldsymbol{U}(x,t)$$

(see Exercise 5.2.10). The action of the two-dimensional symmetry group $T(\gamma) = T_1(\gamma_1)T_2(\gamma_2)$ on the stationary solution $\boldsymbol{U}_0 := \phi_0(1,0)^{\mathrm{T}}$ generates a two-dimensional manifold of equilibria

$$\mathcal{M}_0 := \{T(\gamma_1)\boldsymbol{U}_0 : \gamma \in \mathbb{R} \times [0, 2\pi)\}.$$

For $\epsilon > 0$ the symmetry T_2 is broken, and the profile ϕ_0 splits into $\phi_\pm$. It is natural to ask about the fate of $\mathcal{M}_0$, which apparently breaks into two distinct one-dimensional manifolds,

$$\mathcal{M}_\pm := \{T_1(\gamma)\phi_\pm : \gamma_1 \in \mathbb{R}\}.$$

We use perturbation methods to characterize the spectrum of the linearization associated with each of these manifolds. We first identify the spectrum of the linearization about $\mathcal{M}_0$. Linearizing (6.1.8) about $\boldsymbol{U}_0$ yields the operator

$$\mathcal{L}_0 := \begin{pmatrix} \mathcal{L}_1^0 & 0 \\ 0 & \mathcal{L}_2^0 \end{pmatrix}, \tag{6.1.12}$$

where

$$\mathcal{L}_1^0 = \partial_x^2 - 1 + 3\phi_0^2, \quad \mathcal{L}_2^0 = \partial_x^2 - 1 + \phi_0^2. \tag{6.1.13}$$

The suboperators are Sturmian, and since $\{T_1'\boldsymbol{U}_0, T_2'\boldsymbol{U}_0\} \in \ker(\mathcal{L}_0)$ we readily verify that

$$\mathcal{L}_1^0(\partial_x\phi_0) = 0, \quad \mathcal{L}_2^0\phi_0 = 0. \tag{6.1.14}$$

From Exercise 6.1.2 we infer that the only spectrum of $\mathcal{L}_0$ on the imaginary axis is the zero eigenvalue, with $\mathrm{m_a}(0) = \mathrm{m_g}(0) = 2$. In addition, there is one unstable eigenvalue, associated with the ground state of $\mathcal{L}_0^1$, and the rest of the spectrum is bounded away from the imaginary axis. Changes in the number of unstable point spectrum associated with the linearization about the manifold $\mathcal{M}_\pm$ hinge upon the fate of these two eigenvalues. Since the translational symmetry is not broken, one eigenvalue must remain at zero: what is the fate of the other eigenvalue due to the "broken symmetry"?

Before addressing this issue, we verify that there are no other equilibria that bifurcate from $\mathcal{M}_0$ by searching for a stationary solution $\boldsymbol{U}$ near $T_2(\gamma_2)\boldsymbol{U}_0$ as

$$\boldsymbol{U} = T_2(\gamma_2)\boldsymbol{U}_0 + \epsilon\boldsymbol{U}_1 + \mathcal{O}(\epsilon^2).$$

Substituting this expression into (6.1.8), and requiring $\partial_t\boldsymbol{U} = 0$, yields the order of $\mathcal{O}(\epsilon)$ balance

$$\mathcal{L}_{\gamma_2}\boldsymbol{U}_1 = \begin{pmatrix} \sin\theta & 2\cos\theta \\ 0 & -\sin\theta \end{pmatrix} T_2(\gamma_2)\boldsymbol{U}_0,$$

where $\mathcal{L}_{\gamma_2}$ refers to the linearization about $T_2(\gamma_2)\boldsymbol{U}_0$. Using the commutative relation (4.2.5), i.e., $\mathcal{L}_{\gamma_2}T(\gamma_2) = T(\gamma_2)\mathcal{L}_0$, and acting on both sides of the previous equation by $T_2(-\gamma_2)$, yields

$$\begin{aligned}\mathcal{L}_0[T_2(-\gamma_2)\boldsymbol{U}_1] &= T_2(-\gamma_2)\begin{pmatrix}\sin\theta & 2\cos\theta\\ 0 & -\sin\theta\end{pmatrix}T_2(\gamma_2)\boldsymbol{U}_0,\\ &= \begin{pmatrix}\sin(\theta+2\gamma_2)\\ -2\sin\gamma_2\sin(\theta+\gamma_2)\end{pmatrix}\phi_0.\end{aligned}$$

We can solve for $T_2(-\gamma_2)\boldsymbol{U}_1$ if and only if the right-hand side is orthogonal to $\ker(\mathcal{L}_0)$. Since ϕ_0 is homoclinic we deduce that $\langle\phi_0,\partial_x\phi_0\rangle = 0$: the orthogonality of the right-hand side to the translational symmetry eigenfunction $T_1'\boldsymbol{U}_0$ is immediate. The orthogonality to the rotational symmetry eigenfunction $T_2'\boldsymbol{U}_0 = \phi_0(0,1)^{\mathrm{T}}$ requires $\sin\gamma_2\sin(\theta+\gamma_2) = 0$; that is, $\gamma_2 = 0$ or $\gamma_2 = -\theta$. In these two cases we solve for $\boldsymbol{U}_1$ as

$$\gamma_2 = 0: \ \boldsymbol{U}_1 = \sin\theta(\mathcal{L}_1^0)^{-1}\phi_0\boldsymbol{v}_+; \quad \gamma_2 = -\theta: \ \boldsymbol{U}_1 = -\sin\theta(\mathcal{L}_1^0)^{-1}\phi_0\boldsymbol{v}_-.$$

At the moment we have the formal expansions for the steady state,

$$\boldsymbol{U}\sim\phi_0\boldsymbol{v}_+ + \epsilon\sin\theta(\mathcal{L}_1^0)^{-1}\phi_0\boldsymbol{v}_+, \quad \boldsymbol{U}\sim\phi_0\boldsymbol{v}_+ - \epsilon\sin\theta(\mathcal{L}_1^0)^{-1}\phi_0\boldsymbol{v}_-.$$

We can compare these expansions to the expansion of $\boldsymbol{U}_\pm$ about $\epsilon = 0$, which takes the form

$$\boldsymbol{U}_\pm\sim\phi_0\boldsymbol{v}_+ \pm\epsilon\frac{1}{2}\sin\theta(\phi_0 + x\partial_x\phi_0)\boldsymbol{v}_\pm.$$

Verifying that

$$\mathcal{L}_1^0(\phi_0 + x\partial_x\phi_0) = 2\phi_0 \quad\Rightarrow\quad (\mathcal{L}_1^0)^{-1}\phi_0 = \frac{1}{2}(\phi_0 + x\partial_x\phi_0),$$

we see the two expressions match, and we have formally confirmed that the two solutions that bifurcate from $\boldsymbol{U}_0$ are precisely the solutions $\boldsymbol{U}_\pm$ determined above.

We now address the stability of the $\boldsymbol{U}_+$ branch for small ϵ. Linearizing about $\boldsymbol{U}_+$ yields the operator

$$\mathcal{L} := \begin{pmatrix}\mathcal{L}_1^\epsilon & -2\epsilon\mu\\ 0 & \mathcal{L}_2^\epsilon\end{pmatrix},$$

where the Sturmian suboperators take the form

$$\mathcal{L}_1^\epsilon = \partial_x^2 - A_+^2 + 3\phi_+^2, \quad \mathcal{L}_2^\epsilon = \partial_x^2 - A_-^2 + \phi_+^2.$$

Using (6.1.9) to expand $\phi_\pm$ we find that the operator $\mathcal{L}$ takes the form (6.0.1) with $\mathcal{L}_1$ given by

$$\mathcal{L}_1 = \begin{pmatrix}\sin\theta(-1+3\phi_0^2+3x\phi_0\partial_x\phi_0) & -2\mu\\ 0 & \sin\theta(1+\phi_0^2+x\phi_0\partial_x\phi_0)\end{pmatrix}. \tag{6.1.15}$$

Since the kernel of $\mathcal{L}_0$ is semi-simple, we are within the framework of Chapter 6.1. Setting

$$\begin{aligned}\boldsymbol{\Psi}_1 &= \begin{pmatrix}\partial_x\phi_0\\0\end{pmatrix}, \quad \boldsymbol{\Psi}_2 = \begin{pmatrix}0\\\phi_0\end{pmatrix},\\ \boldsymbol{\Psi}_1^{\mathrm{a}} &= \frac{1}{\langle\partial_x\phi_0,\partial_x\phi_0\rangle}\begin{pmatrix}\partial_x\phi_0\\0\end{pmatrix}, \quad \boldsymbol{\Psi}_2^{\mathrm{a}} = \frac{1}{\langle\phi_0,\phi_0\rangle}\begin{pmatrix}0\\\phi_0\end{pmatrix},\end{aligned} \tag{6.1.16}$$

the matrix $\boldsymbol{L}_1$ given in (6.1.5) becomes

$$\boldsymbol{L}_1 = \begin{pmatrix}\langle\mathcal{L}_1\boldsymbol{\Psi}_1,\boldsymbol{\Psi}_1^{\mathrm{a}}\rangle & 0\\ 0 & \langle\mathcal{L}_1\boldsymbol{\Psi}_2,\boldsymbol{\Psi}_2^{\mathrm{a}}\rangle\end{pmatrix},$$

with the off-diagonal terms are zero by parity considerations.

As a result of Exercise 6.1.3 we see

$$\langle\mathcal{L}_1\boldsymbol{\Psi}_1,\boldsymbol{\Psi}_1^{\mathrm{a}}\rangle = 0, \quad \langle\mathcal{L}_1\boldsymbol{\Psi}_2,\boldsymbol{\Psi}_2^{\mathrm{a}}\rangle = 2\sin\theta,$$

so that $\sigma_{\mathrm{pt}}(\boldsymbol{L}_1) = \{0, 2\sin\theta\}$, and the kernel of $\mathcal{L}_0$ perturbs into two eigenvalues. One of the eigenvalues remains at zero, which is expected from the translational symmetry. The associated eigenvector of $\boldsymbol{L}_1$ for the nonzero eigenvalue is given $(0,1)^{\mathrm{T}}$. The broken rotational symmetry drives the other eigenvalue according to

$$\lambda = \epsilon 2\sin\theta + \mathcal{O}(\epsilon^2).$$

If $\sin\theta > 0$ the eigenvalue enters the right-half complex plain and the manifold $\mathcal{M}_+$ is exponentially unstable for $\epsilon > 0$ sufficiently small; furthermore, the linearization has two real unstable eigenvalues. Conversely, if $\sin(\theta) < 0$, then the manifold is still unstable for $\epsilon > 0$, but the linearization has now only one unstable $\mathcal{O}(1)$ real eigenvalue. We further deduce from (6.1.1) that at leading order the "broken-rotational" eigenfunction is

$$\boldsymbol{V} = \boldsymbol{\Psi}_2 + \mathcal{O}(\epsilon) = \phi_0\begin{pmatrix}0\\1\end{pmatrix} + \mathcal{O}(\epsilon).$$

Upon noting the form of T_2', we see that this eigenfunction corresponds to a rotation of the $\epsilon = 0$ wave $\boldsymbol{U}_0$ in the $\boldsymbol{U}$ plane.

6.1.2 Example: Spatially Periodic Waves of gKdV

The general theory developed in Chapter 5.2.1 for the orbital stability of solitary waves can also be applied to spatially periodic structures. As an application of the perturbation methods we investigate the orbital stability of small-amplitude spatially periodic waves that bifurcate from the constant solution of the generalized Korteweg–de Vries equation. The stability of spatially periodic equilibria has been the subject of a great deal of recent research, e.g., see[**14, 15, 66–68, 88, 89, 133, 217, 218**] and the references therein.

In the traveling coordinate $\xi = x - ct$ with speed $c > 0$ the gKdV equation becomes (5.2.5). Finding spatially periodic solutions of this equation is straightforward from a phase plane analysis, for which they correspond to periodic orbits of the traveling-wave ordinary differential equation (ODE) (5.2.6). The Hamiltonian structure of the traveling-wave ODE makes the existence of these closed orbits self-evident: they are the dotted-red orbits appearing in Fig. 5.2. We wish to determine their period and to study the corresponding linearized operators, which is easiest in the small-amplitude limit corresponding to the small limit cycles enclosing the fixed point at $u \equiv a := (c(p+1))^{1/p}$. This motivates the change of variables $u = a + \epsilon v$ for $\epsilon \ll 1$, for which v satisfies

$$\partial_t v = \partial_\xi \left(-\partial_\xi^2 v - cpv - \frac{\epsilon}{2} a^{p-1} p v^2 + O(\epsilon^2) \right), \quad v(\xi,0) = v_0(\xi). \tag{6.1.17}$$

It is significant that the sign of one of the linear terms has flipped, changing from cu to $-cpv$, which gives the dominant linear term, $\partial_\xi^2 + cp$, a nontrivial kernel spanned by spatially periodic solutions. We rescale (6.1.17), and generalize the form of the nonlinearity to obtain a more interesting bifurcation problem, considering instead

$$\partial_t u = \partial_\xi \left(-\omega \partial_\xi^2 u - u - \epsilon u^{p+1} \right), \quad u(\xi,0) = u_0(\xi), \tag{6.1.18}$$

for $\omega, \epsilon > 0$ and $p \in \mathbb{N}_+$.

Remark 6.1.1. The Eq. (6.1.18) can also be obtained directly from the gKdV, (5.2.3), by considering *left*-going waves $\xi = x + ct$ with $c > 0$ and rescaling the amplitude. However, in the context of classical cnoidal waves it is natural to think of the periodic solutions as right-going waves that bifurcate out of the spatially constant solution, as outlined above.

The stability problem for (6.1.18) is naturally posed on the Hilbert space of $2L$-periodic functions $Y = H^1_{\text{per}}[-L,+L] \subset X = L^2[-L,+L]$ with usual L^2 inner product

$$\langle f, g \rangle = \int_{-L}^{+L} f(x)\overline{g(x)}\,dx.$$

We assume that $\epsilon \ll 1$, and while it may seem intuitive to take the periodicity as a free parameter, it is computationally (and notationally) more convenient to fix $L = \pi$ and treat ω as the tuning parameter to obtain the desired period.

Remark 6.1.2. It is worthwhile to make a clarifying comment about periodic function spaces. Setting

$$\psi_j(x) = \frac{1}{\sqrt{2\pi}} e^{ijx}, \quad j \in \mathbb{Z},$$

by Plancherel's equality we have $f \in L^2[-\pi,+\pi]$ if and only if

$$f = \sum_{j=-\infty}^{+\infty} f_j \psi_j,\ f_j = \langle f, \psi_j \rangle; \quad \|f\|_{L^2}^2 = \sum_{j=-\infty}^{+\infty} |f_j|^2 < \infty.$$

As for the norms $H^k(\mathbb{R})$ in Chapter 2.2.1, the usual norm for the space $H^k_{\text{per}}[-\pi,+\pi]$ is equivalent to

$$\|f\|_{H^k_{\text{per}}}^2 = \sum_{j=-\infty}^{+\infty} (1 + j^{2k})|f_j|^2,$$

moreover for $u \in H^k_{\text{per}}[-\pi,\pi]$ we have $\partial_x^j u(-\pi) = \partial_x^j u(\pi)$, for $j = 0,\dots,k-1$. In the literature one will often see the notation $L^2_{\text{per}}[-\pi,+\pi]$, which serves to reinforce that the boundary conditions of the underlying problem are 2π-spatially periodic.

In order to apply Theorem 5.2.11 we first introduce the Hamiltonian

$$\mathcal{H}(u) := \int_{-\pi}^{+\pi} \left(\frac{1}{2}\omega(\partial_\xi u)^2 - \frac{1}{2}u^2 - \frac{1}{p+2}\epsilon u^{p+2} \right) d\xi,$$

for which the gKdV equation (6.1.18) takes Hamiltonian form

$$\partial_t u = \partial_\xi \mathcal{H}'(u), \quad u(0) = u_0.$$

We observe that the skew operator $\mathcal{J} = \partial_\xi$ in (6.1.18) has a nontrivial kernel $\ker(\mathcal{J}) = \text{span}\{1\} \subset H^1_{\text{per}}[-\pi,+\pi]$. Using (5.2.57) the nontrivial kernel generates the conserved quantity

$$Q_2(u) := \langle u, 1 \rangle;$$

in other words, the flow preserves the mean of the initial data. Following the discussion and notation preceding the statement of Theorem 5.2.11 in Chapter 5.2.2, we introduce

$$Y_0 = \{u \in H^1_{\text{per}}[-\pi,+\pi] : \langle 1, u \rangle = 0\} \subset \ker(\mathcal{J})^\perp =: X_0 \subset L^2([-\pi,\pi]).$$

The nontrivial flow generated by the gKdV takes place on an affine translate of this codimension-one subspace. The operator $\mathcal{J}$ is Fredholm of index zero, and $\mathcal{J}^{-1} \in \mathcal{B}(X_0, Y_0)$. Since $\mathcal{J} = T'$, where $T(\gamma)u(\xi,t) = u(\xi+\gamma,t)$ denotes the spatial translational symmetry associated with gKdV, we have

$\mathcal{J}^{-1}T' = \mathcal{I} : Y_0 \mapsto Y_0$ is both self-adjoint and bounded. The conserved quantity generated by this symmetry is the total mass,

$$Q_1(u) = \langle u, u \rangle.$$

We now turn to the construction of the critical points. We find the 2π-periodic solutions via a perturbation expansion, using the Poincaré–Lindstedt method. We integrate the steady-state version of (6.1.18) once, which leads to the equilibrium ODE for ϕ,

$$\omega \partial_\xi^2 \phi + \phi + \epsilon \phi^{p+1} = a_0, \quad a_0 \in \mathbb{R}. \tag{6.1.19}$$

For $\epsilon = 0$ the system has 2π-periodic solutions for $\omega = 1$. For $\epsilon > 0$ we adjust the free parameter ω to account for the impact of the weak nonlinearity on the spatial period. The leading order expansions of ϕ and ω,

$$\phi(\xi) = \phi_0(\xi) + \epsilon \phi_1(\xi) + \mathcal{O}(\epsilon^2), \quad \omega = 1 + \epsilon \omega_1 + \mathcal{O}(\epsilon^2),$$

yield the following system of equations:

$$\begin{aligned} \mathcal{O}(1): &\quad \mathcal{L}_0 \phi_0 = -a_0 \\ \mathcal{O}(\epsilon): &\quad \mathcal{L}_0 \phi_1 = 2\omega_1 \partial_\xi^2 \phi_0 + \phi_0^{p+1}, \end{aligned}$$

where the dominant linear operator is $\mathcal{L}_0 := -(\partial_\xi^2 + 1)$. Restricting our attention to even solutions, the $\mathcal{O}(1)$ equation yields

$$\phi_0(\xi) = a_0 + b_0 \cos \xi, \quad b_0 \in \mathbb{R} \backslash \{0\}. \tag{6.1.20}$$

By the Fredholm alternative an even 2π-periodic solution ϕ_1 to the $\mathcal{O}(\epsilon)$ equation can be found if and only if

$$2\omega_1 \partial_\xi^2 \phi_0 + \phi_0^{p+1} \in \ker(\mathcal{L}_0)^\perp = \mathrm{span}\{\cos \xi, \sin \xi\}^\perp.$$

Since ϕ_0 is even in ξ, this solvability condition determines ω_1 via

$$\omega_1 = \frac{1}{2\pi b_0} \langle \phi_0^{p+1}(\xi), \cos \xi \rangle. \tag{6.1.21}$$

Remark 6.1.3. While the solution ϕ of (6.1.19) is not a critical point of this Hamiltonian (unless $a_0 = 0$), it is an equilibrium of the form considered in Hypothesis 5.2.5(f'). The stability of equilibria of augmented traveling-wave equations is discussed in Exercise 5.2.9. We may form an augmented Lagrangian that has the full symmetry group by adding $Q_2(u)$ to $\mathcal{H}(u)$; however, this has no impact on $\mathcal{L}$ in (6.1.22) and we omit this step.

In order to determine the orbital stability of this wave, we must investigate the spectrum of the second variation of the Hamiltonian, which takes the perturbative form

$$\begin{aligned}\mathcal{L} &= -\omega\partial_\xi^2 - 1 - \epsilon(p+1)\phi^p(\xi),\\ &= \underbrace{-\partial_\xi^2 - 1}_{\mathcal{L}_0} + \underbrace{\epsilon\left(-\omega_1\partial_\xi^2 - (p+1)\phi_0^p(\xi)\right)}_{\mathcal{L}_1} + O(\epsilon^2).\end{aligned} \tag{6.1.22}$$

We use perturbation theory to compute the spectrum of $\mathcal{L}$. Since $\mathcal{L}_0$ is a constant coefficient operator, it is easy to verify that on $H^1_{\text{per}}[-\pi,+\pi]$ its eigenspaces are $\text{span}\{\cos(k\xi),\sin(k\xi)\}$ with associated eigenvalues $\lambda_k = k^2 - 1$ for $k = 0,1,\ldots$. In particular, $\lambda_0 = -1$ is a simple eigenvalue, and $\lambda_1 = 0$ with $\ker(\mathcal{L}_0) = \text{span}\{\cos\xi,\sin\xi\}$, so that $\text{n}(\mathcal{L}_0) = 1$ and $\text{z}(\mathcal{L}_0) = 2$: the rest of the spectrum is on the positive real axis and uniformly bounded away from zero. With respect to the double eigenvalue at zero, for $\epsilon > 0$ at least one eigenvalue, which in an abuse of notation we label λ_0, will remain at the origin due to the translational symmetry. The fate of the other eigenvalue, which we label λ_1, is determined using the perturbative theory leading to (6.1.6). For the problem at hand the matrix $\boldsymbol{L}_1 \in \mathbb{R}^{2\times 2}$ is given by

$$\boldsymbol{L}_1 = \begin{pmatrix} \langle\mathcal{L}_1\cos\xi,\cos\xi\rangle & \langle\mathcal{L}_1\cos\xi,\sin\xi\rangle \\ \langle\mathcal{L}_1\sin\xi,\cos\xi\rangle & \langle\mathcal{L}_1\sin\xi,\sin\xi\rangle \end{pmatrix} = \begin{pmatrix} \langle\mathcal{L}_1\cos\xi,\cos\xi\rangle & 0 \\ 0 & \langle\mathcal{L}_1\sin\xi,\sin\xi\rangle \end{pmatrix}.$$

The off-diagonal terms are zero due to parity considerations. The diagonal terms can be calculated directly from (6.1.20), (6.1.21), and (6.1.22), from which we deduce that $\langle\mathcal{L}_1\sin\xi,\sin\xi\rangle = 0$ as anticipated, while for $p \geq 1$,

$$\langle\mathcal{L}_1\cos\xi,\cos\xi\rangle = \begin{cases} -(p-1)a_0 f(a_0,b,p), & p \text{ odd} \\ -g(a_0,b,p), & p \text{ even}, \end{cases} \tag{6.1.23}$$

where the functions f and g are strictly positive for all values of (a_0,b,p). The small (generically) nonzero eigenvalue has the expansion

$$\lambda_1(\epsilon) = \epsilon\langle\mathcal{L}_1\cos\xi,\cos\xi\rangle + \mathcal{O}(\epsilon^2), \tag{6.1.24}$$

with corresponding eigenfunction

$$\psi_1 = \cos\xi + \mathcal{O}(\epsilon).$$

If $p = 1$, then $\langle\mathcal{L}_1\cos\xi,\cos\xi\rangle = 0$, and determining the fate of the nonzero eigenvalue requires going to the next order in the perturbation expansion. On the other hand, for $p \geq 3$ and odd we know that the eigenvalue depends continuously on a_0, and changes sign in an $\mathcal{O}(\epsilon)$ neighborhood of $a_0 = 0$, forming a two-dimensional kernel as it passes through zero.

In light of Exercise 6.1.7, we have shown that for $\epsilon > 0$ sufficiently small,

$$\mathrm{z}(\mathcal{L}) = \begin{cases} 2, & p \text{ odd}, a_0 = \mathcal{O}(\epsilon) \\ 1, & \text{otherwise}, \end{cases} \quad \text{and} \quad \mathrm{n}(\mathcal{L}) = \begin{cases} 1, & p \text{ odd}, a_0 < 0 \\ 2, & p \text{ odd}, a_0 > 0 \\ 2, & p \text{ even}. \end{cases}$$

Equivalently we have established that

$$\mathrm{z}(\mathcal{L}) = \begin{cases} 2, & \lambda_1(\epsilon) = 0 \\ 1, & \lambda_1(\epsilon) \neq 0, \end{cases} \quad \text{and} \quad \mathrm{n}(\mathcal{L}) = \begin{cases} 1, & \lambda_1(\epsilon) \geq 0 \\ 2, & \lambda_1(\epsilon) < 0. \end{cases} \tag{6.1.25}$$

We investigate the case $\mathrm{z}(\mathcal{L}) = 1$, i.e., $\lambda_1(\epsilon) \neq 0$. By Theorem 5.2.11 the manifold $\mathcal{M}_T(\phi)$ of translates of the periodic wave ϕ is $H^1_{\mathrm{per}}[-\pi, +\pi]$ orbitally stable to same mass and mean-zero perturbations if the matrix $\boldsymbol{D}$ given in (5.2.53) is nonsingular and satisfies

$$\mathrm{n}(\mathcal{L}) = \mathrm{n}(\boldsymbol{D}).$$

For the problem at hand the constraint matrix $\boldsymbol{D} \in \mathbb{R}^{2\times 2}$ takes the form

$$\boldsymbol{D} = \begin{pmatrix} \langle \mathcal{L}^{-1}1, 1\rangle & \langle \mathcal{L}^{-1}1, \phi\rangle \\ \langle \mathcal{L}^{-1}1, \phi\rangle & \langle \mathcal{L}^{-1}\phi, \phi\rangle \end{pmatrix}. \tag{6.1.26}$$

It remains to calculate the negative index of the matrix $\boldsymbol{D}$. The original characterization (5.2.54) of $\mathcal{L}^{-1}\phi$ through $\partial_c \phi$ [see (5.2.36)] is complicated in this problem by the dependence of the period of ϕ for the unscaled problem upon the wave speed c. Instead, we exploit the singular nature of $\mathcal{L}^{-1}$ as $\epsilon \to 0^+$. Any $\psi \in \ker(\mathcal{L})^\perp$ can be decomposed as

$$\psi = \alpha(\epsilon)\psi_1(\epsilon) + \psi^\perp, \tag{6.1.27}$$

where $\psi^\perp \in (\ker(\mathcal{L}) \cup \mathrm{span}\{\psi_1(\epsilon)\})^\perp$, and $\alpha = \langle \psi, \psi_1\rangle / \|\psi_1\|^2$. From the orthogonality of ψ_1 with $\psi^\perp$, and the fact that $\psi_1(\epsilon)$ is an eigenfunction for $\mathcal{L}$, we have

$$\langle \mathcal{L}^{-1}\psi, \psi\rangle = \frac{\alpha^2(\epsilon)}{\lambda_1(\epsilon)}\|\psi_1\|^2 + \langle \mathcal{L}^{-1}\psi^\perp, \psi^\perp\rangle. \tag{6.1.28}$$

The action of $\mathcal{L}^{-1}$ on $(\ker(\mathcal{L}) \cup \mathrm{span}\{\psi_1(\epsilon)\})^\perp$ is smooth in ϵ, since $\mathcal{L}$ has no small eigenvalues on that space. Moreover, α is smooth in ϵ, and if $\langle \mathcal{L}_1 \cos\xi, \cos\xi\rangle \neq 0$ then λ_1 is proportional to ϵ at leading order. Consequently, if $\alpha(0) \neq 0$, then the ψ_1 term in (6.1.28) dominates as $\epsilon \to 0^+$ and the inner product is $O(\epsilon^{-1})$, while if $\alpha(0) = 0$ then the ψ_1 term is $\mathcal{O}(\epsilon)$ and the $\psi^\perp$ inner product is $\mathcal{O}(1)$ and dominates.

Now, when $\epsilon = 0$ we have not only $\langle 1, \psi_1(0)\rangle = 0$, but also that the wave is given by $\phi = a_0 + b_0\psi_1(0)$. Comparing with (6.1.27) with $\psi = \phi$ we see that $\alpha(0) = b_0$, from which we deduce via (6.1.28) that

$$\langle \mathcal{L}^{-1}\phi, \phi\rangle = \frac{b_0^2\|\psi_1\|^2}{\lambda_1} + \mathcal{O}(1) = \frac{b_0^2\|\psi_1\|^2}{\langle \mathcal{L}_1 \cos\xi, \cos\xi\rangle}\epsilon^{-1} + \mathcal{O}(1),$$

and

$$\langle \mathcal{L}^{-1}1, 1\rangle = \langle \mathcal{L}^{-1}1^{\perp}, 1^{\perp}\rangle + \mathcal{O}(\epsilon) = \langle \mathcal{L}_0^{-1}1, 1\rangle + \mathcal{O}(\epsilon) = -2\pi + \mathcal{O}(\epsilon)$$

Applying the decomposition to the off-diagonal terms in $\boldsymbol{D}$ shows that they are at most $\mathcal{O}(1)$. Since $b_0 \neq 0$, the dominance of $\boldsymbol{D}_{22}$ implies that $\sigma(\boldsymbol{D}) = \{\langle \mathcal{L}^{-1}1, 1\rangle + \mathcal{O}(\epsilon), \langle \mathcal{L}^{-1}\phi, \phi\rangle + \mathcal{O}(1)\}$, and hence

$$\mathrm{n}(\boldsymbol{D}) = \begin{cases} 1 & \lambda_1 > 0, \\ 2 & \lambda_1 < 0. \end{cases} \tag{6.1.29}$$

In conclusion, comparing (6.1.25) with (6.1.29) we deduce for $a_0 \neq 0$ that $\mathrm{n}(\mathcal{L}) = \mathrm{n}(\boldsymbol{D})$, and moreover $\mathrm{z}(\mathcal{L}) = 1$ with $\ker(\mathcal{L}) = \mathrm{span}\{T'\phi\}$. This satisfies the conditions of Theorem 5.2.11 and establishes the orbital stability of the translational manifold of 2π-periodic solutions in the H^1 norm. We emphasize that the stability is with respect to mean-zero perturbations from $H^1_{\mathrm{per}}[-\pi, \pi]$, i.e., $\langle u(0), 1\rangle = \langle \phi, 1\rangle$, in the system (6.1.18).

Remark 6.1.4. In the limit $\epsilon \ll 1$ the off-diagonal terms in $\boldsymbol{D}$ are unimportant, and the negative index $\mathrm{n}(\boldsymbol{D})$ tracks the sign of the bifurcating eigenvalue, λ_1, of $\mathcal{L}$. However, as ϵ grows it is entirely possible that the off-diagonal terms, which gauge the impact of the coupling between the kernel of $\mathcal{J}$ and the constraint induced by the conserved charge, could change the negative index of $\boldsymbol{D}$ and induce a bifurcation in the existence problem for the periodic wave. See Bronski and Johnson [**41**], Bronski et al. [**43**] for further discussion of this problem

Exercises

Exercise 6.1.1. Assume $\mathcal{L}_0 - \lambda_0\mathcal{I}$ is Fredholm of index zero. Use the Fredholm alternative to show there exists a basis $\{\psi_1^{\mathrm{a}}, \dots, \psi_m^{\mathrm{a}}\}$ of $\ker(\mathcal{L}_0^{\mathrm{a}} - \overline{\lambda_0}\mathcal{I})$ satisfying the normalization $\langle \psi_i, \psi_j^{\mathrm{a}}\rangle = \delta_{ij}$ if and only if $\mathrm{m_a}(\lambda_0) = \mathrm{m_g}(\lambda_0) = m$.

Exercise 6.1.2. Consider the operator $\mathcal{L}_0$ given by

$$\mathcal{L}_0 := \begin{pmatrix} \mathcal{L}_1^0 & 0 \\ 0 & \mathcal{L}_2^0 \end{pmatrix},$$

where

$$\mathcal{L}_1^0 = \partial_x^2 - 1 + 3\phi_0^2, \quad \mathcal{L}_2^0 = \partial_x^2 - 1 + \phi_0^2,$$

and $\phi_0(x)$ is even and defined as the homoclinic solution to (6.1.14). The kernel of $\mathcal{L}_0$ was computed in (6.1.14). Compute the rest of the spectrum of $\mathcal{L}_0$. In particular, verify that $\mathcal{L}_0$ is Fredholm of index zero, that there is precisely one real-valued unstable eigenvalue, and that the essential spectrum is uniformly bounded away from the imaginary axis.

Exercise 6.1.3. Consider the perturbed operator $\mathcal{L}_1$ given in (6.1.15). For the vector-valued functions given in (6.1.16), use integration by parts, and the identity

$$(\partial_x\phi_0)^2 = \phi_0^2 - \frac{1}{2}\phi_0^4,$$

to show that

$$\langle \mathcal{L}_1\Psi_1, \Psi_1^{\mathrm{a}}\rangle = 0, \quad \langle \mathcal{L}_1\Psi_2, \Psi_2^{\mathrm{a}}\rangle = 2\sin\theta.$$

Exercise 6.1.4. In Chapter 6.1.1 the small eigenvalues were located for the solution associated with $\boldsymbol{U}_+$. Determine the location of the $\mathcal{O}(\epsilon)$ eigenvalue for the solution associated with $\boldsymbol{U}_-$. In addition, derive a leading order expression for the eigenfunction associated with this small eigenvalue.

Exercise 6.1.5. Consider the eigenvalue problem given by

$$\mathcal{L}\boldsymbol{v} = \lambda\boldsymbol{v}; \quad \boldsymbol{v} = \begin{pmatrix} v_{\mathrm{r}} \\ v_{\mathrm{i}} \end{pmatrix}, \quad \mathcal{L} = \begin{pmatrix} 0 & \mathcal{L}_2 \\ -\mathcal{L}_1 & -2\mu \end{pmatrix},$$

where for $A_\pm^2 = 1 \pm \sin\theta$,

$$\mathcal{L}_1 = -\partial_x^2 + A_+^2 - 3\phi_\theta^2, \quad \mathcal{L}_2 = -\partial_x^2 + A_-^2 - \phi_\theta^2.$$

Here $\theta \in (-\pi/2, \pi/2)$, and $0 < \mu \le 1$ is chosen so that $\mu = \cos\theta$, and the $\phi_\theta(x)$ is the solution of

$$\partial_x^2\phi_\theta - A_+^2\phi_\theta + \phi_\theta^3 = 0,$$

which is homoclinic to zero. This eigenvalue problem was discussed in Chapter 4.5. When $\theta = 0$ we have that

$$\ker(\mathcal{L}) = \mathrm{span}\left\{\begin{pmatrix} \partial_x\phi_\theta \\ 0 \end{pmatrix}, \begin{pmatrix} -2\mathcal{L}_1^{-1}(\phi_\theta) \\ \phi_\theta \end{pmatrix}\right\},$$

while for $\theta \neq 0$ we have

$$\ker(\mathcal{L}) = \mathrm{span}\left\{\begin{pmatrix} \partial_x\phi_\theta \\ 0 \end{pmatrix}\right\}.$$

Determine to leading order the location of the nonzero $\mathcal{O}(|\theta|)$ eigenvalue.

Exercise 6.1.6. Consider the system of coupled Allen–Cahn equations

$$\begin{aligned} \partial_t u &= \partial_x^2 u - u + u^3 + \epsilon v, \\ \partial_t v &= \partial_x^2 v - v + v^3 + \epsilon d^2 u. \end{aligned}$$

For $\epsilon = 0$ the system has the two-parameter symmetry group consisting of uniform translation and a relative shift,

$$T(\gamma_1, \gamma_2)\begin{pmatrix} u \\ v \end{pmatrix}(x) = \begin{pmatrix} u(x+\gamma_1-\gamma_2) \\ v(x+\gamma_1+\gamma_2) \end{pmatrix},$$

and has a corresponding manifold of equilibria

$$\mathcal{M}_T = \left\{ T(\gamma)\begin{pmatrix} \phi_0 \\ \phi_0 \end{pmatrix} : \gamma \in \mathbb{R}^2 \right\},$$

where ϕ is the heteroclinic solution of

$$\partial_x^2\phi_0 - \phi_0 + \phi_0^3 = 0,$$

which satisfies $\phi_0(x) \to \pm 1$ as $x \to \pm\infty$, and it is odd about $x = 0$.

(a) For $\epsilon = 0$ show that the equilibria are linearly stable. In particular, show that conditions of Corollary 4.3.6 hold, and deduce that $\mathcal{M}_T$ is asymptotically orbitally stable in $H^1(\mathbb{R})$.

(b) Show for $\epsilon \neq 0$ that there are precisely two one-parameter families of equilibria that bifurcate from $\mathcal{M}_T$. In particular, show that they are of the form

$$\mathcal{M}_T^{\pm} := \left\{ T_1(\gamma_1)\begin{pmatrix} \phi_\pm \\ \pm d\phi_\pm \end{pmatrix} : \gamma_1 \in \mathbb{R} \right\},$$

where T_1 is the usual translational symmetry and $\phi_\pm(x;\epsilon)$ is the unique, odd heteroclinic solution of

$$\partial_x^2\phi_\pm - (1 \pm \epsilon d)\phi_\pm + \phi_\pm^3 = 0.$$

(c) For $0 < \epsilon \ll 1$ determine the linear stability of each of the equilibria, and the orbital stability of $\mathcal{M}_T^{\pm}$. In order to identify the sign of the eigenvalue corresponding to the broken shift-symmetry it is helpful to rewrite $\phi_\pm$ as a rescaled version of ϕ_0.

Exercise 6.1.7. Verify the calculations leading to (6.1.23) for $p \geq 2$. In addition, for $p = 1$ show that the small nonzero eigenvalue is positive for all values of (a_0, b).

Exercise 6.1.8. Verify the formulae (6.1.29) for the case $a_0 \neq 0$ and $p = 1$, i.e., when $\langle \mathcal{L}_1 \cos\xi, \cos\xi \rangle = 0$.

6.2 Perturbation of a Geometrically Simple Eigenvalue

Tractable applications of perturbation theory to geometrically simple eigenvalues of differential operators arise in Hamiltonian systems. The perturbation analysis takes a simplified form in the Hamiltonian case, which we present in Chapter 7. However, for completeness we include a discussion of the generic case. For example, the perturbation analysis of the eigenvalues of the NLS linearization presented in Chapter 7.2.2 can also be addressed through the techniques of this section.

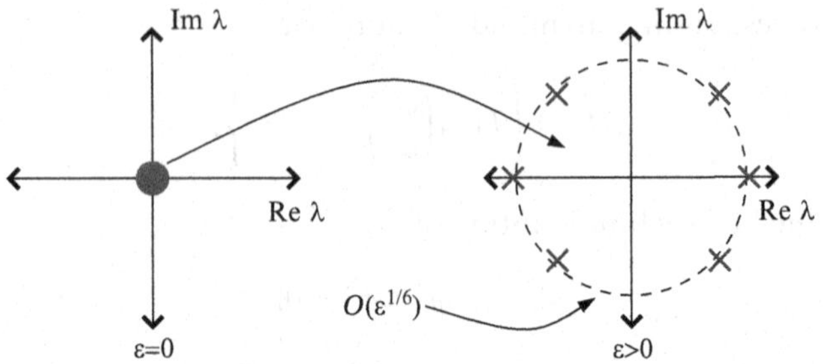

Fig. 6.2 A cartoon of the unfolding of a geometrically simple eigenvalue presented in Chapter 6.2. When $\epsilon = 0$ we assume $\lambda = 0$ is a geometrically simple eigenvalue of $\mathcal{L}_0$ with $\mathrm{m_a}(0) = 6$. For $\epsilon > 0$ the perturbed problem will generically—i.e., for $\eta \neq 0$ as defined in (6.2.2)—have six eigenvalues that lie to leading order on a circle of radius $\mathcal{O}(\epsilon^{1/6})$. (Color figure online.)

We consider the expansion (6.0.1) for which the operator $\mathcal{L}_0 - \lambda_0\mathcal{I}$ is Fredholm of index zero and has $\lambda_0 \in \sigma_{\mathrm{pt}}(\mathcal{L}_0)$ with $\mathrm{m_g}(\lambda_0) = 1$ and $\mathrm{m_a}(\lambda_0) = p \geq 2$. Without loss of generality, we assume that $\lambda_0 = 0$. From the Fredholm theory we know that $\mathcal{L}_0$ has a Jordan chain which terminates at p elements, and $\mathrm{gker}(\mathcal{L}_0) = \mathrm{span}\{\psi_1,\ldots,\psi_p\}$, where for $j = 1,\ldots,p$,

$$\mathcal{L}_0\psi_j = \psi_{j-1}, \quad \psi_0 = 0.$$

Moreover, $\dim[\ker(\mathcal{L}_0^{\mathrm{a}})] = 1$, and since $\psi_j \in R(\mathcal{L}_0)$ for $j = 1,\ldots,p-1$, it follows from the Fredholm alternative that $\psi_j \in \ker(\mathcal{L}_0^{\mathrm{a}})^{\perp}$ for these j. That is, for $\psi^{\mathrm{a}} \in \ker(\mathcal{L}_0^{\mathrm{a}})$ the generalized eigenfunctions satisfy

$$\langle\psi_j,\psi_1^{\mathrm{a}}\rangle = 0, \quad j = 1,\ldots,p-1; \quad \langle\psi_p,\psi_1^{\mathrm{a}}\rangle \neq 0.$$

In order to track the eigenvalues for the perturbed problem, we expand the perturbed eigenvalue and eigenfunction of $\mathcal{L}$ as

$$\begin{aligned}\lambda &= \epsilon^{1/p}\lambda_1 + \epsilon^{2/p}\lambda_2 + \cdots + \epsilon\lambda_p + \mathcal{O}\left(\epsilon^{(p+1)/p}\right),\\ v &= v_0 + \epsilon^{1/p}v_1 + \epsilon^{2/p}v_2 + \cdots + \epsilon v_p + \mathcal{O}\left(\epsilon^{(p+1)/p}\right).\end{aligned} \tag{6.2.1}$$

Plugging this expansion into the eigenvalue problem

$$(\mathcal{L}_0 + \epsilon\mathcal{L}_1 + \cdots)v = \lambda v,$$

and equating orders of $\epsilon^{j/p}$ we obtain

$$\mathcal{O}\left(\epsilon^{j/p}\right): \quad \mathcal{L}_0 v_j = \sum_{i=1}^{j} \lambda_i v_{j-i}, \quad j = 0,\dots,p-1,$$

$$\mathcal{O}(\epsilon): \quad \mathcal{L}_0 v_p = \sum_{i=1}^{p} \lambda_i v_{p-i} - \mathcal{L}_1 v_0.$$

This upper-triangular system is solvable, since we can make the right-hand side orthogonal to $\ker(\mathcal{L}_0^{\mathrm{a}}) = \mathrm{span}\{\psi^{\mathrm{a}}\}$. Indeed, setting $v_0 = \psi_1$, then for $j = 2,\dots,p$ the higher-order corrections can be expressed as

$$v_{j-1} = \lambda_1^{j-1}\psi_j + \tilde{v}_{j-1}, \quad \tilde{v}_{j-1} \in \mathrm{span}\{\psi_1,\dots,\psi_{j-1}\} \subset \ker(\mathcal{L}_0^{\mathrm{a}})^{\perp}.$$

In particular, note that for $j = 0,\dots,p-2$ the solutions satisfy $v_j \in \ker(\mathcal{L}_0^{\mathrm{a}})^{\perp}$, while $v_{p-1} = \lambda_1^{p-1}\psi_p + \tilde{v}_{p-1}$, where $\tilde{v}_{p-1} \in \ker(\mathcal{L}_0^{\mathrm{a}})^{\perp}$. The $\mathcal{O}(\epsilon)$ equation is solvable if and only if

$$0 = \left\langle \sum_{i=1}^{p} \lambda_i v_{p-i} - \mathcal{L}_1 v_0, \psi_1^{\mathrm{a}} \right\rangle = \left\langle \lambda_1 v_{p-1} - \mathcal{L}_1 v_0, \psi_1^{\mathrm{a}} \right\rangle = \lambda_1^p \langle \psi_p, \psi_1^{\mathrm{a}} \rangle - \langle \mathcal{L}_1 \psi_1, \psi_1^{\mathrm{a}} \rangle,$$

where in the second equality we used the fact that the remainder of the summation lies within $\ker(\mathcal{L}_0^{\mathrm{a}})^{\perp}$, and in the third equality we used the expressions for v_0 and v_{p-1}. There are p solutions to this equation,

$$\lambda_1 = \eta^{1/p} \mathrm{e}^{2\pi \mathrm{i}\ell/p}, \quad \ell = 0,\dots,p-1; \quad \eta := \frac{\langle \mathcal{L}_1 \psi_1, \psi_1^{\mathrm{a}} \rangle}{\langle \psi_p, \psi_1^{\mathrm{a}} \rangle}. \tag{6.2.2}$$

For $\eta \neq 0$ it is the case that at leading order the eigenvalues are evenly spaced on the circle of radius $\mathcal{O}(\epsilon^{1/p})$, i.e., on the circle $\{\lambda : |\lambda| = (\eta\epsilon)^{1/p}\}$ (see Fig. 6.2). If $\eta = 0$, then the perturbed eigenvalues are in the disk $\{\lambda : |\lambda| \leq \mathcal{O}(\epsilon^{2/p})\}$. The interested reader can consult [**144**, **162**] for information on more general expansions.

Chapter 7
Point Spectrum: Linear Hamiltonian Systems

Hamiltonian systems are about balance, with the energy and other invariants preserved under the flow. For a spatially localized critical point of a Hamiltonian system, the balance is reflected in the symmetry of the spectrum, which typically pins the essential spectrum to the imaginary axis in unweighted spaces. The mechanism for bifurcation in Hamiltonian systems thus falls upon the point spectrum. Moreover, as foreshadowed in the discussion of Chapter 6, most examples of a complex point spectrum, such as nontrivial Jordan blocks, arise in the Hamiltonian context. This complex linear structure, with the associated possibility of bifurcation, adds to the richness of Hamiltonian dynamics.

In Chapter 5.2.2 we showed that a Hamiltonian system, of the form (5.2.26) with an N dimensional symmetry group could be reformulated so that traveling solutions, ϕ_c, associated with the underlying symmetries became critical points of the Lagrangian Λ, introduced in (5.2.45). Moreover, in Theorem 5.2.11 the orbital stability of the symmetry manifold $\mathcal{M}_T(\boldsymbol{c})$ of the boosted Hamiltonian system,

$$\partial_t u = \mathcal{J}\frac{\delta\Lambda}{\delta u}(u), \tag{7.0.1}$$

was related to the difference in the negative eigenvalue count of the second variational derivative, $\mathcal{L}$, of Λ at ϕ_c, and the negative eigenvalue count of the constraint matrix $\boldsymbol{D} \in \mathbb{R}^{(N+M)\times(N+M)}$, defined in (5.2.53). Indeed the orbital stability could be guaranteed through an analysis of the self-adjoint operator $\mathcal{L}$ if the equality,

$$\mathrm{n}(\mathcal{L}) - \mathrm{n}(\boldsymbol{D}) = 0,$$

holds. In this case, ϕ_c is a constrained local minimizer of Λ and the level sets of Λ in a neighborhood of ϕ_c are bounded in the admissible space. Irrespective of the properties of the flow (7.0.1) on these level sets the solution u that starts sufficiently close to ϕ_c cannot stray too far away.

T. Kapitula and K. Promislow, *Spectral and Dynamical Stability of Nonlinear Waves*, Applied Mathematical Sciences 185, DOI 10.1007/978-1-4614-6995-7_7,

In this chapter we investigate the spectrum of the full linearization, $\mathcal{J}\mathcal{L}$, of (7.0.1) about a critical point, ϕ_c, of Λ. An immediate result is that $\sigma_{\mathrm{pt}}(\mathcal{J}\mathcal{L})$ enjoys the same fourfold Hamiltonian symmetry of the ODE systems in Proposition 5.1.2:

Proposition 7.0.1. *Consider the linear operator $\mathcal{J}\mathcal{L}$ associated with the linearization of the real Hamiltonian system (7.0.1) about a critical point ϕ_c. The point spectrum $\sigma_{\mathrm{pt}}(\mathcal{J}\mathcal{L})$ is symmetric with respect to the real and imaginary axes. That is, if $\lambda \in \sigma_{\mathrm{pt}}(\mathcal{J}\mathcal{L})$, then the quartet $\{\pm\lambda, \pm\overline{\lambda}\} \subset \sigma_{\mathrm{pt}}(\mathcal{J}\mathcal{L})$.*

It is a nontrivial consequence of Theorem 5.2.11 and Proposition 7.0.1, that for the case $\mathrm{n}(\mathcal{L}) = \mathrm{n}(\boldsymbol{D})$ the point spectrum $\sigma_{\mathrm{pt}}(\mathcal{J}\mathcal{L})$ must lie entirely on the imaginary axis. What happens, however, if $\mathrm{n}(\mathcal{L}) > \mathrm{n}(\boldsymbol{D})$? In this case the level sets of Λ need not be bounded in the admissible directions, and the Hamiltonian flow may possibly lead to instability. For perturbations that leave the essential spectrum in place, the bifurcation from stability to instability must occur via pairs of real eigenvalues, leaving the origin or quartets of complex eigenvalues bifurcating off of the imaginary axis. In either of these cases the spectral analysis of the linearization of a Hamiltonian system is ideally suited for both perturbation and index methods.

In this chapter we study the structure of the Hamiltonian eigenvalue problem

$$\mathcal{J}\mathcal{L}v = \lambda v, \tag{7.0.2}$$

where we supplement Hypothesis 5.2.5 with

(j) the essential spectrum of $\mathcal{J}\mathcal{L}$ is a strict subset of the imaginary axis; that is, there exists $\omega_0 > 0$ such that

$$\sigma_{\mathrm{ess}}(\mathcal{J}\mathcal{L}) \subset (-\mathrm{i}\infty, -\mathrm{i}\omega_0] \cup [\mathrm{i}\omega_0, \mathrm{i}\infty). \tag{7.0.3}$$

Remark 7.0.2. We emphasize that since $\lambda \in \mathbb{C}$, the eigenvalue problem is for complex-valued v, for which

$$\langle \mathcal{J}v, v\rangle = -\langle v, \mathcal{J}v\rangle = -\overline{\langle \mathcal{J}v, v\rangle} \quad \Rightarrow \quad \langle \mathcal{J}v, v\rangle \in \mathrm{i}\mathbb{R},$$

and

$$\langle \mathcal{L}v, v\rangle = \langle v, \mathcal{L}v\rangle = \overline{\langle \mathcal{L}v, v\rangle} \quad \Rightarrow \quad \langle \mathcal{L}v, v\rangle \in \mathbb{R}.$$

This dichotomy drives much of the analysis presented herein.

In Chapter 7.1 we address index theorems that determine the number of complex quartets or real pairs of eigenvalues. These are expressed in terms of the Hamiltonian–Krein index, which uses the Krein index (signature) of the purely imaginary eigenvalues. In Chapter 7.2 we consider two classes of perturbations, parameterized by ϵ: those that break symmetries but preserve the Hamiltonian structure, and those that break the Hamiltonian structure but preserve the symmetries. Under the assumption that the matrix $\boldsymbol{D}$ is nonsingular for $\epsilon = 0$, the unperturbed linearization satisfies $\dim[\ker(\mathcal{J}\mathcal{L}]) = N$ and $\dim[\text{gker}(\mathcal{J}\mathcal{L})] = 2N$, with N Jordan chains of length two. If the perturbed problem remains Hamiltonian but retains only k symmetries, then $2(N-k)$ of these eigenvalues will generically be nonzero. Following Chapter 6.2 we show that these $N-k$ eigenvalue pairs will scale like $\mathcal{O}(\epsilon^{1/2})$ for $\epsilon > 0$. If the perturbation is not Hamiltonian, but does not break any of the symmetries, then generically the Jordan chain structure of the kernel will break apart for $\epsilon > 0$ and the perturbed linearization will satisfy $\dim[\ker(\mathcal{J}\mathcal{L})] = \dim[\text{gker}(\mathcal{J}\mathcal{L})] = N$. We show that generically N nonzero eigenvalues are formed from the broken Hamiltonian structure, with the dislodged eigenvalues scaling like $\mathcal{O}(\epsilon)$.

Exercises

Exercise 7.0.1. Consider a Hamiltonian $\mathcal{H}$ that has a critical point $\phi \in C^\infty(\mathbb{R})$ for which the second variation, $\mathcal{L}$, of $\mathcal{H}$ at ϕ_c is an exponentially asymptotic $2n$th-order *self-adjoint* differential operator of the form (3.1.1). The function ϕ_c is also an equilibrium of the Hamiltonian system

$$\partial_t u = \mathcal{J}\frac{\delta \mathcal{H}}{\delta u},$$

for any skew operator $\mathcal{J} = b_1\partial_x + \cdots + b_{2d+1}\partial_x^{2d+1}$ with $b_{2j+1} \in \mathbb{R}$ for $j = 0, \ldots, d$. Show that $\sigma_{\text{ess}}(\mathcal{J}\mathcal{L}) \subset \mathrm{i}\mathbb{R}$ in the space $H^{2d+2n+1}(\mathbb{R})$,

7.1 The Krein Signature and the Hamiltonian–Krein Index

The symmetry of Proposition 7.0.1 implies that Hamiltonian eigenvalue problems have three types of nonzero eigenvalues: those on the imaginary axis, those belonging to an unstable quartet, and those belonging to an unstable real pair.

Definition 7.1.1 (Real and complex Krein indices). We call the number of positive real eigenvalues of $\mathcal{J}\mathcal{L}$ (up to multiplicity) the *real Krein index*, denoted by k_{r}, and the number of eigenvalues of $\mathcal{J}\mathcal{L}$ with a nonzero imaginary part and a positive real part (again up to multiplicity), the *complex Krein index*, denoted by k_{c}.

The goal of this section is to determine bounds on the number of eigenvalues of each of these three types, and to develop conditions on which the imaginary eigenvalues can transition to unstable pairs and quartets. Proposition 7.0.1 shows that k_{c} is even, since each quartet contributes twice to the count. If either k_{r} or k_{c} is nonzero, then the equilibria that produced the linearization is linearly unstable.

The bifurcation from linear stability to instability occurs when eigenvalues leave the imaginary axis. Those bifurcations that occur at the origin can be analyzed via standard perturbation theory, which we consider in Chapter 7.2. Krein [**178**] was the first to understand that the nonzero imaginary point spectra could be separated into two groups: those with the potential to make trouble, and those of a purer nature. We characterize these in terms of the Krein matrix.

Definition 7.1.2 (Negative Krein index). Let Hypothesis 5.2.5(a)–(j) hold with $\mathcal{L} = \delta^2\mathcal{H}/\delta u^2$. Let $\lambda \in \mathrm{i}\mathbb{R}\backslash\{0\}$ be a purely imaginary nonzero eigenvalue with associated generalized eigenspace $\mathbb{E}^\lambda$ and basis given by $\{v_1^\lambda,\dots,v_k^\lambda\}$. We form the Hermitian Krein matrix $\mathbf{H}^\lambda \in \mathbb{C}^{k\times k}$,

$$\boldsymbol{H}_{ij}^\lambda = \langle v_i^\lambda, \mathcal{L}v_j^\lambda\rangle, \tag{7.1.1}$$

which is the linear operator (matrix) induced by the bilinear form $\langle \mathcal{L}v, v\rangle$ restricted to the finite-dimensional space $v \in \mathbb{E}^\lambda$. For $\lambda \in (\mathrm{i}\mathbb{R}\backslash\{0\}) \cap \sigma_{\mathrm{pt}}(\mathcal{J}\mathcal{L})$ we introduce the *negative Krein index*

$$k_{\mathrm{i}}^-(\lambda) := \mathrm{n}(\mathbb{E}^\lambda), \tag{7.1.2}$$

and define the total negative Krein index

$$k_{\mathrm{i}}^- = k_{\mathrm{i}}^-(\mathcal{J}\mathcal{L}) := \sum_{\mathrm{Re}\,\lambda=0} k_{\mathrm{i}}^-(\lambda). \tag{7.1.3}$$

If $k_{\mathrm{i}}^-(\lambda) \geq 1$, then the eigenvalue λ is said to have a *negative Krein signature*; otherwise, it has a *positive Krein signature.*

Remark 7.1.3. Since $k_{\mathrm{i}}^-(\overline{\lambda}) = k_{\mathrm{i}}^-(\lambda)$ (see Exercise 7.1.1), the negative Krein index, k_{i}^-, is an even number.

Alternatively, the negative Krein index of an imaginary eigenvalue is the dimension of the intersection of the associated generalized eigenspace with the negative cone of $\mathcal{L}$; see Definition 5.2.9. Our first observation is that all of the eigenspaces associated with eigenvalues having a nonzero real part also have a nontrivial intersection with the negative cone. Indeed, fix a nonzero $\lambda \in \sigma_{\mathrm{pt}}(\mathcal{J}\mathcal{L})$ with a generalized eigenspace $\mathbb{E}^\lambda$. Since $\lambda \neq 0$, after applying the Fredholm alternative to the eigenvalue problem (7.0.2) we see that any eigenvalue v must satisfy $v \in \mathrm{gker}(\mathcal{L}\mathcal{J})^\perp$; that is, $\mathbb{E}^\lambda \subset \mathrm{gker}(\mathcal{L}\mathcal{J})^\perp$. Furthermore, from Hypothesis 5.2.5(b'), (e), and (g) we may calculate that

$$\ker(\mathcal{L}\mathcal{J}) = \mathcal{J}^{-1}\ker(\mathcal{L}) \oplus \ker(\mathcal{J}) = S \subset \ker(\mathcal{L})^\perp,$$

where the constraint space S is defined in (5.2.52). We conclude that the eigenspaces associated with nonzero eigenvalues must reside in $S^\perp$.

As in Chapter 5.3 we denote the orthogonal projection onto $S^\perp$ by Π, and the associated constrained form of $\mathcal{L}$ by $\mathcal{L}_\Pi$. Since $E^\lambda \subset S^\perp$, the operator $\mathcal{J}$

has a bounded inverse on $\mathbb{E}^\lambda$. Inverting $\mathcal{J}$ in the eigenvalue problem (7.0.2), yields the general reformulation for the eigenfunction

$$\mathcal{L}v = \lambda \mathcal{J}^{-1} v + \psi,$$

for some $\psi \in \ker(\mathcal{J})$. Taking the X inner product with v we obtain the equalities

$$\langle \mathcal{L}v, v\rangle = \lambda \langle \mathcal{J}^{-1}v, v\rangle + \langle \psi, v\rangle = \lambda \langle \mathcal{J}^{-1}v, v\rangle,$$

where the last equality follows since $v \in \ker(\mathcal{J})^\perp$. Now, $\mathcal{J}^{-1}$ is skew on $\ker(\mathcal{J})^\perp$, and we deduce that $\langle \mathcal{J}^{-1}v, v\rangle \in \mathrm{i}\mathbb{R}$, while $\langle \mathcal{L}v, v\rangle \in \mathbb{R}$ since $\mathcal{L}$ is self-adjoint. This leads to the following Krein dichotomy: either

$$\lambda \in \mathrm{i}\mathbb{R} \quad \text{or} \quad \langle \mathcal{L}v, v\rangle = \langle \mathcal{J}^{-1}v, v\rangle = 0. \tag{7.1.4}$$

This is a telling deduction, for any eigenfunction of $\mathcal{J}\mathcal{L}$ corresponding to an eigenvalue with a nonzero real part must have nontrivial intersection with both the positive and negative eigenspaces of $\mathcal{L}$. In other words, the eigenspace $\mathbb{E}^\lambda$ has a nontrivial intersection with the negative cone of $\mathcal{L}$. Since by assumption the negative cone has finite dimension, the following index is then expected to be finite:

Definition 7.1.4. The Hamiltonian–Krein index is given by the sum

$$K_{\mathrm{Ham}} := k_{\mathrm{r}} + k_{\mathrm{c}} + k_{\mathrm{i}}^-,$$

of the real, complex, and negative Krein indices, defined in Definitions 7.1.1 and 7.1.2.

The dimension of the negative cone of $\mathcal{L}$ is simply $\mathrm{n}(\mathcal{L})$, but the arguments above suggests the intersection of the negative cone with $S^\perp$ is more interesting. In fact, the real quantity of interest is the dimension of the negative cone of $\mathcal{L}_\Pi$, which is $\mathrm{n}(\mathcal{L}_\Pi)$. This leads us full circle, as the index Theorem 5.3.2 implies

$$\mathrm{n}(\mathcal{L}_\Pi) = \mathrm{n}(\mathcal{L}) - \mathrm{n}(\boldsymbol{D}),$$

subject to the assumption that the constraint matrix $\boldsymbol{D}$ of (5.2.53) is nonsingular. Moreover, this assumption on the constraint matrix implies that $\mathrm{gker}(\mathcal{J}\mathcal{L})$ consists of N Jordan chains of length two, each generated by a symmetry. We have the following spectral result: if $\mathrm{n}(\mathcal{L}_\Pi) = 0$, then since $\mathcal{L}_\Pi$ has no nontrivial negative cone the Hamiltonian–Krein index, K_{Ham}, is 0. This argument relates the orbital stability of Theorem 5.2.11 to the spectral stability of $\sigma_{\mathrm{pt}}(\mathcal{J}\mathcal{L})$. Even more significantly, it implies that each of the purely imaginary point spectra have a positive Krein signature.

On the other hand, if $\mathrm{n}(\mathcal{L}) - \mathrm{n}(\boldsymbol{D}) \geq 1$, then it may be that $K_{\mathrm{Ham}} \geq 1$. However, even in this case the wave can still be spectrally stable if $k_{\mathrm{i}}^- = K_{\mathrm{Ham}}$ and $k_{\mathrm{r}} = k_{\mathrm{c}} = 0$; that is, the potentially unstable negative cone induced by the dimension mismatch between the negative spaces of $\mathcal{L}$ and $\boldsymbol{D}$ can be stored in the potentially troublesome negative Krein eigenvalues.

The following theorem shows that the relationship between the negative index of the constrained operator, $\mathrm{n}(\mathcal{L}_\Pi)$, and the Hamiltonian–Krein index is sharp (see Fig. 7.1). The proof of this theorem, in full generality, is out side the scope of this book. Interested readers should consult [**56, 66, 127, 157, 159, 222**] and the references therein.

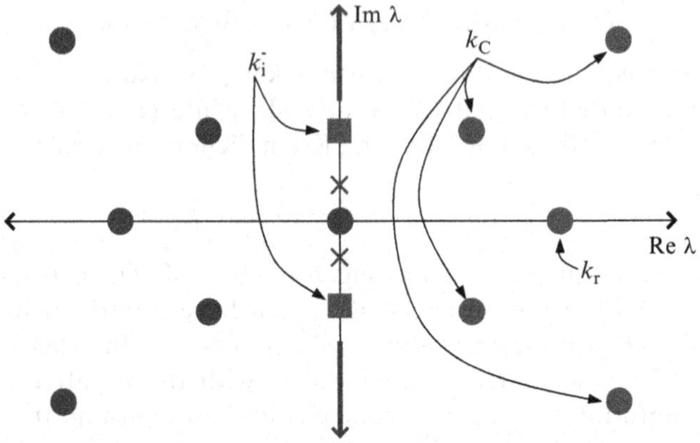

Fig. 7.1 The spectrum of $\mathcal{JL}$. The purely imaginary eigenvalues with a negative Krein signature are denoted by (*red*) squares. The Hamiltonian–Krein index K_{Ham} of Theorem 7.1.5 counts the total number of eigenvalues in the right-half plane (*red* circles) together with the total negative Krein index of $\mathcal{JL}$. This figure depicts the case $K_{\mathrm{Ham}} = 7$ with $k_i^- = 2$, $k_c = 4$, and $k_r = 1$. (Color figure online.)

Theorem 7.1.5. *(Hamiltonian–Krein index theorem) Consider the eigenvalue problem (7.0.2). Suppose that Hypothesis 5.2.5(a) and (b')–(j) are satisfied. If the constraint matrix* **D***, defined in (5.2.53), is nonsingular then the Hamiltonian–Krein index satisfies*

$$K_{\mathrm{Ham}} = \mathrm{n}(\mathcal{L}) - \mathrm{n}(\boldsymbol{D}).$$

Remark 7.1.6. An equivalent statement of the stability Theorem 5.2.11 is that a critical point of a Hamiltonian system is orbitally stable if the constraint matrix nonsingular and the Hamiltonian–Krein index satisfies $K_{\mathrm{Ham}} = 0$.

Remark 7.1.7. Since k_c and k_i^- are both even, if the Hamiltonian–Krein index K_{Ham} is odd, then $k_r \geq 1$ and the critical point ϕ of the Hamiltonian is linearly unstable. This instability result was first observed by Grillakis et al. [**110**].

Remark 7.1.8. The invertibility of the skew operator, $\mathcal{J}$, plays a much more subtle role in the Hamiltonian–Krein index Theorem 7.1.5 than in the stability Theorem 5.2.11. The proof of the Hamiltonian–Krein index requires the stronger assumption Hypothesis 5.2.5(b'). This is nontrivial; indeed, this assumption fails for the gKdV equation since ∂_x does not have a bounded inverse on any finite codimension subspace of $L^2(\mathbb{R})$. This requirement may be remedied by working on algebraically weighted subsets of $L^2(\mathbb{R})$, but then the well-posedness of the gKdV becomes delicate (see Bona et al. [**36**] for the original treatment of this issue for a general class of KdV-type equations).

7.1.1 A Finite-Dimensional Version of Theorem 7.1.5

The proof of Theorem 7.1.5 in an infinite-dimensional setting is beyond the scope of this book; however, many of the key ideas can be conveyed in a relatively simple proof for the finite-dimensional setting. This section can be omitted without impinging upon the material that follows. The proof sketched below is based upon the approach of [**127**] for compact operators. A proof of the Hamiltonian–Krein index theorem in the context of Theorem 7.1.5 can be found in [**157**, **159**]; this proof is based upon the results of Grillakis [**108**]. An alternate proof based on ideas arising from indefinite inner products (leading to the Pontryagin spaces) can be found in [**56**, **222**].

We start with the eigenvalue problem

$$\boldsymbol{J}\boldsymbol{L}\boldsymbol{x} = \lambda \boldsymbol{x} \quad \Rightarrow \quad \boldsymbol{L}\boldsymbol{x} = \lambda \boldsymbol{J}^{-1}\boldsymbol{x}, \tag{7.1.5}$$

where $\boldsymbol{L} \in \mathbb{R}^{n\times n}$ is symmetric and invertible, and $\boldsymbol{J} \in \mathbb{R}^{n\times n}$ is skew-symmetric and invertible with respect to the standard inner product on $\mathbb{C}^n$,

$$\boldsymbol{x}\cdot\boldsymbol{y} = \sum_{j=1}^{n} x_j\overline{y_j}.$$

The matrices enjoy the property that for all $\boldsymbol{x} \in \mathbb{C}^n$,

$$\mathrm{Re}(\boldsymbol{J}^{-1}\boldsymbol{x}\cdot\boldsymbol{x}) = \mathrm{Im}(\boldsymbol{L}\boldsymbol{x}\cdot\boldsymbol{x}) = 0.$$

In the context at hand there are no constraints; indeed, $\boldsymbol{J}$ is assumed invertible, so the constraint space S is trivial and there is no matrix $\boldsymbol{D}$. The statement of Theorem 7.1.5 reduces to $K_{\mathrm{Ham}} = \mathrm{n}(\boldsymbol{L})$. As a further simplification, we assume that all eigenvalues are algebraically simple, i.e., there are no nontrivial Jordan blocks. For each $\lambda \in \sigma(\boldsymbol{J}\boldsymbol{L})$ we denote the basis of the n_λ-dimensional eigenspace $\mathbb{E}^\lambda$ by $\{\boldsymbol{x}_1^\lambda,\dots,\boldsymbol{x}_{n_\lambda}^\lambda\}$.

Theorem 7.1.9. *Consider the finite-dimensional Hamiltonian eigenvalue problem (7.1.5) under the assumption that the skew-symmetric matrix* $\boldsymbol{J}$ *and the symmetric matrix* $\boldsymbol{L}$ *are both invertible. If all of the eigenvalues are algebraically simple, then the Hamiltonian–Krein index given in Definition 7.1.4 satisfies*

$$K_{\mathrm{Ham}} = \mathrm{n}(\boldsymbol{L}).$$

The idea behind the proof of Theorem 7.1.9 is fairly simple: rewrite the matrix $\boldsymbol{L}$ in the basis comprised of the eigenvectors for $\boldsymbol{JL}$. Surprisingly, $\boldsymbol{L}$ is block-diagonal in this basis, and an analysis of the structure of these blocks yields the result. We begin with the following simple observation:

Proposition 7.1.10. *Suppose that $\boldsymbol{x}_\lambda \in \mathbb{E}^\lambda$ and $\boldsymbol{x}_\sigma \in \mathbb{E}^\sigma$. If $\lambda + \overline{\sigma} \neq 0$, then*

$$\boldsymbol{L}\boldsymbol{x}_\lambda \cdot \boldsymbol{x}_\sigma = 0.$$

Proof. Since $\boldsymbol{L}\boldsymbol{x}_\lambda = \lambda \boldsymbol{J}^{-1}\boldsymbol{x}_\lambda$ and $\boldsymbol{L}\boldsymbol{x}_\sigma = \lambda \boldsymbol{J}^{-1}\boldsymbol{x}_\sigma$, we have

$$\boldsymbol{L}\boldsymbol{x}_\lambda \cdot \boldsymbol{x}_\sigma = \lambda \boldsymbol{J}^{-1}\boldsymbol{x}_\lambda \cdot \boldsymbol{x}_\sigma,$$

and

$$\boldsymbol{L}\boldsymbol{x}_\lambda \cdot \boldsymbol{x}_\sigma = \boldsymbol{x}_\lambda \cdot \boldsymbol{L}\boldsymbol{x}_\sigma = \overline{\sigma}\boldsymbol{x}_\lambda \cdot \boldsymbol{J}^{-1}\boldsymbol{x}_\sigma = -\overline{\sigma}\boldsymbol{J}^{-1}\boldsymbol{x}_\lambda \cdot \boldsymbol{x}_\sigma.$$

So long as $\lambda + \overline{\sigma} \neq 0$, equating the two lines above implies that

$$\boldsymbol{J}^{-1}\boldsymbol{x}_\lambda \cdot \boldsymbol{x}_\sigma = 0,$$

which yields the desired result. □

Recall that eigenvalues for (7.1.5) come in the quartets $\{\pm\lambda, \pm\overline{\lambda}\}$. As a consequence of Proposition 7.1.10 it is natural to look at the action of $\boldsymbol{L}$ on the sum of subspaces, $\mathbb{I}_\lambda := \mathbb{E}^\lambda \oplus \mathbb{E}^{-\overline{\lambda}}$. Since $\boldsymbol{e}_\lambda \in \mathbb{I}_\lambda$ implies $\overline{\boldsymbol{e}_\lambda} \in \mathbb{I}_{\overline{\lambda}}$, we may assume that $\boldsymbol{e}_{\overline{\lambda}} = \overline{\boldsymbol{e}_\lambda}$. By assumption the eigenvectors of $\boldsymbol{JL}$ form a basis of $\mathbb{C}^n$, so any $\boldsymbol{x} \in \mathbb{C}^n$ has a decomposition

$$\boldsymbol{x} = \sum_{\lambda \in \sigma(\boldsymbol{JL})} \boldsymbol{e}_\lambda, \tag{7.1.6}$$

where the vectors $\boldsymbol{e}_\lambda$ are uniquely defined. These observations lead to the block-diagonalization of $\boldsymbol{L}$.

Proposition 7.1.11. *For $\boldsymbol{x}$ decomposed as in (7.1.6),*

$$\boldsymbol{L}\boldsymbol{x} \cdot \boldsymbol{x} = \sum_{\lambda \in \sigma(\boldsymbol{JL})} \boldsymbol{L}\boldsymbol{e}_\lambda \cdot \boldsymbol{e}_\lambda, \quad \boldsymbol{e}_\lambda \in \mathbb{I}_\lambda.$$

From Proposition 7.1.11 we see that it is sufficient to characterize the bilinear form $\boldsymbol{L}\boldsymbol{e}_\lambda \cdot \boldsymbol{e}_\lambda$ in order to understand $\boldsymbol{L}\boldsymbol{x} \cdot \boldsymbol{x}$. Moreover, since

$$\overline{\boldsymbol{L}\boldsymbol{e}_\lambda \cdot \boldsymbol{e}_\lambda} = \boldsymbol{L}\overline{\boldsymbol{e}_\lambda} \cdot \overline{\boldsymbol{e}_\lambda} = \boldsymbol{L}\boldsymbol{e}_{\overline{\lambda}} \cdot \boldsymbol{e}_{\overline{\lambda}}, \tag{7.1.7}$$

it is sufficient to understand the bilinear form for $\operatorname{Im}\lambda \geq 0$. The key step is to decompose each vector $\boldsymbol{e}_\lambda \in \mathbb{I}_\lambda$ through the basis of $\mathbb{E}^\lambda$ and $\mathbb{E}^{-\overline{\lambda}}$ as

$$\boldsymbol{e}_\lambda = \sum_{j=1}^{n_\lambda} c_j^\lambda \boldsymbol{x}_j^\lambda + \sum_{j=1}^{n_\lambda} c_j^{-\overline{\lambda}} \boldsymbol{x}_j^{-\overline{\lambda}},$$

so that the bilinear form $\boldsymbol{L}\boldsymbol{e}_\lambda \cdot \boldsymbol{e}_\lambda$ can be written as

$$\boldsymbol{L}\boldsymbol{e}_\lambda \cdot \boldsymbol{e}_\lambda = \boldsymbol{L}|_{I_\lambda}\boldsymbol{c}_\lambda \cdot \boldsymbol{c}_\lambda,$$

where $\boldsymbol{c}_\lambda = (c_1^\lambda, \dots, c_{n^\lambda}^\lambda, c_1^{-\overline{\lambda}}, \dots, c_{n^\lambda}^{-\overline{\lambda}})^{\mathrm{T}}$. For $\operatorname{Re}\lambda \neq 0$ the Hermitian matrix $\boldsymbol{L}|_{\mathbb{I}_\lambda} \in \mathbb{C}^{2n_\lambda \times 2n_\lambda}$ has entries

$$(\boldsymbol{L}|_{\mathbb{I}_\lambda})_{jk} = \begin{cases} \boldsymbol{L}\boldsymbol{x}_j^\lambda \cdot \boldsymbol{x}_k^\lambda, & j,k = 1,\dots,n_\lambda \\ \boldsymbol{L}\boldsymbol{x}_{j-n^\lambda}^{-\overline{\lambda}} \cdot \boldsymbol{x}_k^\lambda, & j = n_\lambda + 1,\dots,2n_\lambda,\ k = 1,\dots,n_\lambda \\ \boldsymbol{L}\boldsymbol{x}_{j-n^\lambda}^{-\overline{\lambda}} \cdot \boldsymbol{x}_{k-n^\lambda}^{-\overline{\lambda}}, & j,k = n_\lambda + 1,\dots,2n_\lambda. \end{cases}$$

Furthermore, as a consequence of Proposition 7.1.10 the upper-left and lower-right blocks of $\boldsymbol{L}|_{\mathbb{I}_\lambda}$ are zero; hence, the matrix takes the form

$$\boldsymbol{L}|_{\mathbb{I}_\lambda} = \begin{pmatrix} \boldsymbol{0}_{n_\lambda} & \boldsymbol{L}_{\mathrm{c}}^\lambda \\ (\boldsymbol{L}_{\mathrm{c}}^\lambda)^{\mathrm{a}} & \boldsymbol{0}_{n_\lambda} \end{pmatrix}. \tag{7.1.8}$$

On the other hand, if $\lambda \in \mathrm{i}\mathbb{R}$ the matrix simply becomes

$$(\boldsymbol{L}|_{I_\lambda})_{jk} = \boldsymbol{L}\boldsymbol{x}_j^\lambda \cdot \boldsymbol{x}_k^\lambda, \quad j,k = 1,\dots,n_\lambda,$$

and no other reduction is possible. In either case, we can rewrite the bilinear form in Proposition 7.1.11 as

$$\boldsymbol{L}\boldsymbol{x} \cdot \boldsymbol{x} = \sum_{\lambda \in \sigma(\boldsymbol{JL})} \boldsymbol{L}|_{\mathbb{I}_\lambda}\boldsymbol{c}_\lambda \cdot \boldsymbol{c}_\lambda, \tag{7.1.9}$$

where each submatrix $\boldsymbol{L}|_{\mathbb{I}_\lambda}$ is Hermitian.

Proposition 7.1.12. *The matrix* $\boldsymbol{L}|_{\mathbb{I}_\lambda}$ *is nonsingular.*

Proof. Suppose not. Let $\boldsymbol{c}_{\mathrm{ker}}^\lambda \in \ker(\boldsymbol{L}|_{\mathbb{I}_\lambda})$ be nontrivial, so that $\boldsymbol{L}|_{\mathbb{I}_\lambda}\boldsymbol{c}_{\mathrm{ker}}^\lambda \cdot \boldsymbol{c}_\lambda = 0$ for any $\boldsymbol{c}_\lambda$. Defining the vector

$$\boldsymbol{e}_{\mathrm{ker}}^\lambda = \sum_{j=1}^{n_\lambda} c_{\mathrm{ker},j}^\lambda \boldsymbol{x}_j^\lambda + \sum_{j=1}^{n_\lambda} c_{\mathrm{ker},j}^{-\overline{\lambda}} \boldsymbol{x}_j^{-\overline{\lambda}},$$

we have $\boldsymbol{L}\boldsymbol{e}_{\mathrm{ker}}^\lambda \cdot \boldsymbol{e}_\lambda = 0$ for any $\boldsymbol{e}_\lambda \in \mathbb{I}_\lambda$. From the generalized orthogonality result of Proposition 7.1.10 and the decomposition (7.1.6) we deduce that $\boldsymbol{L}\boldsymbol{e}_{\mathrm{ker}}^\lambda \cdot \boldsymbol{x} = 0$ for any $\boldsymbol{x} \in \mathbb{C}^n$. Since $\boldsymbol{L}$ is nonsingular this can hold if and only if $\boldsymbol{e}_{\mathrm{ker}}^\lambda = \boldsymbol{0}$, which contradicts the linear independence of the eigenvectors. □

With these basic results in hand, we proceed to the main result. From the decomposition (7.1.9) we deduce

$$\mathrm{n}(\boldsymbol{L}) = \sum_{\lambda \in \sigma(\boldsymbol{JL})} \mathrm{n}(\boldsymbol{L}|_{\mathbb{I}_\lambda}). \tag{7.1.10}$$

For $\lambda \in \sigma(JL) \cap \mathbb{R}$ set

$$k_{\mathrm{r}}(\lambda) := \mathrm{n}(L|_{\mathbb{I}_\lambda}),$$

and for $\lambda \in \sigma(JL)$ with nonzero real and imaginary parts set

$$k_{\mathrm{c}}(\lambda) := \mathrm{n}(L|_{\mathbb{I}_\lambda}).$$

Finally, for purely imaginary eigenvalues, $\lambda \in \sigma(JL) \cap \mathrm{i}\mathbb{R}$, we have the negative Krein index

$$k_{\mathrm{i}}^-(\lambda) := \mathrm{n}(L|_{\mathbb{I}_\lambda}).$$

As a consequence of (7.1.7) we have $\mathrm{n}(L|_{\mathbb{I}_\lambda}) = \mathrm{n}(L|_{\mathbb{I}_{-\overline{\lambda}}})$, and in particular

$$k_{\mathrm{c}}(\overline{\lambda}) = k_{\mathrm{c}}(\lambda), \quad k_{\mathrm{i}}^-(\overline{\lambda}) = k_{\mathrm{i}}^-(\lambda);$$

hence, the numbers

$$k_{\mathrm{c}} := \sum k_{\mathrm{c}}(\lambda), \quad k_{\mathrm{i}}^- := \sum k_{\mathrm{i}}^-(\lambda)$$

are even. On the other hand, the number

$$k_{\mathrm{r}} := \sum k_{\mathrm{r}}(\lambda)$$

can be odd or even. With this notation (7.1.10) can be rewritten as

$$\mathrm{n}(L) = k_{\mathrm{r}} + k_{\mathrm{c}} + k_{\mathrm{i}}^-. \tag{7.1.11}$$

It remains to connect the numbers $k_{\mathrm{r}}(\lambda)$ and $k_{\mathrm{c}}(\lambda)$ and the multiplicities of the associated eigenvalues. Since $\mathrm{Re}\,\lambda \neq 0$, as a consequence of the form of the matrix in (7.1.8), we have for $\mathrm{Re}\,\lambda > 0$

$$\mathrm{n}(L|_{\mathbb{I}_\lambda}) = n_\lambda = \dim(\mathbb{E}^\lambda).$$

This means that $k_{\mathrm{r}}(\lambda) = \dim(\mathbb{E}^\lambda)$ and $k_{\mathrm{c}}(\lambda) = \dim(\mathbb{E}^\lambda)$; in particular, k_{r} equals the total number of positive real eigenvalues (including multiplicity), while k_{c} equals the total number of eigenvalues in the open right-half of the complex plane with nonzero imaginary part. Combining the definition of the Hamiltonian–Krein index with the equality (7.1.11) finishes the proof.

Remark 7.1.13. In this proof we assumed that L had no kernel. If L has a nontrivial kernel with $J^{-1}\ker(L) \subset \ker(L)^\perp$, then by considering the eigenvalue problem on the appropriate subspace we recover the full result of Theorem 7.1.5. In addition, the only place that we really used the fact that eigenvalues were all algebraically simple is in the proof of Proposition 7.1.10. Since this restriction can be removed as a consequence of Exercise 7.1.2, the result is general.

7.1.2 Krein Signature and Bifurcation

Information about the bifurcation of imaginary eigenvalues into an unstable spectrum can also be extracted from the Krein dichotomy (7.1.4) (see Fig. 7.2). It is of particular interest to know that if two pair of nonzero imaginary eigenvalues collide, will they form a complex quartet or remain upon the imaginary axis. The former case is called a Hamiltonian–Hopf bifurcation because the resulting complex eigenvalues induce an oscillatory instability in the underlying flow. As we see, a Hamiltonian–Hopf bifurcation can occur only if the two pairs of eigenvalues have opposing Krein signatures.

Proposition 7.1.14. *Let the operator $\mathcal{L} = \mathcal{L}(\epsilon) : Y \subset X \mapsto Y^*$ be a smooth function of the bifurcation parameter $\epsilon \in \mathbb{R}$ with the assumptions of this section holding uniformly for ϵ in a neighborhood of $\epsilon_0 \in \mathbb{R}$. The collision of two pairs of simple, nonzero eigenvalues $\pm \mathrm{i}d_1(\epsilon)$ and $\pm \mathrm{i}d_2(\epsilon)$ of $\mathcal{J}\mathcal{L}(\epsilon)$ at the nonzero points $\pm \mathrm{i}d_0 \in (-\mathrm{i}\omega_0, \mathrm{i}\omega_0)$ as $\epsilon \to \epsilon_0$ can result in a nontrivial Jordan block and a quartet of complex eigenvalues of nonzero real part only if the eigenvalues $\mathrm{i}d_1$ and $\mathrm{i}d_2$ have opposite Krein signature for all $|\epsilon - \epsilon_0|$ sufficiently small.*

Proof. By symmetry we may focus on the action on the positive imaginary axis $(0, \mathrm{i}\omega)$. Denoting the eigenfunctions of id_1 and id_2 by v_1 and v_2, we may assume that the two eigenvalues are distinct for $\epsilon < \epsilon_0$ and take the common value $\mathrm{i}d_0$ with $d_0 > 0$ at $\epsilon = \epsilon_0$, forming a complex pair $\{\lambda, -\overline{\lambda}\}$ with $\mathrm{Re}\,\lambda > 0$ for $\epsilon > \epsilon_0$. At the bifurcation point there is a two-dimensional eigenspace, $\mathbb{E}^{\mathrm{i}d_0}$, which breaks up, possibly nonanalytically, into the two one-dimensional spaces spanned by v_1 and v_2 for $\epsilon \neq \epsilon_0$. In particular, for $\epsilon > \epsilon_0$ the Krein dichotomy (7.1.4) implies that

$$\langle \mathcal{L} v_j, v_j \rangle = 0, \quad j = 1, 2,$$

and by continuity this equality must hold for $\epsilon = \epsilon_0$. In particular, the Krein matrix satisfies $\mathrm{tr}(\mathbf{H}^{\mathrm{i}d_0}) = 0$, so that the eigenvalues of $\mathbf{H}^{\mathrm{i}d_0}$, which are real, sum to zero. If both eigenvalues of $\mathbf{H}^{\mathrm{i}d_0}$ were zero, then since $\mathcal{L}$ is self-adjoint its restriction to $\mathbb{E}^{\mathrm{i}d_0}$ would be zero, which implies that the eigenvalue $\lambda = \mathrm{i}d_0 = 0$, in contradiction to the assumption that $d_0 > 0$. Thus $\mathbf{H}^{\mathrm{i}d_0}$ has one positive and one negative eigenvalue, in particular $\mathrm{n}(\mathbf{H}^{\mathrm{i}d_0}) = 1$. Finally, since the eigenvalues of $\mathcal{L}$ constrained to $\mathrm{span}\{v_1, v_2\}$ are continuous in ϵ, and since both eigenvalues are nonzero at $\epsilon = \epsilon_0$, then for $0 < \epsilon_0 - \epsilon$ sufficiently small, one of the pairs $\{\pm id_1\}$ and $\{\pm id_2\}$ has a negative Krein signature, while the other pair has a positive signature.

To address the Jordan block structure, suppose that for $\epsilon = \epsilon_0$ the eigenvalue id_0 has algebraic multiplicity two and geometric multiplicity one, so that

$$(\mathcal{J}\mathcal{L} - \mathrm{i}d_0\mathcal{I})v_1 = 0, \quad (\mathcal{J}\mathcal{L} - \mathrm{i}d_0\mathcal{I})v_2 = v_1.$$

The eigenvalue equation for v_1 can be rewritten as

$$0 = (\mathcal{L}\mathcal{J} - \mathrm{i}d_0\mathcal{I})(\mathcal{J}^{-1}v_1) = -(\mathcal{J}\mathcal{L} - \mathrm{i}d_0\mathcal{I})^{\mathrm{a}}\mathcal{J}^{-1}v_1.$$

Consequently, $\ker((\mathcal{J}\mathcal{L} - \mathrm{i}d_0\mathcal{I})^{\mathrm{a}}) = \mathrm{span}\{\mathcal{J}^{-1}v_1\}$ and the Fredholm alternative implies that the Jordan chain equation for v_2 has a solution if and only if

$$0 = \langle \mathcal{J}^{-1}v_1, v_1\rangle = -\frac{\mathrm{i}}{d_0}\langle \mathcal{L}v_1, v_1\rangle.$$

The Krein matrix $\mathbf{H}^{\mathrm{i}d_0}$ is then of the form

$$\mathbf{H}^{\mathrm{i}d_0} = \begin{pmatrix} 0 & a \\ \bar{a} & b \end{pmatrix},$$

where $a \in \mathbb{C}$ and $b \in \mathbb{R}$. A short calculation shows that either $\mathrm{n}(\mathbf{H}^{\mathrm{i}d_0}) = \mathrm{p}(\mathbf{H}^{\mathrm{i}d_0}) = 1$ or the matrix is identically zero. However, the latter possibility is excluded since the collision is at the nonzero value $\mathrm{i}d_0$. The matrix $\mathbf{H}^{\mathrm{i}d_0}$ must have one positive and one negative eigenvalue, and for ϵ sufficiently close to, but not equal to, ϵ_0, the matrix must break up into two one-dimensional spaces of opposite Krein signature. □

Remark 7.1.15. The collision of three of more eigenvalues was considered in Vougalter and Pelinovsky [**275**]. They showed that if the collision leads to a Jordan chain, then it must be the case that before the collision, half of the eigenvalues had a positive Krein signature, whereas the other half had a negative Krein signature. Indeed, this formation of a Jordan chain upon collision is the generic situation. It was shown by MacKay [**196**] that it is a codimension-three phenomena for the collision of eigenvalues of opposite signature to lead to an algebraically simple eigenvalue.

7.1.3 The Jones–Grillakis Instability Index

The Hamiltonian–Krein index result of Theorem 7.1.5 can be refined if $\mathcal{J}$ is canonical and $\mathcal{L}$ is diagonal, i.e.,

$$\mathcal{J} = \begin{pmatrix} 0 & \mathcal{I} \\ -\mathcal{I} & 0 \end{pmatrix}, \quad \mathcal{L} = \begin{pmatrix} \mathcal{L}_+ & 0 \\ 0 & \mathcal{L}_- \end{pmatrix}, \tag{7.1.12}$$

where $\mathcal{L}_\pm$ are self-adjoint. This structure arises in the study of the nonlinear Schrodinger (NLS) equation and its extension to coupled systems of NLS equations. In this context it is possible to compute a lower bound for k_{r}, and when the lower bound is nonzero the underlying equilibria is spectrally unstable.

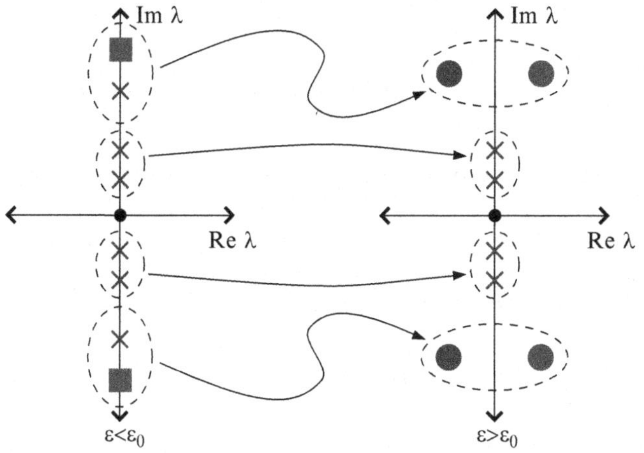

Fig. 7.2 The collision of simple eigenvalues on the imaginary axis. In the left panel there are four pairs of purely imaginary eigenvalues: one pair has a negative Krein signature ((*red*) squares), and the other three pairs have a positive Krein signature ((*blue*) crosses). The dotted oval denotes the pairs of eigenvalues that collide when $\epsilon = \epsilon_0$. The collision of eigenvalues with opposite sign (generically) leads to a quartet of eigenvalues with nonzero real and imaginary parts, while during the collision of eigenvalues of the same Krein signature, the eigenvalues remain semi-simple and stay on the imaginary axis. (Color figure online.)

Since $\mathcal{J}$ has no kernel,

$$\dim[\ker(\mathcal{J}\mathcal{L})] = \dim[\ker(\mathcal{L})] = \dim[\ker(\mathcal{L}_+)] + \dim[\ker(\mathcal{L}_-)].$$

In light of Hypothesis 5.2.5(g) on the kernel of $\mathcal{L}$, there exists a natural number $k \le N = \dim[\ker(\mathcal{L})]$ such that

$$T_j'\phi_c = \begin{pmatrix} 0 \\ \phi_j^- \end{pmatrix}, \quad j = 1,\ldots,k; \; T_j'\phi_c = \begin{pmatrix} \phi_{j-k}^+ \\ 0 \end{pmatrix}, \quad j = k+1,\ldots,N.$$

The assumption Hypothesis 5.2.5(g) that the kernel of $\mathcal{L}$ is spanned by its symmetry elements implies that

$$\ker(\mathcal{L}_+) = \operatorname{span}\{\phi_1^+,\ldots,\phi_{N-k}^+\}, \quad \ker(\mathcal{L}_-) = \operatorname{span}\{\phi_1^-,\ldots,\phi_k^-\}. \tag{7.1.13}$$

As well, the constraint space S takes the from

$$S = \mathcal{J}^{-1}\ker(\mathcal{L}) = \operatorname{span}\left\{\begin{pmatrix} -\phi_j^- \\ 0 \end{pmatrix}\right\}_{j=1}^{k} \bigcup \left\{\begin{pmatrix} 0 \\ \phi_{j-k}^+ \end{pmatrix}\right\}_{j=k+1}^{N}.$$

The relation (5.2.54),

$$\mathcal{L}\partial_{c_i}\phi_c = -\mathcal{J}^{-1}T_i'\phi_c,$$

implies that

$$\partial_{c_j}\phi_c = \begin{pmatrix} \psi_j^+ \\ 0 \end{pmatrix}, \quad j = 1,\ldots,k; \quad \partial_{c_{j+k}}\phi_c = \begin{pmatrix} 0 \\ -\psi_j^- \end{pmatrix}, \quad j = 1,\ldots,N-k,$$

where the functions $\psi_j^\pm \in X$ are related to the $\phi_j^\pm$ through $\mathcal{L}_\pm$ via,

$$\mathcal{L}_+\psi_j^+ = \phi_j^-, \quad j = 1,\ldots,k; \quad \mathcal{L}_-(-\psi_j^-) = \phi_j^+, \quad j = 1,\ldots,N-k. \tag{7.1.14}$$

The constraint matrix $\boldsymbol{D}$, from (5.2.55), becomes block-diagonal of the form

$$\boldsymbol{D} = \begin{pmatrix} \boldsymbol{D}_+ & \boldsymbol{0} \\ \boldsymbol{0} & \boldsymbol{D}_- \end{pmatrix},$$

where the elements of the sub-blocks are expressed as

$$\begin{aligned} (\boldsymbol{D}_+)_{ij} &= \langle \psi_i^+, \mathcal{L}_+\psi_j^+ \rangle, \quad i,j = 1,\ldots k, \\ (\boldsymbol{D}_-)_{ij} &= \langle \psi_i^-, \mathcal{L}_-\psi_j^- \rangle, \quad i,j = 1,\ldots N-k. \end{aligned} \tag{7.1.15}$$

The negative indices of $\mathcal{L}$ and $\boldsymbol{D}$ can be subdivided as

$$\mathrm{n}(\mathcal{L}) = \mathrm{n}(\mathcal{L}_+) + \mathrm{n}(\mathcal{L}_-), \quad \mathrm{n}(\boldsymbol{D}) = \mathrm{n}(\boldsymbol{D}_+) + \mathrm{n}(\boldsymbol{D}_-),$$

and through this subdivision the instability criterion of Theorem 7.1.5 can be strengthened.

Theorem 7.1.16. *In addition to the assumptions of Theorem 7.1.5, assume that $\mathcal{J},\mathcal{L}$ are of the form of (7.1.12), with the N-dimensional kernel of $\mathcal{L}$ given by (7.1.13) while the basis of the complementary constraint space S satisfies the relations (7.1.14). The number of real pairs of eigenvalues of $\mathcal{J}\mathcal{L}$ then satisfies the lower bound*

$$k_{\mathrm{r}} \geq \left| [\mathrm{n}(\mathcal{L}_+) - \mathrm{n}(\boldsymbol{D}_+)] - [\mathrm{n}(\mathcal{L}_-) - \mathrm{n}(\boldsymbol{D}_-)] \right|.$$

In particular, if $|\mathrm{n}(\mathcal{L}_+) - \mathrm{n}(\mathcal{L}_-)| \geq N+1$, then $k_{\mathrm{r}} \geq 1$.

The proof of this theorem is beyond the scope of this book. Theorem 7.1.16 was first established by Jones [**135**], Jones and Moloney [**136, 137**] using dynamical systems arguments. Motivated by this, Grillakis [**107**] proved the result using a functional analytic approach. A proof based upon the index Theorem 5.3.2 was presented in Kapitula and Promislow [**148**]. The instability criterion of Theorem 7.1.16 has seen substantial application (e.g., see [**156, 180, 285, 286**] and the references therein).

Remark 7.1.17. As a consequence of the index Theorem 5.3.2, we have that

$$\mathrm{n}(\mathcal{L}_+) - \mathrm{n}(\boldsymbol{D}_+) = \mathrm{n}(\mathcal{L}_+|_{\Pi_+}), \quad \mathrm{n}(\mathcal{L}_-) - \mathrm{n}(\boldsymbol{D}_-) = \mathrm{n}(\mathcal{L}_-|_{\Pi_-}),$$

where the constraint space for the operator $\mathcal{L}_\pm$ is $\ker(\mathcal{L}_\mp)$, respectively. Since $\dim[\mathrm{gker}(\mathcal{L})] = 2N$, the Jordan chain structure implicit in (7.1.14) guarantees that $\ker(\mathcal{L}_-) \perp \ker(\mathcal{L}_+)$. Consequently, Theorem 7.1.16 provides a lower bound on the number of real pairs of eigenvalues in terms of the mismatch in the dimension of the negative spaces of the two constrained operators,

$$k_\mathrm{r} \geq |\mathrm{n}(\mathcal{L}_+|_{\Pi_+}) - \mathrm{n}(\mathcal{L}_-|_{\Pi_-})|.$$

This formulation is equivalent to the presentation in Grillakis [**107**].

Exercises

Exercise 7.1.1. Show that $k_\mathrm{i}^-(\overline{\lambda}) = k_\mathrm{i}^-(\lambda)$.

Exercise 7.1.2. Prove Proposition 7.1.10 under the assumption that each of the eigenspaces $\mathbb{E}^\lambda$ and $\mathbb{E}^\sigma$ contain generalized eigenvectors, i.e., each eigenspace possesses a nontrivial Jordan block.

Exercise 7.1.3. Suppose that $\mathcal{L} = \mathcal{L}(\epsilon)$ is a smooth function of the bifurcation parameter ϵ, and that when $\epsilon = 0$, $\lambda = \mathrm{i}d_0$ is an algebraically simple eigenvalue of $\mathcal{L}(0)$ which is not contained in the essential spectrum. Further suppose that the matrix $\boldsymbol{E}^{\mathrm{i}d_0}$ is either positive or negative definite.

(a) Show that for ϵ sufficiently small, the eigenvalues that perturb from $\mathrm{i}d_0$ must remain on the imaginary axis.
(b) Show that the eigenvalues of $\mathcal{L}(\epsilon)$ near $\mathrm{i}d_0$ perturb smoothly and each has a constant Krein signature for ϵ sufficiently small.

Exercise 7.1.4. In Chapter 4.5 the analysis of equilibria of the parametrically forced NLS Eq. (4.5.1) led to the eigenvalue problem

$$\mathcal{L}_+\mathcal{L}_-v = -z^2 v,$$

for $z \in \mathbb{C}$ and $v \in L^2(\mathbb{R})$. Here the operators $\mathcal{L}_\pm$ are given by

$$\mathcal{L}_+ = -\partial_x^2 + A_+^2 - 3U_\theta^2, \quad \mathcal{L}_- = -\partial_x^2 + A_-^2 - U_\theta^2,$$

where the equilibrium profile

$$U_\theta(x) = \sqrt{2}A_+\phi(A_+x), \quad A_\pm^2 := 1 \pm \sin\theta, \quad -\pi/2 < \theta < \pi/2,$$

is a rescaling of the homoclinic solution ϕ of

$$\partial_x^2\phi - \phi + \phi^3 = 0.$$

The goal is to show that if $\theta < 0$, then $k_\mathrm{r} = 1$.

(a) Introducing $u = z^{-1}\mathcal{L}_-v$, show that the eigenvalue problem can be written in the canonical framework of (7.1.12).

(b) Use the equation for ϕ to calculate the negative indexes of $\mathcal{L}_\pm$. In particular, show that $\mathrm{n}(\mathcal{L}_-)$ changes value at $\theta = 0$.
(c) Show that for $\theta < 0$, then $k_\mathrm{r} = 1$, and $k_\mathrm{c} = k_\mathrm{i}^- = 0$. What can be concluded about $k_\mathrm{r}, k_\mathrm{c}, k_\mathrm{i}^-$ for $\theta > 0$?

Exercise 7.1.5. This exercise proves Theorem 7.1.16 for matrices under the assumption that $\boldsymbol{L}$ has no kernel. Suppose that for the eigenvalue problem (7.1.5),

$$\boldsymbol{J} = \begin{pmatrix} \boldsymbol{0}_n & \boldsymbol{I}_n \\ -\boldsymbol{I}_n & \boldsymbol{0}_n \end{pmatrix}, \quad \boldsymbol{J} = \begin{pmatrix} \boldsymbol{L}_1 & \boldsymbol{0}_n \\ \boldsymbol{0}_n & \boldsymbol{L}_2 \end{pmatrix},$$

where $\boldsymbol{L}_j$ are symmetric and invertible. Further suppose that all of the eigenvalues are algebraically simple.

(a) Set

$$\boldsymbol{T} := \begin{pmatrix} \boldsymbol{I}_n & \boldsymbol{0}_n \\ \boldsymbol{0}_n & -\boldsymbol{I}_n \end{pmatrix}.$$

Show that if $\boldsymbol{x}_\lambda \in \mathbb{E}^\lambda$, then $\boldsymbol{T}\boldsymbol{x}_\lambda \in \mathbb{E}^{-\lambda}$.
(b) Set

$$\sigma^+(\boldsymbol{JL}) := \{\lambda \in \sigma(\boldsymbol{JL}) : \operatorname{Im}\lambda > 0\} \cup \{\lambda \in \sigma(\boldsymbol{JL}) : \operatorname{Im}\lambda = 0 \text{ and } \operatorname{Re}\lambda > 0\}.$$

Write $\boldsymbol{x}_\lambda = (\boldsymbol{u}_\lambda, \boldsymbol{v}_\lambda)^\mathrm{T}$. Show that the each of the sets

$$\{\boldsymbol{u}_\lambda : \lambda \in \sigma^+(\boldsymbol{JL})\}, \quad \{\boldsymbol{v}_\lambda : \lambda \in \sigma^+(\boldsymbol{JL})\}$$

form a basis of $\mathbb{C}^n$.
(c) Show that if $\lambda \pm \overline{\sigma} \neq 0$, then

$$\boldsymbol{L}_1 \boldsymbol{u}_\lambda \cdot \boldsymbol{u}_\sigma = \boldsymbol{L}_2 \boldsymbol{v}_\lambda \cdot \boldsymbol{v}_\sigma = 0$$

(compare with Proposition 7.1.10).
(d) Show that the matrix $\boldsymbol{L}|_{\mathbb{I}_\lambda}$ as written in (7.1.8) for $\operatorname{Re}\lambda \neq 0$ takes the form

$$(\boldsymbol{L}_\mathrm{c}^\lambda)_{jk} = \boldsymbol{L}_1 \boldsymbol{u}_j^\lambda \cdot \overline{\boldsymbol{u}_k^\lambda} = -\boldsymbol{L}_2 \boldsymbol{v}_j^\lambda \cdot \overline{\boldsymbol{v}_k^\lambda}.$$

(e) If $\operatorname{Re}\lambda = 0$, show that

$$(\boldsymbol{L}|_{I_\lambda})_{jk} = \boldsymbol{L}_1 \boldsymbol{u}_j^\lambda \cdot \overline{\boldsymbol{u}_k^\lambda} + \boldsymbol{L}_2 \boldsymbol{v}_j^\lambda \cdot \overline{\boldsymbol{v}_k^\lambda}.$$

(f) Show that

$$k_\mathrm{r} + k_\mathrm{c} + k_\mathrm{i}^- = \mathrm{n}(\boldsymbol{L}_1) + \mathrm{n}(\boldsymbol{L}_2).$$

(g) For $\lambda \in \sigma^+(\boldsymbol{JL})$ set

$$(\boldsymbol{S}_1)_{jk} := \boldsymbol{L}_1 \boldsymbol{u}_j \cdot \boldsymbol{u}_k, \quad (\boldsymbol{S}_2)_{jk} := \boldsymbol{L}_2 \boldsymbol{v}_j \cdot \boldsymbol{v}_k,$$

where each $\boldsymbol{u}_1,\ldots,\boldsymbol{u}_n$ is the first component of an eigenvector, and each $\boldsymbol{v}_1,\ldots,\boldsymbol{v}_n$ is the second component of an eigenvector. For a self-adjoint matrix $\boldsymbol{S}$ we introduce the negative cone $C_-(\boldsymbol{S}) = \{\boldsymbol{v} : \boldsymbol{S}\boldsymbol{v}\cdot\boldsymbol{v} < 0\} \cup \{\boldsymbol{0}\}$. Show that

$$k_\mathrm{c} + k_\mathrm{i}^- = 2\dim[C_-(\boldsymbol{S}_1) \cap C_-(\boldsymbol{S}_2)].$$

(h) Show that $k_\mathrm{r} \geq |\mathrm{n}(\boldsymbol{L}_1) - \mathrm{n}(\boldsymbol{L}_2)|$, and find an exact equality for k_r.

(i) Show that

$$\dim[C_-(\boldsymbol{S}_1) \cap C_-(\boldsymbol{S}_2)] = \dim[C_-(\boldsymbol{L}_1) \cap C_-(\boldsymbol{L}_2^{-1})],$$

and rewrite the results of (g) and (h) in terms of the matrices $\boldsymbol{L}_1$ and $\boldsymbol{L}_2$ only.

7.2 Symmetry-Breaking Perturbations

Returning to the general Hamiltonian framework, we impose the additional assumption that $\mathcal{J} : X \to X$ has a bounded inverse, so that the kernel of the operator $\mathcal{J}\mathcal{L}$ arises from the symmetry group T, while its generalized kernel, with dimension $2N$, is comprised of the functions

$$\mathrm{gker}(\mathcal{J}\mathcal{L}) = \mathrm{span}\{T_1'\phi_c,\ldots,T_N'\phi_c,\partial_{c_1}\phi_c,\ldots,\partial_{c_N}\phi_c\}.$$

In particular, $\mathrm{gker}(\mathcal{J}\mathcal{L})$ is comprised of N Jordan chains of length two:

$$\mathcal{L}T_j'\phi_c = 0, \quad \mathcal{J}\mathcal{L}(-\partial_{c_j}\phi_c) = T_j'\phi_c, \quad j = 1,\ldots,N. \tag{7.2.1}$$

We investigate two types of perturbations to this framework. The first class of perturbations preserve the Hamiltonian structure; however, only $0 \leq k \leq N-1$ of the symmetries will persist. We will see that this generically produces $2(N-k)$ small but nonzero eigenvalues for the linearized operator. The second class of perturbations breaks the Hamiltonian structure, but all of the symmetries remain. In this problem the N-dimensional kernel remains unperturbed, but the N eigenvalues in the Jordan chains will generically perturb away from zero.

7.2.1 *Hamiltonian Perturbation*

We consider a general perturbed Hamiltonian framework

$$\partial_t u = \mathcal{J}\frac{\delta\mathcal{H}}{\delta u}(u,\epsilon), \quad \mathcal{H}(u,\epsilon) = \mathcal{H}_0(u) + \epsilon\mathcal{H}_1(u), \tag{7.2.2}$$

where both $\mathcal{H}_0, \mathcal{H}_1 : Y \subset X \mapsto Y^*$ are at least C^3 in the Y norm; see Chapter 2.2.3. We assume that the unperturbed Hamiltonian, $\mathcal{H}_0$, has an N-dimensional symmetry group T and a critical point ϕ_c that generates a manifold of critical points

$$\mathcal{M}_T(\phi_c) := \{T(\gamma)\phi_c : \gamma \in \mathbb{R}^N\}.$$

We assume that symmetries $T_1, \ldots, T_k$ persist under $\mathcal{H}_1$, but that $T_{k+1}, \ldots, T_N$ do not. We wish to find criteria that determine which critical points of $\mathcal{M}_T$ persist under the perturbation.

Arguing formally, we fix a choice γ of the symmetry parameter and consider a smooth expansion of the perturbed critical point,

$$u = T(\gamma)\phi_c + \epsilon u_1 + \mathcal{O}(\epsilon^2).$$

Plugging this expansion into the critical point equation

$$0 = \frac{\delta \mathcal{H}}{\delta u}(u, \epsilon) = \frac{\delta \mathcal{H}_0}{\delta u}(u) + \epsilon \frac{\delta \mathcal{H}_1}{\delta u}(u),$$

and collecting powers of ϵ yields

$$\mathcal{O}(1): \quad \frac{\delta \mathcal{H}_0}{\delta u}(T(\gamma)\phi_c) = 0$$

$$\mathcal{O}(\epsilon): \quad \mathcal{L}_0 u_1 = -\frac{\delta}{\delta u}\mathcal{H}_1(T(\gamma)\phi_c), \quad \mathcal{L}_0 := \frac{\delta^2 \mathcal{H}_0}{\delta u^2}(T(\gamma)\phi_c).$$

By assumption the $\mathcal{O}(1)$ equation is satisfied for all $\gamma \in \mathbb{R}^N$. The $\mathcal{O}(\epsilon)$ equation has a solution only if the right-hand side is orthogonal to the kernel of $\mathcal{L}_0$,

$$\frac{\delta}{\delta u}\mathcal{H}_1(T(\gamma)\phi_c) \in \ker(\mathcal{L}_0)^\perp = \operatorname{span}\{T_1' T(\gamma)\phi_c, \ldots, T_N' T(\gamma)\phi_c\}^\perp,$$

which is equivalent to the equations

$$0 = \left\langle \frac{\delta}{\delta u}\mathcal{H}_1(T(\gamma)\phi_c), T_j' T(\gamma)\phi_c \right\rangle = \partial_{\gamma_j}\mathcal{H}_1(T(\gamma)\phi_c), \quad j = 1, \ldots, N.$$

Since $\mathcal{H}_1$ preserves the first k symmetries, the equations for $j = 1, \ldots, k$, must hold and we may assume, without loss of generality, that $\gamma_1 = \cdots = \gamma_k = 0$. It remains to determine $\gamma_{k+1}, \ldots, \gamma_N$ for which,

$$\partial_{\gamma_j}\mathcal{H}_1(T(\gamma)\phi_c) = 0, \quad j = k+1, \ldots, N. \tag{7.2.3}$$

This motivates the definition of the reduced symmetry group

$$T_{\text{red}}(\gamma_{\text{red}}) = T_{k+1}(\gamma_{k+1}) \cdots T_N(\gamma_N), \quad \gamma_{\text{red}} = (\gamma_{k+1}, \ldots, \gamma_N)^{\mathrm{T}} \in \mathbb{R}^{N-k}, \tag{7.2.4}$$

Definition 7.2.1. We say that a critical point $T(\gamma^*)\phi_c$ of $\mathcal{H}_0$ persists if there exists a unique family of critical points $\phi_c(\epsilon)$ of $\mathcal{H}(\cdot,\epsilon)$ that is smooth in ϵ in the Y-norm and satisfies

$$\phi_c(\epsilon) = T_{\text{red}}(\gamma^*_{\text{red}})\phi_c + \epsilon\phi_{c,1}(\epsilon)$$

for $\|\phi_{c,1}(\epsilon)\|_Y$ uniformly bounded for ϵ sufficiently small.

The *reduced Hamiltonian* $\mathcal{H}_{\text{red}} : \mathbb{R}^{N-k} \to \mathbb{R}$ is the perturbed Hamiltonian

$$\mathcal{H}_{\text{red}}(\gamma_{\text{red}}) := \mathcal{H}_1(T_{\text{red}}(\gamma_{\text{red}})\phi_c). \tag{7.2.5}$$

evaluated over the *reduced manifold*

$$\mathcal{M}_{T_{\text{red}}} := \{T_{\text{red}}(\gamma_{\text{red}})\phi_c : \gamma_{\text{red}} \in \mathbb{R}^{N-k}\}.$$

The graph of the reduced Hamiltonian over $\mathbb{R}^{N-k}$ is called the *reduced energy surface*, and it plays a central role in the persistence and stability of critical points

From (7.2.3) we see that points on $\mathcal{M}_T$ that persist under perturbation correspond to critical points of the reduced energy surface. This formal calculation is the basis for a rigorous one, modulo an assumption of nondegeneracy.

Lemma 7.2.2. *We assume that the skew operator $\mathcal{J} : X \to X$ has a bounded inverse and the Hamiltonian $\mathcal{H} = \mathcal{H}(u,\epsilon)$ from (7.2.2) is C^3 in Y-norm and satisfies Hypothesis 5.2.5 (a)–(h),and (j) [see (7.0.3)], uniformly in ϵ. Then a critical point $T_{\text{red}}(\gamma^*_{\text{red}})\phi_c$ of $\mathcal{H}_0$ for $\gamma^*_{\text{red}} \in \mathbb{R}^{N-k}$ persists under perturbation by $\mathcal{H}_1$ if it is a nondegenerate critical point of the reduced Hamiltonian, i.e.,*

$$\nabla\mathcal{H}_{\text{red}}(\gamma^*_{\text{red}}) = \mathbf{0} \quad \text{and} \quad \det\left[\nabla^2_{\gamma_{\text{red}}}\mathcal{H}_{\text{red}}\right] \neq 0.$$

7.2.1.1 Perturbations of K_{Ham}

From Theorem 7.1.5 the location of the small eigenvalues of the perturbed linear Hamiltonian operator depend in good measure upon the fate of the small eigenvalues of the second variation of the perturbed Hamiltonian. Not surprisingly, the nature of the reduced-energy surface at the nondegenerate critical point γ^*_{red} gives the desired information about the location of the small eigenvalues of the self-adjoint problem. Introducing $\boldsymbol{H}_{\text{red}}$, the Hessian matrix of the reduced Hamiltonian,

$$\boldsymbol{H}_{\text{red}} := \nabla^2_{\gamma_{\text{red}}}\mathcal{H}_{\text{red}}, \tag{7.2.6}$$

we have the following result.

Lemma 7.2.3. *Assume that γ^*_{red} is a nondegenerate critical point of the reduced Hamiltonian, let $\phi_c(\epsilon)$ be the associated persistent solution, and let*

$$\mathcal{L}(\epsilon) := \frac{\delta^2 \mathcal{H}}{\delta u^2}(\phi_c(\epsilon)),$$

be the second variation of the perturbed Hamiltonian at $\phi_c(\epsilon)$. Then for $\epsilon > 0$ sufficiently small

$$\mathrm{n}(\mathcal{L}(\epsilon)) - \mathrm{n}(\mathcal{L}(0)) = \mathrm{n}(\boldsymbol{H}_{\text{red}}). \tag{7.2.7}$$

That is, the number of $\mathcal{O}(\epsilon)$ negative eigenvalues of the second variation of the perturbed Hamiltonian equals the number of negative eigenvalues of the second variation of the reduced Hamiltonian.

Remark 7.2.4. For non-Hamiltonian systems the $\mathcal{O}(\epsilon)$ eigenvalues of the linearized operator will generically depend upon the $\mathcal{O}(\epsilon)$ correction to the critical point (e.g., see Chapter 7.2.2). For perturbations that preserve the Hamiltonian structure, however, the perturbing eigenvalues are independent of the correction term to ϕ_c at leading order.

Remark 7.2.5. It follows from (7.2.7) that if the critical point γ^*_{red} is a local minimum of the reduced energy, then all of the perturbed eigenvalues of $\mathcal{L}$ move to the right as ϵ increases from zero. If the critical point is a saddle point on the reduced-energy surface, then each negative eigenvalue associated with the critical point generates exactly one $\mathcal{O}(\epsilon)$ negative eigenvalue of $\mathcal{L}$.

Proof. We present a formal argument; a more rigorous justification can be found in [**157**, Section 4]. Since the operator $\mathcal{L}$ is self-adjoint, each eigenvalue is semi-simple and to leading order the small eigenvalues can be found using the perturbation results of Chapter 6.1. Setting $u_0 := T_{\text{red}}(\gamma^*_{\text{red}})\phi_c$ we expand the persistent critical point as

$$u = u_0 + \epsilon u_1 + \mathcal{O}(\epsilon^2), \tag{7.2.8}$$

so that the second variation of the energy has the expansion, see Chapter 2.2.3

$$\frac{\delta^2 \mathcal{H}}{\delta u^2}(u,\epsilon) = \underbrace{\frac{\delta^2 \mathcal{H}_0}{\delta u^2}(u_0)}_{\mathcal{L}_0} + \epsilon \underbrace{\left(\frac{\delta^2 \mathcal{H}_1}{\delta u^2}(u_0) + \frac{\delta^3 \mathcal{H}_0}{\delta u^3}(u_0)[u_1]\right)}_{\mathcal{L}_1} + \mathcal{O}(\epsilon^2). \tag{7.2.9}$$

By our hypotheses

$$\ker(\mathcal{L}_0) = \mathrm{span}\{T_1' u_0, \ldots, T_N' u_0\};$$

and without loss of generality we assume that this is an orthonormal basis. Since $\mathcal{L}_0$ is self-adjoint, the perturbation matrix $\boldsymbol{L}_1$ given in (6.1.5) has the entries

$$(\boldsymbol{L}_1)_{ij} = \langle T_i' u_0, \mathcal{L}_1 T_j' u_0 \rangle.$$

The form of $\mathcal{L}_1$ in (7.2.9) can be used to simplify the expression for $\boldsymbol{L}_1$. The second term in $\mathcal{L}_1$ involves the third variation of $\mathcal{H}_0$. Since $\mathcal{H}_0$ is invariant under the full set of symmetries, we may derive the relations

$$0 = \frac{\partial^2 \mathcal{H}_0}{\partial \gamma_i \partial \gamma_j}(T(\gamma)u) = \left\langle T_i' u, \frac{\delta^2 \mathcal{H}_0}{\delta u^2}(u) T_j' u \right\rangle. \tag{7.2.10}$$

Plugging the expansion (7.2.8) for u and the expression (7.2.9) for $\delta^2\mathcal{H}/\delta u^2$ into (7.2.10), and equating orders of ϵ, yields

$$\mathcal{O}(1): \quad 0 = \left\langle T_i' u_0, \mathcal{L}_0 T_j' u_0 \right\rangle,$$

$$\mathcal{O}(\epsilon): \quad 0 = \left\langle T_i' u_1, \mathcal{L}_0 T_j' u_0 \right\rangle + \left\langle T_i' u_0, \mathcal{L}_0 T_j' u_1 \right\rangle + \left\langle T_i' u_0, \frac{\delta^3 \mathcal{H}_0}{\delta u^3}(u_0)[u_1] T_j' u_0 \right\rangle.$$

The $\mathcal{O}(1)$ equation is satisfied since $T_j' u_0 \in \ker(\mathcal{L}_0)$. For the same reason, the first two terms of the $\mathcal{O}(\epsilon)$ equation are zero, and we deduce that

$$\left\langle T_i' u_0, \frac{\delta^3 \mathcal{H}_0}{\delta u^3}(u_0)[u_1] T_j' u_0 \right\rangle = 0; \quad i, j = 1, \ldots, N. \tag{7.2.11}$$

With this equality, u_1 drops out of the expression for $\boldsymbol{L}_1$, which simplifies to

$$(\boldsymbol{L}_1)_{ij} = \left\langle \frac{\delta^2 \mathcal{H}_1}{\delta u^2}(u_0) T_i' u_0, T_j' u_0 \right\rangle = \frac{\partial^2 \mathcal{H}_1}{\partial \gamma_i \partial \gamma_j}(\gamma_{\mathrm{red}}^*).$$

Recalling (7.2.6) we have the expression

$$\boldsymbol{L}_1 = \begin{pmatrix} \boldsymbol{0}_k & \boldsymbol{0} \\ \boldsymbol{0} & \boldsymbol{H}_{\mathrm{red}} \end{pmatrix}, \tag{7.2.12}$$

Since the eigenvalues of $\boldsymbol{L}_1$ give the leading order motion of the broken symmetries under perturbation, and since by the nondegenericity assumption none of the eigenvalues of $\boldsymbol{H}_{\mathrm{red}}$ are zero, the result (7.2.7) follows. □

Let us investigate the implications of Lemma 7.2.3 to the Hamiltonian–Krein index count of Theorem 7.1.5. Under the hypothesis of this section, for $\epsilon = 0$ the constraint matrix takes the form $\boldsymbol{D}_{ij} = \langle \mathcal{L}_0 \partial_{c_i} \phi_c, \partial_{c_i} \phi_c \rangle$, and moreover when $\epsilon = 0$ the Hamiltonian–Krein index satisfies

$$K_{\mathrm{Ham}}(0) := k_{\mathrm{r}} + k_{\mathrm{c}} + k_{\mathrm{i}}^- = \mathrm{n}(\mathcal{L}_0) - \mathrm{n}(\boldsymbol{D}).$$

Since the perturbed Hamiltonian has only k symmetries, to leading order the reduced constraint matrix $\boldsymbol{D}_{\text{red}} \in \mathbb{R}^{k\times k}$ for the perturbed problem is given by

$$(\boldsymbol{D}_{\text{red}})_{i,j} = \langle \mathcal{L}_0 \partial_{c_i} \phi_c, \partial_{c_i} \phi_c \rangle, \quad i,j = 1,\ldots,k,$$

i.e., $\boldsymbol{D}_{\text{red}}$ is the top left $k \times k$ block of $\boldsymbol{D}$. Using Lemma 7.2.3 we see that for $\epsilon > 0$ sufficiently small the Hamiltonian–Krein index for the perturbed problem is given by

$$K_{\text{Ham}}(\epsilon) = \mathrm{n}(\mathcal{L}_0) + \mathrm{n}(\boldsymbol{H}_{\text{red}}) - \mathrm{n}(\boldsymbol{D}_{\text{red}}). \tag{7.2.13}$$

We have established the following instability criterion.

Lemma 7.2.6. *Under a Hamiltonian symmetry-breaking perturbation, for $\epsilon > 0$ sufficiently small the difference between the unperturbed and the perturbed Hamiltonian–Krein index satisfies*

$$K_{\text{Ham}}(\epsilon) - K_{\text{Ham}}(0) = \mathrm{n}(\boldsymbol{H}_{\text{red}}) + \mathrm{n}(\boldsymbol{D}) - \mathrm{n}(\boldsymbol{D}_{\text{red}}), \tag{7.2.14}$$

which affords the lower bound

$$K_{\text{Ham}}(\epsilon) - K_{\text{Ham}}(0) \geq \min\{\mathrm{n}(\boldsymbol{H}_{\text{red}}), \mathrm{n}(\boldsymbol{D}) - \mathrm{n}(\boldsymbol{D}_{\text{red}})\}. \tag{7.2.15}$$

Proof. The result (7.2.14) follows from (7.2.13) since $K_{\text{Ham}}(0) = \mathrm{n}(\mathcal{L}_0) - \mathrm{n}(\boldsymbol{D})$. Since $\boldsymbol{D}_{\text{red}}$ is a submatrix of $\boldsymbol{D}$, we have that $\mathrm{n}(\boldsymbol{D}_{\text{red}}) \leq \mathrm{n}(\boldsymbol{D})$, and the inequality (7.2.15) follows from (7.2.14). □

Remark 7.2.7. The negative index of the constraint matrix, $\mathrm{n}(\boldsymbol{D})$, can be thought of as the number of negative (lower energy) directions on the Hamiltonian surface that are blocked by the constraints of the unperturbed problem. Thus, the difference $\mathrm{n}(\boldsymbol{D}) - \mathrm{n}(\boldsymbol{D}_{\text{red}})$ indicates the number of unstable directions that are accessible to the system due to the breaking the symmetries.

7.2.1.2 Perturbations to the Kernel of $\mathcal{JL}$

We return to the eigenvalue problem for $\mathcal{JL}$ and determine to leading order the location of the small eigenvalues that arise from the symmetry breaking. Using the expansion of $\mathcal{L}$ we rewrite it as

$$\mathcal{JL}u = \lambda u \quad \Rightarrow \quad (\mathcal{L}_0 + \epsilon\mathcal{L}_1)u = \lambda \mathcal{J}^{-1} u. \tag{7.2.16}$$

Since each eigenvalue is part of a Jordan chain of length two, the results of Chapter 6.2 motivate expansions of the form

$$\lambda = \sqrt{\epsilon}\,\lambda_1 + \mathcal{O}(\epsilon), \quad u = u_0 + \sqrt{\epsilon}\,u_1 + \epsilon u_2 + \mathcal{O}(\epsilon^{3/2}),$$

where $u_0 \in \text{span}\{T_1'\phi_c, \dots, T_N'\phi_c\} = \ker(\mathcal{L}_0)$. Plugging the expansion in (7.2.16), at $\mathcal{O}(\epsilon^{1/2})$ we obtain the relation

$$\mathcal{L}_0 u_1 = \lambda_1 \mathcal{J}^{-1} u_0. \tag{7.2.17}$$

Using the relation (7.2.1), for any $\boldsymbol{v} = (v_1, \dots, v_N)^{\mathrm{T}} \in \mathbb{C}^N$ we solve (7.2.17),

$$u_0 = \sum_{j=1}^{N} v_j T_j' \phi_c \quad \Rightarrow \quad u_1 = -\lambda_1 \sum_{j=1}^{N} v_j \partial_{c_j} \phi_c \in \text{gker}(\mathcal{J}\mathcal{L}).$$

Using (7.2.17) to simplify the $\mathcal{O}(\epsilon)$ terms in the expansion yields

$$\mathcal{L}_0 u_2 = \lambda_1 \mathcal{J}^{-1} u_1 - \mathcal{L}_1 u_0 = -\sum_{j=1}^{N} v_j \left(\lambda_1^2 \mathcal{J}^{-1} \partial_{c_j} \phi_c + \mathcal{L}_1 T_j' \phi_c \right). \tag{7.2.18}$$

The equation (7.2.18) is solvable for u_2 if and only if the right-hand side is orthogonal to $\ker(\mathcal{L}_0)$. Recalling the definition (5.2.53) of the constraint matrix $\boldsymbol{D}$, we first see

$$\langle \mathcal{J}^{-1} \partial_{c_j} \phi_c, T_\ell' \phi_c \rangle = -\langle \partial_{c_j} \phi_c, \mathcal{J}^{-1} T_\ell' \phi_c \rangle = \boldsymbol{D}_{j\ell}.$$

Regarding the second term, we have already seen that it is simply an entry in the matrix $\boldsymbol{L}_1$ given in (7.2.12). Thus, we may write the solvability condition as a matrix eigenvalue problem for λ_1,

$$(\boldsymbol{L}_1 + \lambda_1^2 \boldsymbol{D}) \boldsymbol{v} = \boldsymbol{0}. \tag{7.2.19}$$

Lemma 7.2.8. *Under a Hamiltonian perturbation that breaks k symmetries, the eigenvalues and eigenfunctions of the broken symmetries satisfy the expansions*

$$\lambda = \sqrt{\epsilon}\,\lambda_1 + \mathcal{O}(\epsilon), \quad u = \sum_{j=1}^{N} v_j \left(T_j' \phi_c + \sqrt{\epsilon}\,\lambda_1 \partial_{c_j} \phi_c \right) + \mathcal{O}(\epsilon),$$

where $\lambda_1, \boldsymbol{v}$ are found by solving the matrix eigenvalue problem (7.2.19).

The result of Lemma 7.2.8 can be used to develop a lower bound on the number, k_{r}, of purely real eigenvalue pairs that bifurcate out of the kernel as the symmetries are broken. Setting $\boldsymbol{z} = \lambda_1 \boldsymbol{D} \boldsymbol{v}$ the system (7.2.19) transforms to

$$\underbrace{\begin{pmatrix} \boldsymbol{0}_N & \boldsymbol{I}_N \\ -\boldsymbol{I}_N & \boldsymbol{0}_N \end{pmatrix}}_{\boldsymbol{J}} \underbrace{\begin{pmatrix} \boldsymbol{L}_1 & \boldsymbol{0}_N \\ \boldsymbol{0}_N & \boldsymbol{D}^{-1} \end{pmatrix}}_{\boldsymbol{L}_{\text{diag}}} \begin{pmatrix} \boldsymbol{v} \\ \boldsymbol{z} \end{pmatrix} = \lambda_1 \begin{pmatrix} \boldsymbol{v} \\ \boldsymbol{z} \end{pmatrix}, \tag{7.2.20}$$

which is a finite-dimensional version of that discussed in Theorem 7.1.16, where the role of $\mathcal{L}_+$ is played by $\boldsymbol{L}_1$, and that of $\mathcal{L}_-$ is played by $\boldsymbol{D}^{-1}$. Thus, we begin by noting that

$$\mathrm{n}(\boldsymbol{L}_1) = \mathrm{n}(\boldsymbol{H}_{\mathrm{red}}), \quad \mathrm{n}(\boldsymbol{D}^{-1}) = \mathrm{n}(\boldsymbol{D}).$$

We must now compute the submatrices $\boldsymbol{D}_\pm$ of (7.1.15). Since $\boldsymbol{D}^{-1}$ has no kernel, $\ker(\boldsymbol{D}^{-1}) = \{\boldsymbol{0}\}$, and since $\boldsymbol{H}_{\mathrm{red}}$ is nonsingular, upon recalling (7.2.12) we have $\ker(\boldsymbol{L}_1) = \mathrm{span}\{\boldsymbol{e}_1,\ldots,\boldsymbol{e}_k\}$, where $\boldsymbol{e}_j \in \mathbb{C}^N$ denotes the jth unit vector. The k-dimensional constraint space S is of the form

$$S = \mathcal{J}^{-1}\ker(\boldsymbol{L}_{\mathrm{diag}}) = \mathrm{span}\{(\boldsymbol{0},\boldsymbol{e}_1)^{\mathrm{T}},\ldots,(\boldsymbol{0},\boldsymbol{e}_k)^{\mathrm{T}}\},$$

so that the matrix $\boldsymbol{D}_+$ defined in (7.1.15) is empty (zero-dimensional). We now compute $\boldsymbol{D}_-$. Using (7.1.14) the generalized eigenfunctions satisfy,

$$\boldsymbol{D}^{-1}(-\psi_j^-) = \boldsymbol{e}^j \quad \Rightarrow \quad \psi_j^- = -\boldsymbol{D}\boldsymbol{e}_j,$$

for $j = 1,\ldots,k$. Consequently, for $i,j = 1,\ldots,k$

$$(\boldsymbol{D}_-)_{ij} = \boldsymbol{D}\boldsymbol{e}_i \cdot \boldsymbol{D}^{-1}(\boldsymbol{D}\boldsymbol{e}_j) = \boldsymbol{e}_i \cdot \boldsymbol{D}\boldsymbol{e}_j \quad \Rightarrow \quad \boldsymbol{D}_- = \boldsymbol{D}_{\mathrm{red}}.$$

In addition to the lower bound of Lemma 7.2.6, we may use Theorem 7.1.16 to establish:

Corollary 7.2.9. *Consider a Hamiltonian perturbation that breaks k symmetries. For $\epsilon > 0$ sufficiently small, the number of pairs of purely real broken-symmetry eigenvalues satisfies the lower bound*

$$k_{\mathrm{r}}(\epsilon) - k_{\mathrm{r}}(0) \geq |\mathrm{n}(\boldsymbol{H}_{\mathrm{red}}) - [\mathrm{n}(\boldsymbol{D}) - \mathrm{n}(\boldsymbol{D}_{\mathrm{red}})]|. \tag{7.2.21}$$

Moreover, if the critical point is a minimizer of the reduced Hamiltonian, i.e., $\mathrm{n}(\boldsymbol{H}_{\mathrm{red}}) = 0$, then all of the unstable small eigenvalues are real-valued, and

$$k_{\mathrm{r}} = \mathrm{n}(\boldsymbol{D}) - \mathrm{n}(\boldsymbol{D}_{\mathrm{red}}). \tag{7.2.22}$$

Proof. The result (7.2.21) is a direct consequence of Theorem 7.1.16. To see the second result, we invoke Lemma 7.2.6 which implies that

$$K_{\mathrm{Ham}}(\epsilon) - K_{\mathrm{Ham}}(0) = \mathrm{n}(\boldsymbol{D}) - \mathrm{n}(\boldsymbol{D}_{\mathrm{red}}),$$

whereas (7.2.21) gives the lower bound

$$k_{\mathrm{r}}(\epsilon) - k_{\mathrm{r}}(0) \geq \mathrm{n}(\boldsymbol{D}) - \mathrm{n}(\boldsymbol{D}_{\mathrm{red}}).$$

Combining these two results yields (7.2.22). □

7.2.1.3 Example: Hamiltonian Perturbation of NLS

We apply the theory developed in Chapter 7.2.1 to a perturbed NLS equation of the form

$$\mathrm{i}\partial_t u + \partial_x^2 u - \omega u + |u|^2 u = \epsilon(V(x)u + \overline{u}). \tag{7.2.23}$$

We introduce $\boldsymbol{u} = (\mathrm{Re}\, u, \mathrm{Im}\, u)^{\mathrm{T}} = (u_1, u_2)^{\mathrm{T}}$ so that the vector system has the Hamiltonian structure

$$\partial_t \boldsymbol{u} = \mathcal{J}\frac{\delta \mathcal{H}}{\delta \boldsymbol{u}}(\boldsymbol{u}), \quad \mathcal{H}(\boldsymbol{u}) = \mathcal{H}_0(\boldsymbol{u}) + \epsilon\mathcal{H}_1(\boldsymbol{u}),$$

where $\mathcal{J}$ and the unperturbed Hamiltonian are as given in Exercise 5.2.10, while the perturbing Hamiltonian takes the form $\mathcal{H}_1 = \mathcal{H}_{\mathrm{trans}} + \mathcal{H}_{\mathrm{rot}}$ where

$$\mathcal{H}_{\mathrm{trans}}(\boldsymbol{u}) = \int_{\mathbb{R}} V(x)|\boldsymbol{u}|^2\,\mathrm{d}x, \quad \mathcal{H}_{\mathrm{rot}}(\boldsymbol{u}) = \int_{\mathbb{R}} \left(u_1^2 - u_2^2\right)\mathrm{d}x.$$

When $\epsilon = 0$ solutions to (7.2.23) are invariant under the symmetries of translation and complex rotation given in Exercise 5.2.10(a), and the manifold $\mathcal{M}_T(\Phi_\omega)$ generated by the image of

$$\Phi_\omega(x) = (\phi_\omega, 0)^T,$$

under the symmetry group T is orbitally stable (see Exercise 5.2.10). In particular, for $\epsilon = 0$ we have $k_{\mathrm{r}}(0) = k_{\mathrm{c}}(0) = k_{\mathrm{i}}^-(0) = 0$.

For $\epsilon > 0$ we assume (for the moment) that some equilibria on $\mathcal{M}_T$ will persist in a perturbed form,

$$\boldsymbol{u}(x) = T(\gamma^*)\Phi_\omega(x) + \epsilon\boldsymbol{u}_1 + \mathcal{O}(\epsilon^2), \quad T(\gamma) := T_1(\gamma_1)T_2(\gamma_2),$$

for a discrete set of values of (γ_1^*, γ_2^*). For $\epsilon > 0$ we anticipate that the spectral stability of the persistent equilibria will depend upon the choice of (γ_1^*, γ_2^*). Both of the symmetries are broken for $\epsilon > 0$: the term $\mathcal{H}_{\mathrm{trans}}$ breaks the translational symmetry, while $\mathcal{H}_{\mathrm{rot}}$ breaks the rotational symmetry. From Lemma 7.2.8 we anticipate that four small eigenvalues of $\mathcal{O}(\sqrt{\epsilon})$ will bifurcate from the origin.

From Lemma 7.2.2 we know the persisting equilibria correspond to nondegenerate critical points of the reduced Hamiltonian. For this perturbation of NLS the reduced Hamiltonian is given by

$$\mathcal{H}_{\mathrm{red}}(\gamma) := \mathcal{H}_1(T(\gamma)\Phi_\omega) = \underbrace{\int_{\mathbb{R}} V(x)\phi_\omega^2(x+\gamma_1)\,\mathrm{d}x}_{\mathcal{H}_{\mathrm{trans}}(\gamma_1)} + \underbrace{\cos 2\gamma_2 \int_{\mathbb{R}} \phi_\omega^2(x)\,\mathrm{d}x}_{\mathcal{H}_{\mathrm{rot}}(\gamma_2)},$$

and its critical points solve

$$\mathcal{H}'_{\text{trans}}(\gamma_1)=0,\quad \mathcal{H}'_{\text{rot}}(\gamma_2)=-2\sin 2\gamma_2\|\phi_\omega\|_2^2=0.$$

The location of the persistent equilibria satisfies

$$\mathcal{H}'_{\text{trans}}(\gamma_1)=-\int_{\mathbb{R}} V'(x)\phi_\omega^2(x+\gamma_1)\,\mathrm{d}x=0,$$

while its rotational angle is of the form $\gamma_2=k\pi/2$ for $k\in\mathbb{N}$. At these critical points the Hessian of the reduced Hamiltonian is diagonal,

$$\boldsymbol{H}_{\text{red}}=\operatorname{diag}(\mathcal{H}''_{\text{trans}}(\gamma_1),\mathcal{H}''_{\text{rot}}(\gamma_2)),\tag{7.2.24}$$

and the reduced Hamiltonian is nonsingular if and only if

$$\mathcal{H}''_{\text{trans}}(\gamma_1)=\int_{\mathbb{R}} V''(x)\phi_\omega^2(x+\gamma_1)\,\mathrm{d}x\neq 0.$$

This last condition guarantees the persistence of the equilibria through Lemma 7.2.2, and is henceforth assumed.

To apply the spectral index results of Lemma 7.2.6 and Corollary 7.2.9 requires that we evaluate the constraint matrix $\boldsymbol{D}$. In Exercise 5.2.10 we saw that for $\epsilon=0$ the eigenvalue problem for $\boldsymbol{v}=(v_1,v_2)^{\mathrm{T}}$ takes the form

$$\mathcal{J}\mathcal{L}_0\boldsymbol{v}:=\begin{pmatrix}0&1\\-1&0\end{pmatrix}\begin{pmatrix}\mathcal{L}_+&0\\0&\mathcal{L}_-\end{pmatrix}\boldsymbol{v}=\lambda\boldsymbol{v},$$

where

$$\mathcal{L}_+=-\partial_x^2+\omega-3\phi_\omega^2,\quad \mathcal{L}_-=-\partial_x^2+\omega-\phi_\omega^2,$$

and ϕ_ω solves

$$-\partial_x^2\phi_\omega+\omega\phi_\omega-\phi_\omega^3=0.\tag{7.2.25}$$

This is the canonical form studied in Chapter 7.1.3, so we know that $\boldsymbol{D}$ will be a diagonal matrix. It is straightforward to calculate that

$$\ker(\mathcal{L}_+)=\operatorname{span}\{\partial_x\phi_\omega\},\quad \ker(\mathcal{L}_-)=\operatorname{span}\{\phi_\omega\},$$

that is, the kernel for $\mathcal{L}_+$ is generated by the spatial translation symmetry, and the kernel for $\mathcal{L}_-$ is generated by the rotation symmetry. From (7.1.14) the generalized eigenfunctions $\psi^\pm$ are the solutions to

$$\mathcal{L}_+\psi^+=\phi_\omega,\quad \mathcal{L}_-(-\psi^-)=\partial_x\phi_\omega.$$

As in Exercise 5.2.10(e) for the NLS (the $\epsilon=0$ problem), we verify that

$$\boldsymbol{D}_+=-\frac{1}{2}\partial_\omega\|\phi_\omega\|_2^2<0,\quad \boldsymbol{D}_-=\frac{1}{4}\|\phi_\omega\|_2^2>0.$$

and in particular

$$n(\boldsymbol{D}) = n(\boldsymbol{D}_+) = n(-\partial_\omega\|\phi_\omega\|_2^2) = 1.$$

Since the operators $\mathcal{L}_\pm$ are Sturmian, and $\phi_\omega > 0$ while $\partial_x\phi_\omega$ has a single zero, it follows that

$$n(\mathcal{L}_+) = 1, \quad n(\mathcal{L}_-) = 0,$$

so that for the unperturbed problem we the Hamiltonian–Krein index satisfies

$$K_{\text{Ham}}(0) = n(\mathcal{L}) - n(\boldsymbol{D}) = 0,$$

which verifies the spectral stability.

In order to determine the stability of the persistent solution, we invoke Lemma 7.2.6 and Corollary 7.2.9. We know that $K_{\text{Ham}}(0) = k_{\text{r}}(0) = 0$ and $n(\boldsymbol{D}) = n(\boldsymbol{D}_+) = 1$, while the breaking of both symmetries implies that $\boldsymbol{D}_{\text{red}}$ has dimension zero. Applying these observations to (7.2.14) and (7.2.21) we find

$$K_{\text{Ham}}(\epsilon) = n(\boldsymbol{H}_{\text{red}}) + 1, \quad k_{\text{r}}(\epsilon) \geq |n(\boldsymbol{H}_{\text{red}}) - 1|.$$

Thus, if the unperturbed critical point is a minimizer of the reduced Hamiltonian, i.e., $n(\boldsymbol{H}_{\text{red}}) = 0$, then for $\epsilon > 0$ we have the counterintuitive result that the wave will be spectrally unstable with $K_{\text{Ham}}(\epsilon) = k_{\text{r}}(\epsilon) = 1$. If the persistent critical point corresponds to a maximum of the reduced Hamiltonian, then $n(\boldsymbol{H}_{\text{red}}) = 2$, so that $K_{\text{Ham}} = 3$, while $k_{\text{r}}(\epsilon) \geq 1$. Moreover, since there are only four small eigenvalues, and two are real, there cannot be a complex quad, and either $k_{\text{r}}(\epsilon) = 2$ and $k_{\text{i}}^-(\epsilon) = 0$, or $k_{\text{r}}(\epsilon) = 1$ and $k_{\text{i}}^-(\epsilon) = 2$. The perturbed wave can be spectrally stable only if it corresponds to a saddle point of the reduced Hamiltonian. In this case there are three possibilities: (a) $k_{\text{r}}(\epsilon) = 2$, (b) $k_{\text{i}}^-(\epsilon) = 2$, or (c) $k_{\text{c}}(\epsilon) = 1$. A specific calculation is required to distinguish between these possibilities, and this is left as Exercise 7.2.3.

Many Hamiltonian systems are functions of complex-valued variables $u : \mathbb{R} \to \mathbb{C}$. In previous sections we have separated complex-valued u into real and imaginary parts. However, it is computationally expedient to view the real-valued Hamiltonian $\mathcal{H}$ as a function of u and its complex conjugate $\overline{u}$. An equivalent formulation of the Hamiltonian system is given by

$$\partial_t u = \mathcal{J}\frac{\delta\mathcal{H}}{\delta\overline{u}}(u), \tag{7.2.26}$$

where u is complex and the $\overline{u}$ variation is defined via

$$\lim_{\delta\to 0}\frac{\mathcal{H}(u,\overline{u}+\delta\overline{v}) - \mathcal{H}(u,\overline{v})}{\delta} = \left\langle\frac{\delta\mathcal{H}}{\delta\overline{u}}(u), v\right\rangle.$$

For example, the Hamiltonian

$$\mathcal{H}(u) := \int_{\mathbb{R}} |\partial_x u|^2 + \frac{1}{2}|u|^4 \,\mathrm{d}x = \int_{\mathbb{R}} \partial_x u \overline{\partial_x u} + \frac{1}{2} u^2 \overline{u}^2 \,\mathrm{d}x,$$

has $\overline{u}$ variation

$$\frac{\delta \mathcal{H}}{\delta \overline{u}} = -\partial_x^2 u + u^2 \overline{u} = -\partial_x^2 u + |u|^2 u.$$

The advantage to this formulation is that the resulting system is for a complex scalar, rather than a two-dimensional real vector. Exercise 7.2.5 uses this formulation.

7.2.2 Non-Hamiltonian Perturbations

We consider a perturbed Hamiltonian system of the form

$$\partial_t u = \mathcal{J} \frac{\delta \mathcal{H}_0}{\delta u}(u) + \epsilon \mathcal{N}(u), \tag{7.2.27}$$

where $\mathcal{N} : Y \mapsto Y^*$ is formally smooth in u. When $\epsilon = 0$ we assume that Hypothesis 5.2.5 holds; in particular, there is an N-dimensional symmetry operator T and the skew operator $\mathcal{J}$ is boundedly invertible on all of X. Moreover, we assume that the perturbation does not break any of the symmetries; that is, for all $\gamma \in \mathbb{R}^N$,

$$T(\gamma)\mathcal{N}(u) = \mathcal{N}(T(\gamma)u). \tag{7.2.28}$$

With the full symmetry group in force the persistence problem for $\mathcal{M}_T(\phi_c)$ may be considered for $\gamma = \mathbf{0}$. Expanding the perturbed critical point as

$$\phi = \phi_c + \epsilon \phi_1 + \epsilon^2 \phi_2 + \mathcal{O}(\epsilon^3), \tag{7.2.29}$$

the equation for equilibria

$$\frac{\delta \mathcal{H}_0}{\delta u}(u) + \epsilon \mathcal{J}^{-1} \mathcal{N}(u) = 0,$$

yields the system of equations

$$\begin{aligned}
\mathcal{O}(1):&\quad \frac{\delta \mathcal{H}_0}{\delta u}(\phi_c) = 0 \\
\mathcal{O}(\epsilon):&\quad \mathcal{L}_0 \phi_1 = -\mathcal{J}^{-1}\mathcal{N}(\phi_c), \quad \mathcal{L}_0 := \frac{\delta^2 \mathcal{H}_0}{\delta u^2}(\phi_c) \\
\mathcal{O}(\epsilon^2):&\quad \mathcal{L}_0 \phi_2 = -\frac{\delta^3 \mathcal{H}_0}{\delta u^3}(\phi_c)[\phi_1, \phi_1] - \mathcal{J}^{-1}\mathcal{N}'(\phi_c)\phi_1.
\end{aligned} \tag{7.2.30}$$

The $\mathcal{O}(1)$ equation is satisfied for any critical point ϕ_c of the unperturbed problem. The linearization $\mathcal{L}_0$ has an N-dimensional kernel, so the $\mathcal{O}(\epsilon)$ equation is solvable if and only if

$$\mathcal{J}^{-1}\mathcal{N}(\phi_c) \in \ker(\mathcal{L}_0)^\perp = \operatorname{span}\{T_1'\phi_c, \dots, T_N'\phi_c\}^\perp.$$

This imposes N conditions on $\boldsymbol{c}$,

$$\left\langle \mathcal{J}^{-1}\mathcal{N}(\phi_c), T_j'\phi_c \right\rangle = -\left\langle \mathcal{N}(\phi_c), \mathcal{J}^{-1}T_j'\phi_c \right\rangle = 0, \tag{7.2.31}$$

for $j = 1, \dots, N$. Assuming that (7.2.31) holds for some $\boldsymbol{c}$, then the correction term is given by

$$\phi_1 = \mathcal{L}_0^{-1}\left(\mathcal{J}^{-1}\mathcal{N}(\phi_c)\right) \in \ker(\mathcal{L}_0)^\perp. \tag{7.2.32}$$

Proceeding formally, we assume that the higher-order corrections can be chosen to satisfy the corresponding solvability conditions.

While the non-Hamiltonian perturbation may impact many parts of the spectrum of the associated linearized operator, in this section we focus on the fate of the $2N$-dimensional generalized kernel of the unperturbed linearization $\mathcal{L}_0$. The linearization about the persistent solution leads to an eigenvalue problem of the form

$$\mathcal{J}\mathcal{L}v = \lambda v, \tag{7.2.33}$$

where the operator $\mathcal{L}$ has the expansion

$$\mathcal{L} = \mathcal{L}_0 + \epsilon\mathcal{J}^{-1}\mathcal{L}_1 + \epsilon^2\mathcal{J}^{-1}\mathcal{L}_2 + \mathcal{O}(\epsilon^3). \tag{7.2.34}$$

The correction term

$$\mathcal{L}_1 = \mathcal{N}'(\phi_c) + \mathcal{J}\frac{\delta^3\mathcal{H}_0}{\delta u^3}(\phi_c)[\phi_1], \tag{7.2.35}$$

is not necessarily self-adjoint. Indeed, in many applications the operator $\mathcal{L}_1$ is dissipative in the sense that it moves the essential spectrum off the imaginary axis and into the left-half complex plane. However, since none of the symmetries are broken, $\mathcal{L}$ will have an N-dimensional kernel,

$$\ker(\mathcal{L}) = \operatorname{span}\{T_1'\phi, \dots, T_N'\phi\}. \tag{7.2.36}$$

Significantly, the expansion (7.2.29) of ϕ induces an expansion for each element of the kernel,

$$T_j'\phi = T_j'\phi_c + \epsilon T_j'\phi_1 + \mathcal{O}(\epsilon^2). \tag{7.2.37}$$

The fate of the N-dimensional set of eigenvalues associated with the generalized eigenfunctions is determined by the following lemma.

Lemma 7.2.10. *Consider the system (7.2.27), where the perturbation $\mathcal{N}(u)$ satisfies the symmetry condition (7.2.28). Assuming the existence of a persistent solution ϕ with an expansion (7.2.29), then the associated linearized problem (7.2.33) will have an N-dimensional kernel given by (7.2.36) and N small eigenvalues $\lambda = \epsilon\lambda_1 + O(\epsilon^2)$. The leading order term λ_1 solves the characteristic equation*

$$\det(\mathbf{\Gamma} - \lambda_1 \mathbf{D}) = 0, \tag{7.2.38}$$

*where the constraint matrix, **D**, is defined in (5.2.53), and $\mathbf{\Gamma} \in \mathbb{R}^{N\times N}$ has entries*

$$\mathbf{\Gamma}_{ij} = -\langle T_j'\phi_1 + \mathcal{L}_1 \partial_{c_j}\phi_c, \mathcal{J}^{-1}T_i'\phi_c\rangle. \tag{7.2.39}$$

Remark 7.2.11. Comparing Lemma 7.2.8 and Lemma 7.2.10 we see that if the perturbation of the original system is Hamiltonian, but breaks symmetries, then the fate of the small eigenvalues that bifurcate out of the origin is determined by the structure of the reduced-energy surface—no knowledge of the $\mathcal{O}(\epsilon)$ correction of the persistent critical point is needed. On the other hand, if the perturbation is not Hamiltonian, but breaks no symmetries, then the location of the small eigenvalues explicitly depends upon the $\mathcal{O}(\epsilon)$ correction to the persistent equilibrium.

Proof. To locate the N small but nonzero eigenvalues, we employ an asymptotic expansion for the eigenvalue and eigenfunction

$$\lambda = \epsilon\lambda_1 + \mathcal{O}(\epsilon^2), \quad v = v_0 + \epsilon v_1 + \mathcal{O}(\epsilon^2)$$

(e.g., see [**197**]). Plugging these expansions into the eigenvalue problem, (7.2.33), yields:

$$\begin{aligned} \mathcal{O}(1):&\ \ \mathcal{L}_0 v_0 = 0 \\ \mathcal{O}(\epsilon):&\ \ \mathcal{L}_0 v_1 = \lambda_1 \mathcal{J}^{-1} v_0 - \mathcal{J}^{-1}\mathcal{L}_1 v_0 \\ \mathcal{O}(\epsilon^2):&\ \ \mathcal{L}_0 v_2 = \lambda_2 \mathcal{J}^{-1} v_0 + \lambda_1 \mathcal{J}^{-1} v_1 - \mathcal{J}^{-1}\mathcal{L}_1 v_1 - \mathcal{J}^{-1}\mathcal{L}_2 v_0. \end{aligned} \tag{7.2.40}$$

Any $\boldsymbol{\alpha} = (\alpha_1, \ldots, \alpha_N) \in \mathbb{C}^N$ generates a solution to the $\mathcal{O}(1)$ equation,

$$v_0 = \sum_{j=1}^{N} \alpha_j T_j' \phi_c.$$

The solvability condition for the $\mathcal{O}(\epsilon)$ equation requires that

$$\lambda_1 \mathcal{J}^{-1} v_0 - \mathcal{J}^{-1}\mathcal{L}_1 v_0 \in \ker(\mathcal{L}_0)^{\perp}.$$

Now, as a consequence of the existence of the N Jordan chains of length two for the unperturbed problem it is the case that $\mathcal{J}^{-1}v_0 \in \ker(\mathcal{L}_0)^{\perp}$. This is initially alarming, as the free parameter λ_1 has dropped out of the problem. However, the term $\mathcal{J}^{-1}\mathcal{L}_1 v_0$ also lies in $\ker(\mathcal{L}_0)^{\perp}$. Indeed, using the expansion (7.2.37) of $T_j'\phi \in \ker(\mathcal{L}_0)$ and the expansion (7.2.34) for $\mathcal{L}$ we see that the identity

$$\mathcal{L}T_j'\phi = 0, \quad j = 1,\ldots,N,$$

is equivalent to the following relations

$$\begin{aligned} \mathcal{O}(1): &\quad \mathcal{L}_0 T_j'\phi_c = 0 \\ \mathcal{O}(\epsilon): &\quad \mathcal{L}_0 T_j'\phi_1 = -\mathcal{J}^{-1}\mathcal{L}_1 T_j'\phi_c \\ \mathcal{O}(\epsilon^2): &\quad \mathcal{L}_0 T_j'\phi_2 = -\mathcal{J}^{-1}\mathcal{L}_1 T_j'\phi_1 - \mathcal{J}^{-1}\mathcal{L}_2 T_j'\phi_c. \end{aligned} \tag{7.2.41}$$

Specifically, the persistence assumption implies, at $\mathcal{O}(\epsilon)$, that the operator $\mathcal{J}^{-1}\mathcal{L}_1 : \ker(\mathcal{L}_0) \mapsto \ker(\mathcal{L}_0)^\perp$. Since $v_0 \in \ker(\mathcal{L}_0)$, we deduce that $\mathcal{J}^{-1}\mathcal{L}_1 v_0 \in \ker(\mathcal{L}_0)^\perp$. Thus, we may solve the $\mathcal{O}(\epsilon)$ expression of (7.2.40) for v_1. Recalling (5.2.54),

$$\mathcal{J}\mathcal{L}_0(-\partial_{c_j}\phi_c) = T_j'\phi_c, \quad j = 1,\ldots,N,$$

and using the $\mathcal{O}(\epsilon)$ equation in (7.2.41), we obtain the expression,

$$v_1 = -\sum_{j=1}^{N} \alpha_j \left(\lambda_1 \partial_{c_j}\phi_c + \mathcal{L}_0^{-1}\mathcal{J}^{-1}\mathcal{L}_1 T_j'\phi_c\right) = \sum_{j=1}^{N} \alpha_j \left(T_j'\phi_1 - \lambda_1 \partial_{c_j}\phi_c\right). \tag{7.2.42}$$

Unfortunately, determining λ_1 requires an examination of the $\mathcal{O}(\epsilon^2)$ terms in the perturbation expansion. Substitution of (7.2.42) into the $\mathcal{O}(\epsilon^2)$ equation of (7.2.41) yields the expression for v_2,

$$\begin{aligned} \mathcal{L}_0 v_2 = \sum_{j=1}^{N} \alpha_j \Big(&\lambda_2 \mathcal{J}^{-1} T_j'\phi_c - [\mathcal{J}^{-1}\mathcal{L}_2 T_j'\phi_c + \mathcal{J}^{-1}\mathcal{L}_1 T_j'\phi_1] \\ &+ \lambda_1 [\mathcal{J}^{-1} T_j'\phi_1 + \mathcal{J}^{-1}\mathcal{L}_1 \partial_{c_j}\phi_c] - \lambda_1^2 \mathcal{J}^{-1}\partial_{c_j}\phi_c\Big). \end{aligned} \tag{7.2.43}$$

We may solve for v_2 only if the right-hand side is orthogonal to $\ker(\mathcal{L}_0)$. Again, we have $\mathcal{J}^{-1}T_j'\phi_c \in \ker(\mathcal{L}_0)^\perp$, so that the λ_2 term drops out of the solvability condition. In addition, from the $\mathcal{O}(\epsilon^2)$ equation of (7.2.41) we know that

$$\mathcal{J}^{-1}\mathcal{L}_2 T_j'\phi_c + \mathcal{J}^{-1}\mathcal{L}_1 T_j'\phi_1 \in \ker(\mathcal{L}_0)^\perp,$$

which leaves only the λ_1 and λ_1^2 terms on the right-hand side of (7.2.43). Taking the inner product of this equation with $T_\ell'\phi_c$, and dividing by $\lambda_1 \neq 0$, yields the system

$$(\boldsymbol{\Gamma} - \lambda_1 \boldsymbol{D})\boldsymbol{\alpha} = \boldsymbol{0}.$$

The constraint matrix $\boldsymbol{D}$ arises from the λ_1^2 terms via the relations

$$\langle \mathcal{J}^{-1}\partial_{c_j}\phi_c, T_\ell'\phi_c\rangle = -\langle \partial_{c_j}\phi_c, \mathcal{J}^{-1}T_\ell'\phi_c\rangle = \langle \partial_{c_j}\phi_c, \mathcal{L}_0 \partial_{c_\ell}\phi_c\rangle = \boldsymbol{D}_{j\ell},$$

and the matrix $\boldsymbol{\Gamma}$, defined in (7.2.39), arises from the λ_1 terms. Of course, the existence of nonzero solutions $\boldsymbol{\alpha}$ is equivalent to (7.2.38). □

7.2.2.1 Example: Non-Hamiltonian Perturbation of NLS

As an application of Lemma 7.2.10 we consider the perturbed NLS equation given by

$$\mathrm{i}\partial_t u + \partial_x^2 u - \omega u + |u|^2 u = \mathrm{i}\epsilon\left(d_1\partial_x^2 u + d_2 u + d_3|u|^2 u + d_4|u|^4 u\right). \tag{7.2.44}$$

The $\mathcal{O}(\epsilon)$-term on the right-hand side is the Ginzburg–Landau perturbation of the NLS. In the context of pulse propagation in optical fibers, u is the envelope of the electric field, $d_1 > 0$ describes spectral filtering, $d_2 < 0$ accounts for linear loss in the fiber, $d_3 > 0$ describes the nonlinear amplification in the fiber, and $d_4 < 0$ accounts for nonlinear saturation (e.g., see [**13, 97, 151, 175, 268**] and the references therein).

Separating u into its real and imaginary parts, $u = u_\mathrm{r} + \mathrm{i}u_\mathrm{i}$, the system can be rewritten in terms of the vector $\boldsymbol{u} = (u_\mathrm{r}, u_\mathrm{i})^\mathrm{T}$,

$$\partial_t \boldsymbol{u} = \mathcal{J}(-\partial_x^2\boldsymbol{u} + \omega\boldsymbol{u} - |\boldsymbol{u}|^2\boldsymbol{u}) + \epsilon\left(d_1\partial_x^2\boldsymbol{u} + d_2\boldsymbol{u} + d_3|\boldsymbol{u}|^2\boldsymbol{u} + d_4|\boldsymbol{u}|^4\boldsymbol{u}\right),$$

where $\mathcal{J}$ is the canonical skew matrix corresponding to $-\mathrm{i}$. For $\epsilon \geq 0$ the equation is invariant under the translational and rotational symmetries (see Exercise 5.2.10 for further discussion); however, when $\epsilon > 0$ the system is no longer Hamiltonian.

We first examine the persistence problem, following the approach of (7.2.29)–(7.2.32). When $\epsilon = 0$ the NLS Hamiltonian has the critical point

$$\Phi_\omega(x) = \begin{pmatrix} \phi_\omega(x) \\ 0 \end{pmatrix}, \quad \phi_\omega(x) = \sqrt{2\omega}\,\mathrm{sech}(\sqrt{\omega}\,x),$$

where $\omega = \omega(\boldsymbol{c})$ (see Exercise 5.2.10(b)). The operator $\mathcal{L}_0$, the second variation of the NLS Hamiltonian at Φ_ω, has a kernel spanned by the symmetries

$$T_1'\Phi_\omega = \begin{pmatrix} \partial_x\phi_\omega \\ 0 \end{pmatrix}, \quad T_2'\Phi_\omega = \begin{pmatrix} 0 \\ \phi_\omega \end{pmatrix}.$$

The residual generated by the perturbation $\mathcal{N}$ takes the form

$$\mathcal{N}(\Phi_\omega) = \begin{pmatrix} d_1\partial_x^2\phi_\omega + d_2\phi_\omega + d_3\phi_\omega^3 + \phi_\omega^5 \\ 0 \end{pmatrix},$$

and since $\mathcal{J}^{-1} = -\mathcal{J}$, the two persistence equations from (7.2.31) reduce to

$$0 = \langle \mathcal{N}(\phi_c), \mathcal{J}^{-1} T_1' \Phi_\omega \rangle$$
$$0 = \langle \mathcal{N}(\phi_c), \mathcal{J}^{-1} T_2' \Phi_\omega \rangle = -d_1 \langle \partial_x^2 \phi_\omega, \phi_\omega \rangle - d_2 \langle \phi_\omega, \phi_\omega \rangle - d_3 \langle \phi_\omega^3, \phi_\omega \rangle - d_4 \langle \phi_\omega^5, \phi_\omega \rangle.$$

The first equation is trivially satisfied, and writing $\phi_\omega = \sqrt{\omega}\phi_0(\sqrt{\omega}x)$, the second equation can be expressed as

$$\omega^2 d_4 \|\phi_0\|_{L^6}^6 + \omega\left(d_3 \|\phi_0|_{L^4}^4 - d_1 \|\partial_x \phi_0\|_{L^2}^2\right) + d_2 \|\phi_0\|_{L^2}^2 = 0.$$

The values of $\omega = \omega_\pm$ that persist under the perturbation satisfy

$$\omega_\pm = \frac{d_1 \|\partial_x \phi_0\|_{L^2}^2 - d_3 \|\phi_0|_{L^4}^4 \pm \sqrt{\left(d_1 \|\partial_x \phi_0\|_{L^2}^2 - d_3 \|\phi_0|_{L^4}^4\right)^2 - 4 d_2 d_4 \|\phi_0\|_{L^6}^6 \|\phi_0\|_{L^2}^2}}{2 d_1 \|\partial_x \phi_0\|_{L^2}^2}.$$

As long as $d_1 \|\partial_x \phi_0\|_{L^2}^2 > d_3 \|\phi_0|_{L^4}^4$ and the discriminant under the square root is positive, then there are two persistent equilibria that are distinguished by their amplitudes, $\omega_+ > \omega_- > 0$. The linearization, $\mathcal{J}\mathcal{L}_0$, of NLS about the critical point Φ_ω takes the form

$$\mathcal{L}_0 = \text{diag}(\mathcal{L}_+, \mathcal{L}_-); \quad \mathcal{L}_+ = -\partial_x^2 + \omega - 3\phi_\omega^2, \quad \mathcal{L}_- = -\partial_x^2 + \omega - \phi_\omega^2.$$

From the expression (7.2.32) we determine that the first-order correction to the wave is given by

$$\Phi_1 = \mathcal{L}_0^{-1}\left(\mathcal{J}^{-1}\mathcal{N}(\phi_c)\right) = \begin{pmatrix} 0 \\ \phi_1 \end{pmatrix}, \ \phi_1 := \mathcal{L}_-^{-1}\left(d_1 \partial_x^2 \phi_\omega + d_2 \phi_\omega + d_3 \phi_\omega^3 + d_4 \phi_\omega^5\right).$$

To invoke Lemma 7.2.10 and locate the small eigenvalues arising from the non-Hamiltonian perturbation we must evaluate the constraint matrix $\boldsymbol{D}$ and the matrix $\boldsymbol{\Gamma}$. The constraint matrix for NLS was computed in Chapter 7.2.1.3, where it was found that $\boldsymbol{D} = \text{diag}(\boldsymbol{D}_+, \boldsymbol{D}_-)$, with $\boldsymbol{D}_+ < 0$ and $\boldsymbol{D}_- > 0$. To compute $\boldsymbol{\Gamma}$ we recall from (5.2.66) that the generalized eigenfunctions have the form

$$\partial_{c_1} \Phi_c = \begin{pmatrix} 0 \\ x\phi_\omega/2 \end{pmatrix}, \quad \partial_{c_2} \Phi_c = \begin{pmatrix} -\partial_\omega \phi_\omega \\ 0 \end{pmatrix}.$$

It is straightforward to verify that the perturbed operator $\mathcal{L}_1$ takes the diagonal form $\mathcal{L}_1 = \text{diag}(\mathcal{L}_1^+, \mathcal{L}_1^-)$, where

$$\mathcal{L}_1^+ = -2\phi_\omega \phi_1 + d_1 \partial_x^2 + d_2 + 3 d_3 \phi_\omega^2 + 5 d_4 \phi_\omega^4$$
$$\mathcal{L}_1^- = 2\phi_\omega \phi_1 + d_1 \partial_x^2 + d_2 + d_3 \phi_\omega^2 + d_4 \phi_\omega^4.$$

The suboperators $\mathcal{L}_1^\pm$ preserve even/odd parity, and hence Φ_1 is even about $x = 0$. Moreover, $T_1'\Phi_1$ is odd about $x = 0$, and $T_2'\Phi_1$ is even about $x = 0$. Examining the formula (7.2.39) for $\boldsymbol{\Gamma}_{ij}$ we see that $\boldsymbol{\Gamma}_{12} = \boldsymbol{\Gamma}_{21} = 0$, while the diagonal entries take the form

$$\mathbf{\Gamma}_{11} = \langle -\partial_x \phi_1 + \mathcal{L}_1^-(x\phi_\omega/2), \partial_x \phi_\omega \rangle, \quad \mathbf{\Gamma}_{22} = \langle -\phi_1 + \mathcal{L}_1^+(\partial_\omega \phi_\omega), \phi_\omega \rangle.$$

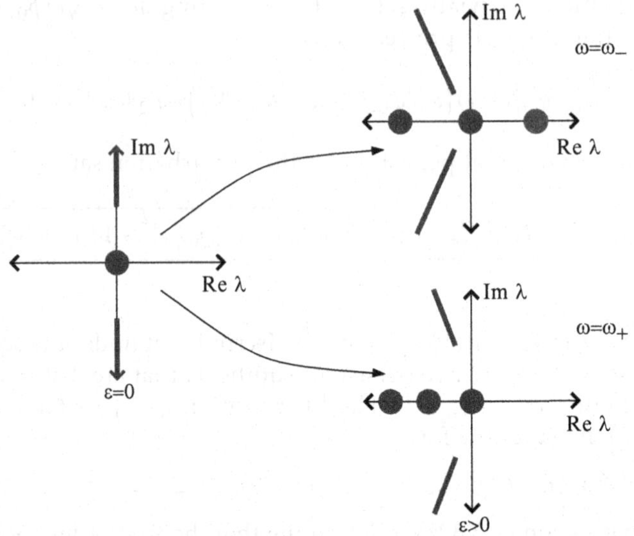

Fig. 7.3 The evolution of the essential spectrum and kernel of NLS operator $\mathcal{JL}$ in (7.2.33) under the Ginzburg–Landau perturbation in (7.2.44). The essential spectrum is marked by thick (*blue*) lines, and the small eigenvalues near the origin are marked by (*blue* for stable, *red* for unstable) filled circles. The left panel depicts the spectrum for the unperturbed problem. The two right panels depict the spectrum for the two persistent solutions associated with $\omega_\pm$. The perturbation moves the essential spectrum into the left-half complex plane, and breaks the four-dimensional generalized kernel into a two-dimensional kernel and two small eigenvalues. (Color figure online.)

From Lemma 7.2.10 we see that the two small nonzero eigenvalues satisfy

$$\text{sign}(\lambda_{1,1}) = -\text{sign}(\mathbf{\Gamma}_{11}), \quad \text{sign}(\lambda_{1,2}) = \text{sign}(\mathbf{\Gamma}_{22}).$$

Although we will not perform the calculation here, it was demonstrated in [**150**, Section 2] that

$$\text{sign}(\mathbf{\Gamma}_{11}) = \begin{cases} -1, & \omega = \omega_- \\ +1, & \omega = \omega_+, \end{cases} \quad \text{sign}(\mathbf{\Gamma}_{22}) = -\text{sign}(d_1) < 0. \tag{7.2.45}$$

We conclude that the persistent equilibria Φ_{ω_-} is spectrally unstable, with one positive and one negative real eigenvalue of $\mathcal{O}(\epsilon)$, while the equilibria Φ_{ω_+} has both $\mathcal{O}(\epsilon)$ eigenvalues real-valued and negative (see Fig. 7.3). Since the essential spectrum is strictly in the left-half complex plane, we would like to conclude that Φ_{ω_+} is asymptotically orbitally stable. However, before this conclusion can be reached, it must be demonstrated that the perturbation does not cause eigenvalues to be ejected from the essential spectrum, particularly from the branch point $\lambda = \mathrm{i}\omega$, which lies on the imaginary axis for $\epsilon = 0$. Such an ejected eigenvalue could well move into the right-half complex plane, even as the essential spectrum retreats into the left-half. The study of eigenvalues being ejected from the branch point of the essential spectrum, known as an *edge bifurcation*, is a particular focus of Chapter 9.

Exercises

Exercise 7.2.1. Prove Lemma 7.2.2. *Hint*: This is the implicit function theorem, in the guise of an equivariant Lyapunov–Schmidt reduction, see [**54**].

Exercise 7.2.2. Consider the eigenvalue problem (7.2.16). Show that if $\lambda = \mathrm{i}\sqrt{\epsilon}\,\lambda_1 + \mathcal{O}(\epsilon) \in \mathrm{i}\mathbb{R}$ is a simple eigenvalue, so that $\lambda_1 = \mathrm{i}\lambda_1$ in (7.2.19), then the Krein index of the eigenvalue is given by

$$\operatorname{sign}(\boldsymbol{v} \cdot \boldsymbol{L}_1 \boldsymbol{v}) = \operatorname{sign}(\boldsymbol{v} \cdot \boldsymbol{D}\boldsymbol{v}),$$

where $\boldsymbol{v} \in \mathbb{C}^N$ is the eigenvector in (7.2.19) associated with λ_1.

Exercise 7.2.3. For the eigenvalue problem associated with Chapter 7.2.1.3, use Lemma 7.2.8 to calculate the leading order corrections to the $\mathcal{O}(\sqrt{\epsilon})$ eigenvalues, and determine the relationship between their location (real, imaginary, complex) and the nature of the energy curves $\mathcal{H}_{\mathrm{trans}}(\gamma_1)$ and $\mathcal{H}_{\mathrm{rot}}(\gamma_2)$. If the perturbed critical point corresponds to a saddle point of the reduced Hamiltonian, what condition determines the spectral stability?

Exercise 7.2.4. Consider the perturbed NLS of the form

$$\mathrm{i}\partial_t u + \partial_x^2 u - \omega u + |u|^2 u = \epsilon V(x) u,$$

where the potential $V(x)$ is the symmetric sum of delta functions, i.e.,

$$V(x) = \delta(x + L) + \delta(x - L).$$

Suppose that $L \gg 1$, i.e., the potential wells are well-separated. Show that the persistent waves are given by $\phi_\omega(x+\gamma)+\mathcal{O}(\epsilon)$, where the pulse is centered either on a well, or symmetrically between the two wells. That is, $\gamma \in \{\pm L, 0\}$. Determine the spectral stability of each of the perturbed waves.

Exercise 7.2.5. Consider the linearly coupled NLS of the form

$$\begin{aligned}
\mathrm{i}\partial_t u_1 + \partial_x^2 u_1 - \omega u_1 + |u_1|^2 u_1 &= \epsilon u_2 \\
\mathrm{i}\partial_t u_2 + \partial_x^2 u_2 - \omega u_2 + |u_2|^2 u_2 &= \epsilon u_1.
\end{aligned}$$

This problem is discussed in some detail in [**147**].

(a) Determine the Hamiltonian for the unperturbed problem, and show that the $\mathcal{O}(\epsilon)$ correction for the perturbed problem is given by

$$\mathcal{H}_1(\boldsymbol{u}) = \int_{\mathbb{R}} (u_1\overline{u}_2 + \overline{u}_1 u_2)\,\mathrm{d}x.$$

(b) Determine the four symmetries for $\epsilon = 0$, and the two remaining symmetries for $\epsilon > 0$.
(c) Modulo symmetries, one persistent critical point is given at $\epsilon = 0$ by

$$\phi_\omega(x) = \sqrt{2\omega}\,\mathrm{sech}(\sqrt{\omega}x)\begin{pmatrix}1\\1\end{pmatrix}.$$

Find all of the perturbations of ϕ_ω that persist for $\epsilon > 0$.
(d) Consider the perturbed waves found in (c). It is the case that $\mathrm{m_a}(0) = 8$ when $\epsilon = 0$, and $\mathrm{m_a}(0) = 4$ for $\epsilon > 0$. Locate the four $\mathcal{O}(\epsilon^{1/2})$ eigenvalues, and if any of these are purely imaginary, determine their Krein signature.

Exercise 7.2.6. Verify the sign of $\boldsymbol{\Gamma}_{11}, \boldsymbol{\Gamma}_{22}$ in (7.2.45).

Exercise 7.2.7. For the problem discussed in Chapter 7.2.2.1 compute $\sigma_{\mathrm{ess}}(\mathcal{L})$ for $\epsilon > 0$, and show that Fig. 7.3 is a faithful representation of that curve. In particular, verify that the essential spectrum for the linearization about Φ_{ω_-} is closer to the origin than that of Φ_{ω_+}.

Exercise 7.2.8. For the problem discussed in Chapter 7.2.2.1 it can be shown that if $d_2 = \mathcal{O}(1)$, then any eigenvalues ejected from the essential spectrum remain in the left-half plane, and the pulse Φ_{ω_+} is spectrally stable for $\epsilon > 0$ sufficiently small. Use Theorem 4.3.5 to prove in this case that the manifold $\mathcal{M}_T(\Phi_{\omega_+})$ is asymptotically orbitally stable.

7.3 Additional Reading

A beautiful overview of the spectral theory for Hamiltonian eigenvalue problems is given by Kollár and Miller [177].

The Hamiltonian–Krein index has also been developed for quadratic Hermitian pencils, which are nonlinear eigenvalue problems of the form $(\lambda^2\mathcal{M}_2 + \lambda\mathcal{M}_1 + \mathcal{M}_0)v = 0$; examples of this theory can be found in Bronski et al. [44], Chugunova and Pelinovsky [55], Kapitula et al. [161], Kollár [176]. The index has recently been extended to problems where the skew operator does not have a bounded inverse by Kapitula and Stefanov [155], Pelinovsky [224].

Spectral calculations for discrete Hamiltonian systems can be found in Kevrekidis and Pelinovsky [170], Lukas et al. [194], Pelinovsky and Kevrekidis [226], Pelinovsky and Sakovich [228], Pelinovsky et al. [232, 233].

For examples of applications of Hamiltonian index theory to the analysis of spectra of linear operators, the interested reader can consult Jackson and Weinstein [**130**], Johansson and Kivshar [**132**], Kivshar et al. [**174**], Ostrovskaya et al. [**214**], Pelinovsky and Kivshar [**227**], Pelinovsky and Yang [**231**], Skryabin [**264**].

Chapter 8
The Evans Function for Boundary-Value Problems

Previously we gathered information about a point spectrum either perturbatively, as in Chapter 6, or in cases where the linear operator has special structure, as arises from symmetries (Chapter 4.2) and in Hamiltonian systems (Chapter 7). We would like a more general tool that allows us to make nonperturbative statements about point spectra, or to investigate point spectra emerging from the absolute spectra of a linearized operator.

The remainder of this book focuses on the development of the *Evans function*, a complex-valued function of the spectral parameter, akin to the characteristic polynomial for matrices, with the property that its zeros correspond to a point spectrum. While the framework we present for the Evans function is motivated by dynamical systems, the function itself is closely related to the transmission coefficient developed within the mathematical physics community. We start with the conceptually and algebraically simplest problem: Sturm-Liouville operators with separated boundary conditions on a bounded domain. The associated eigenvalue problem is understood (see Chapter 2.3) up to the ordering of the eigenvalues and the number of nodal values of the associated eigenfunction. The Evans function gives precise information about the location of the eigenvalues, but more significantly, it places the eigenvalue problem in a natural context that renders transparent its extension to the real line.

8.1 Sturm–Liouville Operators

We construct and study the Evans function for the Sturm–Liouville problem

$$\mathcal{L}p := \partial_x^2 p + a_1(x)\partial_x p + a_0(x)p = \lambda p, \tag{8.1.1}$$

on the interval $[-1,1]$ with the separated boundary conditions of the form

T. Kapitula and K. Promislow, *Spectral and Dynamical Stability of Nonlinear Waves*, Applied Mathematical Sciences 185, DOI 10.1007/978-1-4614-6995-7_8,

$$(b_1^-, b_2^-)\begin{pmatrix} p \\ \partial_x p \end{pmatrix}(-1) = 0, \quad (b_1^+, b_2^+)\begin{pmatrix} p \\ \partial_x p \end{pmatrix}(+1) = 0, \tag{8.1.2}$$

where the boundary condition vectors $\boldsymbol{b}^{\mathrm{a}\pm} := (b_1^\pm, b_2^\pm)$ in (8.1.2) have norm one while the spatial coefficients $a_1(x)$ and $a_0(x)$ are assumed to be piecewise smooth and real-valued. Chapter 2.3 shows that this problem has a countable number of simple, real-valued eigenvalues. Our goal is to construct the Evans function, $E(\lambda)$, which is entire in the spectral parameter λ and has zeros that coincide with eigenvalues up to multiplicity.

The construction of the Evans function has a geometric motivation, which has much in common with the analysis of the spectrum of the exponentially asymptotic linear operators of Chapter 3. We rewrite the eigenvalue problem as a dynamical system, and look at the eigenvalue problem not as an existence problem, but as an intersection problem for distinguished subspaces associated with the dynamical system. This geometric perspective not only allows a simple generalization to the unbounded domain and to higher-order systems, but it affords insight that is not easily motivated by the classical formulation of the eigenvalue problem.

Set $\boldsymbol{Y} = (p, \partial_x p)^{\mathrm{T}}$, and write (8.1.1) as the first-order system

$$\partial_x \boldsymbol{Y} = \boldsymbol{A}(x,\lambda)\boldsymbol{Y}, \quad \boldsymbol{A}(x,\lambda) := \begin{pmatrix} 0 & 1 \\ \lambda - a_0(x) & -a_1(x) \end{pmatrix}. \tag{8.1.3}$$

It is convenient to introduce

$$\boldsymbol{b}^\pm := (-b_2^\pm, b_1^\pm)^{\mathrm{T}},$$

which are perpendicular to the respective boundary vectors $\boldsymbol{b}^{\mathrm{a}\pm}$ and the boundary spaces $\mathbb{B}_{\pm 1} = \mathrm{span}\{\boldsymbol{b}^\pm\}$. Solutions $\boldsymbol{Y}(x,\lambda)$ to (8.1.3) that correspond to eigenfunctions for the eigenvalue problem satisfy the boundary conditions

$$\boldsymbol{Y}(-1,\lambda) \in \mathbb{B}_{-1}, \quad \boldsymbol{Y}(+1,\lambda) \in \mathbb{B}_{+1}. \tag{8.1.4}$$

Our first step towards the Evans function is the construction of a basis for the space of solutions (8.1.3) which satisfy the left and right boundary conditions, respectively. Such solutions are defined up to a scalar normalization. As we shall see in the sequel, there are several natural choices for the scalar normalization that are convenient in different settings. At the moment, we fix the normalization, defining

$$\boldsymbol{J}^\pm(x,\lambda) := \boldsymbol{\Phi}(x,\lambda)\boldsymbol{\Phi}(\pm 1,\lambda)^{-1}\boldsymbol{b}^\pm. \tag{8.1.5}$$

We call the $\boldsymbol{J}^\pm$ *Jost solutions*, in reference to the eponymous functions that play a fundamental role in scattering theory on the line; see Chapter 9.3.2 for a discussion of this connection. For the bounded domain problem the Jost solutions satisfy the initial conditions

$$J^{\pm}(\pm 1, \lambda) = b^{\pm} \quad \Rightarrow \quad J^{\pm}(\pm 1, \lambda) \in \mathbb{B}_{\pm 1}.$$

The Jost solutions provide a map of the vectors $b^{\pm}$ from $x = \pm 1$ to $x = 0$ under the action of the flow generated by (8.1.3) (see Fig.8.1).

The significance of the Jost solutions lies in their linear independence, or lack thereof. As J^{-} and J^{+} span the space of solutions of (8.1.3), which satisfy, respectively, the left and the right boundary conditions, $\lambda = \lambda_0$ can be an eigenvalue of (8.1.1) if and only if the vectors $\{J^{-}(0, \lambda_0), J^{+}(0, \lambda_0)\}$ are linearly dependent. This follows since the associated solution $Y(\cdot, \lambda_0)$ of (8.1.3) must satisfy both boundary conditions, and hence $Y(0, \lambda_0)$ must be a multiple of both $J^{\pm}$ at $x = 0$. That is, at an eigenvalue, the two Jost solutions coincide up to a scalar multiple, and we denote the appropriately normalized vector solution the *Jost eigenfunction*. These observations motivate the following definition:

Definition 8.1.1. The Evans matrix $E(\lambda) \in \mathbb{C}^{2\times 2}$ has columns $J^{\pm}(0, \lambda)$:

$$E(\lambda) := \left(J^{-}, J^{+}\right)(0, \lambda), \tag{8.1.6}$$

and the Evans function $E(\lambda)$ is its determinant

$$E(\lambda) = \det E(\lambda), \tag{8.1.7}$$

where the Jost solutions $J^{\pm}(x, \lambda)$ are given in (8.1.5).

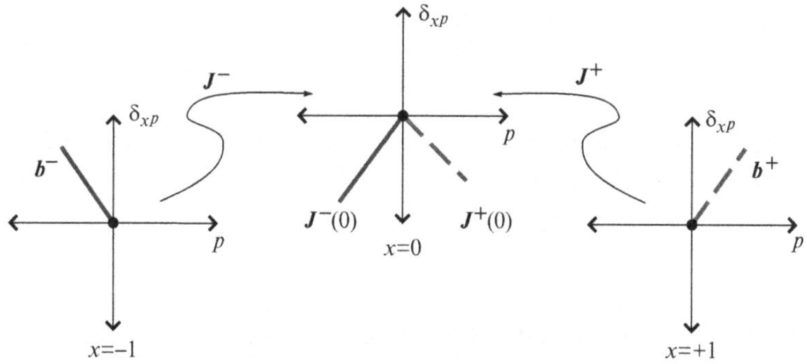

Fig. 8.1 The map induced by the Jost solutions, under the formulation (8.1.5). In this depiction, $J^{\pm}(0, \lambda_0)$ are linearly independent, and $E(\lambda_0) \neq 0$. (Color figure online.)

Remark 8.1.2. The Evans function is a particular Wronskian of the underlying ordinary differential equation (ODE). Its first use in this context can be found in the work of Evans and his analysis of the spectral problem associated with traveling-wave solutions of the Hodgkin–Huxley system; Evans [**77**, **78**, **79**, **80**]. The term *Evans function* was first used in this context by

Jones in his seminal work [134], which addresses the spectral stability of pulse solutions to the singularly perturbed Fitzhugh–Nagumo system.

Calculations with the Evans function require an understanding of the adjoint system. We define the adjoint eigenvalue problem to (8.1.1) as

$$\mathcal{L}^{\mathrm{a}} q := \partial_x^2 q - \partial_x(a_1(x)q) + a_0(x)q = \overline{\lambda} q, \tag{8.1.8}$$

subject to the adjoint boundary conditions

$$(\boldsymbol{b}^{\pm})^{\mathrm{T}} \begin{pmatrix} a_1(\pm 1)q(\pm 1) - \partial_x q(\pm 1) \\ q(\pm 1) \end{pmatrix} = 0, \tag{8.1.9}$$

To exploit the structure relating the eigenvalue problem to its adjoint we vectorize the adjoint eigenvalue problem not through the map $(q, q_x)^T$ but rather through the transformation

$$\boldsymbol{Z} := \begin{pmatrix} a_1(x)q(x) - \partial_x q(x) \\ q(x) \end{pmatrix}. \tag{8.1.10}$$

We also introduce the following notation. For a matrix $\boldsymbol{B} \in \mathbb{C}^{m \times n}$ we denote its Hermitian, or complex-conjugate transpose, by $\boldsymbol{B}^{\mathrm{H}}$, while $\boldsymbol{B}^{-\mathrm{H}}$ denotes the Hermitian of $\boldsymbol{B}^{-1}$. For vectors $\boldsymbol{v}, \boldsymbol{w} \in \mathbb{C}^n$ we define the complex inner product

$$\boldsymbol{v} \cdot \boldsymbol{w} := \boldsymbol{w}^{\mathrm{H}} \boldsymbol{v},$$

which induces the Hermitian as its adjoint

$$\boldsymbol{v} \cdot (\boldsymbol{B}\boldsymbol{w}) = (\boldsymbol{B}\boldsymbol{w})^{\mathrm{H}} \boldsymbol{v} = \boldsymbol{w}^{\mathrm{H}} \boldsymbol{B}^{\mathrm{H}} \boldsymbol{w} = (\boldsymbol{B}^{\mathrm{H}} \boldsymbol{v}) \cdot \boldsymbol{w}.$$

With this notation the choice (8.1.10) for the vectorization yields a vector version of the adjoint eigenvalue problem (8.1.8) with the appealing form

$$\partial_x \boldsymbol{Z} = -\boldsymbol{A}(x, \lambda)^{\mathrm{H}} \boldsymbol{Z}, \tag{8.1.11}$$

subject to the adjoint boundary conditions $\boldsymbol{Z}(\pm 1) \in \mathbb{B}^{\mathrm{a}}_{\pm 1}$, where, consistent with (8.1.9), we define the adjoint boundary spaces

$$\mathbb{B}^{\mathrm{a}}_{\pm 1} := \mathrm{span}\{\boldsymbol{b}^{\mathrm{a}\pm}\}. \tag{8.1.12}$$

The Jost solutions for the adjoint problem, $\boldsymbol{J}^{\mathrm{a}\pm}$, which we call the *adjoint Jost solutions*, span the solution space of (8.1.11) subject to $\mathbb{B}^{\mathrm{a}}_{\pm 1}$. For the moment we impose the normalization $\boldsymbol{J}^{\mathrm{a}\pm}(\pm 1) = \boldsymbol{b}^{\mathrm{a}\pm}$, although other scalar normalizations will also be considered in the sequel.

The utility of the adjoint problem is that it gives a convenient formulation for the inverse of the fundamental matrix solution (FMS) for the original problem.

Lemma 8.1.3. *If $\boldsymbol{\Phi} = (\boldsymbol{Y}_1, \boldsymbol{Y}_2)$ is a FMS for the system (8.1.3), then its inverse is the Hermitian of a FMS for the adjoint problem. Specifically there exists $\boldsymbol{B} \in \mathbb{C}^{2\times 2}$ which is constant in x and invertible such that*

$$\boldsymbol{\Phi}(x)^{-1} = \boldsymbol{B}\,(\boldsymbol{\Phi}^{\mathrm{a}}(x))^{\mathrm{H}}, \tag{8.1.13}$$

where $\boldsymbol{\Phi}^{\mathrm{a}} = (\boldsymbol{Z}_1, \boldsymbol{Z}_2)$ is a FMS for the adjoint problem. In particular, if $\boldsymbol{\Phi}$ and $\boldsymbol{\Phi}^{\mathrm{a}}$ are principal FMSs at $x = x_0$, then $\boldsymbol{B} = \boldsymbol{I}_2$.

Proof. The key observation is that if $\boldsymbol{Y}$ solves (8.1.3) and $\boldsymbol{Z}$ solves the adjoint problem, then their inner product is constant in x,

$$\partial_x(\boldsymbol{Y}\cdot\boldsymbol{Z}) = (\partial_x\boldsymbol{Y})\cdot\boldsymbol{Z} + \boldsymbol{Y}\cdot(\partial_x\boldsymbol{Z}) = (\boldsymbol{A}\boldsymbol{Y})\cdot\boldsymbol{Z} - \boldsymbol{Y}\cdot(\boldsymbol{A}^{\mathrm{H}}\boldsymbol{Z}) = 0.$$

Since for the two FMSs $\boldsymbol{\Phi}$ and $\boldsymbol{\Phi}^{\mathrm{a}}$ we have

$$\left[(\boldsymbol{\Phi}^{\mathrm{a}})^{\mathrm{H}}\boldsymbol{\Phi}\right]_{ij} = \boldsymbol{Z}_i^{\mathrm{H}}\boldsymbol{Y}_j = \boldsymbol{Y}_j\cdot\boldsymbol{Z}_i, \tag{8.1.14}$$

we deduce that

$$\partial_x\left((\boldsymbol{\Phi}^{\mathrm{a}})^{\mathrm{H}}\boldsymbol{\Phi}\right) = 0 \quad \Rightarrow \quad (\boldsymbol{\Phi}^{\mathrm{a}})^{\mathrm{H}}\boldsymbol{\Phi} = \boldsymbol{B}.$$

Since both FMSs are invertible, $\boldsymbol{B}$ is constant and invertible, and the identity (8.1.13) follows. □

The following Proposition establishes key properties of the Evans function.

Proposition 8.1.4. *The Evans function associated with the Sturm-Liouville problem (8.1.1) with boundary conditions (8.1.2) is entire. Furthermore, $\lambda = \lambda_0$ is a simple eigenvalue of $\mathcal{L}$ if and only if $\dim[\ker(\boldsymbol{E}(\lambda_0))] = 1$. In this case*

$$E(\lambda_0) = 0, \quad E'(\lambda_0) = \langle p_0, q_0\rangle \det(\boldsymbol{J}_0^{\mathrm{a}}, \boldsymbol{J}_0)(0) \neq 0. \tag{8.1.15}$$

Here p_0 is the eigenfunction, q_0 is the adjoint eigenfunction, and the Jost and adjoint Jost eigenfunctions are given by

$$\boldsymbol{J}_0(x) = \begin{pmatrix} p_0(x) \\ \partial_x p_0(x) \end{pmatrix}, \quad \boldsymbol{J}_0^{\mathrm{a}}(x) = \begin{pmatrix} a_1(x)q_0(x) - \partial_x q_0(x) \\ q_0(x) \end{pmatrix}. \tag{8.1.16}$$

Moreover, the adjoint eigenfunction q_0 has been is scaled so that $|\boldsymbol{J}_0^{\mathrm{a}}(0)| = 1$.

Proof. Since $\boldsymbol{A}(x,\lambda)$ is entire in λ for fixed x, so is the fundamental matrix solution, and hence the Jost solutions are entire in λ for fixed x (see Lemma 2.1.4). Since the Evans function is an algebraic combination of entire functions, it is entire. From Exercise 8.1.1 we see that $E(\lambda_0) = 0$ if and only if λ_0 is an eigenvalue and $\dim[\ker(\boldsymbol{E}(\lambda_0))] = 1$. We only need to calculate the derivative at an eigenvalue, λ_0. Since $E(\lambda_0) = 0$, the Jost solutions are linearly dependent, satisfying $\boldsymbol{J}^-(0,\lambda_0) = c\boldsymbol{J}^+(0,\lambda_0)$ for some $c \in \mathbb{C}$. Without loss of generality, we may rescale the Jost solutions so that $c = 1$, which implies

$$\boldsymbol{J}^-(x,\lambda_0) = \boldsymbol{J}^+(x,\lambda_0) \equiv \boldsymbol{J}_0(x), \quad \boldsymbol{J}_0(x) := \begin{pmatrix} p_0(x) \\ \partial_x p_0(x) \end{pmatrix}, \tag{8.1.17}$$

where p_0 is the eigenfunction of $\mathcal{L}$ associated with λ_0, and $\boldsymbol{J}_0$ is its vectorized form. We have that

$$\begin{aligned} E'(\lambda_0) &= \det(\partial_\lambda \boldsymbol{J}^-, \boldsymbol{J}^+)(0,\lambda_0) + \det(\boldsymbol{J}^-, \partial_\lambda \boldsymbol{J}^+)(0,\lambda_0) \\ &= \det(\partial_\lambda(\boldsymbol{J}^- - \boldsymbol{J}^+)(0,\lambda_0), \boldsymbol{J}_0(0)). \end{aligned} \tag{8.1.18}$$

To evaluate the derivative we must determine $\partial_\lambda \boldsymbol{J}^\pm(0,\lambda_0)$. The Jost solutions satisfy the ODE (8.1.3): differentiating it with respect to λ, and noting that the boundary conditions are λ-independent yields

$$\partial_x(\partial_\lambda \boldsymbol{J}^\pm) = \boldsymbol{A}(x,\lambda_0)\partial_\lambda \boldsymbol{J}^\pm + \partial_\lambda \boldsymbol{A}(x,\lambda_0)\boldsymbol{J}_0, \quad \partial_\lambda \boldsymbol{J}^\pm(\pm 1,\lambda_0) = \boldsymbol{0}.$$

Variation of parameters yields the two solutions

$$\partial_\lambda \boldsymbol{J}^\pm(x,\lambda_0) = \boldsymbol{\Phi}(x,\lambda_0)\int_{\pm 1}^{x} \boldsymbol{\Phi}(s,\lambda_0)^{-1}\partial_\lambda \boldsymbol{A}(s,\lambda_0)\boldsymbol{J}_0(s)\,\mathrm{d}s,$$

so that the difference, evaluated at $x = 0$, takes the form

$$\partial_\lambda(\boldsymbol{J}^- - \boldsymbol{J}^+)(0,\lambda_0) = \boldsymbol{\Phi}(0,\lambda_0)\int_{-1}^{+1} \boldsymbol{\Phi}(s,\lambda_0)^{-1}\partial_\lambda \boldsymbol{A}(s,\lambda_0)\boldsymbol{J}_0(s)\,\mathrm{d}s. \tag{8.1.19}$$

Using the form of $\boldsymbol{J}_0$ and $\boldsymbol{A}$ given in (8.1.17) and (8.1.3), respectively, we rewrite (8.1.19) as

$$\partial_\lambda(\boldsymbol{J}^- - \boldsymbol{J}^+)(0,\lambda_0) = \boldsymbol{\Phi}(0,\lambda_0)\int_{-1}^{+1} \boldsymbol{\Phi}(s,\lambda_0)^{-1}\begin{pmatrix} 0 \\ p_0(s) \end{pmatrix}\mathrm{d}s. \tag{8.1.20}$$

To simplify the expression (8.1.20) we first choose the FMS in the form $\boldsymbol{\Phi} = (\boldsymbol{Y}_1, \boldsymbol{J}_0)$ where $\boldsymbol{J}_0$ is generated by the eigenfunction and $\boldsymbol{Y}_1$ is any other linearly independent solution. As a consequence of Lemma 8.1.3 we know that the rows of $\boldsymbol{\Phi}^{-1}$ correspond to ODE adjoint solutions: write $\boldsymbol{\Phi}^{-1} = (\boldsymbol{Z}_1, \boldsymbol{Z}_2)^{\mathrm{H}}$, where each $\boldsymbol{Z}_i$ solves the adjoint ODE. Since $\boldsymbol{\Phi}^{-1}\boldsymbol{\Phi} = \boldsymbol{I}_2$ we know that

$$\boldsymbol{Z}_1 \cdot \boldsymbol{Y}_1 = 1,\ \boldsymbol{Z}_1 \cdot \boldsymbol{J}_0 = 0; \quad \boldsymbol{Z}_2 \cdot \boldsymbol{Y}_1 = 0,\ \boldsymbol{Z}_2 \cdot \boldsymbol{J}_0 = 1.$$

The choices for $\boldsymbol{Z}_1$ are limited; it solves the adjoint ODE and is orthogonal to $\boldsymbol{J}_0$ for all x. Since the boundary spaces, $\mathbb{B}_{\pm 1}$, and adjoint boundary spaces, $\mathbb{B}^{\mathrm{a}}_{\pm 1}$, are orthogonal, and since $\boldsymbol{J}_0(\pm 1) \in \mathbb{B}_{\pm 1}$, it follows that $\boldsymbol{Z}_1(\pm 1) \in \mathbb{B}^{\mathrm{a}}_{\pm 1}$. In particular, $\boldsymbol{Z}_1$ is an adjoint eigenfunction and up to normalization satisfies $\boldsymbol{Z}_1 = \alpha \boldsymbol{J}_0^{\mathrm{a}}$, where $\boldsymbol{J}_0^{\mathrm{a}}$ is given in (8.1.16). For $\boldsymbol{Y}_1$, we decompose its value at $x = 0$ as $\boldsymbol{Y}_1(0) = c_1^{\mathrm{a}} \boldsymbol{J}_0^{\mathrm{a}}(0) + c_2^{\mathrm{a}} \boldsymbol{J}_0(0)$. Since $\boldsymbol{Z}_1 = \alpha \boldsymbol{J}_0^{\mathrm{a}}$ and $\boldsymbol{J}_0^{\mathrm{a}} \cdot \boldsymbol{J}_0 = 0$, the first orthogonality condition implies that

$$(c_1^{\mathrm{a}})^{-1} = \alpha \boldsymbol{J}_0^{\mathrm{a}}(0) \cdot \boldsymbol{J}_0^{\mathrm{a}}(0) = \alpha,$$

and we are free to choose $c_1^{\mathrm{a}} = \alpha = 1$, and $c_2^{\mathrm{a}} = 0$ (the second equality above follows from the assumed normalization $|\boldsymbol{J}_0^{\mathrm{a}}(0)| = 1$). This yields the FMS $\boldsymbol{\Phi}$ with initial value $\boldsymbol{\Phi}(0, \lambda_0) = \left(\boldsymbol{J}_0^{\mathrm{a}}, \boldsymbol{J}_0\right)(0)$, for which $\boldsymbol{\Phi}^{-1} = (\boldsymbol{J}_0^{\mathrm{a}}, \boldsymbol{Z}_2)^{\mathrm{H}}$, where the form of the second row, $\boldsymbol{Z}_2^{\mathrm{H}}$, will be immaterial.

For any $\boldsymbol{f} \in \mathbb{C}^2$ we have

$$\boldsymbol{\Phi}(x, \lambda_0)^{-1} \boldsymbol{f}(x) = \begin{pmatrix} (\boldsymbol{J}_0^{\mathrm{a}})^{\mathrm{H}}(x) \\ \boldsymbol{Z}_2^{\mathrm{H}}(x) \end{pmatrix} \boldsymbol{f}(x) = \begin{pmatrix} \boldsymbol{f}(x) \cdot \boldsymbol{J}_0^{\mathrm{a}}(x) \\ \boldsymbol{f}(x) \cdot \boldsymbol{Z}_2(x, \lambda_0) \end{pmatrix}.$$

Using the form of $\boldsymbol{J}_0^{\mathrm{a}}$ from (8.1.16), the integrand of (8.1.20) reduces to

$$\boldsymbol{\Phi}(s, \lambda_0)^{-1} \begin{pmatrix} 0 \\ p_0(s) \end{pmatrix} = \begin{pmatrix} p_0(s)\overline{q_0(s)} \\ \tilde{c}(s) \end{pmatrix},$$

where the function $\tilde{c}(s)$ is immaterial. Consequently, (8.1.20) becomes

$$\partial_\lambda (\boldsymbol{J}^- - \boldsymbol{J}^+)(0, \lambda_0) = \left(\boldsymbol{J}_0^{\mathrm{a}}(0), \boldsymbol{J}_0(0)\right) \int_{-1}^{+1} \begin{pmatrix} p_0(s)\overline{q_0(s)} \\ \tilde{c}(s) \end{pmatrix} \mathrm{d}s = \langle p_0, q_0 \rangle \boldsymbol{J}_0^{\mathrm{a}}(0) + c \boldsymbol{J}_0(0).$$

Plugging this expression and (8.1.8) into (8.1.18) yields

$$E'(\lambda_0) = \langle p_0, q_0 \rangle \det\left(\boldsymbol{J}_0^{\mathrm{a}}, \boldsymbol{J}_0\right)(0) + c \det\left(\boldsymbol{J}_0, \boldsymbol{J}_0\right)(0). \tag{8.1.21}$$

As $\det\left(\boldsymbol{J}_0, \boldsymbol{J}_0\right)(0) = 0$ the second term drops out and we obtain (8.1.15).

We conclude by demonstrating that if the eigenvalue is simple, then $E'(\lambda_0) \neq 0$. By its construction we know that $\det\left(\boldsymbol{J}_0^{\mathrm{a}}, \boldsymbol{J}_0\right)(0) = \det \boldsymbol{\Phi}(0, \lambda_0) \neq 0$; thus, if the derivative is zero it must that $\langle p_0, q_0 \rangle = 0$. From Exercise 8.1.1 we know that $\mathrm{m_g}(\lambda_0) = 1$, and hence $\mathrm{m_a}(\lambda_0) \geq 2$ if and only if there is a generalized eigenfunction p_1, which solves

$$(\mathcal{L} - \lambda_0) p_1 = p_0.$$

By the Fredholm alternative

$$(\mathcal{L} - \lambda_0) p_1 = p_0 \text{ is solvable} \quad \Leftrightarrow \quad p_0 \in \ker(\mathcal{L}^{\mathrm{a}} - \overline{\lambda_0})^{\perp} = \mathrm{span}\{q_0\}^{\perp},$$

thus $\mathrm{m_a}(\lambda_0) \geq 2$ if and only if $\langle p_0, q_0 \rangle = 0$. In particular, if λ_0 is simple, then $\langle p_0, q_0 \rangle \neq 0$ and $E'(\lambda_0) \neq 0$. □

Remark 8.1.5. In Proposition 8.1.4 the adjoint Jost solution is scaled to have length one at $x = 0$. For a more general choice of the adjoint eigenfunction the derivative formula in (8.1.15) is divided by $|J_0^{\mathrm{a}}(0)|^2 > 0$. Since the Evans function is only defined up to a multiplicative constant, this normalization has little impact on the applications that follow.

Remark 8.1.6. In general, the Jost solutions, and hence the Evans function, are only defined up to multiplicative constants. Differing normalizations of the Jost solutions produce an Evans function with the same zeros, including multiplicity. Typically, the location of the zeros is the relevant information. However, in some applications, such as the orientation index (see Chapter 9.4), the sign of the Evans function is important. Indeed, in this application the ratio of the sign of the Evans function for λ near zero and near infinity are determined, and the multiplicative constant is unimportant, although care must be taken to insure that the normalizations at the two values of λ do not induce a sign change.

Example 8.1.7. Consider (8.1.1) with $a_1(x) = a_0(x) \equiv 0$. The principal fundamental matrix solution at $x = 0$ is given by

$$\boldsymbol{\Phi}(x,\lambda) = \begin{pmatrix} \cosh(\sqrt{\lambda}x) & \sinh(\sqrt{\lambda}x)/\sqrt{\lambda} \\ \sqrt{\lambda}\sinh(\sqrt{\lambda}x) & \cosh(\sqrt{\lambda}x) \end{pmatrix}.$$

Since

$$\cosh(\sqrt{\lambda}x) = \sum_{n=0}^{\infty} \frac{\lambda^n x^{2n}}{(2n)!}, \quad \sinh(\sqrt{\lambda}x) = \sqrt{\lambda}\sum_{n=0}^{\infty} \frac{\lambda^n x^{2n+1}}{(2n+1)!},$$

it is clear that $\boldsymbol{\Phi}(x,\lambda)$ is entire in λ for each fixed value of x. The Jost solutions satisfy

$$\boldsymbol{J}^{\pm}(0,\lambda) = \boldsymbol{\Phi}(\pm 1,\lambda)^{-1}\boldsymbol{b}^{\pm},$$

so that the Evans function is given by

$$E(\lambda) = \det(\boldsymbol{J}^-,\boldsymbol{J}^+)(0,\lambda) = \det(\boldsymbol{b}^-,\boldsymbol{b}^+)\cosh(2\sqrt{\lambda}) + (b_1^- b + 1^+ - b_2^- b_2^+\lambda)\frac{\sinh(2\sqrt{\lambda})}{\sqrt{\lambda}}.$$

In particular, under the transformation $\lambda = -\gamma^2 \in \mathbb{R}^-$ the Evans function takes the more familiar expression

$$E(\gamma) = \det(\boldsymbol{b}^-,\boldsymbol{b}^+)\cos(2\gamma) + (b_1^- b + 1^+ + b_2^- b_2^+\gamma^2)\frac{\sin(2\gamma)}{\gamma},$$

that traditionally arises for the Sturm–Liouville eigenvalue problem in an undergraduate course in partial differential equations [**111**].

Following (3.1.5), we demonstrate that the resolvent $(\mathcal{L} - \lambda)^{-1}$ is well-defined and bounded for all $\lambda \notin \sigma_{\mathrm{pt}}(\mathcal{L})$. In particular, $\rho(\mathcal{L}) = \mathbb{C}\backslash\sigma_{\mathrm{pt}}(\mathcal{L})$ and $\mathcal{L}$ has no essential spectrum.

Lemma 8.1.8. *Consider the operator (8.1.1) with boundary conditions (8.1.2). Then* $\sigma(\mathcal{L}) = \sigma_{\mathrm{pt}}(\mathcal{L})$ *and* $\sigma_{\mathrm{ess}}(\mathcal{L}) = \emptyset$. *Furthermore, all of the eigenvalues are simple, and the only accumulation point of the spectrum is* $\lambda = \infty$.

Proof. As a consequence of Proposition 8.1.4 we know that eigenvalues correspond to the zeros of an entire function, and moreover, from Exercise 8.1.6 the eigenvalues are simple. Since the zeros of an entire function are isolated, there is no accumulation point except possibly at $\lambda = \infty$. Suppose that $\lambda \notin \sigma_{\mathrm{pt}}(\mathcal{L})$, and consider the problem

$$(\mathcal{L} - \lambda)u = f.$$

The equivalent vectorized problem is

$$\partial_x \boldsymbol{Y} = \boldsymbol{A}(x,\lambda)\boldsymbol{Y} + \boldsymbol{F}, \quad \boldsymbol{F} = \begin{pmatrix} 0 \\ f \end{pmatrix}.$$

Let the Jost solutions be the columns of the fundamental solution, i.e.,

$$\boldsymbol{\Phi}(x,\lambda) = (\boldsymbol{J}^-, \boldsymbol{J}^+)(x,\lambda):$$

since by assumption $E(\lambda) \neq 0$, this matrix is nonsingular for all x. For any $c_- \in \mathbb{C}$, the function

$$\boldsymbol{y}^-(x,\lambda) = c_- \boldsymbol{J}^-(x,\lambda) + \boldsymbol{\Phi}(x,\lambda) \int_{-1}^{x} \boldsymbol{\Phi}(s,\lambda)^{-1} \boldsymbol{F}(s)\,\mathrm{d}s,$$

solves the ODE and satisfies the boundary condition at $x = -1$. Similarly for any $c_+ \in \mathbb{C}$,

$$\boldsymbol{y}^+(x,\lambda) = -c_+ \boldsymbol{J}^+(x,\lambda) + \boldsymbol{\Phi}(x,\lambda) \int_{+1}^{x} \boldsymbol{\Phi}(s,\lambda)^{-1} \boldsymbol{F}(s)\,\mathrm{d}s,$$

solves the ODE and satisfies the boundary condition at $x = +1$. The function

$$\boldsymbol{y}(x,\lambda) = \begin{cases} \boldsymbol{y}^-(x,\lambda), & x < 0 \\ \boldsymbol{y}^+(x,\lambda), & x > 0, \end{cases}$$

satisfies the ODE on $x < 0$ and $x > 0$ as well as both BCs. If $\boldsymbol{y}(x,\lambda)$ is continuous at $x = 0$, then it is the solution of the resolvent equation. Continuity requires that the vector $\boldsymbol{c} = (c_-, c_+)^{\mathrm{T}}$ satisfy

$$\boldsymbol{c} = \int_{-1}^{+1} \boldsymbol{\Phi}(s,\lambda)^{-1} \boldsymbol{F}(s)\,\mathrm{d}s.$$

Since the columns of the matrix are smooth, and the interval of integration is bounded, we clearly have the inequality

$$|c| \leq C(\lambda)\|\boldsymbol{F}\|_2 = \|f\|_2;$$

consequently, the solution is continuous and satisfies the estimate,

$$\|\boldsymbol{y}(\cdot,\lambda)\|_2 \leq C(\lambda)\|f\|_2.$$

In particular, the solution u is $C^1[-1,1]$ and we deduce that if $\lambda \notin \sigma_{\mathrm{pt}}(\mathcal{L})$, then for any $f \in L^2([-1,1])$ the system $(\mathcal{L}-\lambda)u = f$ has a unique solution satisfying $\|u\|_{H^1} \leq C\|f\|_2$. In particular, λ is in the resolvent set, $\rho(\mathcal{L})$. □

Exercises

Exercise 8.1.1. Show for the Evans function $E(\lambda)$ defined in Definition 8.1.1 that $E(\lambda_0) = 0$ if and only $\lambda_0 \in \sigma_{\mathrm{pt}}(\mathcal{L})$. Moreover, use the Evans function construction to show that $\mathrm{m_g}(\lambda_0) = 1$ for any $\lambda_0 \in \sigma_{\mathrm{pt}}(\mathcal{L})$.

Exercise 8.1.2. Derive the adjoint boundary conditions as those on $q \in H^2([-1,1])$ required to insure that

$$\langle \mathcal{L}p, q\rangle = \langle p, \mathcal{L}^{\mathrm{a}} q\rangle,$$

for all $p \in H^2([-1,1])$ satisfying (8.1.2).

Exercise 8.1.3. Show that if the Evans function is nonzero for $\lambda \in \mathbb{C}$, then:

(a) $\boldsymbol{\Phi}(x,\lambda) := \left(\boldsymbol{J}^+(x,\lambda), \boldsymbol{J}^-(x,\lambda)\right)$ is a FMS for (8.1.3)
(b) The adjoint Jost solutions $\{\boldsymbol{J}^{\mathrm{a}\pm}(\cdot,\lambda)\}$ can be normalized so that

$$\boldsymbol{\Phi}(x,\lambda)^{-1} = \left(\boldsymbol{J}^{\mathrm{a}-}(x,\lambda), \boldsymbol{J}^{\mathrm{a}+}(x,\lambda)\right).$$

Exercise 8.1.4. Show that $\langle q_0, p_0\rangle \neq 0$ by determining the relation between the eigenfunction p_0 and the adjoint eigenfunction q_0. Deduce that all eigenvalues of $\mathcal{L}$ are simple, in agreement with the Sturm–Liouville theory. *Hint: $\mathcal{L}$ can be symmetrized (made self-adjoint).*

Exercise 8.1.5. Construct the Evans function for the operator $\mathcal{L}$ of (8.1.1) when

$$a_1(x) \equiv 0, \quad a_0(x) = \begin{cases} a_0^-, & -1 \leq x < 0 \\ a_0^+, & 0 \leq x \leq 1. \end{cases}$$

Exercise 8.1.6. Construct the Evans function for the operator $\mathcal{L}$ of (8.1.1) when $a_1(x) \equiv 0$ and $a_0(x) = a_0\delta(x)$, for $a_0 \in \mathbb{C}$. Is it possible to show, either analytically or numerically, that $E(\lambda)$ has zeros with nonzero imaginary part?

8.2 Higher-Order Operators

The construction of the Evans function naturally extends to the nth-order operator and associated eigenvalue problem first discussed in Chapter 3. Specifically we consider the eigenvalue problem

$$\mathcal{L}p := \partial_x^n p + a_{n-1}(x)\partial_x^{n-1} p + \cdots + a_1(x)\partial_x p + a_0(x)p = \lambda p. \tag{8.2.1}$$

where the coefficients of $\mathcal{L}$ are assumed to be piecewise smooth and the boundary conditions take the form

$$\boldsymbol{B}_- \begin{pmatrix} p \\ \partial_x p \\ \vdots \\ \partial_x^{n-1} p \end{pmatrix}(-1) = \boldsymbol{0}, \quad \boldsymbol{B}_+ \begin{pmatrix} p \\ \partial_x p \\ \vdots \\ \partial_x^{n-1} p \end{pmatrix}(+1) = \boldsymbol{0}, \tag{8.2.2}$$

where the boundary matrices $\boldsymbol{B}_- \in \mathbb{R}^{k\times n}$ and $\boldsymbol{B}_+ \in \mathbb{R}^{(n-k)\times n}$ have full rank.

Setting $\boldsymbol{Y} = (p, \partial_x p, \ldots, \partial_x^{n-1} p)^{\mathrm{T}}$, the vectorized version of the eigenvalue problem becomes the first-order ODE

$$\partial_x \boldsymbol{Y} = \boldsymbol{A}(x,\lambda)\boldsymbol{Y}, \quad \boldsymbol{A}(x,\lambda) = \begin{pmatrix} 0 & 1 & \ldots & 0 & 0 \\ 0 & 0 & \ldots & 0 & 0 \\ \vdots & \vdots & \vdots\vdots\vdots & \vdots & \vdots \\ 0 & 0 & \ldots & 0 & 1 \\ \lambda - a_0(x) & -a_1(x) & \ldots & -a_{n-2}(x) & -a_{n-1}(x) \end{pmatrix}, \tag{8.2.3}$$

subject to the boundary conditions

$$\begin{aligned} \boldsymbol{Y}(-1,\lambda) &\in \mathbb{B}_{-1} := \ker(\boldsymbol{B}_-) = \operatorname{span}\{\boldsymbol{b}_1^-, \ldots, \boldsymbol{b}_{n-k}^-\} \\ \boldsymbol{Y}(+1,\lambda) &\in \mathbb{B}_{+1} := \ker(\boldsymbol{B}_+) = \operatorname{span}\{\boldsymbol{b}_1^+, \ldots, \boldsymbol{b}_k^+\}. \end{aligned} \tag{8.2.4}$$

The definitions of the Jost solutions and Evans function are natural extensions of the second-order operator case Definition 8.1.1. For each $\lambda \in \mathbb{C}$ let $\boldsymbol{\Phi}(x,\lambda) \in \mathbb{C}^{n\times n}$ be a fundamental matrix solution to (8.2.3) and define the Jost solutions by

$$\boldsymbol{J}_j^-(x,\lambda) = \boldsymbol{\Phi}(x,\lambda)\boldsymbol{\Phi}(-1,\lambda)^{-1}\boldsymbol{b}_j^-, \quad j = 1, \ldots, n-k, \tag{8.2.5}$$

(the flow of a vector in $\mathbb{B}_{-1}$ from $x = -1$ to $x = 0$) and

$$\boldsymbol{J}_j^+(x,\lambda) = \boldsymbol{\Phi}(x,\lambda)\boldsymbol{\Phi}(+1,\lambda)^{-1}\boldsymbol{b}_j^+, \quad j = 1, \ldots, k \tag{8.2.6}$$

(the flow of a vector in $\mathbb{B}_{+1}$ from $x = +1$ to $x = 0$). For fixed x the Jost solutions inherit the λ-analyticity of the fundamental matrix solution. Moreover, $\lambda_0 \in \mathbb{C}$ is an eigenvalue of $\mathcal{L}$ subject to (8.2.1) if and only if

$$\dim\left[\operatorname{span}\{J_1^-(0,\lambda_0),\ldots,J_{n-k}^-(0,\lambda_0)\}\cap\operatorname{span}\{J_1^+(0,\lambda_0),\ldots,J_k^-(0,\lambda_0)\}\right]\geq 1. \tag{8.2.7}$$

These observations motivate the following definition of the Evans function.

Definition 8.2.1. The Evans matrix $\boldsymbol{E}(\lambda)\in\mathbb{C}^{n\times n}$ is the matrix with columns composed of the Jost solutions evaluated at $x=0$:

$$\boldsymbol{E}(\lambda):=\left(J_1^-,\ldots,J_{n-k}^-,J_1^+,\cdots,J_k^+\right)(0,\lambda), \tag{8.2.8}$$

and the Evans function $E(\lambda)$ is its determinant

$$E(\lambda)=\det\boldsymbol{E}(\lambda), \tag{8.2.9}$$

where the Jost solutions are defined in (8.2.5) and (8.2.6).

Remark 8.2.2. The Jost solutions are only defined up to the choice of basis of $\mathbb{B}_{\pm 1}$ in (8.2.4). However, different choices of basis vectors lead to formulations of the Evans function with the same zeros. The choice of basis can even depend upon λ. Indeed, let $\boldsymbol{C}_-(\lambda)\in\mathbb{C}^{(n-k)\times(n-k)}$ and $\boldsymbol{C}_+(\lambda)\in\mathbb{C}^{k\times k}$ be entire in λ and nonsingular. We may define a new basis for $\mathbb{B}_{\pm 1}$ via

$$\left(\boldsymbol{c}_1^-,\ldots,\boldsymbol{c}_{n-k}^-\right)(\lambda)=\left(\boldsymbol{b}_1^-,\ldots,\boldsymbol{b}_{n-k}^-\right)\boldsymbol{C}_-(\lambda),$$

and

$$\left(\boldsymbol{c}_1^+,\ldots,\boldsymbol{c}_k^+\right)(\lambda)=\left(\boldsymbol{b}_1^+,\ldots,\boldsymbol{b}_k^+\right)\boldsymbol{C}_+(\lambda).$$

The associated Jost solutions for these new bases are given by

$$\tilde{J}_j^-(x,\lambda)=\boldsymbol{\Phi}(x,\lambda)\boldsymbol{\Phi}(-1,\lambda)^{-1}\boldsymbol{c}_j^-(\lambda),\quad j=1,\ldots,n-k,$$

and

$$\tilde{J}_j^+(x,\lambda)=\boldsymbol{\Phi}(x,\lambda)\boldsymbol{\Phi}(+1,\lambda)^{-1}\boldsymbol{c}_j^+(\lambda),\quad j=1,\ldots,k.$$

In terms of the reformulated Jost solutions, the associated Evans function takes the form

$$\tilde{E}(\lambda)=\det\left(\tilde{J}_1^-,\ldots,\tilde{J}_{n-k}^-,\tilde{J}_1^+,\ldots,\tilde{J}_k^+\right)(0,\lambda).$$

However,

$$\left(\tilde{J}_1^-,\ldots,\tilde{J}_{n-k}^-,\tilde{J}_1^+,\ldots,\tilde{J}_k^+\right)(0,\lambda)=\left(J_1^-,\ldots,J_{n-k}^-,J_1^+,\ldots,J_k^+\right)(0,\lambda)\operatorname{diag}(\boldsymbol{C}_-,\boldsymbol{C}_+)(\lambda),$$

so that the two Evans functions are related via

$$\tilde{E}(\lambda)=\det\boldsymbol{C}_-(\lambda)\det\boldsymbol{C}_+(\lambda)\,E(\lambda).$$

Since $\boldsymbol{C}_\pm(\lambda)$ are entire and nonsingular, the Evans functions are both analytic, and share the same zeros, up to multiplicity.

Recall that the adjoint ODE played an important role in evaluating the Evans function associated with the Sturm–Liouville problem. The adjoint equation associated with (8.2.3) takes the form

$$\partial_x \boldsymbol{Z} = -\boldsymbol{A}(x,\lambda)^{\mathrm{H}} \boldsymbol{Z} \tag{8.2.10}$$

where $\boldsymbol{Z}$ satisfies the adjoint boundary conditions

$$\boldsymbol{Z}(-1) \in \mathbb{B}^{\mathrm{a}}_{-1} := \mathbb{B}^{\perp}_{-1}, \quad \boldsymbol{Z}(1) \in \mathbb{B}^{\mathrm{a}}_{+1} := B^{\perp}_{+1}. \tag{8.2.11}$$

In particular, Lemma 8.1.3 holds for the system (8.2.3) subject to (8.2.4) and this adjoint system. The following result extends Proposition 8.1.4; we sketch the elements of the proof that are similar.

Proposition 8.2.3. *The Evans function for the eigenvalue problem (8.2.1) with boundary conditions (8.2.2) is entire. Every eigenvalue $\lambda_0 \in \sigma_{\mathrm{pt}}(\mathcal{L})$ satisfies $\mathrm{m_g}(\lambda_0) = \dim[\ker(\boldsymbol{E}(\lambda_0))]$, while $\mathrm{m_a}(\lambda_0) = d$ if and only if λ_0 is a zero of E of multiplicity d. In particular, if the eigenvalue is simple with associated eigenfunction p_0, then*

$$E(\lambda_0) = 0, \quad E'(\lambda_0) = \langle p_0, q_0 \rangle \det \boldsymbol{\Phi}(0,\lambda_0) \neq 0. \tag{8.2.12}$$

Here the adjoint eigenfunction, q_0, is scaled so that the associated adjoint Jost eigenvector, $\boldsymbol{J}_0^{\mathrm{a}}$, satisfies $|\boldsymbol{J}_0^{\mathrm{a}}(0)| = 1$, and $\boldsymbol{\Phi}(x,\lambda_0)$ is the fundamental matrix solution with normalization (8.2.15).

Remark 8.2.4. The following proof of the second statement of Proposition 8.2.3 is given under the additional assumption that there exists a perturbation of the coefficient $a_0(x) \mapsto a_0(x) + \epsilon a_1(x)$ for $0 < \epsilon \ll 1$ for which the perturbed operator has d simple eigenvalues that satisfy

$$\lim_{\epsilon \to 0} \lambda_j(\epsilon) = \lambda_0, \quad |\lambda_j(\epsilon) - \lambda_0| \le o(1),$$

for $j = 1, \ldots, d$. A more rigorous proof for the case of $\mathrm{m_g}(\lambda_0) = 1$ and $\mathrm{m_a}(\lambda_0) = d \ge 2$ is given in Chapter 8.2.1, and a proof for the general case is given in Chapter 8.2.2.

Proof. Suppose that λ_0 is a simple eigenvalue with associated eigenfunction p_0. We know that $E(\lambda_0) = 0$, and we show that $E'(\lambda_0) \neq 0$ and derive its expression in (8.2.12). Since $\lambda_0 \in \sigma_{\mathrm{pt}}(\mathcal{L})$, we know from Exercise 8.2.1 that the Evans matrix satisfies $\dim[\ker(\boldsymbol{E}(\lambda_0))] = 1$, so that the Jost solutions are not linearly independent. In particular, there is a choice of basis of $\mathbb{B}_{\pm 1}$ for which

$$\boldsymbol{J}_1^-(\lambda_0, x) = \boldsymbol{J}_1^+(\lambda_0, x) \equiv \boldsymbol{J}_0(x), \quad \boldsymbol{J}_0(x) := \begin{pmatrix} p_0(x) \\ \partial_x p_0(x) \\ \vdots \\ \partial_x^{n-1} p_0(x) \end{pmatrix}.$$

Differentiating the expression, (8.2.9), for the Evans function with respect to λ we see, as in (8.1.18), that

$$E'(\lambda_0) = \det\left(\partial_\lambda(J_1^- - J_1^+), J_2^-, \ldots, J_{n-k}^-, J_0, J_2^+, \ldots, J_k^+\right)(0, \lambda_0). \tag{8.2.13}$$

Following the argument leading to (8.1.20), the form (8.2.3) of $A(x, \lambda)$ yields

$$\partial_\lambda(J_1^- - J_1^+)(0, \lambda_0) = \Phi(0, \lambda_0) \int_{-1}^{+1} \Phi(s, \lambda_0)^{-1} \begin{pmatrix} 0 \\ \vdots \\ 0 \\ 0 \\ p_0(s) \end{pmatrix} ds. \tag{8.2.14}$$

Since λ_0 is a simple eigenvalue of $\mathcal{L}$, it follows that $\overline{\lambda_0}$ is a simple eigenvalue of $\mathcal{L}^a$ and there exists a unique solution $J_0^a(x)$ of the adjoint equation (8.2.10) that satisfies the adjoint BCs (8.2.11). Recalling the Evans matrix defined in (8.2.8) we have $\dim[\ker(E(\lambda_0))] = 1$, since if not there would be two linearly independent combinations of Jost solutions that would solve both the left and the right boundary conditions, and hence $\lambda = \lambda_0$ would not be a simple eigenvalue of $\mathcal{L}$. Moreover, for $j = 1, \ldots, n-k$ the quantity $J_j^-(x) \cdot J_0^a(x)$ is both independent of x and is zero at $x = -1$, while for $j = 1, \ldots, k$ the quantity $J_j^+(x) \cdot J_0^a(x)$ is both independent of x and is zero at $x = 1$. Consequently the vector $J_0^a(0)$, together with the range of $E(\lambda_0)$, spans $\mathbb{C}^n$, and we may form a FMS with the initial condition

$$\Phi(0, \lambda_0) = \left(J_0^a, J_2^-, \ldots, J_{n-k}^-, J_0, J_2^+, \ldots, J_k^+\right)(0, \lambda_0). \tag{8.2.15}$$

From Lemma 8.1.3 we know that $\Phi^{-1} = (Z_1, \cdots, Z_n)^{\mathrm{H}}$, where each Z_j solves the adjoint equation, but not necessarily the adjoint boundary conditions. Moreover, since $Y_j \cdot Z_k$ is independent of x and $\Phi\Phi^{-1} = I_n$, we have the conditions $Z_k \cdot Y_j = \delta_{jk}$ where δ is the Kronecker delta. From these conditions and the normalization of J_0^a it follows that the first row of Φ^{-1} is given by $\overline{J_0^a(s)}^{\mathrm{T}}$. The last entry of the adjoint eigenfunction, J_0^a, denoted q_0, lies in $\ker(\mathcal{L}^a - \overline{\lambda_0})$. Using this and the form of (8.2.15) to evaluate (8.2.14) yields

$$\partial_\lambda(J_1^- - J_1^+)(0, \lambda_0) = \langle p_0, q_0\rangle J_0^a(0) + \sum_{j=2}^{n-k} c_j^- J_j^-(0, \lambda_0) + c_0 J_0(0) + \sum_{j=2}^{k} c_j^+ J_j^+(0, \lambda_0) \tag{8.2.16}$$

for some constants $c_0, c_j^\pm$. However, the value of the constants is immaterial as only the first term contributes to the Evans function. Indeed, plugging the expression of (8.2.16) into (8.2.13) yields the formula (8.2.12) for $E'(\lambda_0)$. Moreover, the derivative must be nonzero if the eigenvalue is simple, since the Fredholm theory implies that $\langle p_0, q_0\rangle \neq 0$, while $\det \Phi(0, \lambda_0) \neq 0$ by its construction.

In the case that the eigenvalue has algebraic multiplicity $\mathrm{m_a}(\lambda_0) = d \geq 2$, we introduce a perturbation

$$a_0(x) \mapsto a_0(x) + \epsilon a_1(x), \quad 0 \leq \epsilon \ll 1, \tag{8.2.17}$$

of the coefficient $a_0(x)$ in the operator $\mathcal{L}$ defined in (8.2.1). In Chapter 6 we saw that for $0 < \epsilon \ll 1$ the eigenvalue of multiplicity d will (generically) break up into d simple eigenvalues; furthermore, these eigenvalues, denoted $\lambda_j(\epsilon)$ for $j = 1, \ldots, d$ will satisfy

$$\lim_{\epsilon \to 0} \lambda_j(\epsilon) = \lambda_0, \quad |\lambda_j(\epsilon) - \lambda_0| \leq \mathcal{O}(\epsilon^{1/d}). \tag{8.2.18}$$

Relaxing the convergence of $|\lambda_j - \lambda_0|$ to be merely $o(1)$ as $\epsilon \to 0$, we show

$$E(\lambda_0) = E'(\lambda_0) = \cdots = E^{(d-1)}(\lambda_0) = 0, \quad E^{(d)}(\lambda_0) \neq 0,$$

i.e., the zero of the Evans function is of order d.

Construct a positively oriented circle in the complex plane,

$$C_\delta(\lambda_0) := \{\lambda \in \mathbb{C} : |\lambda - \lambda_0| = \delta\}$$

with δ sufficiently small that its contains no eigenvalues of $\mathcal{L}$ other than λ_0 when $\epsilon = 0$. For $0 < \epsilon \ll 1$ sufficiently small, by assumption there will be precisely d simple eigenvalues in the interior of $C_\delta(\lambda_0)$. The matrix $\boldsymbol{A}(x, \lambda, \epsilon)$, as given in (8.2.3), is entire in λ for fixed (x, ϵ), and analytic in ϵ at $\epsilon = 0$ for fixed (x, λ). Consequently, the FMS $\boldsymbol{\Phi}(x, \lambda, \epsilon)$ and the Evans function inherit this analyticity. The winding number $W(\epsilon)$ for the Evans function,

$$W(\epsilon) = \frac{1}{2\pi \mathrm{i}} \oint_{C_\delta(\lambda_0)} \frac{\partial_\lambda E(\lambda, \epsilon)}{E(\lambda, \epsilon)} \, \mathrm{d}\lambda,$$

is continuous, integer-valued, and counts the total number of zeros (including multiplicity) of $E(\lambda, \epsilon)$ contained in the interior of $C_\delta(\lambda_0)$ (e.g., see [**203**]). By assumption the eigenvalues of $\mathcal{L}(\epsilon)$ are simple, so that $W(\epsilon) = d$ for $0 < \epsilon \ll 1$, and by continuity we deduce that $W(0) = d$, for any $\delta > 0$ sufficiently small. In particular, $E(\lambda, 0)$ has a zero of multiplicity d at $\lambda = \lambda_0$. □

Example 8.2.5. The equations of motion for the transverse vibrations of a spatially uniform elastic bar supported on an elastic foundation, as derived in [**205**, Chapter 10.2], take the form

$$\begin{aligned} &\frac{m}{EI} \partial_t^2 w + \frac{k}{EI} w + \partial_x^4 w = 0 \\ &w(\pm 1, t) = 0, \quad \partial_x^2 w(\pm 1, t) = 0. \end{aligned} \tag{8.2.19}$$

Here $w(x, t)$ is the total deflection of the bar, i.e., the sum of the angle of rotation due to bending and the angle of distortion due to shear, $m > 0$ is the mass per unit length, $EI > 0$ is the moment of inertia about the neutral axis, and $k > 0$ plays the role of a distributed spring, corresponding to a Hookean restoring force

$$F_{\text{rest}}(x,t) = -kw(x,t).$$

We have scaled the length of the bar to be two. The boundary conditions correspond to a hinged setting at both ends, attached to a uniformly distributed elastic support.

The PDE (8.2.19) is linear with constant coefficients, so we can solve it via separation of variables. Setting $w(x,t) = e^{\gamma t}p(x)$ and substituting into (8.2.19) yields the boundary-value problem

$$\begin{aligned} \partial_x^4 p + \frac{k}{EI}p &= -\frac{m}{EI}\gamma^2 p \\ p(\pm 1) = 0, \quad \partial_x^2 p(\pm 1) &= 0. \end{aligned} \tag{8.2.20}$$

To simplify notation, we introduce

$$a_0 := \frac{k}{EI}, \qquad \lambda := -\frac{m}{EI}\gamma^2, \tag{8.2.21}$$

which puts (8.2.20) in the form

$$\partial_x^4 p + a_0 p = \lambda p; \quad p(\pm 1) = \partial_x^2 p(\pm 1) = 0. \tag{8.2.22}$$

The eigenvalue problem (8.2.22) is self-adjoint, and the eigenvalues λ are real-valued and relate to γ through (8.2.21). In particular, $\lambda > 0$ corresponds to oscillatory behavior in w, while $\lambda < 0$ generates pairs of solutions w that experience exponential growth and decay, so that instabilities are associated with negative values of λ.

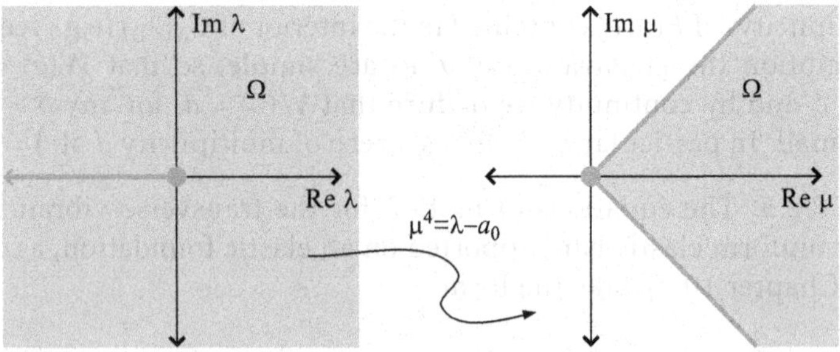

Fig. 8.2 The transformation $\mu^4 := \lambda - a_0$. The solid (*green*) curve is the branch cut, and the (*green*) circle is the branch point. The shaded region Ω corresponds to those values of λ (*left panel*) and μ (*right panel*) for which a zero of the Evans function corresponds to an eigenvalue for (8.2.19). (Color figure online.)

To construct the Evans function associated with (8.2.22), we vectorize (8.2.22). Setting $\boldsymbol{Y} = (p, \partial_x p, \partial_x^2 p, \partial_x^3 p)^{\mathrm{T}}$, the matrix $\boldsymbol{A}(x,\lambda)$ takes the form (8.2.3), while the boundary conditions (8.2.22) become

$$\boldsymbol{B}\boldsymbol{Y}(\pm 1, \lambda) = \boldsymbol{0}, \quad \boldsymbol{B} = \begin{pmatrix} 1 & 0 & 0 & 0 \\ 0 & 0 & 1 & 0 \end{pmatrix}.$$

As a basis for $\mathbb{B}_{\pm 1} := \ker(\boldsymbol{B})$, we choose

$$\boldsymbol{b}_1^{\pm} = \boldsymbol{e}_2, \quad \boldsymbol{b}_2^{\pm} = \boldsymbol{e}_4,$$

where $\boldsymbol{e}_j \in \mathbb{R}^4$ is the jth unit basis vector. To construct the FMS, we set $\mu^4 := \lambda - a_0$ (see Fig. 8.2), and consider μ which satisfy $|\arg(\mu)| < \pi/4$. A FMS takes the form

$$\boldsymbol{\Phi}(x,\lambda) = \begin{pmatrix} \cos\mu x & \sin\mu x & \cosh\mu x & \sinh\mu x \\ -\mu\sin\mu x & \mu\cos\mu x & \mu\sinh\mu x & \mu\cosh\mu x \\ -\mu^2\cos\mu x & -\mu^2\sin\mu x & \mu^2\cosh\mu x & \mu^2\sinh\mu x \\ \mu^3\sin\mu x & \mu^3\cos\mu x & \mu^3\sinh\mu x & \mu^3\cosh\mu x \end{pmatrix}.$$

The Jost solutions, given by (8.2.5) and (8.2.6), have complicated expressions; however, from the definition of the Evans function, (8.2.9), we may calculate that

$$E(\lambda) = -\frac{\sin(2\mu)\sinh(2\mu)}{\mu^2}, \quad \mu^4 = \lambda - a_0.$$

The Evans function has a removable singularity at $\mu = 0$, with analyticity restored by the nonzero value $E(0) = -4 < 0$. The admissible simple zeros, and consequently simple eigenvalues, are given by

$$\mu = \frac{n\pi}{2} \quad \Rightarrow \quad \lambda = a_0 + \left(\frac{n\pi}{2}\right)^4,$$

for $n = 1, 2, \ldots$. Since $a_0 > 0$, all of the eigenvalues are positive, and the system has no linear instabilities.

Remark 8.2.6. There are countably many zeros of the Evans function of the form

$$\sinh(2\mu) = 0 \quad \Rightarrow \quad \mu = \mathrm{i}\frac{n\pi}{2}, \quad n = \pm 1, \pm 2, \ldots.$$

These purely imaginary zeros do not correspond to eigenvalues, since they are not within the set Ω (see Fig. 8.2); indeed, they lie on two different sheets of the Riemann surface defined by $\mu^4 = \lambda - a_0$. The significance of zeros of the Evans function on the Riemann surface associated with the eigenvalue problem is discussed in the Chapter 9.

8.2.1 Rigorous Multiplicity Proof: $\mathrm{m_g}(\lambda_0) = 1^*$

We prove Proposition 8.2.3 in the case that $\mathrm{m_g}(\lambda_0) = 1$ and $\mathrm{m_a}(\lambda_0) = d \geq 2$ without perturbing the operator $\mathcal{L}$. The primary difficulty is to determine a proper description of the bases for $\mathbb{B}_{\pm 1}$. We assume that $\mathrm{m_a}(\lambda_0) = d \geq 2$, so that there exists a Jordan chain

$$(\mathcal{L} - \lambda_0)p_j = p_{j-1}, \quad p_{-1} = 0, \tag{8.2.23}$$

for $j = 0, \dots, d-1$ with each p_j satisfying the boundary conditions, (8.2.2).

We vectorize the problem, and introduce the generalized eigenfunctions,

$$\boldsymbol{J}_{0,j}(x) = \begin{pmatrix} p_j(x) \\ \partial_x p_j(x) \\ \vdots \\ \partial_x^{n-1} p_j(x) \end{pmatrix}, \quad j = 0, \dots, a-1,$$

which, from (8.2.23), satisfy

$$\partial_x \boldsymbol{J}_{0,j} = \boldsymbol{A}(x, \lambda_0)\boldsymbol{J}_{0,j} + \partial_\lambda \boldsymbol{A}_\lambda(x, \lambda_0)\boldsymbol{J}_{0,j-1}, \quad \boldsymbol{J}_{0,-1} = \boldsymbol{0}, \tag{8.2.24}$$

for $j = 0, \dots, d-1$. We choose a λ-dependent basis for $\mathbb{B}_{\pm 1}$ in the following manner. The first basis vectors are defined via

$$\boldsymbol{b}_1^{\pm}(\lambda) = \sum_{k=0}^{d-1} (\lambda - \lambda_0)^k \boldsymbol{J}_{0,k}(\pm 1), \tag{8.2.25}$$

while the remainder, $\boldsymbol{b}_j^-(\lambda)$ for $j = 2, \dots, n-k$, and $\boldsymbol{b}_j^+(\lambda)$ for $j = 2, \dots, k$ are selected to yield an analytic basis $\{\boldsymbol{b}_1^-, \dots, \boldsymbol{b}_{n-k}^-\}$ for $\mathbb{B}_{-1}$ and $\{\boldsymbol{b}_1^+, \dots, \boldsymbol{b}_k^+\}$ for $\mathbb{B}_{+1}$. With the Jost solutions defined according to (8.2.5) and (8.2.6), we have

$$\boldsymbol{J}_1^-(x, \lambda_0) = \boldsymbol{J}_1^+(x, \lambda_0) \equiv \boldsymbol{J}_{0,0}(x),$$

and moreover the following Lemma holds.

Lemma 8.2.7. *With the basis for $\mathbb{B}_{\pm 1}$ chosen as in (8.2.25), we have*

$$\partial_\lambda^k \boldsymbol{J}_1^{\pm}(x, \lambda_0) = k! \boldsymbol{J}_{0,k}(x),$$

for $k = 0, \dots, d-1$.

Proof. The result holds for $k = 0$ and for $k = 1$ since both $\boldsymbol{J}_{0,1}$ and $\partial_\lambda \boldsymbol{J}_1^{\pm}$ solve (8.2.24) with the same initial condition. Arguing inductively, suppose the result holds for $k = k'$ for some $2 \leq k' \leq d-2$. We verify that

$$\partial_x \left(\partial_\lambda^{k'+1} \boldsymbol{J}_1^{\pm} \right) = \boldsymbol{A}(x, \lambda_0) \partial_\lambda^{k'+1} \boldsymbol{J}_1^{\pm} + (k'+1) \partial_\lambda \boldsymbol{A}_\lambda(x, \lambda_0) \partial_\lambda^{k'} \boldsymbol{J}_1^{\pm}$$

$$\partial_\lambda^{k'+1} \boldsymbol{J}_1^{\pm}(\pm 1, \lambda_0) = (k'+1)! \boldsymbol{J}_{0,k'+1}(\pm 1),$$

while from (8.2.24) we see

$$\partial_x[(k'+1)!\boldsymbol{J}_{0,k'+1}] = \boldsymbol{A}(x,\lambda_0)[(k'+1)!\boldsymbol{J}_{0,k'+1}] + (k'+1)\partial_\lambda \boldsymbol{A}(x,\lambda_0)[(k')!\boldsymbol{J}_{0,k'}],$$

with the appropriate initial condition attached. By the inductive assumption

$$\partial_\lambda^{k'} \boldsymbol{J}_1^{\pm}(x,\lambda_0) = (k')!\boldsymbol{J}_{0,k'}(x),$$

and we deduce that $\boldsymbol{J}_{0,k'+1}(x)$ and $\partial_\lambda^{k'+1}\boldsymbol{J}_1^{\pm}(x,\lambda_0)$ satisfy the same ODE with the same initial condition, and hence they coincide. □

We define the Evans function according to (8.2.9),

$$E(\lambda) = \det\left(\boldsymbol{J}_1^-,\dots \boldsymbol{J}_{n-k}^-,\boldsymbol{J}_1^+,\dots,\boldsymbol{J}_k^+\right)(0,\lambda). \tag{8.2.26}$$

Since $\mathcal{L}-\lambda_0$ is Fredholm of index zero, there is a nontrivial $q_0 \in \ker(\mathcal{L}^{\mathrm{a}}-\overline{\lambda_0})$. Let $\boldsymbol{J}_0^{\mathrm{a}}(x)$ be a solution to the vectorized adjoint system that satisfies $\boldsymbol{J}_0^{\mathrm{a}}(\pm 1) \in \mathbb{B}_{\pm 1}^{\perp}$; moreover, it has q_0 as its final entry. As in the proof of Proposition 8.2.3, $\mathcal{L}$ has only a one-dimensional eigenspace, so that $\dim[\ker(\boldsymbol{E}(\lambda_0)] = 1$ and we may construct a FMS $\boldsymbol{\Phi}(x,\lambda_0)$ that satisfies

$$\boldsymbol{\Phi}(0,\lambda_0) = \left(\boldsymbol{J}_0^{\mathrm{a}},\boldsymbol{J}_2^-,\dots,\boldsymbol{J}_{n-k}^-,\boldsymbol{J}_{0,0},\boldsymbol{J}_2^+,\dots,\boldsymbol{J}_k^+\right)(0,\lambda_0). \tag{8.2.27}$$

The adjoint eigenfunction $\boldsymbol{J}_0^{\mathrm{a}}(x)$ is scaled to have length one when $x=0$. By Lemma 8.2.7 we have

$$\partial_\lambda^k(\boldsymbol{J}_1^- - \boldsymbol{J}_1^+)(\lambda_0) = \boldsymbol{0},\quad k=0,\dots,d-1;$$

consequently, for $0 \le k \le d$,

$$E^{(k)}(\lambda_0) = \det\left(\partial_\lambda^k(\boldsymbol{J}_1^- - \boldsymbol{J}_1^+),\boldsymbol{J}_2^-,\dots,\boldsymbol{J}_{n-k}^-,\boldsymbol{J}_{0,0},\boldsymbol{J}_2^+,\dots,\boldsymbol{J}_k^+\right)(0,\lambda_0), \tag{8.2.28}$$

and

$$E^{(k)}(\lambda_0) = 0,\quad 0 \le k \le d-1.$$

To determine $E^{(d)}(\lambda_0)$ it remains to evaluate $\partial_\lambda^d(\boldsymbol{J}_1^- - \boldsymbol{J}_1^+)(0,\lambda_0)$.

From Lemma 8.2.7 and the form of $\boldsymbol{b}_1^{\pm}(\lambda)$ in (8.2.25) we have

$$\partial_x(\partial_\lambda^d \boldsymbol{J}_1^{\pm}) = \boldsymbol{A}(x,\lambda_0)\partial_\lambda^d \boldsymbol{J}_1^{\pm} + d!\,\partial_\lambda \boldsymbol{A}(x,\lambda_0)\boldsymbol{J}_{0,d-1}(x),\quad \partial_\lambda^d \boldsymbol{J}_1^{\pm}(\pm 1,\lambda_0) = \boldsymbol{0}. \tag{8.2.29}$$

Variation of parameters applied to (8.2.29) yields the solution

$$\partial_\lambda^d \boldsymbol{J}_1^{\pm}(x,\lambda_0) = d!\,\boldsymbol{\Phi}(x,\lambda_0)\int_{\pm 1}^{x} \boldsymbol{\Phi}(s,\lambda_0)^{-1}\partial_\lambda \boldsymbol{A}(s,\lambda_0)\boldsymbol{J}_{0,d-1}(s)\,\mathrm{d}s,$$

and subtracting $\boldsymbol{J}^+$ and $\boldsymbol{J}^-$ yields

$$\partial_\lambda^d(J_1^- - J_1^+)(0,\lambda_0) = d!\,\mathbf{\Phi}(0,\lambda_0)\int_{-1}^{+1}\mathbf{\Phi}(s,\lambda_0)^{-1}\partial_\lambda \mathbf{A}(s,\lambda_0)\mathbf{J}_{0,d-1}(s)\,\mathrm{d}s. \quad (8.2.30)$$

Following the argument of Proposition 8.2.3, with $\mathbf{J}_0$ replaced by $\mathbf{J}_{0,d-1}$, we obtain

$$E^{(d)}(\lambda_0) = d!\langle p_{d-1}, q_0\rangle \det \mathbf{\Phi}(0,\lambda_0).$$

Since the Jordan chain terminates at length d, the Fredholm alternative implies that $\langle p_{d-1}, q_0\rangle \neq 0$, and hence $E^{(d)}(\lambda_0) \neq 0$.

We have established the forward direction of the following theorem.

Theorem 8.2.8. *An eigenvalue $\lambda = \lambda_0$ of $\mathcal{L}$ satisfies $\mathrm{m_g}(\lambda_0) = 1$ and $\mathrm{m_a}(\lambda_0) = d$ if and only if the Evans matrix satisfies $\dim[\ker(\mathbf{E}(\lambda_0))] = 1$ and the Evans function constructed in (8.2.9) satisfies*

$$E(\lambda_0) = E'(\lambda_0) = \cdots = E^{(d-1)}(\lambda_0) = 0, \quad E^{(d)}(\lambda_0) = d!\langle p_{d-1}, q_0\rangle \det \mathbf{\Phi}(0,\lambda_0) \neq 0.$$

Here p_{d-1} is the dth eigenfunction in the Jordan chain [see (8.2.23)] and q_0 is the adjoint eigenfunction. The normalization, $\mathbf{\Phi}(0,\lambda_0)$, for the fundamental matrix solution is defined in (8.2.27).

8.2.2 Rigorous Multiplicity Proof: $\mathrm{m_g}(\lambda_0) \geq 2$*

We sketch the proof of the following result.

Theorem 8.2.9. *Let $\mathcal{L}$ be an operator of the form (8.2.1). An eigenvalue $\lambda_0 \in \sigma_{\mathrm{pt}}(\mathcal{L})$ satisfies $\mathrm{m_g}(\lambda_0) = \dim[\ker(\mathbf{E}(\lambda_0))]$. Moreover, $\mathrm{m_a}(\lambda_0) = d$ if and only if*

$$E(\lambda_0) = E'(\lambda_0) = \cdots = E^{(d-1)}(\lambda_0) = 0,$$

and $E^{(d)}(\lambda_0) \neq 0$. In particular, the value of $E^{(d)}(\lambda_0)$ is given by (8.2.32).

Proof. Assume that $\mathrm{m_g}(\lambda_0) = m$ for some $2 \leq m \leq \min(k, n-k)$, and $\mathrm{m_a}(\lambda_0) = d \geq m$. Then there exists a set of eigenfunctions $p_{j,i}$ for $i = 0,\ldots,m-1$ and $j = 0,\ldots,d_i - 1$ with $\sum d_i = d$ such that

$$(\mathcal{L}-\lambda_0)p_{j,i} = p_{j-1,i}, \quad p_{-1,i} = 0, \quad (8.2.31)$$

where $p_{j,i}(\pm 1)$ satisfy the boundary conditions (see [**94**]). Setting

$$J_{0,j,i} := (p_{j,i}, \partial_x p_{j,i}, \ldots, \partial_x^{n-1} p_{j,i})^{\mathrm{T}},$$

it follows from (8.2.31) that

$$\partial_x \mathbf{J}_{0,j,i} = \mathbf{A}(x,\lambda_0)\mathbf{J}_{0,j,i} + \partial_\lambda \mathbf{A}(x,\lambda_0)\mathbf{J}_{0,j-1,i}, \quad \mathbf{J}_{0,-1,i} = \mathbf{0}.$$

Again, the key is to properly define the bases for $\mathbb{B}_{\pm1}$. For $i = 1,\ldots,m$ set

$$b_i^{\pm}(\lambda) = \sum_{k=0}^{d_i-1}(\lambda-\lambda_0)^k J_{0,k,i-1}(\pm1,\lambda_0),$$

and choose $b_i^{\pm}(\lambda)$ for $i \ge m+1$ so that $\{b_1^-,\ldots,b_k^-\}$ is an analytic basis for $\mathbb{B}_{-1}$, and $\{b_1^+,\ldots,b_{n-k}^-\}$ is an analytic basis for $\mathbb{B}_{+1}$. Defining the Jost solutions in the usual way, the analogue of Lemma 8.2.7 is

$$\partial_\lambda^k J_i^{\pm}(x,\lambda_0) = k!J_{0,k,i}(x), \quad i = 1,\ldots,m,\ k = 0,\ldots,d_i-1.$$

Since $\mathcal{L}-\lambda_0$ is Fredholm of index zero, there is a basis $\{q_0,\ldots,q_{m-1}\}$ of $\ker(\mathcal{L}^{\mathrm{a}}-\overline{\lambda_0})$. Choosing $J_i^{\mathrm{a}}(x)$ to be the associated adjoint eigenfunctions having length one when $x=0$, we have $J_i^{\mathrm{a}}(\pm1) \in \mathbb{B}_{\pm1}^{\perp}$, and we may form a FMS at $\lambda = \lambda_0$ with the normalization

$$\Phi(\lambda_0,0) = \left(J_0^{\mathrm{a}}, J_1^{\mathrm{a}},\ldots,J_{m-1}^{\mathrm{a}}, J_m^-,\ldots,J_{n-k}^-, J_{0,0,0},\ldots,J_{0,0,m-1}, J_m^+,\ldots,J_k^+\right)(0,\lambda_0).$$

The Evans function satisfies

$$E(\lambda_0) = E'(\lambda_0) = \cdots = E^{(a-1)}(\lambda_0) = 0,$$

and

$$E^{(d)}(\lambda_0) = \frac{d!}{\prod_{i=0}^{m-1} d_i!}\det\left(\partial_\lambda^{d_0}(J_1^- - J_1^+),\ldots,\partial_\lambda^{d_{m-1}}(J_m^- - J_m^+),\right.$$
$$\left. J_{m+1}^-,\ldots,J_{n-k}^-, J_1^+,\ldots,J_k^+\right)(0,\lambda_0).$$

As in (8.2.30) we find that

$$\partial_\lambda^{d_{i-1}}(J_1^- - J_1^+)(0,\lambda_0) = d_{i-1}!\,\Phi(0,\lambda_0)\int_{-1}^{+1}\Phi(s,\lambda_0)^{-1}\partial_\lambda A(s,\lambda_0)J_{0,d_{i-1},i}(s)\,\mathrm{d}s,$$

for $i = 1,\ldots,m$. Moreover

$$\Phi(0,\lambda_0)\int_{-1}^{+1}\Phi(s,\lambda_0)^{-1}\partial_\lambda A(s,\lambda_0)J_{0,d_{i-1},i}(s)\,\mathrm{d}s = \sum_{j=0}^{m-1}\langle p_{d_i,i}, q_j\rangle J_j^{\mathrm{a}}(0) + d_i,$$

where

$$d_i \in \mathrm{span}\{J_{m+1}^-,\ldots,J_{n-k}^-, J_{0,0,0},\ldots,J_{0,0,m-1}, J_{m+1}^+,\ldots,J_k^+\}.$$

Substituting and simplifying yields

$$E^{(d)}(\lambda_0) = d!\,\det F(\lambda_0)\,\det\Phi(0,\lambda_0), \tag{8.2.32}$$

where

$$F(\lambda_0) = \begin{pmatrix} \langle p_{d_0,0}, q_0 \rangle & \langle p_{d_0,0}, q_1 \rangle & \cdots & \langle p_{d_0,0}, q_{m-1} \rangle \\ \vdots & \vdots & & \vdots \\ \langle p_{d_{m-1},m-1}, q_0 \rangle & \langle p_{d_{m-1},m-1}, q_1 \rangle & \cdots & \langle p_{d_{m-1},m-1}, q_{m-1} \rangle \end{pmatrix}.$$

The Fredholm alternative implies that $F(\lambda_0)$ is nonsingular; hence, we see that $E^{(d)}(\lambda_0) \neq 0$. □

Exercises

Exercise 8.2.1. Let $\mathcal{L}$ be an operator of the form (8.2.1). If λ_0 is an eigenvalue of $\mathcal{L}$, show that $\mathrm{m_g}(\lambda_0) = \dim[\ker(E(\lambda_0))]$, where the Evans matrix is defined in (8.2.8).

Exercise 8.2.2. If $\lambda_0 \in \sigma_{\mathrm{pt}}(\mathcal{L})$, prove that

$$\mathrm{m_g}(\lambda_0) \leq \min(k, n-k).$$

Exercise 8.2.3. Suppose that $\mathcal{L}$ is an nth-order operator of the form (8.2.1), and that $\lambda_0 \in \mathbb{C}$ is a simple eigenvalue of $\mathcal{L}$ with eigenfunction p_0 and adjoint eigenfunction q_0. Then the vectorized eigenvalue problem, (8.2.3), has eigenfunction J_0, given in terms of p_0. For a system of dimension $n = 3$, give an explicit construction, in terms of q_0, of the adjoint solution $J_0^{\mathrm{a}}(x)$ which solves (8.2.10) subject to (8.2.11). The solution for general n can be found through Pego and Weinstein [**220**, Section 1(e)].

Exercise 8.2.4. Suppose that $\mathrm{m_g}(\lambda_0) = 1$ and $\mathrm{m_a}(\lambda_0) = k \geq 2$, so that there is the Jordan chain

$$(\mathcal{L} - \lambda_0)p_j = p_{j-1}, \quad p_{-1} = 0,$$

for $j = 0, \ldots, k-1$. Give an explicit expression for $a_0(x)$ in (8.2.17) which guarantees that there will be k simple eigenvalues for $\epsilon > 0$ sufficiently small.

Exercise 8.2.5. Consider the eigenvalue problem (8.2.1) with separated boundary conditions (8.2.2), where now $B_- \in \mathbb{R}^{k \times n}$ and $B_+ \in \mathbb{R}^{\ell \times n}$ are of full rank.

(a) Suppose that $k + \ell = n$. Show that $\sigma_{\mathrm{ess}}(\mathcal{L}) = \emptyset$.
(b) Suppose that $k + \ell \geq n + 1$. Show that $\sigma_{\mathrm{ess}}(\mathcal{L}) = \mathbb{C}$; in particular, show that every point $\lambda \in \mathbb{C}$ is an eigenvalue (not isolated) with $\mathrm{m_g}(\lambda) \geq k + \ell - n$.
(c) What can be said if $1 \leq k + \ell \leq n - 1$?

Exercise 8.2.6. Establish the converse direction of Theorem 8.2.8. That is, if $\dim\ker[E(\lambda_0)] = 1$ and λ_0 is a zero of E of multiplicity $d \geq 1$, then $\mathrm{m_g}(\lambda_0) = 1$ and $\mathrm{m_a}(\lambda_0) = d$. *Hint: You must construct the Jordan chain for the vectorized eigenvalue problem. Determine the equation that $J_k(x) := \partial_\lambda^k \left(J_1^- - J_1^+\right)(x, \lambda_0)$ satisfies for $k = 1, \cdots, d-1$. Use (8.2.28) to determine $J_k(0)$.*

Exercise 8.2.7. Establish the converse direction of Theorem 8.2.9. That is, if $\dim[\ker[\boldsymbol{E}(\lambda_0)]] = m$ and λ_0 is a zero of E of multiplicity $d \geq m$, then $\mathrm{m_g}(\lambda_0) = m$ and $\mathrm{m_a}(\lambda_0) = d$.

8.3 Second-Order Systems

The construction of the Evans function for second-order systems with separated boundary conditions requires a proper formulation of the boundary conditions and the adjoint system. We set up the framework and state the main result, as the proofs generalize naturally. We consider the system

$$\mathcal{L}\boldsymbol{p} := \boldsymbol{d}\partial_x^2\boldsymbol{p} + \boldsymbol{a}_1(x)\partial_x\boldsymbol{p} + \boldsymbol{a}_0(x)\boldsymbol{p} = \lambda\boldsymbol{p}, \tag{8.3.1}$$

where $\boldsymbol{a}_1(x), \boldsymbol{a}_0(x) \in \mathbb{R}^{n\times n}$ are piecewise smooth, and $\boldsymbol{d} \in \mathbb{R}^{n\times n}$ is uniformly invertible. The boundary conditions take the form

$$\boldsymbol{B}_-\begin{pmatrix} \boldsymbol{p}(-1) \\ \partial_x\boldsymbol{p}(-1) \end{pmatrix} = \boldsymbol{0}, \quad \boldsymbol{B}_+\begin{pmatrix} \boldsymbol{p}(+1) \\ \partial_x\boldsymbol{p}(+1) \end{pmatrix} = \boldsymbol{0}, \tag{8.3.2}$$

where the boundary matrices $\boldsymbol{B}_- \in \mathbb{R}^{k\times 2n}$ and $\boldsymbol{B}_+ \in \mathbb{R}^{(2n-k)\times 2n}$ have full-rank.

Setting $\boldsymbol{Y} = (\boldsymbol{p}, \partial_x\boldsymbol{p})^{\mathrm{T}}$ the system (8.3.1) is written as a vectorized first-order system

$$\partial_x\boldsymbol{Y} = \boldsymbol{A}(x,\lambda)\boldsymbol{Y}, \quad \boldsymbol{A}(x,\lambda) := \begin{pmatrix} 0 & \boldsymbol{I}_n \\ \boldsymbol{d}^{-1}(\lambda\boldsymbol{I}_n - \boldsymbol{a}_0(x)) & -\boldsymbol{d}^{-1}\boldsymbol{a}_1(x) \end{pmatrix}. \tag{8.3.3}$$

subject to the boundary conditions

$$\begin{aligned} \boldsymbol{Y}(-1,\lambda) &\in \mathbb{B}_{-1} := \ker(\boldsymbol{B}_-) = \mathrm{span}\{\boldsymbol{b}_1^-, \dots, \boldsymbol{b}_{2n-k}^-\} \\ \boldsymbol{Y}(+1,\lambda) &\in \mathbb{B}_{+1} := \ker(\boldsymbol{B}_+) = \mathrm{span}\{\boldsymbol{b}_1^+, \dots, \boldsymbol{b}_k^+\}. \end{aligned} \tag{8.3.4}$$

For each $\lambda \in \mathbb{C}$ let $\boldsymbol{\Phi}(x,\lambda) \in \mathbb{C}^{2n\times 2n}$ represent a fundamental matrix solution of (8.3.3). We define the Jost solutions

$$\boldsymbol{J}_j^-(x,\lambda) := \boldsymbol{\Phi}(x,\lambda)\boldsymbol{\Phi}(-1,\lambda)^{-1}\boldsymbol{b}_j^-, \quad j = 1,\dots,2n-k,$$

and

$$\boldsymbol{J}_j^+(x,\lambda) := \boldsymbol{\Phi}(x,\lambda)\boldsymbol{\Phi}(+1,\lambda)^{-1}\boldsymbol{b}_j^+, \quad j = 1,\dots,k.$$

The Evans function is then

$$E(\lambda) = \det\left(\boldsymbol{J}_1^-, \dots, \boldsymbol{J}_{2n-k}^-, \boldsymbol{J}_1^+, \dots, \boldsymbol{J}_k^+\right)(0,\lambda). \tag{8.3.5}$$

The adjoint eigenvalue problem takes the form

$$\mathcal{L}^{a}\boldsymbol{q} := \boldsymbol{d}^{T}\partial_x^2\boldsymbol{q} - \partial_x[\boldsymbol{a}_1(x)^{T}\boldsymbol{q}] + \boldsymbol{a}_0(x)^{T}\boldsymbol{q} = \overline{\lambda}\boldsymbol{q}, \tag{8.3.6}$$

so that if $\boldsymbol{q}$ solves (8.3.6), then

$$\boldsymbol{J}^{a}(x) = \begin{pmatrix} \boldsymbol{a}_1(x)^{T}\boldsymbol{q}(x) - \boldsymbol{d}^{T}\partial_x\boldsymbol{q}(x) \\ \boldsymbol{d}^{T}\boldsymbol{q}(x) \end{pmatrix}$$

is a solution to the adjoint ODE $\partial_x\boldsymbol{Z} = -\boldsymbol{A}(x,\lambda)^{H}\boldsymbol{Z}$. Lemma 8.1.3 generalizes, and in addition, noting that

$$\partial_\lambda\boldsymbol{A}(x,\lambda)\boldsymbol{Y}(x) = \begin{pmatrix} \boldsymbol{0}_n & \boldsymbol{0}_n \\ \boldsymbol{d}^{-1} & \boldsymbol{0}_n \end{pmatrix}\begin{pmatrix} \boldsymbol{p}(x) \\ \partial_x\boldsymbol{p}(x) \end{pmatrix} = \begin{pmatrix} \boldsymbol{0} \\ \boldsymbol{d}^{-1}\boldsymbol{p}(x) \end{pmatrix},$$

we have the equality

$$\boldsymbol{J}^{a}(x)^{H}\partial_\lambda\boldsymbol{A}(x,\lambda)\boldsymbol{Y}(x) = \boldsymbol{p}(x)\cdot\boldsymbol{q}(x),$$

which generalizes (8.1.21) and (8.2.30). The results for the nth-order problem, in particular Proposition 8.2.3 and Theorem 8.2.8, carry over.

Example 8.3.1. We consider a pair of Sturm–Liouville operators that are coupled through their boundary conditions. Specifically we take system $\boldsymbol{d} = \boldsymbol{I}_2$ and $\boldsymbol{a}_1(x) = \boldsymbol{a}_0(x) \equiv \boldsymbol{0}_2$ in (8.3.1) with boundary condition matrices

$$\boldsymbol{B}_- = \boldsymbol{B}_+ = \begin{pmatrix} 1 & 1 & 0 & 0 \\ 0 & 0 & 0 & 1 \end{pmatrix}.$$

That is, for $\boldsymbol{p} = (p_1, p_2)^{T}$, the boundary conditions are

$$p_1(\pm 1) + p_2(\pm 1) = 0, \quad \partial_x p_2(\pm 1) = 0.$$

Using the example following Proposition 8.1.4 we can conclude that the principal FMS at $x = 0$ is given by

$$\boldsymbol{\Phi}(x,\lambda) = \begin{pmatrix} \cosh(\sqrt{\lambda}x) & 0 & \sinh(\sqrt{\lambda}x)/\sqrt{\lambda} & 0 \\ 0 & \cosh(\sqrt{\lambda}x) & 0 & \sinh(\sqrt{\lambda}x)/\sqrt{\lambda} \\ \sqrt{\lambda}\sinh(\sqrt{\lambda}x) & 0 & \cosh(\sqrt{\lambda}x) & 0 \\ 0 & \sqrt{\lambda}\sinh(\sqrt{\lambda}x) & 0 & \cosh(\sqrt{\lambda}x) \end{pmatrix}.$$

Since

$$\mathbb{B}_{\pm 1} := \ker(\boldsymbol{B}_\pm) = \text{span}\{\underbrace{(-1,1,0,0)^{T}}_{\boldsymbol{b}_1^{\pm}}, \underbrace{(0,0,1,0)^{T}}_{\boldsymbol{b}_2^{\pm}}\},$$

the Jost solutions are given by

$$\boldsymbol{J}_j^{\pm}(x,\lambda) = \boldsymbol{\Phi}(x,\lambda)\boldsymbol{\Phi}(\pm 1,\lambda)^{-1}\boldsymbol{b}_j^{\pm}, \quad j = 1,2,$$

and the Evans function reduces to

$$E(\lambda) = -4\cosh^2(\sqrt{\lambda})\sinh^2(\sqrt{\lambda}).$$

Introducing the Riemann map $\lambda = -\gamma^2$, the Evans function takes a more convenient form

$$E(\gamma) = -4\cos^2(\gamma)\sin^2(\gamma),$$

with its zeros given by

$$\gamma = \frac{2\ell+1}{2}\pi \quad \Rightarrow \quad \lambda = -\left(\frac{2\ell+1}{2}\pi\right)^2, \quad \ell = 0, 1, 2, \ldots,$$

and

$$\gamma = \ell\pi \quad \Rightarrow \quad \lambda = -(\ell\pi)^2, \quad \ell = 0, 1, 2, \ldots.$$

The eigenvalues are all real-valued, as in the scalar Sturm–Liouville problem; however, the coupling of the two problems via the boundary conditions turns each zero into a double zero. Each eigenvalue is semi-simple with multiplicity two.

Exercises

Exercise 8.3.1. Consider a linear operator of the form (8.3.1) with coefficients $\boldsymbol{d} = \boldsymbol{I}_2$, $\boldsymbol{a}_1(x) \equiv \boldsymbol{0}_2$, and

$$\boldsymbol{a}_0(x) = \begin{cases} \mathrm{diag}(a_0, 0), & -1 \le x < 0 \\ \mathrm{diag}(-a_0, 0), & 0 \le x \le 1, \end{cases}$$

where $a_0 \in \mathbb{R}$. Construct the Evans function for the boundary conditions

$$\boldsymbol{B}_- = \boldsymbol{B}_+ = \begin{pmatrix} 1 & b & 0 & 0 \\ 0 & 0 & 0 & 1 \end{pmatrix},$$

where $b \in \mathbb{R}$ is arbitrary. Investigate the movement of the zeros of the Evans function as (a_0, b) vary, and determine regions in the parameter space for which there are zeros with nonzero imaginary part.

Exercise 8.3.2. Consider a linear operator of the form (8.3.1) with coefficients $\boldsymbol{d} = \boldsymbol{I}_2$, $\boldsymbol{a}_1(x) \equiv \boldsymbol{0}_2$, and

$$\boldsymbol{a}_0(x) = a_0\delta(x)\begin{pmatrix} 1 & 0 \\ 0 & 0 \end{pmatrix}, \quad a_0 \in \mathbb{C}.$$

Construct the Evans function for the boundary conditions

$$B_- = B_+ = \begin{pmatrix} 1\,0\,0\,0 \\ 0\,0\,0\,1 \end{pmatrix}.$$

Investigate the movement of the zeros of the Evans function as $a_0 \in \mathbb{C}$ is varied, and determine regions in the complex plane for which there are zeros with nonzero imaginary part.

8.4 The Evans Function for Periodic Problems

Nonseparated boundary conditions couple values at the left and right endpoints. Our discussion focuses on the periodic boundary conditions, which are the most common examples of this class. We consider the eigenvalue problem associated with an nth-order linear differential operator of the form

$$\mathcal{L}p := \partial_x^n p + a_{n-1}(x)\partial_x^{n-1} p + \cdots + a_1(x)\partial_x p + a_0(x)p = \lambda p, \tag{8.4.1}$$

where the coefficients are periodic with a common period, i.e., $a_j(x+\pi) = a_j(x)$ for $j = 0,\ldots,n-1$.

This problem was considered in Chapter 3.3, where the operator $\mathcal{L}$ acted on the whole line, and we saw that on $H^n(\mathbb{R})$ all of the spectrum was essential. However, the essential spectrum could be decomposed into a union of point eigenvalues via a Bloch-wave decomposition that reduces the system to a subproblem on a bounded domain. Indeed, introducing $p = e^{i\mu x}q$ for $-1 < \mu \le 1$, the eigenvalue problem was recast in terms of the operator

$$\mathcal{L}_\mu q := (\partial_x + i\mu)^n q + a_{n-1}(x)(\partial_x + i\mu)^{n-1} q + \cdots + a_1(x)(\partial_x + i\mu)q + a_0(x)q, \tag{8.4.2}$$

acting on $H^n_{\text{per}}([0,\pi])$, and we concluded that $\lambda \in \sigma_{\text{ess}}(\mathcal{L})$ if and only if $\lambda \in \sigma_{\text{pt}}(\mathcal{L}_\mu)$ for some $-1 < \mu \le 1$.

The eigenvalue problem for $\mathcal{L}_\mu$ is equivalent to a spatially rescaled and vectorized version of (8.4.1). Introducing $\mathbf{Y} = (p, \partial_x p, \ldots, \partial_x^{n-1} p)^{\mathrm{T}}$, we rewrite the eigenvalue problem (8.4.1) as a linear system

$$\partial_x \mathbf{Y} = \mathbf{A}(x,\lambda)\mathbf{Y}, \quad \mathbf{A}(x,\lambda) = \begin{pmatrix} 0 & 1 & \ldots & 0 & 0 \\ 0 & 0 & \ldots & 0 & 0 \\ \vdots & \vdots & \vdots\vdots\vdots & \vdots & \vdots \\ 0 & 0 & \ldots & 0 & 1 \\ \lambda - a_0(x) & -a_1(x) & \ldots & -a_{n-2}(x) & -a_{n-1}(x) \end{pmatrix}, \tag{8.4.3}$$

subject to $\mathbf{Y}(\pi,\lambda) = e^{i\mu\pi}\mathbf{Y}(0,\lambda)$. Making the change of variables

$$\mathbf{Y} = e^{i\mu x}\tilde{\mathbf{Y}}$$

and dropping the tilde, the system (8.4.3) reduces to a one-parameter family of periodic boundary-value problems,

$$\partial_x \boldsymbol{Y} = \underbrace{[\boldsymbol{A}(x,\lambda) - \mathrm{i}\mu \boldsymbol{I}_n]}_{\boldsymbol{A}(x,\lambda,\mu)}\boldsymbol{Y}, \quad \boldsymbol{Y}(\pi,\lambda,\mu) = \boldsymbol{Y}(0,\lambda,\mu). \tag{8.4.4}$$

The goal of this section is to construct the Evans function for the Bloch-wave eigenvalue problem (8.4.4). To apply the construction of the preceding sections, we follow [**91**], embedding the system (8.4.4) into a larger problem of dimension $2n$ that converts the periodic boundary conditions to separated ones. This is achieved by coupling (8.4.4) to the trivial system $\partial_x \boldsymbol{Z} = \boldsymbol{0}$ for $\boldsymbol{Z} \in \mathbb{C}^n$, with the purpose of propagating the value of $\boldsymbol{Y}$ on the left boundary to the right boundary, where it can be introduced as a local boundary condition. The new system variable takes the from

$$\boldsymbol{W} := \begin{pmatrix} \boldsymbol{Y} \\ \boldsymbol{Z} \end{pmatrix} \in \mathbb{C}^{2n},$$

which satisfies the larger system

$$\partial_x \boldsymbol{W} = \begin{pmatrix} \boldsymbol{A}(x,\lambda,\mu) & \boldsymbol{0}_n \\ \boldsymbol{0}_n & \boldsymbol{0}_n \end{pmatrix} \boldsymbol{W}. \tag{8.4.5}$$

Letting $\boldsymbol{e}_i \in \mathbb{C}^n$ denote the canonical basis, we set $\boldsymbol{b}_\ell^\pm = (\boldsymbol{e}_\ell, \boldsymbol{e}_\ell)^{\mathrm{T}}$, for $\ell = 1,\dots,n$ and introduce the boundary subspaces for (8.4.5), which mimic periodic conditions

$$\boldsymbol{W}(0) \in \mathbb{B}_0 := \operatorname{span}\{\boldsymbol{b}_1^-,\dots,\boldsymbol{b}_n^-\}, \quad \boldsymbol{W}(\pi) \in \mathbb{B}_\pi := \operatorname{span}\{\boldsymbol{b}_1^+,\dots,\boldsymbol{b}_n^+\}. \tag{8.4.6}$$

The system (8.4.5)–(8.4.6) has separated boundary conditions for which the construction of the Evans function follows from the Jost solutions. Indeed, a FMS $\boldsymbol{\Phi}(x,\lambda,\mu)$ for (8.4.4) induces a FMS

$$\boldsymbol{\Psi}(x,\lambda,\mu) = \begin{pmatrix} \boldsymbol{\Phi}(x,\lambda,\mu) & \boldsymbol{0}_n \\ \boldsymbol{0}_n & \boldsymbol{I}_n \end{pmatrix},$$

for (8.4.5). This motivates the definition of the Jost solutions for $j = 1,\dots,n$ as

$$\boldsymbol{J}_j^-(x,\lambda,\mu) = \boldsymbol{\Psi}(x,\lambda,\mu)\boldsymbol{\Psi}(0,\lambda,\mu)^{-1}\boldsymbol{b}_j^- = \begin{pmatrix} \boldsymbol{\Phi}(x,\lambda,\mu)\boldsymbol{\Phi}(0,\lambda,\mu)^{-1}\boldsymbol{e}_j \\ \boldsymbol{e}_j \end{pmatrix}, \tag{8.4.7}$$

and

$$\boldsymbol{J}_j^+(x,\lambda,\mu) = \boldsymbol{\Psi}(x,\lambda,\mu)\boldsymbol{\Psi}(\pi,\lambda,\mu)^{-1}\boldsymbol{b}_j^+ = \begin{pmatrix} \boldsymbol{\Phi}(x,\lambda,\mu)\boldsymbol{\Phi}(\pi,\lambda,\mu)^{-1}\boldsymbol{e}_j \\ \boldsymbol{e}_j \end{pmatrix}. \tag{8.4.8}$$

The Evans function is defined as

$$E(\lambda,\mu) = \det\left(J_1^-,\ldots,J_n^-,J_1^+,\ldots,J_n^+\right)(\pi/2,\lambda,\mu), \tag{8.4.9}$$

and inherits the following properties from Proposition 8.2.3.

Lemma 8.4.1. *The Evans function as defined in (8.4.9) is an entire function of λ for fixed μ and of μ for fixed λ. It has at most a countable number of zeros, with a zero at (λ,μ) if and only if λ is an eigenvalue of $\mathcal{L}_\mu$ defined in (8.4.2). Moreover, the multiplicity of the zero of the Evans function is equal to the algebraic multiplicity of the eigenvalue of $\mathcal{L}_\mu$ at λ.*

The definition (8.4.9) naturally sets up the application of Proposition 8.2.3; however, in Chapter 3.3 we showed that there is a nontrivial solution to (8.4.4) if and only if

$$\det(\boldsymbol{\Phi}(\pi,\lambda,\mu) - \boldsymbol{I}_n) = 0, \tag{8.4.10}$$

where $\boldsymbol{\Phi}(x,\lambda,\mu)$ is the principal FMS to (8.4.4) at $x = 0$ (see (3.3.5)). This more intuitive definition of the Evans function is equivalent to (8.4.9).

Lemma 8.4.2. *Let $\boldsymbol{\Phi}(x,\lambda,\mu)$ be a fundamental matrix solution to (8.4.4). The Evans function defined by*

$$E(\lambda,\mu) = \det\left(\boldsymbol{\Phi}(\pi,\lambda,\mu)\boldsymbol{\Phi}(0,\lambda,\mu)^{-1} - \boldsymbol{I}_n\right), \tag{8.4.11}$$

shares the properties stated in Lemma 8.4.1 for the definition (8.4.9).

Proof. From the definition, (8.4.7)–(8.4.8), of the Jost solutions we write

$$\left(J_1^-,\ldots,J_n^-,J_1^+,\ldots,J_n^+\right)(\pi/2,\lambda,\mu) = \begin{pmatrix} \boldsymbol{\Phi}(\pi/2,\lambda,\mu)\boldsymbol{\Phi}(0,\lambda,\mu)^{-1} & \boldsymbol{\Phi}(\pi/2,\lambda,\mu)\boldsymbol{\Phi}(\pi,\lambda,\mu)^{-1} \\ \boldsymbol{I}_n & \boldsymbol{I}_n \end{pmatrix}.$$

The matrix on the right can be factored

$$\begin{pmatrix} \boldsymbol{\Phi}(\pi/2,\lambda,\mu)\boldsymbol{\Phi}(0,\lambda,\mu)^{-1} & \boldsymbol{\Phi}(\pi/2,\lambda,\mu)\boldsymbol{\Phi}(\pi,\lambda,\mu)^{-1} \\ \boldsymbol{I}_n & \boldsymbol{I}_n \end{pmatrix} = \begin{pmatrix} \boldsymbol{\Phi}(\pi/2,\lambda,\mu)\boldsymbol{\Phi}(\pi,\lambda,\mu)^{-1} & \boldsymbol{0}_n \\ \boldsymbol{0}_n & \boldsymbol{I}_n \end{pmatrix} \begin{pmatrix} \boldsymbol{\Phi}(\pi,\lambda,\mu)\boldsymbol{\Phi}(0,\lambda,\mu)^{-1} & \boldsymbol{I}_n \\ \boldsymbol{I}_n & \boldsymbol{I}_n \end{pmatrix};$$

consequently, we rewrite the Evans function in (8.4.9) as

$$E(\lambda,\mu) = \det\left(\boldsymbol{\Phi}(\pi/2,\lambda,\mu)\boldsymbol{\Phi}(\pi,\lambda,\mu)^{-1}\right)\det\left(\boldsymbol{\Phi}(\pi,\lambda,\mu)\boldsymbol{\Phi}(0,\lambda,\mu)^{-1} - \boldsymbol{I}_n\right).$$

Since $\boldsymbol{\Phi}(\pi/2,\lambda,\mu)\boldsymbol{\Phi}(\pi,\lambda,\mu)^{-1}$ is nonsingular and analytic in both λ and μ, its determinant is analytic and nonzero. Consequently, we can renormalize the Evans function to remove the first term in the product without modifying the properties of Lemma 8.4.1. □

8.4.1 Application: Spectral Properties

The Evans function for the periodic problem is a convenient tool to prove spectral properties for the Bloch-wave decomposition (8.4.2). We first establish that the spectra of $\mathcal{L}$ consists of closed curves.

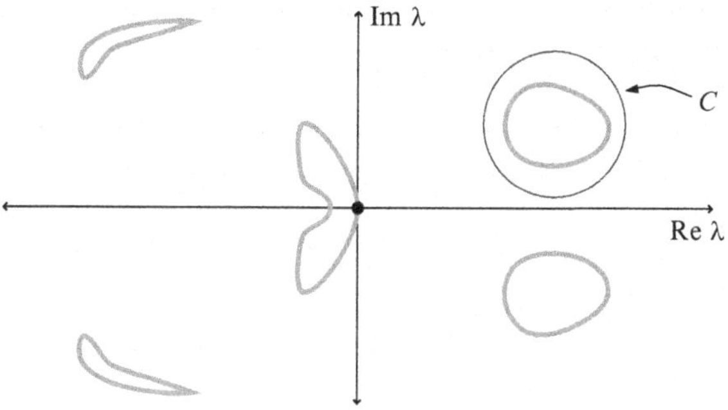

Fig. 8.3 A typical essential spectrum, denoted by the thick (*green*) curves, of an operator $\mathcal{L}$ of the form (8.4.1) subject to periodic boundary conditions. The contour C of Theorem 8.4.3 is depicted by the thin (*blue*) curve. (Color figure online.)

Theorem 8.4.3. *Consider the operator $\mathcal{L}$ defined in* (8.4.1). *Let $C \subset \mathbb{C}$ be a simple closed curve oriented in the positive sense, which does not intersect $\sigma(\mathcal{L})$. Then the winding number*

$$W(\mu) = \frac{1}{2\pi \mathrm{i}} \oint_C \frac{\partial_\lambda E(\lambda, \mu)}{E(\lambda, \mu)} \, \mathrm{d}\lambda, \tag{8.4.12}$$

is constant for $\mu \in (-1, 1]$. Moreover, if $W(0) = 1$, then the spectra inside of C forms a smooth, closed curve.

Proof. Let $C \subset \mathbb{C}$ be a positively oriented simple closed curve that does not intersect $\sigma(\mathcal{L})$ (see Fig. 8.3). Let $W(\mu)$ be the winding number defined in (8.4.12), and further suppose that $W(\mu_0) = m$. That is, there are m eigenvalues (counting multiplicity) of $\mathcal{L}_{\mu_0}$, defined in (8.4.4), inside C. Since the integrand is analytic in both λ and μ, we have that $W(\mu)$ is analytic in μ as long as the curve C does not intersect any spectra. As the winding number is integer-valued, it must be constant in μ as long as a zero of the Evans function does not intersect the curve C, which is precisely excluded by the assumption $C \cap \sigma(\mathcal{L}) = \emptyset$.

If in addition $W(0) = 1$, then $W(\mu) = 1$ for all $\mu \in (-1, 1]$; consequently, for each μ there is a unique $\lambda = \lambda(\mu)$ for which $E(\lambda(\mu), \mu) \equiv 0$. Since the eigenvalue $\lambda(\mu)$ of $\mathcal{L}_\mu$ is simple we deduce from (8.2.12) that $\partial_\lambda E(\lambda(\mu), \mu) \neq 0$. The implicit function theorem implies that the curve $\lambda = \lambda(\mu)$ lies in $C^\infty[-1, 1]$. To see that the curve is closed we show that $\lambda(-1) = \lambda(+1)$. This follows since the eigenvalue problems at $\mu = \pm 1$ are identical and have a single solution. Indeed, system (8.4.3) with boundary condition $\mathbf{Y}(\pi,\lambda) = e^{i\mu\pi}\mathbf{Y}(0,\lambda)$, is equivalent to (8.4.4), which has periodic boundary conditions. For $\mu = \pm 1$ the boundary condition factors agree: $e^{-i\pi} = e^{+i\pi} = -1$, and the result follows. □

The next result concerns the eigenvalue problem when the spatial interval is multiplied by an integral amount, but the coefficients remain π-periodic. That is, we consider the eigenvalue problem for $\mathcal{L}$ subject to periodic boundary conditions on $0 \le x \le k\pi$ for some integer $k \ge 2$; this introduces additional eigenfunctions that are not periodic at $x = \pi$ but are periodic at $x = k\pi$. We call this the $k\pi$-periodic problem.

Corollary 8.4.4. *Consider the eigenvalue problem* (8.4.1) *subject to periodic boundary conditions on* $[0,\pi]$. *Let* $C \subset \mathbb{C}$ *be a positively oriented, simple closed curve that does not intersect* $\sigma(\mathcal{L})$. *If* $W(0) = m$ *for the* π*-periodic Bloch-wave problem, then for each* $-1/k < \mu \le 1/k$ *the corresponding* $k\pi$*-periodic Bloch-wave problem satisfies* $W(\mu) = km$.

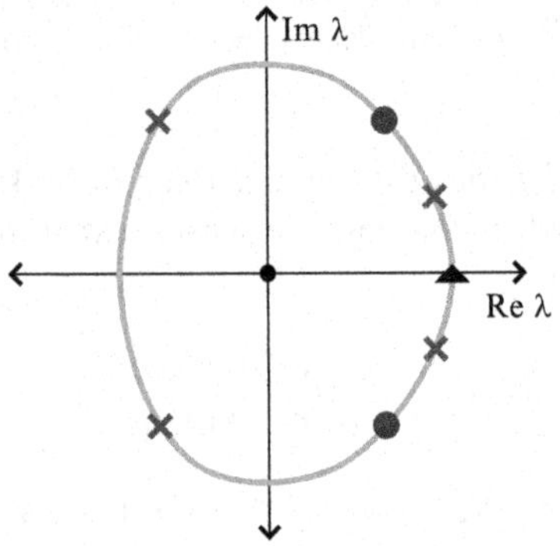

Fig. 8.4 A curve of essential spectrum of $\mathcal{L}$ for the π periodic problem for which $W(0) = 1$. For a generic value of $\mu \in (-1,1]$, the eigenvalue of $\mathcal{L}_\mu$ for $k = 1$, which is common to each $k\pi$ problem, is denoted by a *black* triangle. The additional eigenvalues of the 3π problem are denoted by (*blue*) circles, and those of the 5π problem are denoted by (*red*) crosses. (Color figure online.)

Proof. The linearity and the π periodicity of the coefficients of the vectorized π-eigenvalue problem (8.4.4) guarantee that a solution $\mathbf{y}(x)$ that satisfies $\mathbf{Y}(\pi,\lambda) = e^{i\mu\pi}\mathbf{Y}(0,\lambda)$ will also satisfy $\mathbf{Y}(k\pi,\lambda) = e^{ik\mu\pi}\mathbf{Y}(0,\lambda)$. Consequently, $\mathbf{Y}$ satisfies the η-Bloch wave problem on $k\pi$ if and only if it solves the corresponding π problem,

$$\mathbf{Y}' = \mathbf{A}(x,\lambda)\mathbf{Y}, \quad \mathbf{Y}(\pi,\lambda) = e^{i\mu\pi}\mathbf{Y}(0,\lambda); \quad \mu = \eta + 2\frac{j}{k},\ j = 0,\dots,k-1. \tag{8.4.13}$$

That is, for each η associated with the $k\pi$-periodic problem, there correspond precisely k values of μ associated with the π-periodic problem.

To complete the proof, we take C as in Corollary 8.4.4 and denote the Evans function for the π-periodic problem by $E_1(\lambda,\mu)$ and for the $k\pi$ problem by $E_k(\lambda,\eta)$, with corresponding winding numbers, $W_1(\mu)$ and $W_k(\eta)$. Assume that $W_1(0) = m$; then Theorem 8.4.3 implies that $W_1(\mu) = m$ for $-1 < \mu \leq 1$. The equivalence between the $k\pi$-problem and the system (8.4.13) implies that (see Fig. 8.4)

$$W_k(\eta) = \sum_{j=0}^{k-1} W_1(\eta + 2j/k) = km. \qquad \square$$

Example 8.4.5. Consider the parametrically forced Ginzburg–Landau equation

$$\partial_t u = \partial_x^2 u - u + |u|^2 u + \mathrm{i}\left(\epsilon u - \overline{u}\mathrm{e}^{-\mathrm{i}\theta}\right), \tag{8.4.14}$$

where $u \in \mathbb{C}$, $(x,t) \in \mathbb{R} \times \mathbb{R}^+ \mapsto \mathbb{C}$, and the parameters satisfy $-\pi/2 < \theta < \pi/2$ with $0 < \epsilon = \cos\theta < 1$. The system possesses a real-valued periodic equilibria

$$U_\theta(x) = \sqrt{2}A\,\mathrm{dn}(Ax,k), \quad A^2 := \frac{1+\sin\theta}{2-k^2}, \tag{8.4.15}$$

where $\mathrm{dn}(x,k)$ for $0 \leq k < 1$ is the Jacobi elliptic function with period $2K(k)$, and $K(k)$ is the elliptic integral of first kind, i.e.,

$$K(k) = \int_0^1 \frac{\mathrm{d}t}{\sqrt{(1-t^2)(1-k^2t^2)}}.$$

For a discussion of the Jacobi elliptic functions, see, e.g., [**203**, Chapter III.6].

We will see that $\theta = 0$ is a natural bifurcation point for the equilibria. For $\theta > 0$ the equilibria is narrower with a larger amplitude, and for $\theta < 0$ it is wider with a smaller amplitude. This dichotomy also distinguishes the stability of the two equilibria (also see Chapter 10.3.2). Writing $u = U_\theta + u_\mathrm{r} + \mathrm{i}u_\mathrm{i}$ and linearizing in $\boldsymbol{U} = (u_\mathrm{r}, u_\mathrm{i})^\mathrm{T}$, leads to the operator $\mathcal{L}$ and the associated eigenvalue problem of the form

$$\mathcal{L}\begin{pmatrix} u_\mathrm{r} \\ u_\mathrm{i} \end{pmatrix} = \lambda \begin{pmatrix} u_\mathrm{r} \\ u_\mathrm{i} \end{pmatrix}, \quad \mathcal{L} := \begin{pmatrix} \mathcal{L}_+ & -2\epsilon \\ 0 & \mathcal{L}_- \end{pmatrix}. \tag{8.4.16}$$

Here the suboperators are given by

$$\mathcal{L}_+ = \partial_x^2 - (1+\sin\theta) + 3U_\theta^2, \quad \mathcal{L}_- = \partial_x^2 - (1-\sin\theta) + U_\theta^2.$$

The spatial period of the operators is $2K(k)$ instead of π, so the Floquet exponent (Bloch parameter) μ has the domain $-\pi/(2K(k)) < \mu \leq \pi/(2K(k))$.

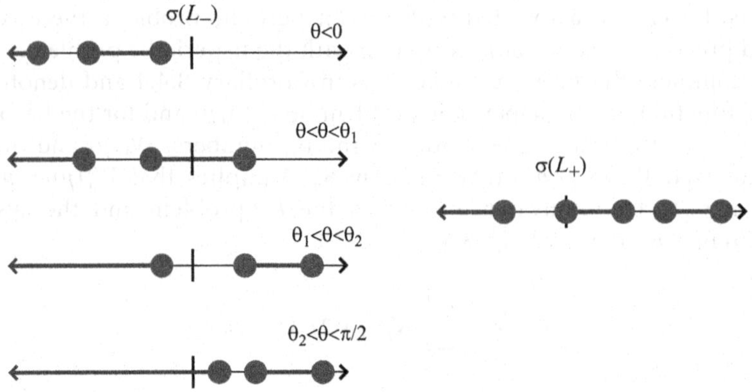

Fig. 8.5 A depiction of $\sigma(\mathcal{L}_-)$ (*left*) and $\sigma(\mathcal{L}_+)$ (*right*). Thick (*blue*) lines corresponds to bands of spectra while the filled (*red*) circles correspond to band edges. The small vertical line denotes $\lambda = 0$. (Color figure online.)

The first step in understanding the full spectrum is to understand the spectrum of $\mathcal{L}_\pm$. Introducing the rescaled independent variable, $y = Ax$, and using the identity $\mathrm{dn}^2(y,k) = 1 - k^2\,\mathrm{sn}^2(y,k)$, where $\mathrm{sn}(y,k)$ is another Jacobi elliptic function, the suboperators can be written as

$$\mathcal{L}_+ = A^2\left(\mathcal{L}_2 + (4+k^2)\right)$$
$$\mathcal{L}_- = A^2\left(\mathcal{L}_1 + \frac{4\sin\theta + (1-\sin\theta)k^2}{1+\sin\theta}\right).$$

where $\mathcal{L}_1 := \partial_y^2 - 2k^2\,\mathrm{sn}^2(y,k)$ and $\mathcal{L}_2 := \partial_y^2 - 6k^2\,\mathrm{sn}^2(y,k)$ are Hill operators. In particular, $\mathcal{L}_2$ is a 1-gap Lamé operator with

$$\sigma(\mathcal{L}_1) = (-\infty, -(1+k^2)] \cup [-1, -k^2],$$

while $\mathcal{L}_2$ is a 2-gap Lamé operator with

$$\sigma(\mathcal{L}_2) = (-\infty, -2(1+k^2+a(k))] \cup [-(4+k^2), -(1+4k^2)] \cup [-(1+k^2), -2(1+k^2-a(k))],$$

where $a(k) := \sqrt{1-k^2+k^4}$, see [**83**]. Consequently, the operators $\tilde{\mathcal{L}_\pm} := \mathcal{L}_\pm/A^2$, have spectra

$$\sigma(\tilde{\mathcal{L}_-}) = \left(-\infty, \frac{-1+(3-2k^2)\sin\theta}{1+\sin\theta}\right] \cup \left[\frac{-(1-k^2)+(3-k^2)\sin\theta}{1+\sin\theta}, \frac{2(2-k^2)\sin\theta}{1+\sin\theta}\right]$$
$$\sigma(\tilde{\mathcal{L}_+}) = (-\infty, 2-k^2-2a(\mathrm{k})] \cup [0, 3(1-\mathrm{k}^2)] \cup [3, 2-\mathrm{k}^2+2\mathrm{a}(\mathrm{k})].$$

Up to a scaling, the spectrum of $\mathcal{L}_+$ is independent of θ. Introducing

$$\theta_1 := \sin^{-1}\left(\frac{1-k^2}{3-k^2}\right), \quad \theta_2 := \sin^{-1}\left(\frac{1}{3-2k^2}\right),$$

the location of $\sigma(\mathcal{L}_-)$ relative to $\lambda = 0$ depends upon the relation of θ to θ_1 and θ_2, as depicted in Fig. 8.5.

Knowledge of the suboperators allows us to determine the spectrum of the full operator $\mathcal{L}$. Recalling (8.4.2), we introduce,

$$\mathcal{L}_{+,\mu} = (\partial_x + \mathrm{i}\mu)^2 - (1 + \sin\theta) + 3U_\theta^2,$$

with a similar formulation for $\mathcal{L}_{-,\mu}$. From knowledge of $\sigma(\mathcal{L}_+)$ we deduce that $0 \notin \sigma(\mathcal{L}_{+,\mu})$ for $\mu > 0$, and hence $\mathcal{L}_{+,\mu}$ in invertible. If $\phi_\pm(\mu)$ are eigenfunctions of the Bloch-wave operators $\mathcal{L}_{\pm,\mu}$ corresponding to eigenvalues $\lambda_\pm(\mu)$, respectively, then both $\lambda_\pm(\mu) \in \sigma(\mathcal{L}_\mu)$, with the eigenvalue/eigenfunction correspondence given by

$$\lambda = \lambda_-(\mu), \quad \boldsymbol{u}_-(\mu) = \begin{pmatrix} 2\epsilon\mathcal{L}_{+,\mu}^{-1}\phi_-(\mu) \\ \phi_-(\mu) \end{pmatrix}; \quad \lambda = \lambda_+(\mu), \quad \boldsymbol{u}_+(\mu) = \begin{pmatrix} \phi_+(\mu) \\ 0 \end{pmatrix}.$$

The value of $\lambda(0)$ can be deduced from the continuity of the curve $\lambda = \lambda(\mu)$ in μ. As expected for an upper triangular operator such as $\mathcal{L}$, we deduce that $\sigma(\mathcal{L}) = \sigma(\mathcal{L}_-) \cup \sigma(\mathcal{L}_+)$. Moreover, $\mathcal{L}$ always has two unstable spectral bands associated to $\sigma(\mathcal{L}_+)$, with additional unstable spectral bands generated by $\sigma(\mathcal{L}_-)$ as θ is increased through zero and θ_2.

Exercises

Exercise 8.4.1. Show that the solutions to (8.4.5) subject to (8.4.6) are in one-to-one correspondence with solutions to (8.4.4).

Exercise 8.4.2. Consider the Sturm–Liouville operator

$$\mathcal{L}p = \partial_x^2 p + a_1(x)p + a_0(x)p, \quad a_j(x+\pi) = a_j(x),$$

acting on the space $H^2_{\mathrm{per}}([0,\pi])$. Show that:

(a) $\sigma_{\mathrm{pt}}(\mathcal{L})$ is a countable set for which the only possible accumulation point is ∞.

(b) $\sigma_{\mathrm{ess}}(\mathcal{L}) = \emptyset$ (compare with Lemma 8.1.8).

Exercise 8.4.3. Consider the nth-order operator introduced in (8.4.1) acting on $H^n_{\mathrm{per}}([0,\pi])$. Show that $\sigma_{\mathrm{ess}}(\mathcal{L}) = \emptyset$.

8.5 Additional Reading

The use of the Evans function or a Bloch-wave decomposition to analyze the spectral stability of periodic waves can be found in Barker et al. [**24**], Bronski and Rapti [**42**], Gallay and Hărăguş [**89**], Hărăguş [**124**, **125**], Hărăguş et al. [**128**], Ivey and Lafortune [**129**].

Chapter 9
The Evans Function for Sturm–Liouville Operators on the Real Line

In this chapter we construct the Evans function for exponentially asymptotic differential operators on unbounded domains. While several key elements of the construction naturally carry over from the bounded domain construction, important new subtleties arise. To focus on these key issues, we restrict our attention to second-order Sturm–Liouville operators. The extension to higher-order linear operators in addressed in Chapter 10.

The most salient obstacle to our construction of the Evans function is a hidden dependence of the boundary conditions upon the eigenvalue parameter, λ. The natural condition, that the solutions of the eigenvalue problem decay to zero at infinity, requires quantification, and is more subtle than it may first appear. Indeed, we will see that the essential elements of the construction require that the coefficients of the linear operator decay at an exponential rate to constant values at $x = \pm\infty$. For the bounded domain eigenvalue problem the vectorized boundary conditions are of the form $\boldsymbol{Y}(x = \pm 1; \lambda) \in \mathbb{B}_{\pm 1}$, where the boundary spaces $\mathbb{B}_{\pm 1}$ are independent of λ. On the unbounded domain the goal is to construct asymptotic spaces $\mathbb{B}_{\pm}(\lambda)$, related to the stable and unstable eigenspaces of the limiting operators $\boldsymbol{A}_{\pm}(\lambda)$, defined in (3.1.4), and to find solutions $\boldsymbol{Y}_{\pm}(\cdot, \lambda)$ of the vectorized eigenvalue problem that approach these spaces as $x \to \pm\infty$. While these eigenspaces are generically analytic in λ, where they change dimension there is a potential loss of analyticity. The restoration of analyticity requires the introduction of branch cuts, branch points, and Riemann surfaces: features that have no analogy in the bounded domain problem. The natural domain of the Evans function is bounded by the essential spectrum of the associated operator. An analytic extension of the Evans function beyond this natural domain requires a subtle modification of the definition of $\mathbb{B}_{\pm}(\lambda)$, while an analysis of the Evans function at the branch points and into the absolute spectrum requires the introduction of an appropriate Riemann surface.

The chapter is organized as follows. We first introduce the spectral projections and the associated Jost solutions associated with the eigenvalue problem. It is these functions that define the boundary condition spaces

T. Kapitula and K. Promislow, *Spectral and Dynamical Stability of Nonlinear Waves*, Applied Mathematical Sciences 185, DOI 10.1007/978-1-4614-6995-7_9,

$\mathbb{B}_\pm(\lambda)$ and the resultant Evans function. Subsequently, we consider three applications: orientation index, edge bifurcations, and large domain limits. The orientation (parity) index detects positive, real (unstable) eigenvalues by comparing the behavior of the Evans function at $\lambda = 0$ and $\lambda = +\infty$. *Edge bifurcations* track eigenvalues as they emerge from and disappear into the absolute spectrum. This type of bifurcation is outside the scope of classical perturbation theory discussed in Chapter 6, because the linear operator is typically no longer Fredholm when the eigenvalue parameter enters the absolute spectrum. Its analysis requires an analytic extension of the Evans function onto the associated Riemann surface. The large domain limit studies the convergence of the spectrum of an operator on an unbounded domain to that of its restriction to large, but finite domains. The convergence is markedly different if the bounded domain problem has separated boundary conditions, for which the bounded domain point spectrum approaches the absolute spectrum of the unbounded problem, or nonseparated boundary conditions, for which the bounded domain point spectrum converges to the essential spectrum corresponding to the unbounded problem in an unweighted space.

9.1 The Whole-Line Eigenvalue Problem

Consider the eigenvalue problem for the Sturm–Liouville operator $\mathcal{L}$,

$$\mathcal{L}p := \partial_x^2 p + a_1(x)\partial_x p + a_0(x)p = \lambda p, \tag{9.1.1}$$

where $\mathcal{L} : H^2(\mathbb{R}) \subset L^2(\mathbb{R}) \mapsto L^2(\mathbb{R})$. The nominal "boundary conditions" can be interpreted as

$$|p(x)| + |\partial_x p(x)| \to 0, \quad x \to \pm\infty. \tag{9.1.2}$$

To obtain meaningful control of these boundary conditions we require that the operator $\mathcal{L}$ be exponentially asymptotic, i.e., there exist asymptotic constants a_0^0, a_1^0 and an exponential decay rate $\alpha > 0$ such that

$$|a_0(x) - a_0^-| + |a_1(x) - a_1^-| \le Ce^{-\alpha|x|},\ x < 0; \quad |a_0(x) - a_0^+| + |a_1(x) - a_1^+| \le Ce^{-\alpha|x|},\ x > 0.$$

Such operators frequently arise as the linearization of an partial differential equation about a front-type equilibria. However, to minimize notational complexity we focus on the special case

$$|a_0(x) - a_0^0| + |a_1(x) - a_1^0| \le Ce^{-\alpha|x|}, \tag{9.1.3}$$

which is typical of operators arising as linearizations about a pulse, or homoclinic, equilibria.

As in Chapter 3, the first step is to rewrite the eigenvalue problem (9.1.1) as a first-order vector system for the unknown $\boldsymbol{Y} = (p, \partial_x p)^{\mathrm{T}}$ that satisfies

$$\partial_x \boldsymbol{Y} = \underbrace{[\boldsymbol{A}_0(\lambda) + \boldsymbol{R}(x)]}_{\boldsymbol{A}(x,\lambda)} \boldsymbol{Y}, \tag{9.1.4}$$

where we have introduced the asymptotic and localized matrices

$$\boldsymbol{A}_0(\lambda) = \begin{pmatrix} 0 & 1 \\ \lambda - a_0^0 & -a_1^0 \end{pmatrix}, \quad \boldsymbol{R}(x) = \begin{pmatrix} 0 & 0 \\ a_0^0 - a_0(x) & a_1^0 - a_1(x) \end{pmatrix}.$$

In particular, from (9.1.3), the localized matrix decays exponentially as $x \to \pm\infty$,

$$|\boldsymbol{R}(x)| \le C\mathrm{e}^{-\alpha|x|}, \quad x \in \mathbb{R}. \tag{9.1.5}$$

The nominal "boundary conditions" (9.1.2) become $|\boldsymbol{Y}(x)| \to 0$ as $x \to \pm\infty$.

The matrix eigenvalues (the eigenvalues of $\boldsymbol{A}_0(\lambda)$) play a key role in the construction. For the case at hand they take the form

$$\mu_1(\lambda) = -\frac{a_1^0}{2} + \sqrt{\lambda - (a_0^0 - (a_1^0)^2/4)}, \quad \mu_2(\lambda) = -\frac{a_1^0}{2} - \sqrt{\lambda - (a_0^0 - (a_1^0)^2/4)}. \tag{9.1.6}$$

It is significant that the matrix eigenvalues are not analytic in λ; indeed, they have a branch point,

$$\lambda_{\mathrm{br}} := a_0^0 - \frac{1}{4}(a_1^0)^2.$$

Choosing the branch cut along the negative real axis, so that $\mathrm{Re}\sqrt{\lambda - \lambda_{\mathrm{br}}} \ge 0$, we have the ordering $\mathrm{Re}\,\mu_1 \ge \mathrm{Re}\,\mu_2$ and the branch cut corresponds to the absolute spectrum of the operator $\mathcal{L}$ (see Definition 3.2.3),

$$\sigma_{\mathrm{abs}}(\mathcal{L}) = \{\lambda \in \mathbb{C} : \lambda = \lambda_{\mathrm{br}} - k^2,\ k \in \mathbb{R}\}.$$

Because the matrix $\boldsymbol{A}(x,\lambda)$ has the same asymptotic limits at $x = \pm\infty$ the operator $\mathcal{L}$ has a single Fredholm border (see Theorem 3.1.13), where one (or possibly two if $a_0^1 = 0$) of the matrix eigenvalues becomes purely imaginary; furthermore, by Theorem 3.1.11 and Remark 3.1.15, the Fredholm border coincides with the essential spectrum and is given by the parabolic curve

$$\sigma_{\mathrm{ess}}(\mathcal{L}) = \{\lambda \in \mathbb{C} : \lambda = a_0^0 - k^2 + \mathrm{i}a_1^0 k,\ k \in \mathbb{R}\}.$$

If we perturb the operator $\mathcal{L}$ so that the coefficient $a_0(x)$ has distinct limits $a_0^+ \ne a_0^-$ at $\pm\infty$, then the Fredholm boundary will split into two curves that form the boundary of the essential spectrum.

As depicted in Fig. 9.1, we can see that for λ to the right of $\sigma_{\mathrm{ess}}(\mathcal{L})$ the matrix eigenvalues satisfy $\mathrm{Re}\,\mu_1(\lambda) > 0 > \mathrm{Re}\,\mu_2(\lambda)$, so that the Fredholm index of $\mathcal{L}-\lambda$ is zero. On the complementary region, for λ to the left of $\sigma_{\rm ess}(\mathcal{L})$, we have either $0 > \operatorname{Re}\mu_1(\lambda) \geq \operatorname{Re}\mu_2(\lambda)$ (for $a_1^0 > 0$) or $\operatorname{Re}\mu_1(\lambda) \geq \operatorname{Re}\mu_2(\lambda) > 0$ (for $a_1^0 < 0$).

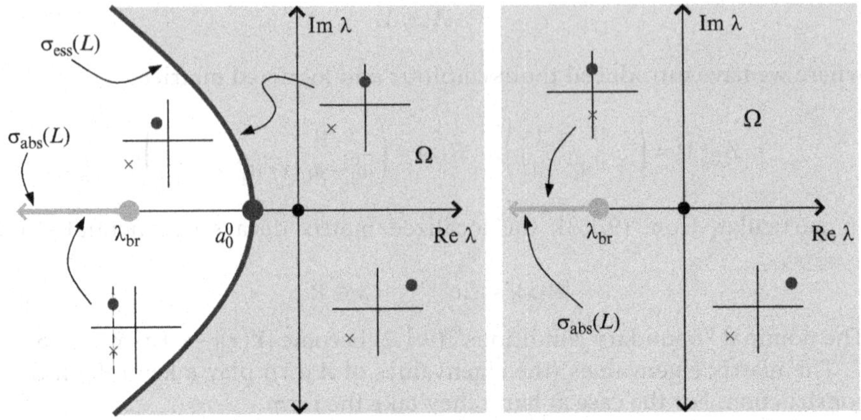

Fig. 9.1 The *left panel* shows the decomposition of the spectrum of $\mathcal{L}$ for the eigenvalue problem (9.2.1) with the matrix eigenvalues depicted in the insets for $a_1^0 > 0$. The *right panel* shows the decomposition when $a_1^0 = 0$. In the natural domain of the Evans function, Ω, which is the the shaded region to the right of the essential spectrum $\sigma_{\rm ess}(\mathcal{L})$, the operator $\mathcal{L}-\lambda$ is Fredholm of index zero, with Morse index $\mathrm{i}(A_0(\lambda)) = 1$, since $\mu_1(\lambda) > 0$ (*blue* circle), and $\mu_2(\lambda) < 0$ (*red* cross). The branch point is always to the left of the essential spectrum, except if $a_1^0 = 0$, when it lies upon it. At the branch point $\mu_1 = \mu_2$, while on the absolute spectrum $\operatorname{Re}\mu_1 = \operatorname{Re}\mu_2$. As $a_0^1 \to 0^+$ the essential spectrum merges with the absolute spectrum so that $\sigma_{\rm abs}(\mathcal{L}) = \sigma_{\rm ess}(\mathcal{L}.)$ (Color figure online.)

The operator $\mathcal{L}$ is Fredholm of index zero on its essential resolvent, that is, for λ to the right of $\sigma_{\rm ess}(\mathcal{L})$, so that on this set $\mathbb{E}_+^s(\lambda) \oplus \mathbb{E}_-^u(\lambda) = \mathbb{C}^2$; see (3.1.13). This condition determines the natural domain of the Evans function

Definition 9.1.1. The connected component of the set $\lambda \in \mathbb{C}$ that contains $\operatorname{Re}\lambda = +\infty$ and for which $\mathcal{L}-\lambda$ has Fredholm index zero is called the *natural domain*, Ω, of the Evans function.

Our analysis will require we extend the Evans function analytically, not only outside of its natural domain Ω, but into the absolute spectrum. This will requires an unfolding of the branch cut, which is accomplished by introducing the Riemann surface through the transformation,

$$\gamma^2 := \lambda - \lambda_{\rm br}. \tag{9.1.7}$$

The Riemann variable γ takes values on a two-sheeted cover of $\mathbb{C}$ (see Fig. 9.2). The benefit of working on the Riemann surface is that the matrix eigenvalues are entire functions of γ,

$$\mu_1(\gamma) = -\frac{a_1^0}{2} + \gamma, \quad \mu_2(\gamma) = -\frac{a_1^0}{2} - \gamma. \tag{9.1.8}$$

Moreover, as depicted in (9.2), under the transformation (9.1.7) the Fredholm boundary is mapped to the vertical line $\mathrm{Re}\,\gamma = |a_1^0|/2$, while the absolute spectrum is mapped to the line $\mathrm{Re}\,\gamma = 0$. The natural domain of the Evans function, Ω, is still to the right of the Fredholm boundary. The natural domain is a subset of the *physical sheet* of the Riemann surface, i.e., the set $\mathrm{Re}\,\gamma > 0$; the set $\mathrm{Re}\,\gamma < 0$ is called the *resonance sheet* of the Riemann surface. In the remainder of this chapter we shall use γ as the spectral parameter with the understanding that it relates to λ through (9.1.7).

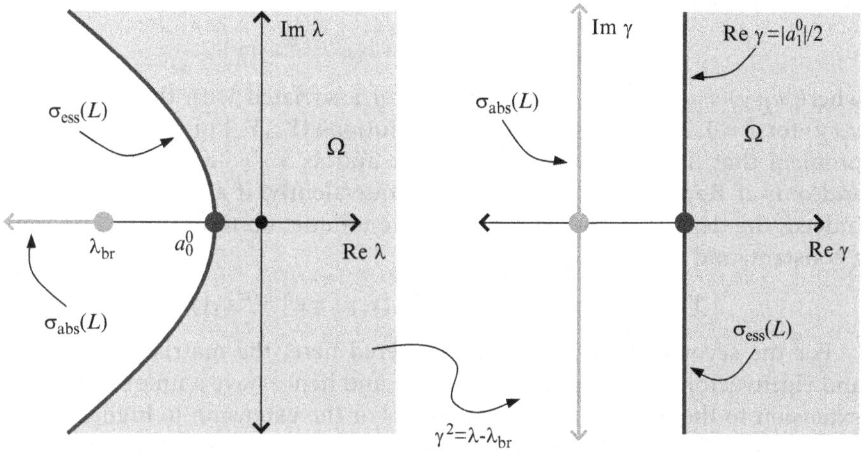

Fig. 9.2 The *left panel* depicts the spectral sets of $\mathcal{L}$ in the λ-plane for the eigenvalue problem (9.1.1) (compare to Fig. 9.1). The *right panel* depicts the spectral sets on the Riemann surface defined by $\gamma^2 = \lambda - \lambda_{\mathrm{br}}$. The natural domain Ω corresponds to the set $\mathrm{Re}\,\gamma > |a_1^0|/2$. The absolute spectrum is mapped onto the imaginary axis, and the matrix eigenvalues, which have a jump across the branch cut, are analytic in γ. The set $\mathrm{Re}\,\gamma > 0$ is called the *physical sheet* of the Riemann surface, and the set $\mathrm{Re}\,\gamma < 0$, which is the second cover of $\mathbb{C}$, is called the *resonance sheet*. (Color figure online.)

9.2 Spectral Projections and the Jost Solutions

Rewritten in terms of the spectral parameter γ, the eigenvalue problem (9.1.4) becomes

$$\partial_x \boldsymbol{Y} = [\boldsymbol{A}_0(\gamma) + \boldsymbol{R}(x)]\boldsymbol{Y}, \quad \boldsymbol{A}_0(\gamma) = \begin{pmatrix} 0 & 1 \\ \gamma^2 - (a_1^0)^2/4 & -a_1^0 \end{pmatrix}. \tag{9.2.1}$$

For $\gamma \in \Omega$ we wish to identify subspaces $\mathbb{B}_\pm(\gamma) \subset \mathbb{C}^2$ with the property that a solution $\boldsymbol{Y}$ of (9.2.1) decays to zero as $x \to \pm\infty$ if and only if $\lim_{x\to\pm\infty} \boldsymbol{Y}(x,\gamma) \in \mathbb{B}_\pm(\gamma)$, respectively. Since the localized term $\boldsymbol{R}(x)$ decays exponentially as

$x \to \pm\infty$, it is natural to exploit the exponential dichotomy enjoyed by the solutions of the asymptotic system (e.g., see [**59**]). Indeed, we will show that for $|x|$ sufficiently large the solutions of (9.2.1) closely shadow those of the asymptotic system

$$\partial_x \boldsymbol{Y} = \boldsymbol{A}_0(\gamma)\boldsymbol{Y}, \quad \boldsymbol{A}_0(\gamma) = \begin{pmatrix} 0 & 1 \\ \gamma^2 - (a_1^0)^2/4 & -a_1^0 \end{pmatrix}, \tag{9.2.2}$$

and we thus anticipate that $\mathbb{B}_-(\gamma) = \mathbb{E}_-^{\mathrm{u}}(\gamma)$ and $\mathbb{B}_+(\gamma) = \mathbb{E}_+^{\mathrm{s}}(\gamma)$.

For γ away from the branch point the asymptotic problem (9.2.2) has the general solution

$$\boldsymbol{Y}(x,\gamma) = c_1 \mathrm{e}^{\mu_1(\gamma)x}\boldsymbol{v}_1(\gamma) + c_2 \mathrm{e}^{\mu_2(\gamma)x}\boldsymbol{v}_2(\gamma),$$

where $\boldsymbol{v}_j(\gamma) = (1, \mu_j(\gamma))^{\mathrm{T}}$ is the eigenvector associated with the eigenvalue $\mu_j(\gamma)$ for $j = 1,2$. It is possible to have solutions $\{\boldsymbol{Y}_+, \boldsymbol{Y}_-\}$ of this asymptotic problem that decay in norm as $x \to +\infty$ and as $x \to -\infty$, respectively, if and only if $\mathrm{Re}\,\mu_2(\gamma) < 0 < \mathrm{Re}\,\mu_1(\gamma)$, or equivalently, if and only if $\gamma \in \Omega$. Indeed, the desired decaying solutions are unique, up to multiplication by a constant, and take the form

$$\boldsymbol{Y}_+(x,\gamma) = \mathrm{e}^{\mu_2(\gamma)x}\boldsymbol{v}_2(\gamma), \quad \boldsymbol{Y}_-(x,\gamma) = \mathrm{e}^{\mu_1(\gamma)x}\boldsymbol{v}_1(\gamma).$$

For the second-order problem considered here, the matrix eigenvalues and eigenvectors are entire functions of γ and hence have a unique analytic extension to the whole Riemann surface. For the extension to higher-order operators in Chapter 10 there will generically be several matrix eigenvalues with positive and negative real parts. As γ varies over the natural domain Ω these matrix eigenvalues may merge and form Jordan blocks, thereby losing analyticity. This potential difficulty is overcome by considering the collective spectral projection onto the stable and unstable eigenspaces of $\boldsymbol{A}_0$, which are analytic on the natural domain.

To construct the spectral projections associated with $\mathbb{E}_-^u$ and $\mathbb{E}_+^s$ of $\boldsymbol{A}_0(\gamma)$ we define positively oriented circles

$$C_{\mathrm{u}} := \{\mu \in \mathbb{C} : |\mu - \mu_1(\gamma)| = \delta\}, \quad C_{\mathrm{s}} := \{\mu \in \mathbb{C} : |\mu - \mu_2(\gamma)| = \delta\},$$

where $\delta > 0$ is chosen sufficiently small so that the curves do not intersect. This is always possible for γ away from the branch point $\gamma = 0$ (see Fig. 9.3). The spectral projections onto the stable and unstable subspaces of the asymptotic matrix, $P^{\mathrm{s}}(\gamma) : \mathbb{C}^2 \mapsto \mathbb{E}^{\mathrm{s}}(\gamma)$ and $P^{\mathrm{u}}(\gamma) : \mathbb{C}^2 \mapsto \mathbb{E}^{\mathrm{u}}(\gamma)$, are defined by the Dunford integral formula,

$$P^{\mathrm{s}}(\gamma)\boldsymbol{v} = \frac{1}{2\pi\mathrm{i}} \oint_{C_{\mathrm{s}}} (\mu \boldsymbol{I}_2 - \boldsymbol{A}_0(\gamma))^{-1}\boldsymbol{v}\,\mathrm{d}\mu,$$

$$P^{\mathrm{u}}(\gamma)\boldsymbol{v} = \frac{1}{2\pi\mathrm{i}} \oint_{C_{\mathrm{u}}} (\mu \boldsymbol{I}_2 - \boldsymbol{A}_0(\gamma))^{-1}\boldsymbol{v}\,\mathrm{d}\mu.$$

The analyticity of $P^{s,u}(\gamma)$ follows immediately from the Cauchy integral theorem and the analyticity of $\boldsymbol{A}_0(\gamma)$. The fact that these functions are projections that commute with $\boldsymbol{A}_0(\gamma)$ follows from the matrix-valued version of the Cauchy residue theorem (also see the brief discussion in Chapter 2.2.4). The projections can be defined and are analytic so long as the matrix eigenvalues μ_1 and μ_2 can be separated; that is, so long as γ is not at the branch point $\gamma = 0$. More concretely, for our 2×2 system the projections take the simple form

$$P^s(\gamma)\boldsymbol{v} = \frac{\boldsymbol{v} \cdot \boldsymbol{v}_2^a}{\boldsymbol{v}_2 \cdot \boldsymbol{v}_2^a} \boldsymbol{v}_2(\gamma), \quad P^u(\gamma)\boldsymbol{v} = \frac{\boldsymbol{v} \cdot \boldsymbol{v}_1^a}{\boldsymbol{v}_1 \cdot \boldsymbol{v}_1^a} \boldsymbol{v}_1(\gamma),$$

where $\boldsymbol{v}_{1,2}^a(\gamma)$ are the adjoint eigenvectors of $\boldsymbol{A}_0(\gamma)$. The normalizing terms $\boldsymbol{v}_j \cdot \boldsymbol{v}_j^a$ for $j = 1,2$ are analytic functions of γ, and nonzero away from the branch point $\gamma = 0$. The spectral projections have the following properties:

(a) They are mutually orthogonal, i.e., $P^s(\gamma)P^u(\gamma) = P^u(\gamma)P^s(\gamma) = 0$,
(b) $P^s(\gamma)\boldsymbol{v} + P^u(\gamma)\boldsymbol{v} = \boldsymbol{v}$ for $\gamma \neq 0$,
(c) $P^{s,u}(\gamma)\boldsymbol{A}_0(\gamma) = \boldsymbol{A}_0(\gamma)P^{s,u}(\gamma) \quad \Rightarrow \quad P^{s,u}(\gamma)e^{\boldsymbol{A}_0(\gamma)x} = e^{\boldsymbol{A}_0(\gamma)x}P^{s,u}(\gamma)$.

Most importantly, we may quantify the decay (or growth) rates of solutions that lie in the range of the spectral projections. Introducing

$$\eta_u(\gamma) := \operatorname{Re}\mu_1(\gamma) = -\frac{a_1^0}{2} + \operatorname{Re}\gamma, \quad \eta_s(\gamma) := \operatorname{Re}\mu_2(\gamma) = -\frac{a_1^0}{2} - \operatorname{Re}\gamma,$$

we remark that $\eta_s \leq \eta_u$ on the physical sheet of the Riemann surface. Adapting Theorem 2.1.23 we have the following exponential dichotomy.

Proposition 9.2.1. *For γ away from the branch point, $\gamma = 0$, the stable and unstable spectral projections associated with $\boldsymbol{A}_0(\gamma)$ induce the following decay estimates upon the matrix exponential*

$$|P^s(\gamma)e^{\boldsymbol{A}_0(\gamma)(x-\tau)}\boldsymbol{v}| = |e^{\boldsymbol{A}_0(\gamma)(x-\tau)}P^s(\gamma)\boldsymbol{v}| \leq Ce^{\eta_s(\gamma)(x-\tau)}|\boldsymbol{v}|, \quad x \geq \tau, \tag{9.2.3}$$

and

$$|P^u(\gamma)e^{\boldsymbol{A}_0(\gamma)(x-\tau)}\boldsymbol{v}| = |e^{\boldsymbol{A}_0(\gamma)(x-\tau)}P^u(\gamma)\boldsymbol{v}| \leq Ce^{\eta_u(\gamma)(x-\tau)}|\boldsymbol{v}|, \quad x \leq \tau. \tag{9.2.4}$$

In particular, if $\gamma \in \Omega$, so that $\eta_s < 0 < \eta_u$, then for fixed x we have

$$\lim_{\tau \to -\infty} |P^s(\gamma)e^{\boldsymbol{A}_0(\gamma)(x-\tau)}\boldsymbol{v}| = 0, \quad \lim_{\tau \to +\infty} |P^u(\gamma)e^{\boldsymbol{A}_0(\gamma)(x-\tau)}\boldsymbol{v}| = 0,$$

and the approach is exponentially fast.

We extend this dichotomy to the full problem, (9.2.1), through the Jost solutions, $\boldsymbol{J}_\pm(\cdot,\gamma)$. For $\gamma \in \Omega$ these are the solutions that decay as either $x \to +\infty$ or as $x \to -\infty$. However, to facilitate the extension of the Jost solutions beyond the natural domain we define them not through their decay properties, but by their approach to the solutions $\mathbf{Y}_\pm(\cdot,\gamma)$ of the asymptotic problem. This permits an extension to all of the physical sheet of the Riemann surface, where $\text{Re}\,\gamma > 0$, and even to the portion of the resonance sheet for which $\text{Re}\,\gamma > -\alpha/2$, where α is the decay parameter for the localized matrix $\mathbf{R}(x)$ in (9.2.1). We call this set the extended domain of the Evans function.

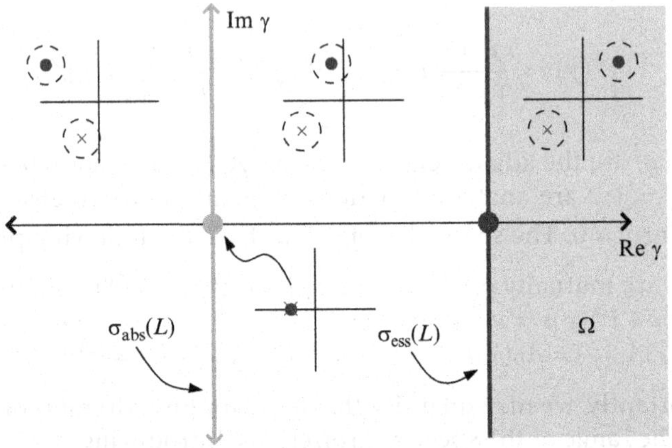

Fig. 9.3 The matrix eigenvalues are given in the inset of each relevant region of the Riemann surface (for $a_1^0 > 0$). Except at the branch point, $\gamma = 0$, the eigenvalue $\mu_1(\gamma)$ (*blue* circle) and the eigenvalue $\mu_2(\gamma)$ (*red* cross) can be separated by small contours of integration used to define the spectral projections. (Color figure online.)

Definition 9.2.2. Given a decay rate $\alpha > 0$ of the exponentially localized matrix $\mathbf{R}(x)$, we define the spectral gap

$$g_\gamma := \eta_u - \eta_s + \alpha. \tag{9.2.5}$$

The extended domain of the Evans function is the set of γ for which the spectral gap is positive, i.e.,

$$\Omega_e := \left\{\gamma \in \mathbb{C} : g_\gamma > 0\right\} = \left\{\gamma \in \mathbb{C} : \text{Re}\,\gamma > -\frac{\alpha}{2}\right\}. \tag{9.2.6}$$

The existence and asymptotic behavior of the Jost solutions on the extended domain is established in the following theorem (also see [**243**, Theorem XI.57]). Recall that $\lambda = \gamma^2 + \lambda_{br}$.

Theorem 9.2.3. *Consider the system (9.2.1) where the exponentially localized matrix $\mathbf{R}(x)$ satisfies the decay estimate (9.1.5) for some $\alpha > 0$. Let $\mathbf{v}_s(\gamma) \in \mathbb{B}_+(\gamma)$ and $\mathbf{v}_u(\gamma) \in \mathbb{B}_-(\gamma)$ be analytic functions of $\gamma \in \Omega_e$ with a continuous limit at $\gamma = 0$. There exist $C, L_0 > 0$ such that for all $\gamma \in \Omega_e$ and for all $L \geq L_0$ there are two Jost solutions $\mathbf{J}^\pm(x,\gamma)$ that solve (9.2.1), are analytic in γ for fixed x, and have the asymptotic form,*

$$
\begin{aligned}
|J^+(x,\gamma) - e^{A_0(\gamma)x}v_s(\gamma)| &\le Ce^{-\alpha L/2}\left(1+g_\gamma^{-1}\right)e^{\eta_s(\gamma)x}|v_s(\gamma)|, \quad x \ge +L \\
|J^-(x,\gamma) - e^{A_0(\gamma)x}v_u(\gamma)| &\le Ce^{-\alpha L/2}\left(1+g_\gamma^{-1}\right)e^{\eta_u(\gamma)x}|v_u(\gamma)|, \quad x \le -L.
\end{aligned} \tag{9.2.7}
$$

Proof. We suppress the γ dependence of $J^\pm(x,\gamma)$ and consider only $J^+(x)$. Considering the ansatz

$$
J^+(x) = e^{A_0 x}v_s + K(x): \tag{9.2.8}
$$

we assume that $v_s \neq 0$, so that $e^{A_0x}v_s$ is a nonzero solution to the asymptotic system. We wish to bound the term $K(x)$, which denotes the correction induced by the exponentially localized matrix $R(x)$. Plugging this ansatz into the eigenvalue problem (9.2.1), we see that the correction term must satisfy the system

$$
\partial_x K = A(x)K + R(x)e^{A_0x}v_s = A_0K + R(x)\left[e^{A_0x}v_s + K(x)\right].
$$

For any $x_0 \in \mathbb{R}$, variation of parameters yields the solution

$$
K(x) = \int_{x_0}^{x} e^{A_0(x-\tau)}R(\tau)\left[e^{A_0\tau}v_s + K(\tau)\right]\mathrm{d}\tau. \tag{9.2.9}
$$

The goal is find a choice of initial data for which $|K(x)|$ decays at an exponential rate as $x \to +\infty$. We consider two cases: (i) when γ is sufficiently far from the branch point, i.e., $|\gamma| > \alpha/4$, we will use the spectral projections to determine the appropriate initial data; and (ii) when γ is close to the branch point, i.e., $|\gamma| \le \alpha/4$, where the decay of the localized perturbation R suffices.

In the first case, the projection operators satisfy $I_2 = P^s + P^u$, and commute with A_0 and its exponential. We use these facts to break the integral into two parts, a stable projection integrated for $\tau \le x$, and an unstable projection integrated for $\tau \ge x$,

$$
\begin{aligned}
K(x) = &\int_{L}^{x} e^{A_0(x-\tau)}P^s R(\tau)\left[e^{A_0\tau}v_s + K(\tau)\right]\mathrm{d}\tau \\
&- \int_{x}^{+\infty} e^{A_0(x-\tau)}P^u R(\tau)\left[e^{A_0\tau}v_s + K(\tau)\right]\mathrm{d}\tau,
\end{aligned} \tag{9.2.10}
$$

where the choice of limits in the integrals, particularly $L \ge 0$, corresponds to a particular choice of initial data. In order to show that $K(x)$ decays as $x \to +\infty$ we use the bounds (9.2.3) and (9.2.4) and the decay of $R(x)$ in (9.1.5) to estimate the P^s terms for $x \ge \tau \ge L$,

$$
\begin{aligned}
|e^{A_0(x-\tau)}P^s R(\tau)e^{A_0\tau}v_s| &\le Ce^{\eta_s x}e^{-\alpha\tau}|v_s| \\
|e^{A_0(x-\tau)}P^s R(\tau)K(\tau)| &\le Ce^{\eta_s x}e^{-\alpha\tau}|e^{-\eta_s\tau}K(\tau)|,
\end{aligned}
$$

and the P^{u} terms for $\tau \geq x \geq L$,

$$|e^{A_0(x-\tau)} P^{\mathrm{u}} \boldsymbol{R}(\tau) e^{A_0 \tau} \boldsymbol{v}_{\mathrm{s}}| \leq C e^{\eta_{\mathrm{u}} x} e^{(\eta_{\mathrm{s}}-\eta_{\mathrm{u}}-\alpha)\tau} |\boldsymbol{v}_{\mathrm{s}}|$$
$$|e^{A_0(x-\tau)} P^{\mathrm{u}} \boldsymbol{R}(\tau) \boldsymbol{K}(\tau)| \leq C e^{\eta_{\mathrm{u}} x} e^{-\alpha\tau} |e^{-\eta_{\mathrm{u}}\tau} \boldsymbol{K}(\tau)|.$$

Applying these estimates to (9.2.10) and simplifying yields

$$\begin{aligned} |\boldsymbol{K}(x)| \leq C \Big(& e^{\eta_{\mathrm{s}} x} \int_L^x e^{-\alpha\tau} (|\boldsymbol{v}_{\mathrm{s}}| + |e^{-\eta_{\mathrm{s}}\tau} \boldsymbol{K}(\tau)|)\, d\tau \\ & + e^{\eta_{\mathrm{u}} x} \int_x^{+\infty} e^{-g_\gamma \tau} (|\boldsymbol{v}_{\mathrm{s}}| + |e^{-\eta_{\mathrm{s}}\tau} \boldsymbol{K}(\tau)|)\, d\tau \Big). \end{aligned} \tag{9.2.11}$$

Multiplying (9.2.11) by $e^{-\eta_{\mathrm{s}} x}$ allows us to rewrite the inequality as

$$\begin{aligned} |e^{-\eta_{\mathrm{s}} x} \boldsymbol{K}(x)| \leq C \Big(& \int_L^x e^{-\alpha\tau} (|\boldsymbol{v}_{\mathrm{s}}| + |e^{-\eta_{\mathrm{s}}\tau} \boldsymbol{K}(\tau)|)\, d\tau \\ & + e^{(g_\gamma - \alpha)x} \int_x^{+\infty} e^{-g_\gamma \tau} (|\boldsymbol{v}_{\mathrm{s}}| + |e^{-\eta_{\mathrm{s}}\tau} \boldsymbol{K}(\tau)|)\, d\tau \Big). \end{aligned} \tag{9.2.12}$$

The structure of the inequality above suggests bounding $\boldsymbol{K}(\tau)$ with its exponentially weighted modulus

$$M := \sup_{\tau \geq L} (e^{-\eta_{\mathrm{s}}\tau} |\boldsymbol{K}(\tau)|).$$

The modulus M is independent of τ, depending only upon L. Taking the sup over $x \geq L$ in (9.2.12) yields

$$M \leq C(|\boldsymbol{v}_{\mathrm{s}}| + M) \sup_{x \geq L} \left(\int_L^x e^{-\alpha\tau}\, d\tau + e^{(g_\gamma - \alpha)x} \int_x^{+\infty} e^{-g_\gamma \tau}\, d\tau \right).$$

For $\gamma \in \Omega_{\mathrm{e}}$ the spectral gap satisfies $g_\gamma > 0$. Evaluating the integrals we obtain the bound, valid for $x \geq L$ and C independent of γ,

$$M \leq C e^{-\alpha L} \left(1 + g_\gamma^{-1}\right) (|\boldsymbol{v}_{\mathrm{s}}| + M).$$

In particular, there exists $L_0 = L_0(\delta)$ such that for $L \geq L_0$ we have $C e^{-\alpha L}(1 + g_\gamma^{-1}) < 1/2$, and we may absorb the M term on the right-hand side into the M term on the left-hand side. Recalling the definition of M, this yields the bound

$$|\boldsymbol{K}(x)| \leq C e^{-\alpha L} \left(1 + g_\gamma^{-1}\right) e^{\eta_{\mathrm{s}} x} |\boldsymbol{v}_{\mathrm{s}}|, \tag{9.2.13}$$

which is slightly stronger than (9.2.7).

For the case when γ is near the branch point, i.e., $|\gamma| \leq \alpha/4$, the spectral gap, g_γ, tends to zero and the spectral projections become poorly scaled as the matrix eigenvalues and eigenvectors approach each other. In this case, we express (9.2.9) in the form

$$K(x) = -\int_x^{+\infty} e^{A_0(x-\tau)} R(\tau)\left[e^{A_0\tau} v_s + K(\tau)\right] d\tau. \tag{9.2.14}$$

Writing $\eta_s = \eta_b - \operatorname{Re}\gamma$ and $\eta_u = \eta_b - \operatorname{Re}\gamma$, where $\eta_b = -a_1^0/2$ is the common value of matrix eigenvalues at the branch point, we have the bounds

$$\left|e^{A_0 s}\right| \le C e^{(\eta_b+|\gamma|)s}, \quad s>0; \quad \left|e^{A_0 s}\right| \le C e^{(\eta_b-|\gamma|)s}, \quad s<0.$$

Since for $x > L$ the exponents in the integrand of (9.2.14) satisfy $x-\tau < 0$ and $\tau > 0$, we may bound $K(x)$ as

$$|K(x)| \le C e^{(\eta_b-|\gamma|)x} \int_x^{+\infty} \left[e^{(-\alpha+2|\gamma|)\tau}|v_s| + e^{(-\alpha-\eta_b+|\gamma|)\tau}|K(\tau)|\right] d\tau, \tag{9.2.15}$$

i.e.,

$$e^{(-\eta_b+|\gamma|)x}|K(x)| \le C e^{(\eta_b-|\gamma|)x} \int_x^{+\infty} \left[e^{(-\alpha+2|\gamma|)\tau}|v_s| + e^{-\alpha\tau} e^{(-\eta_b+|\gamma|)\tau}|K(\tau)|\right] d\tau.$$

Introducing

$$M := \sup_{x\ge L}\left(e^{(-\eta_b+|\gamma|)x}|K(x)|\right),$$

following the same argument as above we find that there exists L_0 such that for $x \ge L \ge L_0$,

$$M \le C e^{(-\alpha+2|\gamma|+\eta_b)x}|v_s|.$$

Hence, for $x \ge L$,

$$|K(x)| \le C e^{(-\alpha+|\gamma|+\eta_b)x}|v_s| \le C e^{-\alpha L/2} e^{\eta_s x}|v_s|,$$

which is consistent with (9.2.7).

The analyticity of the Jost function in γ, away from the branch point, follows from the analyticity of the vector v_s, the analyticity of the spectral projections, and the analyticity of $A_0(\gamma)$, together with the uniform convergence of the integrals. The analyticity at the branch point follows from the fact that v_s is continuous there, and from the uniform convergence of the integral in (9.2.15). □

The crucial assumption in Theorem 9.2.3 is the spectral gap condition implicit in the definition (9.2.6) of the extended domain Ω_e. This gap condition always holds for $\gamma \in \Omega$, and the positivity of the decay rate $\alpha > 0$ allows us to analytically continue the Jost solution across the Fredholm boundary, over the remainder of the physical sheet, and some distance onto the resonance sheet of the Riemann surface. Note that when $a_1^0 = 0$ the Fredholm boundary coincides with the absolute spectrum, and the branch point lies on the boundary of the essential spectrum. If, moreover, $\boldsymbol{R}(x)$ has compact support, then we may take $\alpha = \infty$, and the Jost solutions will be entire on the Riemann surface. Conversely, there are examples, e.g., see Chapter 9.3.2, where the Evans function has a pole precisely on the boundary of the extended domain, Ω_e.

Remark 9.2.4. The notion of Jost solutions as special solutions to linear ODEs arose in the late 1970s, see, e.g., Chadan and Sabatier [**47**], Reed and Simon [**243**] and the references therein.

Remark 9.2.5. For the second-order problem (9.2.1) the matrix eigenvalues and eigenvectors are entire in γ; in particular, they are analytic on the extended domain Ω_e. In this case the Jost solutions $\boldsymbol{J}^{\pm}(x,\gamma)$ are uniquely determined up to the choice of matrix eigenvectors $\boldsymbol{v}_s(\gamma) = \boldsymbol{v}_2(\gamma)$ and $\boldsymbol{v}_u(\gamma) = \boldsymbol{v}_1(\gamma)$, through the asymptotic relation

$$|\boldsymbol{J}^+(x,\gamma) - e^{\mu_2(\gamma)x}\boldsymbol{v}_2(\gamma)| \leq Ce^{-\alpha L/2}|\boldsymbol{v}_2(\gamma)|, \quad x \geq L, \tag{9.2.16}$$

with a corresponding relation for $\boldsymbol{J}^-$. The normalization of the Jost solutions in a neighborhood of the branch point is investigated in Chapter 9.5. For higher-order problems the normalization of the Jost solutions is complicated by the fact that the eigenvalues and eigenvectors may merge, forming Jordon block structures and losing analyticity, even for γ within the natural domain Ω.

As we saw in Chapter 4, is can be surprisingly difficult to obtain estimates on the resolvent $(\mathcal{L} - \lambda)^{-1}$ for large λ. One might hope that a simple scaling argument would show that on the natural domain, for $|\gamma| \gg 1$, the Evans function for the full system approaches the Evans function for the constant coefficient, asymptotic system. This is a bit too much to ask for. Rather, defining a slight subset

$$\Omega_\delta = \left\{\lambda \in \Omega : |\arg(\lambda - \lambda_{br})| < \frac{\pi}{2} - \delta\right\}, \tag{9.2.17}$$

then for any $\delta > 0$ the Jost solutions, after a renormalization, converge to the corresponding asymptotic solutions on the half-line as λ grows large within Ω_δ. By abuse of notation we also denote the image of Ω_δ under the Riemann map, (9.1.7), by Ω_δ.

Corollary 9.2.6. *Fix $\delta > 0$. For all $\gamma \in \Omega_\delta$ with $|\gamma|$ sufficiently large there is a choice of Jost solutions that satisfy*

$$\begin{aligned} |\tilde{\boldsymbol{J}}^+(x,\gamma) - e^{\boldsymbol{A}_0(\gamma)x}\boldsymbol{v}_s(\gamma)| &\leq \frac{C}{|\gamma|}e^{-\alpha x}|\boldsymbol{v}_s(\gamma)|, \quad x \geq 0 \\ |\tilde{\boldsymbol{J}}^-(x,\gamma) - e^{\boldsymbol{A}_0(\gamma)x}\boldsymbol{v}_u(\gamma)| &\leq \frac{C}{|\gamma|}e^{\alpha x}|\boldsymbol{v}_u(\gamma)|, \quad x \leq 0. \end{aligned} \tag{9.2.18}$$

Here the vectors $\boldsymbol{v}_{s,u}(\gamma)$ are chosen as in Theorem 9.2.3.

Remark 9.2.7. The important point of Corollary 9.2.6 is that the renormalized Jost solutions for the full problem are small perturbations of the Jost solutions for the constant coefficient asymptotic problem on the half-line. We have exchanged $L \gg 1$ for $\gamma \in \Omega$ and sufficiently large in magnitude.

Proof. A key observation is that $\gamma \in \Omega_\delta$ implies both that $\eta_{\mathrm{u}}(\gamma) \sim |\gamma|$ and $\eta_{\mathrm{s}}(\gamma) \sim -|\gamma|$ as $|\gamma| \to \infty$. We consider $\boldsymbol{J}^{+}(x,\gamma)$, and revisit the formula (9.2.10) for the correction term $\boldsymbol{K}$ of the Jost function. We set $L = 0$, which amounts to choosing a different initial data, and yields a different Jost function, denoted by $\tilde{\boldsymbol{J}}^{+}$. However, in the case of a second-order system the decaying spaces are one-dimensional, and the renormalization is merely a scalar multiplication.

Our goal is to show that if $|\gamma|$ is sufficiently large, then $|\boldsymbol{K}(x)|$ will be uniformly small. We begin by rewriting the estimates preceding (9.2.11) as

$$|\mathrm{e}^{\boldsymbol{A}_0(x-\tau)} P^{\mathrm{s}} \boldsymbol{R}(\tau)\boldsymbol{v}| \leq C\mathrm{e}^{\eta_{\mathrm{s}} x}\mathrm{e}^{-(\eta_{\mathrm{s}}+\alpha)\tau}|\boldsymbol{v}|, \; x-\tau \geq 0$$
$$|\mathrm{e}^{\boldsymbol{A}_0(x-\tau)} P^{\mathrm{u}} \boldsymbol{R}(\tau)\boldsymbol{v}| \leq C\mathrm{e}^{\eta_{\mathrm{u}} x}\mathrm{e}^{-(\eta_{\mathrm{u}}+\alpha)\tau}|\boldsymbol{v}|, \; x-\tau \leq 0.$$

Introduce the unweighted modulus

$$M := \sup_{\tau \geq 0} |\boldsymbol{K}(\tau)|,$$

and note that for $\gamma \in \Omega_\delta \subset \Omega$ we have $\eta_{\mathrm{s}} < 0 < \eta_{\mathrm{u}}$ and hence $|\mathrm{e}^{\boldsymbol{A}_0 \tau}\boldsymbol{v}_{\mathrm{s}}| \leq C|\boldsymbol{v}_{\mathrm{s}}|$. This affords an estimate analogous to (9.2.11),

$$\begin{aligned} |\boldsymbol{K}(x)| &\leq C\left(\mathrm{e}^{\eta_{\mathrm{s}} x}\int_0^x \mathrm{e}^{-(\eta_{\mathrm{s}}+\alpha)\tau}\,\mathrm{d}\tau + \mathrm{e}^{\eta_{\mathrm{u}} x}\int_x^{+\infty} \mathrm{e}^{-(\eta_{\mathrm{u}}+\alpha)\tau}\,\mathrm{d}\tau\right)(|\boldsymbol{v}_{\mathrm{s}}| + M) \\ &\leq C\mathrm{e}^{-\alpha x}\left(-\frac{1}{\eta_{\mathrm{s}}+\alpha} + \frac{1}{\eta_{\mathrm{u}}+\alpha}\right)(|\boldsymbol{v}_{\mathrm{s}}| + M) \leq C\mathrm{e}^{-\alpha x}\frac{|\boldsymbol{v}_{\mathrm{s}}| + M}{|\gamma|}, \end{aligned} \tag{9.2.19}$$

where the last inequality is where we require that $\gamma \in \Omega_\delta$. Taking the sup over $x > 0$, since $\mathrm{e}^{-\alpha x} \leq 1$ for all $x \geq 0$, the M term on the right-hand side can be absorbed into the M term on the left-hand side for $|\gamma|$ sufficiently large. This yields the bound

$$M \leq C\frac{|\boldsymbol{v}_{\mathrm{s}}|}{|\gamma|},$$

and from (9.2.19) we deduce (9.2.18). □

Exercises

Exercise 9.2.1. Suppose that $\gamma \in \Omega$. Show that for $n \in \mathbb{N}$ the Jost solutions satisfy

$$\lim_{x \to \pm\infty} \partial_\gamma^n \boldsymbol{J}^{\pm}(x,\gamma) = \boldsymbol{0},$$

and that the approach is exponentially fast.

Exercise 9.2.2. Show that the Jost solutions can be constructed under the weaker assumption that $|\boldsymbol{R}(x)| \leq C/(1+|x|)^p$ for some $p > 1$, and find their domain of analyticity in γ.

9.3 The Evans Function

Solutions to the eigenvalue problem (9.1.1) that lie in $H^2(\mathbb{R})$ enjoy the nominal "boundary condition" (9.1.2). For $\gamma \in \Omega$, Theorem 9.2.3 constructs the Jost solutions $\boldsymbol{J}^+(x,\gamma)$, which decays as $x \to +\infty$, and $\boldsymbol{J}^-(x,\gamma)$, which decays as $x \to -\infty$. It is easy to see that if for some $\gamma_0 \in \Omega$ the two Jost solutions $\{\boldsymbol{J}^\pm(x,\gamma_0)\}$ are linearly dependent,

$$a_-\boldsymbol{J}^-(x,\gamma_0) + a_+\boldsymbol{J}^+(x,\gamma_0) = \boldsymbol{0},$$

then $\boldsymbol{J}_0 := \boldsymbol{J}^- = -a_+/a_-\boldsymbol{J}^+$ solves the eigenvalue problem and decays exponentially at both $x = \pm\infty$. Conversely, if we can show that the Jost solution $\boldsymbol{J}^+(x,\gamma)$ spans the linear space of solutions of (9.2.1) that decay at $x = +\infty$, while $\boldsymbol{J}^-(x,\gamma)$ spans the linear space of solutions which decay at $x = -\infty$, then their linear dependence would be *equivalent* to the existence of the eigenvalue γ_0. This task is less immediate than in the case of the bounded domain construction, but the adjoint problem proves quite useful.

As in Chapter 8.1, we introduce a vectorized version of the eigenvalue problem for the adjoint operator $\mathcal{L}^{\mathrm{a}}$, in the form

$$\partial_x \boldsymbol{Z} = -\boldsymbol{A}(x,\gamma)^{\mathrm{H}}\boldsymbol{Z}. \tag{9.3.1}$$

Moreover, as in Lemma 8.1.3, if $\boldsymbol{Y}(x,\gamma)$ and $\boldsymbol{Z}(x,\gamma)$ are solutions to (9.2.1) and the adjoint problem (9.3.1) respectively, then the value of their scalar product is independent of $x \in \mathbb{R}$, i.e.,

$$\partial_x[\boldsymbol{Y}(x,\gamma)\cdot\boldsymbol{Z}(x,\gamma)] = 0. \tag{9.3.2}$$

The matrix eigenvalues for the adjoint system, denoted $\mu_j^{\mathrm{a}}(\gamma)$ for $j = 1,2$, satisfy $\mu_j^{\mathrm{a}}(\gamma) = -\overline{\mu_j(\gamma)}$, so that the natural and extended domains, as well as the branch points, are the same as for the original problem. Moreover, Theorem 9.2.3 applies to the adjoint problem, yielding the existence of the adjoint Jost solutions, $\boldsymbol{J}^{\mathrm{a}\pm}(x,\gamma)$, which for $\gamma \in \Omega$ decay at $x = \pm\infty$ with exponential rates $-\eta_{\mathrm{u}} < 0$ and $-\eta_{\mathrm{s}} > 0$, respectively. Through (9.3.2), the adjoint Jost solutions induce the complementary Jost soutions: solutions of (9.2.1) with asymptotic behavior that is inversely proportional to the adjoint Jost solution.

Lemma 9.3.1. *Under the assumptions of Theorem 9.2.3, for $\gamma \in \Omega_e$ there exists complementary Jost solutions $\boldsymbol{K}^{\pm}(x,\gamma)$ of (9.2.1) that satisfy*

$$\lim_{x\to+\infty} |e^{\eta_u x}\boldsymbol{K}^{+}(x,\gamma)| = 1, \quad \lim_{x\to-\infty} |e^{\eta_s x}\boldsymbol{K}^{-}(x,\gamma)| = 1. \tag{9.3.3}$$

In particular, on the physical sheet, $\mathrm{Re}\,\gamma > 0$, *each of the sets* $\{\boldsymbol{J}^{+},\boldsymbol{K}^{+}\}$ *and* $\{\boldsymbol{J}^{-},\boldsymbol{K}^{-}\}$ *are linearly independent and span the solution space of (9.2.1).*

Proof. We consider $\boldsymbol{K}^{+}(x,\gamma)$ and choose $\boldsymbol{K}^{+}(0,\gamma)$ so that $\boldsymbol{J}^{a+}(0,\gamma)\cdot\boldsymbol{K}^{+}(0,\gamma) = 1$. By (9.3.2) we see that $\boldsymbol{J}^{a+}(x,\gamma)\cdot\boldsymbol{K}^{+}(x,\gamma) = 1$ for all x. Under a normalization analogous to Remark 9.2.5, $\boldsymbol{J}^{a+}(x,\gamma)$ is asymptotically exponential with rate $-\eta_u(\gamma)$ as $x \to +\infty$: it follows that $\boldsymbol{K}^{+}(x,\gamma)$ is asymptotically exponential with rate $\eta_u(\gamma)$ as $x \to +\infty$. The limit (9.3.3) follows from a rescaling. The existence of $\boldsymbol{K}^{-}(x,\gamma)$ follows from similar arguments. Since $\eta_s < \eta_u$ for $\mathrm{Re}\,\gamma > 0$, the linear independence of $\boldsymbol{J}^{+}$ and $\boldsymbol{K}^{+}$ follows from the dichotomy of their exponential limits. □

With Lemma 9.3.1 in hand the equivalence of the linear dependence of the Jost solutions and the existence of an eigenvalue of (9.1.1) reduces to a simple observation. Suppose that $\gamma_0 \in \Omega$ is an eigenvalue of $\mathcal{L}$, and let $\boldsymbol{Y}_0(x)$ denote the corresponding vector-valued eigenfunction—we wish to show that it is proportional to the two Jost solutions, and hence is a Jost eigenfunction. By Lemma 9.3.1 we have two representations of $\boldsymbol{Y}_0$,

$$\boldsymbol{Y}_0 = a_1\boldsymbol{J}^{-}(x,\gamma_0) + a_2\boldsymbol{K}^{-}(x,\gamma_0), \quad \boldsymbol{J}_0 = b_1\boldsymbol{J}^{+}(x,\gamma_0) + b_2\boldsymbol{K}^{+}(x,\gamma_0).$$

Since $\boldsymbol{Y}_0(x)$ decays at both $x = \pm\infty$, from the first representation the limit $x \to -\infty$ implies that $a_2 = 0$, while from the second the limit $x \to \infty$ shows that $b_2 = 0$. This establishes the linear dependence of the Jost solutions and motivates the following definition of the Evans function on its extended domain Ω_e, given in (9.2.6)

Definition 9.3.2. Let $\mathcal{L}$ be an exponentially asymptotic operator of the form (9.1.1). On its extended domain, $\Omega_e \subset \mathbb{C}$, given in (9.2.6), the associated Evans matrix is the concatenation of the Jost solutions, defined in (9.2.7), evaluated at $x = 0$,

$$\boldsymbol{E}(\gamma) := \left(\boldsymbol{J}^{-},\boldsymbol{J}^{+}\right)(0,\gamma), \tag{9.3.4}$$

and the Evans function is its determinant

$$E(\gamma) := \det \boldsymbol{E}(\gamma). \tag{9.3.5}$$

Remark 9.3.3. The Evans function is determined only up to the specification of the matrix eigenvectors, $\{\boldsymbol{v}_1(\gamma),\boldsymbol{v}_2(\gamma)\}$. Different analytic choices of these eigenvectors over the extended domain lead to distinct Evans functions; however, the Evans functions share the same zeros on Ω_e, up to multiplicity.

The zeros of the Evans function on the natural domain Ω correspond to the point spectrum of the operator $\mathcal{L}$.

Lemma 9.3.4. *The Evans function associated with the Sturm–Liouville problem (9.1.1) is analytic on its extended domain, Ω_e. On its natural domain, $\gamma = \gamma_0 \in \Omega$ corresponds to an eigenvalue of $\mathcal{L}$ if and only if $E(\gamma_0) = 0$, with the order of the zero of the Evans function equal to the multiplicity of the eigenvalue. In particular, $\gamma = \gamma_0$ is a simple eigenvalue if and only if*

$$E(\gamma_0) = 0, \quad E'(\gamma_0) = 2\gamma_0 \langle p_0, q_0 \rangle \det(\boldsymbol{J}_0^{\mathrm{a}}, \boldsymbol{J}_0)(0) \neq 0. \tag{9.3.6}$$

Here p_0 and q_0 are the eigenfunction and adjoint eigenfunction associated with the eigenvalue $\lambda_0 = \lambda_{\mathrm{br}} + \gamma_0^2$ of $\mathcal{L}$, and

$$\boldsymbol{J}_0(x) = \begin{pmatrix} p_0(x) \\ \partial_x p_0(x) \end{pmatrix}, \quad \boldsymbol{J}_0^{\mathrm{a}}(x) = \begin{pmatrix} a_1(x) q_0(x) - \partial_x q_0(x) \\ q_0(x) \end{pmatrix},$$

are the eigensolution of (9.2.1) and the adjoint problem (9.3.1), respectively. The adjoint solution is scaled so that $|\boldsymbol{J}_0^{\mathrm{a}}(0)| = 1$.

Proof. In light of the preceding discussion, all that remains to be proven is that within the natural domain Ω, the order of the zero of the Evans function equals the algebraic multiplicity of the eigenvalue. However, this follows from the bounded domain arguments used to establish Proposition 8.1.4, modulo a slight variation required to establish the formula (9.3.6) for $E'(\gamma_0)$, since $'$ now denotes differentiation with respect to γ and not λ. Assuming, without loss of generality, that the Jost solutions are normalized so that $\boldsymbol{J}_0 := \boldsymbol{J}^-(x,\gamma_0) = \boldsymbol{J}^+(x,\gamma_0)$, the key step is to construct $\partial_\gamma \boldsymbol{J}^\pm$, which satisfies

$$\partial_x(\partial_\gamma \boldsymbol{J}^\pm) = \boldsymbol{A}(x,\gamma_0)\partial_\gamma \boldsymbol{J}^\pm + \partial_\gamma \boldsymbol{A}(x,\gamma_0)\boldsymbol{J}_0, \quad \lim_{x\to\pm\infty} \partial_\gamma \boldsymbol{J}^\pm(x,\gamma_0) = \boldsymbol{0}.$$

The existence of solutions of this ODE with the required boundary conditions is established in Exercise 9.2.1. Adapting the arguments from (8.1.17) to (8.1.21) to derive (9.3.6) only requires replacing the λ derivatives with $\partial_\gamma \boldsymbol{A} = 2\gamma \partial_\lambda \boldsymbol{A}$, and evaluating the integrals over the whole line. □

Remark 9.3.5. Recall the discussion in Remark 8.1.5 regarding the scaling of the adjoint Jost solution $\boldsymbol{J}_0^{\mathrm{a}}(x)$ for the Evans function derivative formula. The result also applies here: if the adjoint solution is not scaled, then the derivative expression must be divided by $|\boldsymbol{J}_0^{\mathrm{a}}(0)|^2 > 0$.

Example 9.3.6. Consider the constant coefficient Sturm–Liouville operator $\mathcal{L}$ with $a_1(x) \equiv a_1^0$ and $a_0(x) \equiv a_0^0$. The Jost solutions are given by

$$\boldsymbol{J}^-(x,\gamma) = \mathrm{e}^{\mu_1(\gamma)x}\boldsymbol{v}_1(\gamma), \quad \boldsymbol{J}^+(x,\gamma) = \mathrm{e}^{\mu_2(\gamma)x}\boldsymbol{v}_2(\gamma),$$

where

$$\mu_1(\gamma) = -\frac{a_1^0}{2} + \gamma, \quad v_1(\gamma) = \begin{pmatrix} 1 \\ \mu_1(\gamma) \end{pmatrix}; \quad \mu_2(\gamma) = -\frac{a_1^0}{2} - \gamma, \quad v_2(\gamma) = \begin{pmatrix} 1 \\ \mu_2(\gamma) \end{pmatrix}.$$

The Evans function is

$$E(\gamma) = \det\big(J^-, J^+\big)(0,\gamma) = \mu_2(\gamma) - \mu_1(\gamma) = -2\gamma.$$

The Evans function is zero only at the branch point $\gamma = 0$, and $\mathcal{L}$ has no point spectrum.

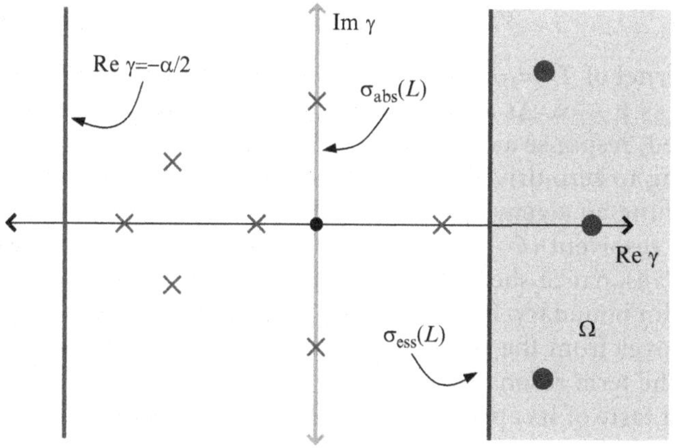

Fig. 9.4 The regions of the Riemann surface for which the Evans function is well-defined (compare with Fig. 9.3). On the natural domain Ω the zeros of the Evans function (*blue* circles) correspond to eigenvalues. For λ in the extended domain Ω_e but outside of Ω the zeros of the Evans function (*red* crosses) correspond to resonance poles. (Color figure online.)

Remark 9.3.7. It is important to observe that zeros of the Evans function outside of its natural domain but within the extended domain do not generically correspond to point spectrum, since the asymptotic properties of the Jost solutions are extended to preserve their analyticity, and not their decay. In $\Omega_e\backslash\Omega$ it is possible to have eigenvalues of $\mathcal{L} - \lambda$ that are not zeros of the Evans function, and to have zeros of the Evans function that are not eigenvalues. It is counterintuitive that the latter are more important. Zeros of the Evans function that cross into the natural domain under perturbation become point spectra of $\mathcal{L}$, because in the natural domain the associated solution for the vectorized system will decay as $x \to \pm\infty$. On the other hand, eigenvalues in $\Omega_e\backslash\Omega$ have eigenfunctions that decay as $x \to \pm\infty$; however, the analytic continuation of these eigenfunctions generally do *not* decay at $x = \pm\infty$ as γ crosses into Ω.

Definition 9.3.8. The zeros of the Evans function that lie in the domain $\Omega_e \backslash \Omega$ are called *resonance poles* (see Fig. 9.4). The associated eigenfunctions, $p_0(x,\gamma)$ and $J_0(x,\gamma)$, are called *resonance functions.*

The term *resonance pole* arises from analogy with the temporal behavior of a forced oscillator

$$\partial_t^2 y + \omega_0^2 y = \sin(\omega t).$$

As the forcing frequency ω approaches the natural frequency ω_0 the system response $y(t)$ grows without bound, because the forcing term lies within the kernel of the operator $\mathcal{T}_0 := \partial_t^2 + \omega_0^2$. When $\omega = \omega_0$, the system is said to be in resonance. If a small damping term is added to the system, $0 < \epsilon \ll 1$,

$$\partial_t^2 y - 2\epsilon \partial_t y + \omega_0^2 y = \sin(\omega t),$$

then the kernel of $\mathcal{T}_\epsilon := \partial_t^2 - \epsilon \partial_t + \omega_0^2$ consists of functions that grow at the rate $\mathcal{O}(e^{\epsilon t})$ as $t \to \infty$. As a consequence the forced system exhibits a large, but bounded, response as ω approaches the resonant frequency ω_0. Tuning the damping to zero drives the amplitude of the resonance response to ∞, just as pushing an eigenvalue out of the essential spectrum will generate a pole of the resolvent $(\mathcal{L} - \lambda)^{-1}$. In many physically important cases $a_1^0 = 0$, so that the resonance sheet of the Riemann surface is on the other side of the Fredholm boundary. In this case the pole of the resolvent (eigenvalue of $\mathcal{L}$) crosses over from the resonance sheet. In fact, the term resonance pole motivates the term resonance sheet.

As a first taste of its applicability and usefulness, we use the Evans function to give an elementary proof of the following well-known fact.

Lemma 9.3.9. *Let $\mathcal{K}$ be a compact subset of $\mathbb{R}^n$ and consider a family of exponentially localized Sturm–Liouville operators $\mathcal{L}(\boldsymbol{p}) = \partial_x^2 + a_1(x,\boldsymbol{p})\partial_x + a_0(x,\boldsymbol{p})$ of the form (9.1.1), which depends smoothly upon parameters $\boldsymbol{p} \in \mathcal{K}$. Suppose that there exists $\alpha_0 > 0$ such that the exponential decay rate satisfies $\inf_{\boldsymbol{p}\in\mathcal{K}} \alpha(\boldsymbol{p}) \geq \alpha_0$. Fix $\delta > 0$, then for each $\boldsymbol{p}$ the operator $\mathcal{L}(\boldsymbol{p})$ has at most a finite number of eigenvalues in Ω_δ, defined in (9.2.17); moreover, there exists an $M > 0$ such that the Evans function $E(\gamma,\boldsymbol{p})$ has no zeros on the set*

$$|\gamma| \geq M, \quad \gamma \in \Omega_\delta.$$

In particular, the number of eigenvalues of $\mathcal{L}(\boldsymbol{p}) \cap \Omega_\delta$, counted according to multiplicity, can increase or decrease only by zeros of the Evans function crossing $\partial\Omega_\delta(\boldsymbol{p})$.

Proof. Fix $\boldsymbol{p} \in \mathcal{K}$, and hereafter suppress the $\boldsymbol{p}$-dependence of solutions. By Corollary 9.2.6 the Jost solutions at $x = 0$ for $\gamma \in \Omega_\delta$ sufficiently large are given by

$$\boldsymbol{J}^+(0,\gamma) = \boldsymbol{v}_s(\gamma) + \mathcal{O}(|\gamma|^{-1}), \quad \boldsymbol{J}^-(0,\gamma) = \boldsymbol{v}_u(\gamma) + \mathcal{O}(|\gamma|^{-1}),$$

where $\boldsymbol{v}_{\mathrm{s,u}}(\gamma)$ are eigenvectors for the matrix $\boldsymbol{A}_0(\gamma)$ (also see Theorem 9.2.3). The Evans function satisfies

$$E(\gamma) = \det(\boldsymbol{J}^-, \boldsymbol{J}^+)(0,\gamma) = \det(\boldsymbol{v}_{\mathrm{s}}, \boldsymbol{v}_{\mathrm{u}})(\gamma) + \mathcal{O}(|\gamma|^{-1}), \tag{9.3.7}$$

where the eigenvectors $\{\boldsymbol{v}_{\mathrm{u}}, \boldsymbol{v}_{\mathrm{s}}\}$ are linearly independent on $\Omega \supset \Omega_\delta$. Hence the Evans function is not zero for $|\gamma| \in \Omega_\delta$ sufficiently large. Since the Evans function is analytic, it has only a finite number of zeros on bounded sets. The compactness of $\mathcal{K}$ and smooth dependence upon $\boldsymbol{p}$ yields the remaining results. □

An explicit determination of the Jost solutions typically requires solving a nonconstant coefficient ODE with nontrivial asymptotic boundary conditions. This renders it unusual, but very informative, when the Evans function has an explicit formulation. In the following examples we derive explicit formulas for the Evans function and use it to track the motion of resonance poles and their transformation into point spectra as they cross into the natural domain Ω.

9.3.1 Example: Square-Well Potential

Consider a Sturm–Liouville operator with piecewise constant coefficients. Specifically, suppose that $a_1(x) \equiv 0$ and

$$a_0(x) = \begin{cases} 0, & |x| > 1, \\ F, & |x| \leq 1. \end{cases}$$

In this case the essential spectrum coincides with the negative real axis, and the Riemann surface reduces to $\lambda = \gamma^2$. Since the coefficients $a_i(x)$ have compact support, we anticipate that the Jost solutions and hence the Evans function will be entire in γ over the Riemann surface. The Jost solutions $\boldsymbol{J}^\pm(x,\gamma)$ are defined in Theorem 9.2.3 via their asymptotic behavior at $x = \pm\infty$, respectively. Since the ODE (9.2.1) is constant coefficient for $|x| > 1$ we readily calculate that

$$\boldsymbol{J}^-(x,\gamma) = \mathrm{e}^{\gamma x}\begin{pmatrix} 1 \\ \gamma \end{pmatrix}, \quad x \leq -1; \qquad \boldsymbol{J}^+(x,\gamma) = \mathrm{e}^{-\gamma x}\begin{pmatrix} 1 \\ -\gamma \end{pmatrix}, \quad x \geq 1.$$

To define the Evans function we must continue $\boldsymbol{J}^\pm$ from $x = \pm 1$ to $x = 0$. We do so through the fundamental matrix solution, which for $|x| < 1$ takes the form

$$\boldsymbol{\Phi}(x,\gamma,F) = \begin{pmatrix} \cosh(\sqrt{\gamma - F}\,x) & \sinh(\sqrt{\gamma - F}\,x)/\sqrt{\gamma - F} \\ \sqrt{\gamma - F}\,\sinh(\sqrt{\gamma - F}\,x) & \cosh(\sqrt{\gamma - F}\,x). \end{pmatrix}$$

It is not immediately apparent that $\boldsymbol{\Phi}$ is entire in γ for fixed x; however, $\cosh(x)$ and $\sinh(x)/x$ are even and the singularities are removable, so the matrix solution is entire in γ for fixed x. The continuation of the Jost solutions to $x = 0$ takes the form

$$\begin{aligned} J^{\pm}(0,\gamma,F) &= \boldsymbol{\Phi}(0,\gamma,F)\boldsymbol{\Phi}(\pm 1,\gamma,F)^{-1}J^{\pm}(\pm 1,\gamma) \\ &= \mathrm{e}^{-\gamma}\boldsymbol{\Phi}(0,\gamma,F)\boldsymbol{\Phi}(\pm 1,\gamma,F)^{-1}\begin{pmatrix} 1 \\ \pm\gamma \end{pmatrix}. \end{aligned}$$

A simple calculation yields the Evans function

$$\begin{aligned} E(\gamma,F) &= \det\left(J^{-},J^{+}\right)(0,\gamma,F), \\ &= -2\mathrm{e}^{-2\gamma}\left(\gamma\cosh(2\sqrt{\gamma-F})+\sqrt{\gamma-F}\sinh(2\sqrt{\gamma-F})+F\frac{\sinh(2\sqrt{\gamma-alF})}{2\sqrt{\gamma-F}}\right). \end{aligned} \tag{9.3.8}$$

The eigenvalue problem (9.1.1) associated to these potentials is self-adjoint, and the spectrum is real, so that the zeros of the Evans function on its natural domain, Ω, must be real. In particular eigenvalues can enter the natural domain from the Fredholm boundary only at the branch point $\gamma = 0$, where the Evans function takes the simple expression

$$E(0,F) = \begin{cases} \sqrt{F}\sin(2\sqrt{F}), & F > 0 \\ -\sqrt{-F}\sinh(2\sqrt{-F}), & F < 0, \end{cases}$$

If the potential F is negative, then $E(0,F) < 0$, and no eigenvalues can enter Ω from the resonance sheet $\mathrm{Re}\,\gamma < 0$. However, if the potential F is positive, then $E(0,F) = 0$ for $F = F_n := n^2\pi^2/4$, $n \in \mathbb{N}$. A numerical examination of the Evans function shows it has real-valued zeros for positive and negative real values of γ for $F > 0$; furthermore, at the critical values $F = F_n$, a zero of E moves from the negative real γ-axis, where it is a resonance pole, to the positive real γ-axis, where it becomes point spectra (see Fig. 9.5). Moreover, the zeros of E that enter at the branch point $\gamma = 0$ cannot escape to $\gamma = +\infty$ and hence must accumulate on the positive real axis. Indeed, for $\gamma > 0$ it is straightforward to see that

$$E(\gamma,F) \sim -2\gamma\mathrm{e}^{-2(\gamma-\sqrt{\gamma})} < 0, \quad \text{for } \gamma \gg 1,$$

has no large zeros on the positive real axis.

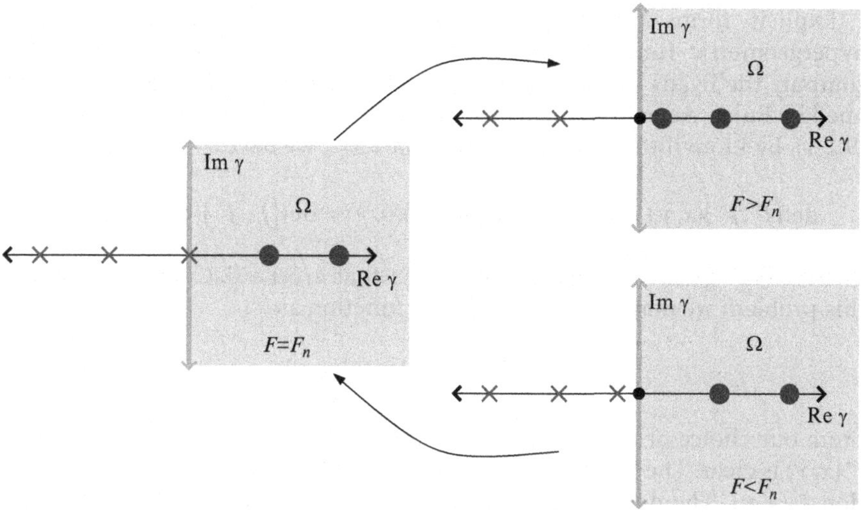

Fig. 9.5 The movement of the resonance poles (*red crosses*) and eigenvalues (*blue* circles), identified via zeros of the Evans function, as the potential F increases through the values $F_n := n^2\pi^2/4$. (*right bottom*) As F approaches F_n from below a resonance pole approaches the Fredholm boundary from the resonance sheet (*left*). At $F = F_n$ the resonance pole enters the Fredholm boundary at the branch point $\gamma = 0$ (*right top*). As F increases beyond F_n, an eigenvalue emerges from the branch point onto the positive real axis. The associated eigenfunction will exhibit slow spatial decay when the eigenvalue is proximal to the Fredholm boundary

9.3.2 Example: Reflectionless Potential

Reflectionless potentials have a distinguished history in scattering problems, where the Jost solutions were originally introduced [**3**, **74**, **207**, **243**]. In particular, the eigenvalue problem (9.1.1) for the case

$$a_1(x) \equiv 0, \quad a_0(x) = F_0 \operatorname{sech}^2(x)\,(F_0 > 0), \tag{9.3.9}$$

plays a special role in the application of the inverse scattering transform to the Korteweg–de Vries equation. Since $|a_0(x)| \le Ce^{-2|x|}$, by Lemma 9.3.4 the Evans function will be analytic on its extended domain $\Omega_e = \{\gamma : \operatorname{Re}\gamma > -1\}$. Under the normalization of Remark 9.2.5 the Jost solutions satisfy

$$\lim_{x\to-\infty} e^{-\gamma x} J^-(x,\gamma) = \begin{pmatrix} 1 \\ \gamma \end{pmatrix}; \quad \lim_{x\to+\infty} e^{\gamma x} J^+(x,\gamma) = \begin{pmatrix} 1 \\ -\gamma \end{pmatrix}, \tag{9.3.10}$$

and the Evans function is given by

$$E(\gamma) = \det\left(J^-, J^+\right)(0,\gamma).$$

Explicit forms for the Jost solutions can be found by using the hypergeometric function (e.g., see [**74**, Chapter 3.2]); however, we will compute the Evans function explicitly by using asymptotic relations developed by Euler. As both of the Jost solutions $\boldsymbol{J}^{\pm}(x,\gamma)$ are solutions of the ODE (9.2.1), by Liouville's formula, see Lemma 2.1.2, we have

$$\det\big(\boldsymbol{J}^-,\boldsymbol{J}^+\big)(x,\gamma) = \mathrm{e}^{\int_0^x a_1(s)\,\mathrm{d}s}\det\big(\boldsymbol{J}^-,\boldsymbol{J}^+\big)(0,\gamma) = \det\big(\boldsymbol{J}^-,\boldsymbol{J}^+\big)(0,\gamma) = E(\gamma).$$

The second equality follows from the fact that $a_1(x) \equiv 0$. Consequently, for this problem we may redefine the Evans function as

$$E(\gamma) = \lim_{x\to+\infty} \det\big(\boldsymbol{J}^-,\boldsymbol{J}^+\big)(x,\gamma). \tag{9.3.11}$$

From our choice of normalization given in (9.3.10) the limit at $x = +\infty$ for $\boldsymbol{J}^+(x,\gamma)$ is clear. The key is to determine the limit at $x = +\infty$ for the Jost function $\boldsymbol{J}^-(x,\gamma)$. The determination of this limit is precisely the 1D scattering problem. From Lemma 9.3.1 we may write $\boldsymbol{J}^-(x,\gamma)$ as a linear combination of $\boldsymbol{J}^+(x,\gamma)$ and $\boldsymbol{K}^+(x,\gamma)$, which for $x \gg 1$ takes the form

$$\mathrm{e}^{-\gamma x}\boldsymbol{J}^-(x,\gamma) \sim t(\gamma)\begin{pmatrix} 1 \\ \gamma \end{pmatrix} + r(\gamma)\mathrm{e}^{-2\gamma x}\begin{pmatrix} 1 \\ -\gamma \end{pmatrix}.$$

Here $t(\gamma)$ is the *transmission coefficient* and $r(\gamma)$ is the *reflection coefficient*, describing the fractions of the incoming wave that are transmitted and reflected, respectively, by the obstacle described by the potential a_0. Evaluating the limit in (9.3.11) we find

$$E(\gamma) = \lim_{x\to+\infty} \det\left(\mathrm{e}^{-\gamma x}\boldsymbol{J}^-(x,\gamma), \mathrm{e}^{\gamma x}\boldsymbol{J}^+(x,\gamma)\right) = -2\gamma t(\gamma). \tag{9.3.12}$$

Therefore, except at $\gamma = 0$ the Evans function and the transmission coefficient give the same information regarding the location and multiplicity of eigenvalues. As a consequence of [**74**, pp. 45–48] it can be shown that

$$t(\gamma) = \frac{\Gamma(\gamma)\Gamma(1+\gamma)}{\Gamma(1/2+\sqrt{\mathrm{F}_0+1/4}+\gamma)\Gamma(1/2-\sqrt{\mathrm{F}_0+1/4}+\gamma)}$$
$$r(\gamma) = \frac{\Gamma(-\gamma)\Gamma(1+\gamma)}{\Gamma(1/2+\sqrt{\mathrm{F}_0+1/4})\Gamma(1/2-\sqrt{\mathrm{F}_0+1/4})},$$

where $\Gamma(z)$ is the well-known *gamma function* (see [**203**, Chapter II.10.52]), a meromorphic extension of the factorial function which enjoys the following properties (among many others):

(a) $1/\Gamma(z)$ is an entire function with simple zeros at $-z \in \mathbb{N}_0$, i.e., the non-positive integers,
(b) $\Gamma(1+z) = z\Gamma(z)$ with $\Gamma(1) = 1$,
(c) $\Gamma(-z)\Gamma(z) = -\pi/z\sin(\pi z)$.

Using the properties (b) and (c) the transmission and reflection coefficients may be simplified to

$$t(\gamma) = \frac{\gamma\Gamma^2(\gamma)}{\Gamma(1/2+\sqrt{F_0+1/4}+\gamma)\Gamma(1/2-\sqrt{F_0+1/4}+\gamma)}$$
$$r(\gamma) = -\frac{\cos(\pi\sqrt{F_0+1/4})}{\sin(\pi\gamma)}.$$

In the special case that $F_0 = \ell(\ell+1)$ for some $\ell \in \mathbb{N}_0$, then $r(\gamma) \equiv 0$, and the potential a_0 is said to be *reflectionless*. If $\ell = 0$, then $t(\gamma) \equiv 1$. If $\ell \in \mathbb{N}$, then property (b) allows the simplification

$$t(\gamma) = \frac{\gamma\Gamma^2(\gamma)}{\Gamma(1+\ell+\gamma)\Gamma(-\ell+\gamma)} = \prod_{j=1}^{\ell} \frac{\gamma - j}{\gamma + j}.$$

The transmission coefficient $t(\gamma)$ reduces to a rational function with simple zeros at $\gamma = 1,\dots,\ell$ and simple poles at $\gamma = -\ell,\dots,-1$. When the potential is not reflectionless, then $\gamma = 0$ is a simple pole, and $\gamma = -n$ is a double pole of the transmission coefficient for $n \in \mathbb{N}$. Furthermore, $t(\gamma) = 0$ for $-1/2 \pm \sqrt{F_0+1/4} - \gamma \in \mathbb{N}_0$, so there are $n_{F_0}+1$ real positive zeros, where n_{F_0} is the unique positive integer such that

$$\sqrt{F_0+1/4} - 3/2 < n_{F_0} < \sqrt{F_0+1/4} - 1/2.$$

The gamma function has a pole with residue one at the origin, so that one may calculate from (9.3.12) that

$$\lim_{\gamma\to 0} E(\gamma) = -2\lim_{\gamma\to 0} \gamma t(\gamma) = -\frac{2}{\pi}\cos(\pi\sqrt{F_0+1/4}). \tag{9.3.13}$$

In other words, $E(0) = 0$ if and only if the potential is reflectionless, and one eigenvalue crosses from the resonance sheet onto the positive real axis as F_0 increases through each of the values $\ell(\ell+1)$.

As a final observation, from the properties of the gamma function we can verify that the Evans function is analytic on the domain $\{\gamma : \mathrm{Re}\,\gamma > -1\}$, with a pole at $\gamma = -1$. This demonstrates that the loss of analyticity of the Evans function is not an artifact of the proof of Lemma 9.3.4, and the constraint imposed on the domain of analyticity by the exponential decay of the potential $a_0(x)$ can be sharp.

Exercises

Exercise 9.3.1. Derive an expression for the Evans function in the case that a_0 is a square-well but $a_1(x) \equiv a_1^0 \neq 0$. The transformation

$$p := e^{-a_1^0 x/2} q,$$

not only simplifies the calculations, but can be used to show that the eigenvalues are real.

9.4 Application: The Orientation Index

In general, it is difficult to explicitly calculate the Evans function. However, it is often the case that we can determine its behavior at $\lambda = 0\,(\gamma = \sqrt{-\lambda_{\mathrm{br}}})$ and in the limit $\lambda \to +\infty\,(\gamma \to +\infty)$. The orientation index uses this information to detect positive real eigenvalues of the underlying operator, which are associated with the temporal instability of the associated equilibria. When the operator arises as the linearization about an equilibrium of a system of the form (4.0.1), which possesses a symmetry group T satisfying (4.2.1), then Lemma 4.2.1 generates $\ker(\mathcal{L})$. Often it can be determined if these elements span the kernel, and hence the degree of the zero of the Evans function at $\lambda = 0$ is known. For a real-valued operator, the Evans function is real when λ (or γ) is real, and thus the sign of the first nonzero derivative of the Evans function determines the sign of the Evans function near zero. If the sign of the Evans function at infinity can be determined, and the signs at λ near zero and at λ near infinity differ, then the Evans function must have an odd, and nonzero, number of real roots (see Fig. 9.6). While the idea behind the orientation index is quite simple, the actual computations can be nontrivial (e.g., see Alexander and Jones [**10**], Pego and Weinstein [220], and the examples in [**9, 31, 38, 96, 134, 143, 179**]).

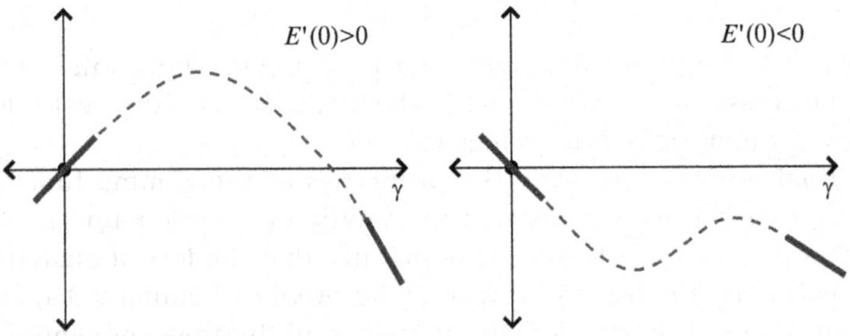

Fig. 9.6 The Evans function for real-valued γ. It is assumed that $\gamma = 0$ is a simple zero, and hence $E'(0) \neq 0$. The known values of E for γ small and large are denoted by the solid (*blue*) curves, and a potential plot for $\gamma = \mathcal{O}(1)$ is denoted by the dashed (*red*) curve. Since the Evans function is smooth, this situation $E'(0) > 0$ and $E(\infty) < 0$ (*left figure*) implies an odd number of positive zeros, while $E'(0) < 0$ (*right figure*) and $E(\infty) < 0$ implies an even number, potentially zero, of positive zeros. (Color figure online.)

To illustrate the orientation index, we reconsider the eigenvalue problem of Chapter 9.3.2, i.e.,

$$\partial_x^2 p + F_0 \operatorname{sech}^2(x) p = \lambda p, \quad F_0 > 0, \tag{9.4.1}$$

which arises from the linearization of the nonlinear Schrödinger equation about its sech pulse solution. On the Riemann surface defined by $\lambda = \gamma^2$ we computed the Evans function for (9.4.1),

$$E(\gamma) = -2\gamma \begin{cases} \dfrac{\gamma \Gamma^2(\gamma)}{\Gamma(1/2 + \sqrt{F_0 + 1/4} + \gamma)\Gamma(1/2 - \sqrt{F_0 + 1/4} + \gamma)}, & F_0 \neq \ell(\ell+1) \\ \displaystyle\prod_{j=1}^{\ell} \frac{\gamma - j}{\gamma + j}, & F_0 = \ell(\ell+1), \end{cases} \tag{9.4.2}$$

where $\ell \in \mathbb{N}_0$. Recall that $E(0) = 0$ if and only if the potential is reflectionless, i.e., $F_0 = \ell(\ell+1)$, in which case

$$E'(0) = 2(-1)^{\ell+1},$$

so that $E'(0) < 0$ for ℓ even, and $E'(0) > 0$ for ℓ odd. Since

$$\lim_{\gamma \to +\infty} \frac{E(\gamma)}{\gamma} = -2,$$

we have $E(\gamma) < 0$ for $\gamma \gg 1$. From the continuity of the Evans function on $\gamma \geq 0$ we deduce that it has an odd number of positive zeros if $E'(0) > 0$, and in particular at least one.

To extend the orientation index construction to the class of Sturm–Liouville operators (9.1.1), we first observe via Lemma 9.3.9 that the Evans function is nonzero for sufficiently large real γ so that it has a fixed sign at $\gamma = +\infty$. Moreover, we assume $a_0^0 < 0$, so that $\lambda = 0$, that is $\gamma = \gamma_0 := \sqrt{-\lambda_{\mathrm{br}}} > 0$, lies within the natural domain of the Evans function (see Fig. 9.2). Under these conditions we make the following definitions.

Definition 9.4.1. The sign of the Evans function for large $\gamma \in \mathbb{R}^+$ is given by

$$E(+\infty) := \lim_{\gamma \to +\infty} \operatorname{sign}[E(\gamma)]. \tag{9.4.3}$$

If $\lambda = 0$, corresponding to $\gamma = \gamma_0$, is an eigenvalue of finite order p, so that $\partial_\gamma^{(p+1)} E(\gamma_0) \neq 0$, then the *orientation index* is defined by

$$\mathbb{O} := \operatorname{sign}\left[E(+\infty)\partial_\gamma^{(p+1)} E(\gamma_0)\right]. \tag{9.4.4}$$

It is clear that if the orientation index is positive, then the associated Evans function has an even number (possibly zero) of zeros on the open interval (γ_0, ∞), while if it is negative, then the Evans function has an odd, necessarily nonzero, number of zeros on (γ_0, ∞). In the context of Sturm–Liouville operators, the orientation index recovers results of the classical Sturmian oscillation theorem.

Theorem 9.4.2 (Orientation index). *Consider the Sturm–Liouville eigenvalue problem (9.1.1) with $a_0^0 < 0$: then $\lambda = 0$, corresponding to $\gamma_0 = \sqrt{-\lambda_{\rm br}}$, is positive and resides in the natural domain of the Evans function. If $\gamma = \gamma_0$ is a simple eigenvalue with associated eigenfunction $p_0(x)$, then the orientation index is given by*

$$\mathbb{O} = \lim_{x \to \infty} (\operatorname{sign}[p_0(-x)p_0(x)]), \tag{9.4.5}$$

where the limit necessarily exists.

Proof. Since $\gamma_0 > 0 \in \Omega$, from Lemma 9.3.4 we know that

$$E'(\gamma_0) = 2\gamma_0 \langle p_0, q_0 \rangle \det\left(J_0^{\rm a}, J_0\right)\Big|_{x=0},$$

where

$$J_0(x) = \begin{pmatrix} p_0(x) \\ \partial_x p_0(x) \end{pmatrix}, \quad J_0^{\rm a}(x) = \begin{pmatrix} a_1(x)q_0(x) - \partial_x q_0(x) \\ q_0(x) \end{pmatrix},$$

and the adjoint eigenfunction, q_0, has been scaled so that $|J_0^{\rm a}(0)| = 1$. Since p_0 is the eigenfunction of (9.1.1) associated with $\gamma = \gamma_0$ ($\lambda = 0$), the adjoint eigenfunction takes the form $q_0 = \rho p_0$, where the weight function $\rho(x)$ symmetrizes $\mathcal{L}$; see (9.5.3). The vectorized adjoint eigenfunction takes the simplified form

$$J_0^{\rm a}(x) = \rho \begin{pmatrix} -\partial_x p_0(x) \\ p_0(x) \end{pmatrix};$$

hence, the assumption leading to the derivative formula for the Evans function is that the eigenfunction $p_0(x)$ has been scaled so that $|J_0(0)| = |J_0^{\rm a}(0)| = 1$. By assumption, $E'(\gamma_0) \neq 0$; in particular, the derivative of the Evans function can be reduced to

$$E'(\gamma_0) = -2\gamma_0 \|p_0\|_\rho^2 < 0$$

where $\|\cdot\|_\rho$ denotes the ρ-weighted norm. Note that if the eigenfunction were not scaled, then the resulting formula for $E'(\gamma_0)$ would be a positive multiple of that given above, which does not affect the sign.

Having established the sign of $E'(\gamma_0)$, it remains to compute $E(+\infty)$. On the one hand, this seems to be a straightforward calculation; indeed, it appears to be the case that we can simply use (9.3.7) in the proof of Lemma 9.3.9. However, the vectors $v_{\rm s,u}(\gamma)$ are not uniquely determined: we must understand how the Jost solutions define an *oriented basis* for $\gamma \in \mathbb{R} \cap \Omega$.

For the matrix eigenvalues we choose the matrix eigenvectors with the normalization

$$\mu_1(\gamma) = -\frac{a_1^0}{2} + \gamma,\ v_1(\gamma) = \begin{pmatrix} 1 \\ \mu_1(\gamma) \end{pmatrix}; \quad \mu_2(\gamma) = -\frac{a_1^0}{2} - \gamma,\ v_2(\gamma) = \begin{pmatrix} 1 \\ \mu_2(\gamma) \end{pmatrix}, \tag{9.4.6}$$

and with the normalization (9.2.16) the Jost solutions satisfy

$$\lim_{x\to-\infty} e^{-\mu_1(\gamma)x} J^-(x,\gamma) = c_-(\gamma)\boldsymbol{v}_1(\gamma), \quad \lim_{x\to+\infty} e^{-\mu_2(\gamma)x} J^+(x,\gamma) = c_+(\gamma)\boldsymbol{v}_2(\gamma). \tag{9.4.7}$$

Here $c_\pm(\gamma)$ are analytic and nonzero on $\Omega_e \cap \mathbb{R}$, but they must be normalized. For each real γ the vectors $\{c_-(\gamma)\boldsymbol{v}_1(\gamma), c_+(\gamma)\boldsymbol{v}_2(\gamma)\}$ form a basis for $\mathbb{R}^2$, for which the orientation satisfies

$$\mathbb{O}_\gamma := \text{sign}[\det(c_-\boldsymbol{v}_1, c_+\boldsymbol{v}_2)(\gamma)] = \text{sign}[-2\gamma c_-(\gamma)c_+(\gamma)] = -\text{sign}[c_-(\gamma)c_+(\gamma)], \tag{9.4.8}$$

where we used (9.4.6) to deduce the first equality. Since $c_\pm(\gamma)$ never changes sign for $\gamma \in \mathbb{R} \cap \Omega$ it follows that $\mathbb{O}_\gamma = \mathbb{O}_{\gamma_0}$ for all real γ in the natural domain.

Turning to $\gamma = \gamma_0$, the formula for $E'(\gamma_0)$ requires the normalization

$$J^-(x,\gamma_0) \equiv J^+(x,\gamma_0) = \begin{pmatrix} p_0(x) \\ \partial_x p_0(x) \end{pmatrix}.$$

Using the limits

$$\lim_{x\to-\infty} e^{-\mu_1(\gamma_0)x} \begin{pmatrix} p_0(x) \\ \partial_x p_0(x) \end{pmatrix} = c_-^* \boldsymbol{v}_1(\gamma_0), \quad \lim_{x\to+\infty} e^{-\mu_2(\gamma_0)x} \begin{pmatrix} p_0(x) \\ \partial_x p_0(x) \end{pmatrix} = c_+^* \boldsymbol{v}_2(\gamma_0),$$

to define $c_\pm^*$, and comparing with (9.4.7), we see that the coefficients $c_\pm(\gamma)$ satisfy $c_\pm(\gamma_0) = c_\pm^*$. From the normalization (9.4.6) of the matrix eigenvectors we find that for $x > 0$ sufficiently large,

$$\text{sign}[c_\pm^*] = \text{sign}[p_0(\pm x)].$$

In particular, from (9.4.8) we deduce that

$$\mathbb{O}_{\gamma_0} = -\text{sign}[c_-^* c_+^*] = -\text{sign}[p_0(-x)p_0(x)], \quad x \gg 1.$$

For $\gamma \gg 1$, equation (9.3.7) from Lemma 9.3.9 yields the reduction

$$E(\gamma) = \det(c_-\boldsymbol{v}_1, c_+\boldsymbol{v}_2)(\gamma) + \mathcal{O}(\gamma^{-1}).$$

In particular, from (9.4.3) and (9.4.8) we determine that

$$E(+\infty) = \lim_{\gamma\to\infty} \text{sign}[\det(c_-\boldsymbol{v}_1, c_+\boldsymbol{v}_2)(\gamma)] = \mathbb{O}_\gamma = \mathbb{O}_{\gamma_0}.$$

Using the definition, (9.4.4), of the orientation index, and the fact that $E'(\gamma_0) < 0$, we find

$$\mathbb{O} = \text{sign}[E(+\infty)E'(\gamma_0)] = -\mathbb{O}_{\gamma_0} = \text{sign}[p_0(-x)p_0(x)], \quad x \gg 1,$$

which establishes (9.4.5). □

Remark 9.4.3. If p_0 is odd, so that $p_0(x)p_0(-x) = -p_0(x)^2 < 0$, then the Evans function has an odd number of positive zeros, whereas if p_0 is even, so that $p_0(x)p_0(-x) = p_0(x)^2 > 0$, then it has an even number of zeros. This is consistent with the Sturm–Liouville theory presented in Theorem 2.3.3.

9.5 Application: Edge Bifurcations

A central application of the Evans function, which distinguishes it from the tools of regular perturbation theory, is its ability to locate (potential) eigenvalues within the essential and absolute spectrum of the operator. Indeed, a key issue in many problems is the possibility of point spectra emerging from the branch point of the absolute spectra under perturbation, a phenomena called an *edge bifurcation*. Examples of edge bifurcations arising within the nonlinear Schrödinger equation are given in Kapitula and Sandstede [**151**, **153**], Li and Promislow [**193**]. There is also an extensive literature on applications arising within mathematical physics [**8**, **86**, **87**, **98**, **131**, **240–243**, **261–263**, **296**], in which the problem is studied via related ideas from classical scattering theory or the Fredholm determinant.

The Evans function is analytic at the branch point, $\gamma_{\text{br}} = 0$, and we assume that it has a zero there as well. However, neither the eigenvalue perturbation results of Chapter 6.1 nor the derivative formula given in Lemma 9.3.4 can be directly applied since the embedded eigenfunction associated with the zero at the branch point does not lie in $L^2(\mathbb{R})$. Moreover, the use of a weighted norm may shift the essential spectrum but will not move the branch point. Resolving this issue requires novel formulas for the partial derivatives of the Evans function within the absolute spectrum.

We fix a framework, considering a linear operator of the form

$$\mathcal{L} := \overbrace{\partial_x^2 + a_{10}(x)\partial_x + a_{00}(x)}^{\mathcal{L}_0} + \epsilon\overbrace{\Big((a_{11}^0 + a_{11}(x))\partial_x + (a_{01}^0 + a_{01}(x))\Big)}^{\mathcal{L}_1}, \tag{9.5.1}$$

where all of the spatially dependent coefficients decay exponentially, i.e..

$$|a_{10}(x)| + |a_{00}(x)| + |a_{11}(x)| + |a_{01}(x)| \le c e^{-\alpha|x|}.$$

The branch point of the operator $\mathcal{L}$ occurs at $\lambda_{\text{br}} = \epsilon a_{01}^0 - \epsilon^2 (a_{11}^0)^2/4$, with the associated Riemann surface given by

$$\lambda := \lambda_{\text{br}} + \gamma^2, \tag{9.5.2}$$

so that the branch point remains at $\gamma = 0$ for $\epsilon \ne 0$. The Evans function is analytic on the extended domain $\{\gamma : \operatorname{Re}\gamma > -\alpha/2\}$. From Chapter 2.3 we recall that the operator $\mathcal{L}_0$ is self-adjoint in the weighted space $L^2_\rho(\mathbb{R})$ with weight

$$\rho(x) = \exp\left[\int_0^x a_{10}(s)\,\mathrm{d}s\right], \tag{9.5.3}$$

which has finite limits $\rho_{\pm\infty} := \lim_{x\to\pm\infty}\rho(x)$.

Since the coefficients of $\mathcal{L}_0$ are asymptotically zero at infinity, the essential and absolute spectrum coincide with the negative real axis. Indeed, if $a_{10}(x)$ tended to a nonzero limit a_{10}^0 then the essential spectrum

$$\sigma_{\text{ess}}(\mathcal{L}_0) = \{\lambda \in \mathbb{C} : \lambda = -k^2 + \mathrm{i}a_{10}^0 k,\ k \in \mathbb{R}\},$$

would lie strictly to the right of the absolute spectrum

$$\sigma_{\text{abs}}(\mathcal{L}_0) = \{\lambda \in \mathbb{C} : \lambda = -(a_{10}^0)^2/4 - k^2,\ k \in \mathbb{R}\},$$

(see Fig. 9.2). In this case any zeros ejected from the branch point would emerge into the resonance sheet, which is less significant than new eigenvalues being created on the physical sheet.

We assume that for $\epsilon = 0$ the Evans function, $E = E(\gamma, \epsilon)$, has a zero at the branch point at $\gamma = 0$. We will show that the zero must be simple, and hence $\partial_\gamma E(0,0) \neq 0$. In particular, since $\mathcal{L}$ depends smoothly upon ϵ, the Evans function is also smooth in ϵ. For ϵ sufficiently small, the implicit function theorem implies that the zero, $\gamma = \gamma(\epsilon)$, of E satisfies (see Fig. 9.7)

$$\gamma(\epsilon) = -\frac{\partial_\epsilon E(0,0)}{\partial_\gamma E(0,0)}\epsilon + \mathcal{O}(\epsilon^2). \tag{9.5.4}$$

In the remainder of this section we evaluate these partial derivatives, obtaining the following theorem.

Theorem 9.5.1 (Edge bifurcation). *Consider the operators $\mathcal{L} = \mathcal{L}(\epsilon)$, introduced in (9.5.1), with the Evans function, $E = E(\gamma, \epsilon)$. If for $\epsilon = 0$ the Evans function has a zero at the branch point $\gamma = 0$, i.e., $E(0,0) = 0$, then the zero is simple and for ϵ sufficiently small it satisfies the following asymptotic expansion*

$$\gamma(\epsilon) = \epsilon\gamma_1 + O(\epsilon^2), \tag{9.5.5}$$

where the leading order motion on the Riemann surface is governed by the term

$$\gamma_1 = \frac{a_{11}^0\left(\rho_{-\infty} - (p_0^+)^2\rho_{+\infty} + 2\langle p_0, \partial_x p_0\rangle_\rho\right)/2 + \langle p_0, \tilde{\mathcal{L}}_1 p_0\rangle_\rho}{\rho_{-\infty} + (p_0^+)^2\rho_{+\infty}}. \tag{9.5.6}$$

Here $p_0 \in L^\infty(\mathbb{R})$ is the embedded eigenvalue constructed in Lemma 9.5.3 with limit value p_0^+, ρ is the weight (9.5.3) with limits $\rho_{\pm\infty}$, and the operator $\tilde{\mathcal{L}}_1$ is the localized part of the operator $\mathcal{L}_1$, i.e.,

$$\tilde{\mathcal{L}}_1 = a_{11}(x)\partial_x + a_{01}(x). \tag{9.5.7}$$

Remark 9.5.2. Generically the zero of the Evans function leaves the branch point at an $\mathcal{O}(\epsilon)$ rate in the γ-plane, which by the definition of the Riemann surface in (9.5.2) corresponds to an $\mathcal{O}(\epsilon^2)$ motion in the λ-plane. Indeed, the zero will be $\mathcal{O}(\epsilon^2)$-close to the branch point λ_{br}. While the branch point of the absolute spectrum generically moves at an $O(\epsilon)$ rate in the λ plane, it is the $\mathcal{O}(\epsilon^2)$ motion of the zero with respect to the branch point that is significant. By comparison, simple eigenvalues within the natural domain of the Evans functions generically perturb at an $\mathcal{O}(\epsilon)$ rate in λ.

9.5.1 The $\epsilon = 0$ Problem

We begin by gathering some preliminary results for the $\epsilon = 0$ case. Setting $\mathbf{Y} = (p, \partial_x p)^{\mathrm{T}}$ and $\lambda = \gamma^2$ the vectorized eigenvalue problem for $\mathcal{L}_0$ takes the form

$$\partial_x \mathbf{Y} = [\mathbf{A}_0(\gamma) + \mathbf{R}(x)]\mathbf{Y}, \tag{9.5.8}$$

where the asymptotic and localized matrices are given by

$$\mathbf{A}_0(\gamma) = \begin{pmatrix} 0 & 1 \\ \gamma^2 & 0 \end{pmatrix}, \quad \mathbf{R}(x) = \begin{pmatrix} 0 & 0 \\ -a_{00}(x) & -a_{10}(x) \end{pmatrix}.$$

For $\operatorname{Re}\gamma > -\alpha/2$ the Jost solutions have the asymptotic behavior

$$\lim_{x \to -\infty} \mathrm{e}^{-\gamma x} \mathbf{J}^-(x,\gamma) = \boldsymbol{v}^-(\gamma), \qquad \lim_{x \to +\infty} \mathrm{e}^{+\gamma x} \mathbf{J}^+(x,\gamma) = \boldsymbol{v}^+(\gamma), \tag{9.5.9}$$

where the eigenvectors

$$\boldsymbol{v}^-(\gamma) = c^-(\gamma)\begin{pmatrix} 1 \\ \gamma \end{pmatrix}, \quad \boldsymbol{v}^+(\gamma) = c^+(\gamma)\begin{pmatrix} 1 \\ -\gamma \end{pmatrix} \tag{9.5.10}$$

are analytic in γ. Here $c^{\pm}(\gamma) \neq 0$ for all γ. The Evans function is given by (9.3.5) and it is analytic for $\operatorname{Re}\gamma > -\alpha/2$. The assumption $E(0) = 0$ implies that $\mathbf{J}^{\pm}(x,0)$ are linearly dependent, and hence from (9.5.9) evaluated at $\gamma = 0$ there exists a solution $\mathbf{J}_0(x)$ of (9.5.8) that converges to a constant at both $\pm\infty$. This embedded "eigenfunction" plays a central role in the perturbation theory, and we collect its properties in the following lemma.

Lemma 9.5.3. *For $\gamma = 0$ assume there is a unique $p_0(x)$ satisfying $\mathcal{L}_0 p_0 = 0$ with the limiting values*

$$\lim_{x \to -\infty} p_0(x) = 1, \quad \lim_{x \to +\infty} p_0(x) = p_0^+, \tag{9.5.11}$$

for some $p_0^+ \in \mathbb{R}$. Then, for the choice of constants $c^-(\gamma) \equiv 1$ and $c^+(0) = p_0^+$ in (9.5.10), the Jost solutions $\mathbf{J}^{\pm}(x,\gamma)$ take a common value at $\gamma = 0$, denoted

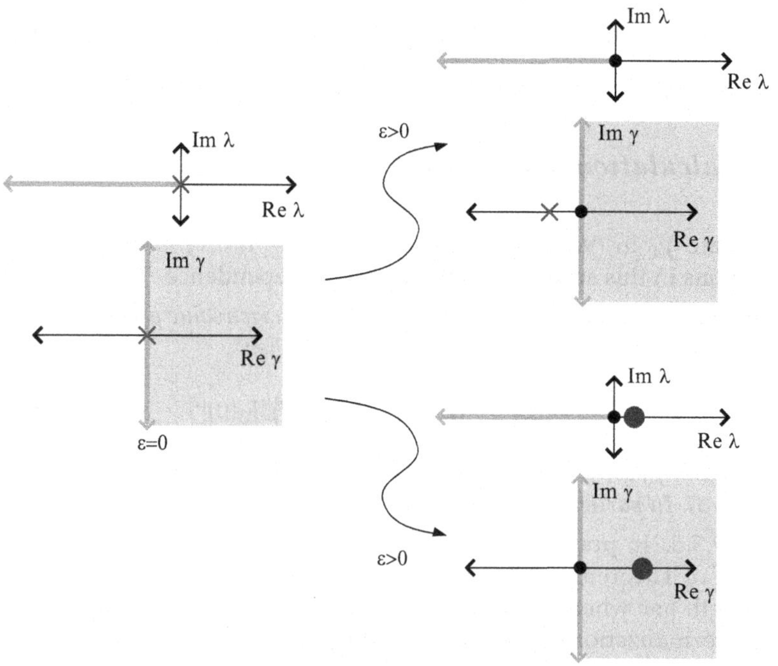

Fig. 9.7 Two distinct unfoldings of a zero of the Evans function at a branch point. The *leftmost panel* depicts the zero of the Evans function with a (*red*) cross at the branch point $\lambda = 0$ in the λ-plane, and a corresponding zero at the branch point $\gamma = 0$ in the γ-plane. The absolute spectrum is denoted by a colored line in the λ-plane, while the physical sheet of the Riemann surface, $\mathrm{Re}\,\gamma > 0$, is shaded. For $\epsilon > 0$ there are two possibilities: in the *right top panel* the zero moves onto the resonance sheet of the Riemann surface, while in the *right bottom panel* the zero moves onto the physical sheet. When the zero moves onto the physical sheet it corresponds to an eigenvalue, and is denoted by a (*blue*) circle. The position of the zero scales differently in the two planes: it is $\mathcal{O}(\epsilon)$ in the γ-plane and $\mathcal{O}(\epsilon^2)$ in the λ-plane. (Color figure online.)

$$J_0(x) = \begin{pmatrix} p_0(x) \\ \partial_x p_0(x) \end{pmatrix}. \tag{9.5.12}$$

In particular, we have the limits

$$\lim_{x\to-\infty} J_0(x) = \begin{pmatrix} 1 \\ 0 \end{pmatrix}, \quad \lim_{x\to+\infty} J_0(x) = \begin{pmatrix} p_0^+ \\ 0 \end{pmatrix},$$

The associated adjoint solution $J_0^{\mathrm{a}}(x)$ *at* $\gamma = 0$ *has the form*

$$J_0^{\mathrm{a}}(x) = \rho(x) \begin{pmatrix} -\partial_x p_0(x) \\ p_0(x) \end{pmatrix},$$

with the limiting values

$$\lim_{x\to-\infty} J_0^{\mathrm{a}}(x) = \rho_- \begin{pmatrix} 0 \\ 1 \end{pmatrix}, \quad \lim_{x\to+\infty} J_0^{\mathrm{a}}(x) = \rho_+ \begin{pmatrix} 0 \\ p_0^+ \end{pmatrix}. \tag{9.5.13}$$

9.5.2 Calculation of $\partial_\gamma E(0,0)$

To evaluate γ_1 in (9.5.5) we must determine $\partial_\gamma E(0,0)$. Since $\epsilon = 0$ in all calculations in this section we suppress the ϵ dependence.

Lemma 9.5.4. *Assume the Evans function for the eigenvalue problem associated with the operator $\mathcal{L}_0$ of (9.5.1) satisfies $E(0) = 0$, then*

$$\partial_\gamma E(0) = -\left(\rho_{-\infty} + (p_0^+)^2 \rho_{+\infty}\right) |\boldsymbol{J}_0(0)|^2 \tag{9.5.14}$$

where p_0 and $\boldsymbol{J}_0$ are defined in (9.5.11) and (9.5.12), while $\rho > 0$ is the weight from (9.5.3). In particular, $\partial_\gamma E(0) < 0$.

Remark 9.5.5. In previous Evans function derivative calculations (e.g., see the proof of Theorem 9.4.2) we have normalized the adjoint Jost solution to have length one when $x = 0$. Here, and in the remainder of Chapter 9.5, we use the normalizations presented in Lemma 9.5.3.

Proof. From the normalization of the Jost solutions introduced in Lemma 9.5.3 we have $\boldsymbol{J}^-(x,0) = \boldsymbol{J}^+(x,0) \equiv \boldsymbol{J}_0(x)$, and hence

$$\partial_\gamma E(0) = \det\left(\partial_\gamma(\boldsymbol{J}^- - \boldsymbol{J}^+), \boldsymbol{J}_0\right)(0,0). \tag{9.5.15}$$

To calculate $\partial_\gamma \boldsymbol{J}^\pm(0,0)$ we examine the remainder term $\boldsymbol{K}^+$ near $\gamma = 0$ (see Chapter 9.2). Taking ∂_γ of (9.2.8) we find

$$\partial_\gamma \boldsymbol{J}^+(0,0) = \partial_\gamma \boldsymbol{v}^+(0) + \partial_\gamma \boldsymbol{K}^+(0,0), \tag{9.5.16}$$

where $|\boldsymbol{K}(x,\gamma)| \sim \mathrm{e}^{-\alpha L}\mathrm{e}^{-\gamma x}$ for $x \geq L \gg 0$. The remainder K satisfies the ODE

$$\partial_x \boldsymbol{K}^+ = \boldsymbol{A}(x,\gamma)\boldsymbol{K}^+ + \boldsymbol{R}(x)[\mathrm{e}^{-\gamma x}\boldsymbol{v}^+(\gamma)].$$

Evaluating the ∂_γ of this ODE at $\gamma = 0$, and using $\partial_\gamma \boldsymbol{A}(x,0) = \boldsymbol{0}_2$, yields,

$$\partial_x(\partial_\gamma \boldsymbol{K}^+) = \boldsymbol{A}(x,0)\partial_\gamma \boldsymbol{K} + \boldsymbol{R}(x)[\partial_\gamma \boldsymbol{v}^+(0) - x\boldsymbol{v}^+(0)]. \tag{9.5.17}$$

We use variation of parameters to solve (9.5.17). Since $|\partial_\gamma \boldsymbol{K}^+(x,0)| \to 0$ as $x \to +\infty$, we find

$$\partial_\gamma \boldsymbol{K}^+(x,0) = \boldsymbol{\Phi}(x) \int_{+\infty}^{x} \boldsymbol{\Phi}^{-1}(s)\boldsymbol{R}(s)[\partial_\gamma \boldsymbol{v}^+(0) - s\boldsymbol{v}^+(0)]\,\mathrm{d}s, \tag{9.5.18}$$

where $\boldsymbol{\Phi}(x)$ is the matrix solution of (9.5.8) for $\gamma = 0$ which at $x = 0$ satisfies

$$\boldsymbol{\Phi}(0) = (\boldsymbol{J}_0^{\mathrm{a}}, \boldsymbol{J}_0)(0). \tag{9.5.19}$$

The integral in (9.5.18) is well-defined since $|\boldsymbol{R}(x)| \to 0$ exponentially fast as $x \to +\infty$, and the fundamental matrix and its inverse can grow at most polynomially in x for $\gamma = 0$. To evaluate the integrand we use the relations

$$\boldsymbol{R}(x) = \boldsymbol{A}(x,0) - \boldsymbol{A}_0(0), \quad \partial_x \boldsymbol{\Phi}^{-1} = -\boldsymbol{\Phi}^{-1}(x)\boldsymbol{A}(x,0)$$

to obtain

$$\begin{aligned} &\boldsymbol{\Phi}(s)^{-1}\boldsymbol{R}(s)[\partial_\gamma \boldsymbol{v}^+(0) - s\boldsymbol{v}^+(0)] \\ &\quad = -\partial_s \boldsymbol{\Phi}^{-1}(s)[\partial_\gamma \boldsymbol{v}^+(0) - s\boldsymbol{v}^+(0)] - \boldsymbol{\Phi}^{-1}(s)\boldsymbol{A}_0(0)[\partial_\gamma \boldsymbol{v}^+(0) - s\boldsymbol{v}^+(0)]. \end{aligned} \tag{9.5.20}$$

At the branch point the eigenvector $\boldsymbol{v}^+$ is part of a Jordan block of $\boldsymbol{A}_0$. Indeed, taking the γ derivative of $\boldsymbol{A}_0(\gamma)\boldsymbol{v}^+(\gamma) = -\gamma \boldsymbol{v}^+(\gamma)$, and using $\partial_\gamma \boldsymbol{A}_0(0) = \boldsymbol{0}_2$, we see that

$$\boldsymbol{A}_0(0)\boldsymbol{v}^+(0) = \boldsymbol{0}, \quad \boldsymbol{A}_0(0)\partial_\gamma \boldsymbol{v}^+(0) = -\boldsymbol{v}^+(0).$$

Thus, the right-hand side of (9.5.20) simplifies to

$$\boldsymbol{\Phi}(s)^{-1}\boldsymbol{R}(s)[\partial_\gamma \boldsymbol{v}^+(0) - s\boldsymbol{v}^+(0)] = -\partial_s \boldsymbol{\Phi}^{-1}(s)[\partial_\gamma \boldsymbol{v}^+(0) - s\boldsymbol{v}^+(0)] + \boldsymbol{\Phi}^{-1}(s)\boldsymbol{v}^+(0),$$

and the expression (9.5.18) takes the form

$$\partial_\gamma \boldsymbol{K}^+(x,0) = -\boldsymbol{\Phi}(x) \lim_{L\to+\infty} \int_L^x \left(\partial_s \boldsymbol{\Phi}^{-1}(s)[\partial_\gamma \boldsymbol{v}^+(0) - s\boldsymbol{v}^+(0)] - \boldsymbol{\Phi}(s)^{-1}\boldsymbol{v}^+(0)\right) \mathrm{d}s. \tag{9.5.21}$$

Integrating the first term in the integral above by parts and observing the cancelation with the second term, yields

$$\partial_\gamma \boldsymbol{K}^+(x,0) = x\boldsymbol{v}^+(0) - \partial_\gamma \boldsymbol{v}^+(0) + \lim_{L\to+\infty} \boldsymbol{\Phi}(x)\boldsymbol{\Phi}(L)^{-1}[\partial_\gamma \boldsymbol{v}^+(0) - L\boldsymbol{v}^+(0)],$$

which evaluated at $x = 0$ reduces to

$$\partial_\gamma \boldsymbol{K}(0,0) = -\partial_\gamma \boldsymbol{v}^+(0) + \lim_{L\to+\infty} \boldsymbol{\Phi}(0)\boldsymbol{\Phi}(L)^{-1}[\partial_\gamma \boldsymbol{v}^+(0) - L\boldsymbol{v}^+(0)].$$

Substituting this expressing into (9.5.16), we obtain

$$\partial_\gamma \boldsymbol{J}^+(0,0) = \lim_{L\to+\infty} \boldsymbol{\Phi}(0)\boldsymbol{\Phi}(L)^{-1}[\partial_\gamma \boldsymbol{v}^+(0) - L\boldsymbol{v}^+(0)].$$

In a similar manner it can be shown that

$$\partial_\gamma \boldsymbol{J}^-(0,0) = \lim_{L\to+\infty} \boldsymbol{\Phi}(0)\boldsymbol{\Phi}(-L)^{-1}[\partial_\gamma \boldsymbol{v}^-(0) - L\boldsymbol{v}^-(0)]$$

(see Exercise 9.5.1), so that

$$\partial_\gamma(\boldsymbol{J}^- - \boldsymbol{J}^+)(0,0) = \boldsymbol{\Phi}(0)\boldsymbol{w}_\infty, \tag{9.5.22}$$

where the limit vector is defined by

$$\boldsymbol{w}_\infty := \lim_{L\to+\infty}\left(\boldsymbol{\Phi}(-L)^{-1}[\partial_\gamma \boldsymbol{v}^-(0) - L\boldsymbol{v}^-(0)] - \boldsymbol{\Phi}(L)^{-1}[\partial_\gamma \boldsymbol{v}^+(0) - L\boldsymbol{v}^+(0)]\right).$$

Write $\boldsymbol{w}_\infty = (w_{\infty 1}, w_{\infty 2})^{\mathrm{T}}$. Returning to (9.5.15), and using (9.5.19) and (9.5.22), we can express the derivative of the Evans function as

$$\begin{aligned}\partial_\gamma E(0) &= \det(\boldsymbol{\Phi}(0)\boldsymbol{w}_\infty, \boldsymbol{J}_0(0)) = \det(w_{\infty 1}\boldsymbol{J}_0^{\mathrm{a}}(0) + w_{\infty 2}\boldsymbol{J}_0(0), \boldsymbol{J}_0(0))\\ &= w_{\infty 1}\det(\boldsymbol{J}_0^{\mathrm{a}}, \boldsymbol{J}_0)(0).\end{aligned} \tag{9.5.23}$$

It remains to evaluate $w_{\infty 1}$. From Lemma 8.1.3, we know that for any $\boldsymbol{v} \in \mathbb{C}^2$,

$$\boldsymbol{\Phi}(x)^{-1}\boldsymbol{v} = \begin{pmatrix} \boldsymbol{v}\cdot \boldsymbol{J}_0^{\mathrm{a}}(x) \\ \star \end{pmatrix}, \tag{9.5.24}$$

where the value of the second entry is immaterial. We thus rewrite $w_{\infty 1}$ as

$$w_{\infty 1} = \lim_{L\to+\infty}\left(\boldsymbol{J}_0^{\mathrm{a}}(-L)\cdot[\partial_\gamma \boldsymbol{v}^-(0) - L\boldsymbol{v}^-(0)] - \boldsymbol{J}_0^{\mathrm{a}}(L)\cdot[\partial_\gamma \boldsymbol{v}^+(0) - L\boldsymbol{v}^+(0)]\right)\boldsymbol{J}^{\mathrm{a}}(0),$$

and since $\boldsymbol{v}^\pm(0) \propto (1,0)^{\mathrm{T}}$, we see from the exponential rate of the convergence in (9.5.13) that

$$\lim_{L\to+\infty} L\boldsymbol{J}_0^{\mathrm{a}}(\pm L)\cdot \boldsymbol{v}^\pm(0) = 0.$$

The remaining terms in $w_{\infty 1}$ can be evaluated from the knowledge of $c^\pm(0)$ and the limiting values in (9.5.13), yielding the expression

$$w_{\infty 1} = \mathrm{e}^{-\int_{-\infty}^{0} a_{10}(s)\,\mathrm{d}s} + (p_0^+)^2\mathrm{e}^{\int_0^{+\infty} a_{10}(s)\,\mathrm{d}s} = \rho_{-\infty} + (p_0^+)^2\rho_{+\infty}.$$

Using Lemma 9.5.3 to evaluate the determinant,

$$\det(\boldsymbol{J}_0^{\mathrm{a}}, \boldsymbol{J}_0)(0,0) = -|\boldsymbol{J}_0(0)|^2, \tag{9.5.25}$$

we arrive at (9.5.14). □

Example 9.5.6. Consider the operator $\mathcal{L}_0$ in the case that $a_{10}(x)$ is odd and $a_{00}(x)$ is even. From the decay assumptions on the potentials a_{i0}, the absolute spectrum of the operator $\mathcal{L}_0$ touches the origin, and while the operator is not self-adjoint in $L^2(\mathbb{R})$ it is self-adjoint in the ρ-weighted space. Specifically, applying the transformation

$$p(x,\gamma) \mapsto p(x,\gamma)/\sqrt{\rho},$$

where $\lambda = \gamma^2$, the eigenvalue problem becomes

$$\partial_x^2 p + a(x)p = \gamma^2 p, \quad a(x) := a_{00}(x) - \frac{1}{4}a_{10}^2(x) - \frac{1}{2}\partial_x a_{10}(x). \tag{9.5.26}$$

Under our assumptions the new potential $a(x)$ is even; furthermore the parity of solutions of the eigenvalue problem is preserved under the transformation. Since $a(x)$ is even, it follows that for any solution $p(x,\gamma)$ of the eigenvalue problem, $p(-x,\gamma)$ is also a solution. This symmetry implies that the Jost solutions $\boldsymbol{J}^{\pm}(x,\gamma)$ enjoy the relationship

$$\boldsymbol{J}^{+}(x,\gamma) = c^{+}(\gamma)\begin{pmatrix} 1 & 0 \\ 0 & -1 \end{pmatrix}\boldsymbol{J}^{-}(x,\gamma), \quad c^{+}(\gamma) \neq 0. \tag{9.5.27}$$

Using the notation

$$\boldsymbol{J}^{-}(x,\gamma) = \begin{pmatrix} p(x,\gamma) \\ \partial_x p(x,\gamma) \end{pmatrix},$$

the Evans function takes the form

$$E(\gamma) = \det\left(\boldsymbol{J}^{-},\boldsymbol{J}^{+}\right)(0,\gamma) = -2c^{+}(\gamma)p(0,\gamma)\partial_x p(0,\gamma).$$

Consequently, $E(\gamma) = 0$ if and only if $p(0,\gamma) = 0$ or $\partial_x p(0,\gamma) = 0$. Since $a(x)$ is even, at a zero of E the first component of the Jost function is either even or odd about $x = 0$.

In the case that $E(0) = 0$ and $p_0(x) = p(x,0)$ is even, it follows that $\partial_x p_0(x)$ is odd and with the standard normalization the Jost solutions satisfy $\boldsymbol{J}^{-}(x,0) = \boldsymbol{J}^{+}(x,0)$, so that $c^{+}(\gamma) \equiv 1$. This implies that $p_0^{+} = 1$ in Lemma 9.5.4 and the expression (9.5.14) reduces to

$$\partial_\gamma E(0) = -2p_0(0)^2|\boldsymbol{J}_0(0)|^2 < 0. \tag{9.5.28}$$

Applying the orientation index Theorem 9.4.2, from (9.4.8)

$$\mathbb{O}_\gamma = \det\left(\boldsymbol{v}^{-},\boldsymbol{v}^{+}\right)(\gamma) = -2\gamma < 0,$$

so that $E(\gamma) < 0$ for $\gamma > 0$ and large. Consequently, when $p(x,0)$ is even the Evans function has an even number of real positive zeros, each of which correspond to eigenvalues of $\mathcal{L}_0$ since the zeros lie on the physical sheet of the Riemann surface. This is consistent with the Sturm–Liouville theory (see Remark 9.4.3), even though $p(x,0)$ is an embedded eigenfunction at the branch point and not in the point spectrum of $\mathcal{L}_0$.

9.5.3 Calculation of $\partial_\epsilon E(0,0)$

In this section we compute $\partial_\epsilon E(0,0)$ for the Evans function associated with the operator $\mathcal{L}$ introduced in (9.5.1). Setting $\boldsymbol{Y} = \boldsymbol{Y}(x,\lambda,\epsilon) = (p, \partial_x p)^{\mathrm{T}}$, the eigenvalue problem for the full operator $\mathcal{L}$ takes the form

$$\partial_x \boldsymbol{Y} = [\boldsymbol{A}_0(\lambda) + \boldsymbol{R}_0(x) + \epsilon \boldsymbol{R}_1(x)]\boldsymbol{Y},$$

where

$$\boldsymbol{A}_0(\lambda) = \begin{pmatrix} 0 & 1 \\ \lambda - \epsilon a_{11}^0 & -\epsilon a_{01}^0 \end{pmatrix}, \quad \boldsymbol{R}_0(x) = \begin{pmatrix} 0 & 0 \\ -a_{10}(x) & -a_{00}(x) \end{pmatrix}$$

and

$$\boldsymbol{R}_1(x) = \begin{pmatrix} 0 & 0 \\ -a_{11}(x) & -a_{01}(x) \end{pmatrix}.$$

To simplify the calculation we transform the eigenvalue problem so that within the λ-plane the branch point, λ_{br}, is independent of ϵ. Within the γ-plane, this has the effect of making the perturbation matrices for the vectorized ODE independent of γ; see (9.5.32). We introduce the transformed variable

$$\boldsymbol{y}(x) := \mathrm{e}^{\epsilon a_{11}^0 x/2}\boldsymbol{Y}(x), \tag{9.5.29}$$

which satisfies the system

$$\partial_x \boldsymbol{y} = \left[\boldsymbol{A}_0(\lambda) + \epsilon \frac{1}{2} a_{11}^0 \boldsymbol{I}_2 + \boldsymbol{R}_0(x) + \epsilon \boldsymbol{R}_1(x)\right]\boldsymbol{y}. \tag{9.5.30}$$

When $\epsilon = 0$ we clearly have $\boldsymbol{y}(x,\lambda,0) = \boldsymbol{Y}(x,\lambda,0)$, but more importantly $\partial_\epsilon \boldsymbol{y}(0,\lambda,0) = \partial_\epsilon \boldsymbol{Y}(0,\lambda,0)$, so that the derivative of the Evans function with respect to ϵ is unaffected by this change of variables. The matrix eigenvalues for the system (9.5.30) satisfy

$$\mu^2 = \lambda - \epsilon a_{01}^0 + \frac{1}{4}\epsilon^2 (a_{11}^0)^2,$$

which implies that the Riemann surface should be defined by

$$\gamma^2 := \lambda - \epsilon a_{01}^0 + \frac{1}{4}\epsilon^2 (a_{11}^0)^2. \tag{9.5.31}$$

The limiting matrix becomes

$$\boldsymbol{A}_0(\lambda) + \frac{1}{2}\epsilon a_1 \boldsymbol{I}_2 = \underbrace{\begin{pmatrix} 0 & 1 \\ \gamma^2 & 0 \end{pmatrix} + \epsilon \frac{1}{2} a_{11}^0 \begin{pmatrix} 1 & 0 \\ 0 & -1 \end{pmatrix} + \epsilon^2 \frac{1}{4}(a_{11}^0)^2 \begin{pmatrix} 0 & 0 \\ -1 & 0 \end{pmatrix}}_{\boldsymbol{A}_0(\gamma) + \epsilon \boldsymbol{A}_1 + \epsilon^2 \boldsymbol{A}_2},$$

so that the transformed system may be written as

$$\partial_x \boldsymbol{y} = \underbrace{\left[\boldsymbol{A}_0(\gamma) + \boldsymbol{R}_0(x) + \epsilon(\boldsymbol{A}_1 + \boldsymbol{R}_1(x)) + \epsilon^2 \boldsymbol{A}_2\right]}_{\boldsymbol{A}(x,\gamma,\epsilon)} \boldsymbol{y}, \tag{9.5.32}$$

where only $\boldsymbol{A}_0$ depends upon γ. It is easy to verify that

$$\ker\left[\boldsymbol{A}_0(0) + \epsilon \boldsymbol{A}_1 + \epsilon^2 \boldsymbol{A}_2\right] = \operatorname{span}\left\{\begin{pmatrix} 1 \\ -\epsilon a_{11}^0/2 \end{pmatrix}\right\}, \tag{9.5.33}$$

which depends upon ϵ.

Lemma 9.5.7. *Consider the Evans function, $E(\gamma,\epsilon)$, associated to the operators $\mathcal{L} = \mathcal{L}(\epsilon)$ introduced in (9.5.1). If for $\epsilon = 0$ the Evans function has a zero at the branch point $\gamma = 0$, i.e., $E(0,0) = 0$, then the partial derivative of the Evans function with respect to ϵ at the branch point is given by the formula*

$$\frac{\partial_\epsilon E(0,0)}{|\boldsymbol{J}_0(0)|^2} = a_{11}^0 \left(\frac{\rho_{-\infty} - (p_0^+)^2 \rho_{+\infty}}{2} + \langle p_0, \partial_x p_0 \rangle_\rho \right) + \langle p_0, \tilde{\mathcal{L}}_1 p_0 \rangle_\rho, \tag{9.5.34}$$

where p_0 and $\boldsymbol{J}_0$ are given in (9.5.11) and (9.5.12), while the weight ρ is defined in (9.5.3) and the localized operator $\tilde{\mathcal{L}}_1$ is given by (9.5.7).

Remark 9.5.8. The expression for $\partial_\epsilon E$ is comprised of two terms. The first term, multiplied by a_{11}^0, reflects the fact that $\sigma_{\text{abs}}(\mathcal{L}) \neq \sigma_{\text{ess}}(\mathcal{L})$ for $\epsilon > 0$ if $a_{11}^0 \neq 0$ (see Fig. 9.8). The second term encodes the impact of the variation in the potential with ϵ, and is related to the formula for the derivative of the Evans function at an eigenvalue. Indeed, if the location of the absolute spectrum is independent of ϵ, i.e., $a_{01}^0 = a_{11}^0 = 0$, then $\tilde{\mathcal{L}}_1 = \mathcal{L}_1$ and the expression for $\partial_\epsilon E(0,0)$ reduces to that for an isolated eigenvalue (see Exercise 9.5.5).

Proof. We suppress the γ-dependence, as all solutions of the system (9.5.32) will be evaluated at $\gamma = 0$. Our approach closely follows that of Lemma 9.5.4, and we present only the main points. Since $E(0) = 0$, when $\gamma = 0$ we choose $c^\pm(0)$ as in Lemma 9.5.3 so that $\boldsymbol{J}^+(x,0) = \boldsymbol{J}^-(x,0) = \boldsymbol{J}_0(x,0)$, and the derivative of the Evans function takes the form

$$\partial_\epsilon E(0) = \det\left(\partial_\epsilon(\boldsymbol{J}^- - \boldsymbol{J}^+), \boldsymbol{J}_0\right)(0,0).$$

The Jost solution $\boldsymbol{J}^+(x,\epsilon)$ to (9.5.32) can be written as

$$\boldsymbol{J}^+(x,\epsilon) = \boldsymbol{v}^+(\epsilon) + \boldsymbol{K}^+(x,\epsilon),$$

where

$$\boldsymbol{v}^+(\epsilon) \in \ker\left[\boldsymbol{A}_0(0) + \epsilon \boldsymbol{A}_1 + \epsilon^2 \boldsymbol{A}_2\right], \quad |\boldsymbol{K}^+(x,\epsilon)| = \mathcal{O}(\mathrm{e}^{-\alpha L}), \ x \geq L.$$

The correction term satisfies the nonhomogeneous ODE

$$\partial_x \boldsymbol{K}^+ = \boldsymbol{A}(x,\epsilon)\boldsymbol{K}^+ + [\boldsymbol{R}_0(x) + \epsilon \boldsymbol{R}_1(x)]\boldsymbol{v}^+(\epsilon),$$

so that when $\epsilon = 0$,

$$\partial_x(\partial_\epsilon \boldsymbol{K}^+) = \boldsymbol{A}(x,0)\partial_\epsilon \boldsymbol{K}^+ + [\boldsymbol{A}_1 + \boldsymbol{R}_1(x)]\boldsymbol{K}^+ + \boldsymbol{R}_1(x)\boldsymbol{v}^+(0) + \boldsymbol{R}_0(x)\partial_\epsilon \boldsymbol{v}^+(0).$$

Since the nonhomogeneous terms decay at an exponential rate as $x \to +\infty$, we can use variation of parameters with $\boldsymbol{\Phi}$ as in (9.5.19) to obtain

$$\partial_\epsilon \boldsymbol{K}^+(x,0) = \boldsymbol{\Phi}(x) \int_{+\infty}^{x} \boldsymbol{\Phi}(s)^{-1}\Big([\boldsymbol{A}_1 + \boldsymbol{R}_1(s)]\boldsymbol{K}^+ + \boldsymbol{R}_1(s)\boldsymbol{v}^+(0) + \boldsymbol{R}_0(s)\partial_\epsilon \boldsymbol{v}^+(0)\Big)\,\mathrm{d}s. \tag{9.5.35}$$

To simplify the integrand, we first use $\boldsymbol{v}^+(0) + \boldsymbol{K}^+(x,0) = \boldsymbol{J}_0(x,0)$, to eliminate $\boldsymbol{K}^+$ so that

$$\begin{aligned}
&[\boldsymbol{A}_1 + \boldsymbol{R}_1(s)]\boldsymbol{K}^+ + \boldsymbol{R}_1(s)\boldsymbol{v}^+(0) + \boldsymbol{R}_0(s)\partial_\epsilon \boldsymbol{v}^+(0) \\
&\quad = \boldsymbol{A}_1[\boldsymbol{J}_0(s) - \boldsymbol{v}^+(0)] + \boldsymbol{R}_1(s)\boldsymbol{J}_0(s) + \boldsymbol{R}_0(s)\partial_\epsilon \boldsymbol{v}^+(0), \\
&\quad = [\boldsymbol{A}_1 + \boldsymbol{R}_1(s)]\boldsymbol{J}_0(s) + \boldsymbol{R}_0(s)\partial_\epsilon \boldsymbol{v}^+(0) - \boldsymbol{A}_1 \boldsymbol{v}^+(0).
\end{aligned}$$

Then, taking ∂_ϵ of the defining relation of $\boldsymbol{v}^+$ we see

$$\Big[\boldsymbol{A}_0(0) + \epsilon \boldsymbol{A}_1 + \epsilon^2 \boldsymbol{A}_2\Big]\boldsymbol{v}^+(\epsilon) \equiv \boldsymbol{0} \quad \Rightarrow \quad \boldsymbol{A}_1 \boldsymbol{v}^+(0) = -\boldsymbol{A}_0 \partial_\epsilon \boldsymbol{v}^+(0),$$

which used in conjunction with the identity $\boldsymbol{A}_0(0) + \boldsymbol{R}_0(s) = \boldsymbol{A}(s,0)$ allows us to rewrite the integrand as

$$[\boldsymbol{A}_1 + \boldsymbol{R}_1(s)]\boldsymbol{K}^+ + \boldsymbol{R}_1(s)\boldsymbol{v}^+(0) + \boldsymbol{R}_0(s)\partial_\epsilon \boldsymbol{v}^+(0) = [\boldsymbol{A}_1 + \boldsymbol{R}_1(s)]\boldsymbol{J}_0(s) + \boldsymbol{A}(s,0)\partial_\epsilon \boldsymbol{v}^+(0).$$

The expression (9.5.35) now takes the form

$$\begin{aligned}
\partial_\epsilon \boldsymbol{K}^+(x,0) = \boldsymbol{\Phi}(x) &\int_{+\infty}^{x} \boldsymbol{\Phi}(s)^{-1}[\boldsymbol{A}_1 + \boldsymbol{R}_1(s)]\boldsymbol{J}_0(s)\,\mathrm{d}s \\
+ \boldsymbol{\Phi}(x) &\int_{+\infty}^{x} \boldsymbol{\Phi}(s)^{-1}\boldsymbol{A}(s,0)\partial_\epsilon \boldsymbol{v}^+(0)\,\mathrm{d}s.
\end{aligned} \tag{9.5.36}$$

Using the relation $\partial_x \boldsymbol{\Phi}^{-1} = -\boldsymbol{\Phi}^{-1}(x)\boldsymbol{A}(x,0)$, we integrate by parts in the second term on the right-hand side of (9.5.36), obtaining

$$\begin{aligned}
\boldsymbol{\Phi}(x) \int_{+\infty}^{x} \boldsymbol{\Phi}(s)^{-1}\boldsymbol{A}(s,0)\partial_\epsilon \boldsymbol{v}^+(0)\,\mathrm{d}s &= -\boldsymbol{\Phi}(x) \lim_{L \to +\infty} \int_{L}^{x} \partial_s \boldsymbol{\Phi}(s)^{-1}\partial_\epsilon \boldsymbol{v}^+(0)\,\mathrm{d}s \\
&= \lim_{L \to +\infty} [\boldsymbol{\Phi}(x)\boldsymbol{\Phi}(L)^{-1} - \boldsymbol{I}_2]\partial_\epsilon \boldsymbol{v}^+(0).
\end{aligned}$$

Substituting this formula into (9.5.36) and evaluating the result at $x = 0$ yields an expression for $\partial_\epsilon \boldsymbol{K}^+(0,0)$. Substituting this result into the expression $\partial_\epsilon \boldsymbol{J}^+(0,0) = \partial_\epsilon \boldsymbol{v}^+(0) + \partial_\epsilon \boldsymbol{K}^+(0,0)$ and observing the cancelation of $\partial_\epsilon \boldsymbol{v}^+(0)$ we obtain,

$$\partial_\epsilon \boldsymbol{J}^+(0,0) = \boldsymbol{\Phi}(0)\int_{+\infty}^{x} \boldsymbol{\Phi}(s)^{-1}[\boldsymbol{A}_1 + \boldsymbol{R}_1(s)]\boldsymbol{J}_0(s)\,\mathrm{d}s + \lim_{L\to+\infty} \boldsymbol{\Phi}(0)\boldsymbol{\Phi}(L)^{-1}\partial_\epsilon \boldsymbol{v}^+(0).$$

In a similar manner it can be shown that

$$\partial_\epsilon \boldsymbol{J}^-(0,0) = \boldsymbol{\Phi}(0)\int_{-\infty}^{x} \boldsymbol{\Phi}(s)^{-1}[\boldsymbol{A}_1 + \boldsymbol{R}_1(s)]\boldsymbol{J}_0(s)\,\mathrm{d}s + \lim_{L\to+\infty} \boldsymbol{\Phi}(0)\boldsymbol{\Phi}(-L)^{-1}\partial_\epsilon \boldsymbol{v}^-(0)$$

(see Exercise 9.5.4), so that

$$\partial_\epsilon(\boldsymbol{J}^- - \boldsymbol{J}^+)(0,0) = \boldsymbol{\Phi}(0)\boldsymbol{w}_\infty,$$

where the limiting vector is defined by

$$\boldsymbol{w}_\infty := \int_{\mathbb{R}} \boldsymbol{\Phi}(s)^{-1}[\boldsymbol{A}_1+\boldsymbol{R}_1(s)]\boldsymbol{J}_0(s)\,\mathrm{d}s + \lim_{L\to+\infty}\left(\boldsymbol{\Phi}(-L)^{-1}\partial_\epsilon \boldsymbol{v}^-(0) - \boldsymbol{\Phi}(L)^{-1}\partial_\epsilon \boldsymbol{v}^+(0)\right).$$

Arguing as in (9.5.23) we find

$$\partial_\epsilon E(0) = w_{\infty 1}\det(\boldsymbol{J}_0^{\mathrm{a}}, \boldsymbol{J}_0)(0) = w_{\infty 1}|\boldsymbol{J}_0(0)|^2, \tag{9.5.37}$$

and applying the formula (9.5.24) to $\boldsymbol{w}$ we obtain the simplified expression

$$w_{\infty 1} = \lim_{L\to+\infty}(\boldsymbol{J}_0^{\mathrm{a}}(-L)\cdot\partial_\epsilon \boldsymbol{v}^-(0) - \boldsymbol{J}_0^{\mathrm{a}}(L)\cdot\partial_\epsilon \boldsymbol{v}^+(0)) + \int_{\mathbb{R}} \boldsymbol{J}_0^{\mathrm{a}}(s)\cdot[\boldsymbol{A}_1 + \boldsymbol{R}_1(s)]\boldsymbol{J}_0(s)\,\mathrm{d}s.$$

Recalling the normalization constants $c^\pm$ of $\boldsymbol{v}^\pm$ from Lemma 9.5.3 and the expression (9.5.33) we have

$$\boldsymbol{v}^-(\epsilon) = \begin{pmatrix} 1 \\ -\epsilon a_{11}^0/2 \end{pmatrix}, \quad \boldsymbol{v}^+(\epsilon) = p_0^+\begin{pmatrix} 1 \\ -\epsilon a_{11}^0/2 \end{pmatrix},$$

and taking ∂_ϵ at $\epsilon = 0$ we obtain

$$\partial_\epsilon \boldsymbol{v}^-(0) = -\frac{1}{2}a_{11}^0\begin{pmatrix} 0 \\ 1 \end{pmatrix}, \quad \partial_\epsilon \boldsymbol{v}^+(0) = -\frac{1}{2}p_0^+ a_{11}^0\begin{pmatrix} 0 \\ 1 \end{pmatrix}.$$

Using the asymptotic formulas (9.5.13) we find

$$\begin{aligned}\lim_{L\to+\infty}&(\boldsymbol{J}_0^{\mathrm{a}}(-L)\cdot\partial_\epsilon \boldsymbol{v}^-(0) - \boldsymbol{J}_0^{\mathrm{a}}(L)\cdot\partial_\epsilon \boldsymbol{v}^+(0)) \\ &= -\frac{1}{2}a_{11}^0\left[\mathrm{e}^{-\int_{-\infty}^0 a_{10}(s)\,\mathrm{d}s} - (p_0^+)^2\mathrm{e}^{\int_0^{+\infty} a_{10}(s)\,\mathrm{d}s}\right].\end{aligned} \tag{9.5.38}$$

To simplify the integral in $w_{\infty 1}$ we recall the formulas (9.5.12) and the definition of A_1 and R_1 to evaluate

$$\boldsymbol{J}_0^{\mathrm{a}}(s)\cdot\boldsymbol{A}_1\boldsymbol{J}_0(s) = -a_{11}^0 q_0(s)\partial_s p_0(s),$$

and

$$\boldsymbol{J}_0^{\mathrm{a}}(s)\cdot\boldsymbol{R}_1(s)\boldsymbol{J}_0(s)\,\mathrm{d}s = -q_0(s)\tilde{\mathcal{L}}_1 p_0(s),$$

where we recall the adjoint solution $q_0(x) := \rho(x)p_0(x)$. Combining these two expressions we see that

$$\int_{\mathbb{R}} \boldsymbol{J}_0^{\mathrm{a}}(s)\cdot[\boldsymbol{A}_1+\boldsymbol{R}_1(s)]\boldsymbol{J}_0(s)\,\mathrm{d}s = -a_{11}^0\langle p_0,\partial_x p_0\rangle_\rho - \langle p_0,(\mathcal{L}_1-\mathcal{L}_1^0)p_0\rangle_\rho. \quad (9.5.39)$$

Inserting (9.5.38) and (9.5.39), along with the expression (9.5.25) for the determinant, into the expression (9.5.37) for the derivative of the Evans function yields (9.5.34). □

Example 9.5.9. (cont.) Recall the examination of the operator $\mathcal{L}$ in the case that $a_{10}(x)$ is odd and $a_{00}(x)$ is even. After a transformation the eigenvalue problem for $\mathcal{L}_0$ takes the form [see (9.5.26)]

$$\partial_x^2 p + a(x)p = \lambda p. \quad a(-x) = a(x),$$

Consider the following perturbation of this problem,

$$\partial_x^2 p + \epsilon[a_{10}^0 + a_{11}(x)]\partial_x p + [a(x) + \epsilon(a_{01}^0 + a_{01}(x))]p = \lambda p.$$

Assume $E(0,0) = 0$, and let p_0, defined in (9.5.11), be even in x. We saw in (9.5.28) that

$$\partial_\gamma E(0,0) = -2p_0(0)^2.$$

Let us now consider $\epsilon \neq 0$ and calculate $\partial_\epsilon E(0,0)$ using Lemma 9.5.7. Since $p_0(x)$ is even it follows that

$$1 = \lim_{x\to-\infty} p_0(x) = \lim_{x\to+\infty} p_0(x) \quad \Rightarrow \quad p_0^+ = 1.$$

In addition $a_{10}(x) \equiv 0$, so that the first term multiplied by a_{11}^0 in (9.5.34) vanishes. As well, the symmetrizing weight $\rho \equiv 1$ and the second term multiplied by a_{11}^0 is also zero due to the even parity,

$$\langle p_0,\partial_x p_0\rangle = \frac{1}{2}\int_{\mathbb{R}} \partial_s[p_0(s)]^2\,\mathrm{d}s = \frac{1}{2}\lim_{L\to+\infty}[p(L,0)^2 - p(-L)^2] = 0.$$

The derivative of the Evans function at the branch point then reduces to

$$\partial_\epsilon E(0,0) = \langle p_0,\tilde{\mathcal{L}}_1 p_0\rangle p_0(0)^2.$$

Substituting these results into (9.5.6) we find that the zero at the branch point perturbs to $\gamma = \epsilon\gamma_1 + O(\epsilon^2)$ where

$$\gamma_1 = \frac{1}{2}\langle p_0, \tilde{\mathcal{L}}_1 p_0\rangle.$$

If $\gamma_1 > 0$ the zero moves onto the physical sheet of the Riemann surface for $\epsilon > 0$. However, if $\gamma_1 < 0$ the zero becomes a resonance pole for $\epsilon > 0$ (compare with Fig. 9.7).

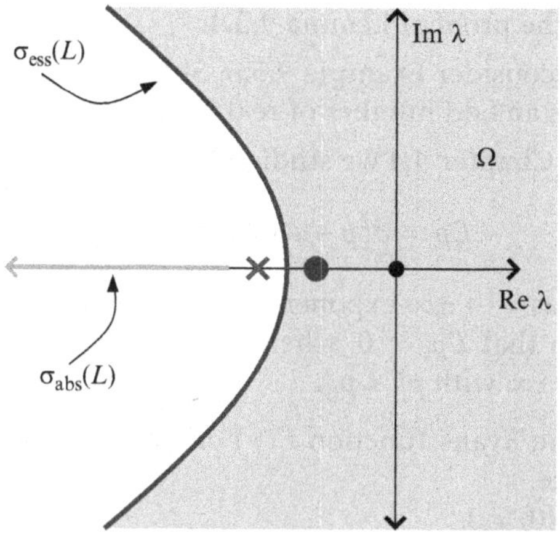

Fig. 9.8 The spectral picture in the λ-plane in the scaling $a_{01}^0 = \mathcal{O}(1) < 0$ for which the branch point, $\lambda_{br} = \epsilon[a_{01}^0 - \epsilon(a_{11}^0)^2/4]$ and the rightmost point of the essential spectrum, $\lambda_{ess} = \epsilon a_{01}^0 + \mathcal{O}(\epsilon^3)$ both move into the left-half complex λ plane for $\epsilon > 0$. If $\gamma_1 > 0$ then the zero of the Evans function moves onto the physical sheet of the Riemann surface, where at leading order it follows the branch point (see Fig. 9.7). Moreover, it becomes an eigenvalue (*blue* circle) only in the case $\gamma_1 > |a_{11}^0|/2$ when it moves to the right of the essential spectrum and onto the natural domain of the Evans function (the shaded region Ω); otherwise it is merely a zero of the Evans function (*red* cross.) (Color figure online.)

It is instructive to examine the relative motion of the zero, $\gamma(\epsilon)$, and the branch point, $\lambda_{br} = \epsilon a_{01}^0 - \epsilon^2(a_{11}^0)^2/4$. Suppose that $\gamma_1 > 0$. Since λ and γ are related by the Riemann relation (9.5.31), we see that in the λ-plane the zero is at

$$\lambda(\epsilon) = \epsilon a_{01}^0 + \epsilon^2\left(\gamma_1^2 - \frac{1}{4}(a_{11}^0)^2\right) + \mathcal{O}(\epsilon^3) = \lambda_{br} + \epsilon^2\gamma_1^2 + \mathcal{O}(\epsilon^3).$$

If $a_{01}^0 = \mathcal{O}(1) < 0$, then the branch point moves to the left and the bifurcating zero satisfies $\mathrm{Re}\,\lambda < 0$, but depending upon the relative size of γ_1 and a_{11}^0 may lie either to the left or the right of the essential spectrum that crosses the real axis at $\epsilon a_{01}^0 + \mathcal{O}(\epsilon^3)$ (see Fig. 9.8). On the other hand, if $a_{01}^0 = \mathcal{O}(\epsilon) < 0$, then rescaling $a_{01}^0 \mapsto \epsilon a_{01}^0 < 0$, we find the zero is located at

$$\lambda(\epsilon) = \epsilon^2\left(\gamma_1^2 + a_{01}^0 - \frac{1}{4}(a_{11}^0)^2\right) + \mathcal{O}(\epsilon^3),$$

and it becomes an eigenvalue with positive real part if $\gamma_1^2 > -a_{01}^0 + (a_{11}^0)^2/4$, while the branch point and essential spectrum both move to the left of zero.

Exercises

Exercise 9.5.1. Derive the equality

$$\partial_\gamma J^-(0,0) = \lim_{L\to+\infty} \Phi(0)\Phi(-L)^{-1}[\partial_\gamma v^-(0) - Lv^-(0)].$$

which appears in the proof of Lemma 9.5.4.

Exercise 9.5.2. Reconsider Example 9.5.6. Show that if $E(0) = 0$ and $p(x,0)$ is odd, then $\mathcal{L}_0$ has an odd number of real positive eigenvalues.

Exercise 9.5.3. In Chapter 4.4 we studied the eigenvalue problem

$$\mathcal{L}p := \partial_x^2 p + a_1(x)\partial_x p = \gamma^2 p,$$

where $a_1(x) \to a_1^\pm$ as $x \to \pm\infty$ exponentially fast with $a_1^- < 0 < a_1^+$. In particular, it was shown that $\mathcal{L}p_0 = 0$, where $p_0(x)$ is monotone increasing, and $p_0(x) \to p_0^\pm$ as $x \to \pm\infty$ with $p_0^+ < p_0^-$.

(a) Construct the Evans function $E(\gamma)$ on an appropriately defined Riemann surface.
(b) Show that $E(0) \neq 0$.

Exercise 9.5.4. For the Jost solution $\boldsymbol{J}^-(x,\epsilon) = \boldsymbol{v}^-(\epsilon) + \boldsymbol{K}^-(x,\epsilon)$, verify the equality

$$\partial_\epsilon \boldsymbol{J}^-(0,0) = \boldsymbol{\Phi}(0)\int_{-\infty}^{x} \boldsymbol{\Phi}(s)^{-1}[\boldsymbol{A}_1 + \boldsymbol{R}_1(s)]\boldsymbol{J}_0(s)\,ds + \lim_{L\to+\infty} \boldsymbol{\Phi}(0)\boldsymbol{\Phi}(-L)^{-1}\partial_\epsilon \boldsymbol{v}^-(0),$$

used in Lemma 9.5.7.

Exercise 9.5.5. Suppose that when $\epsilon = 0$, $\gamma = \gamma_0 \in \sigma_{\rm pt}(\mathcal{L})$ is in the natural domain of the Evans function and is a simple eigenvalue for the operator

$$\mathcal{L} = \partial_x^2 + [a_{10}(x) + \epsilon a_{11}(x)]\partial_x + a_{00}(x) + \epsilon a_{01}(x),$$

where the coefficients are all exponentially localized. For the Evans function $E(\gamma,\epsilon)$ on the Riemann surface $\lambda = \gamma^2$, show that

$$\partial_\epsilon E(\gamma_0,0) = -\langle p_0, \mathcal{L}_1 p_0\rangle_\rho \det\left(\boldsymbol{J}_0^{\rm a}, \boldsymbol{J}_0\right)(0),$$

where p_0 is the eigenfunction of $\mathcal{L}_0$, ρ is the symmetrizing weight, and the adjoint solution has been normalized to have $\boldsymbol{J}_0^{\rm a}(0)| = 1$. Derive an asymptotic expansion for the location of the eigenvalue. Compare your result with the theory presented in Chapter 6.1.

Exercise 9.5.6. Reconsider Example 9.5.9. Compute $\partial_\epsilon E(0,0)$ under the assumption that $p(x,0)$ is odd when $\epsilon = 0$, and derive an expression for the location of the zero λ of the Evans function.

Exercise 9.5.7. Find the small zero of the Evans function for the problem

$$\partial_x^2 p + \epsilon a_{11}^0 \partial_x p + \epsilon [a_{01}^0 + F_0 \operatorname{sech}^4(x)]p = \lambda p.$$

Identify the region(s) in parameter space for which the essential spectrum lies in the left-half complex plane and there is an $\mathcal{O}(\epsilon^2)$ eigenvalue with positive real part.

Exercise 9.5.8. Find the small zero of the Evans function for the problem

$$\partial_x^2 p + \epsilon [a_{10}^0 + a_{10}(x)]\partial_x p + [2\operatorname{sech}^2(x) + \epsilon a_{01}(x)]p = \lambda p,$$

where

$$a_{10}(x) = F_1 \tanh(x)\operatorname{sech}(x), \quad a_{01}(x) = F_0 \operatorname{sech}^4(x)$$

(see Chapter 9.3.2). Identify the region(s) in parameter space for which the essential spectrum lies in the left-half complex plane and there is an $\mathcal{O}(\epsilon^2)$ eigenvalue with positive real part.

9.6 Application: Eigenvalue Problems on Large Intervals with Separated Boundary Conditions

Another task that is well-suited to an Evans function analysis is to identify the relation between the spectrum of an exponentially asymptotic differential operator $\mathcal{L}$ acting on $L^2(\mathbb{R})$ and the spectrum of the same operator on $L^2[-L, L]$, for $L \gg 1$, subject to separated boundary conditions. In particular, how does the choice of boundary conditions impact the convergence of the large interval spectrum to the whole-line spectrum?

While the essential and absolute spectra of exponentially asymptotic operators on $H^2(\mathbb{R})$ have simple characterizations, it is not possible in general to analytically characterize their point spectrum. Often the determination of a point spectrum is performed numerically, on a finite interval, and so the issue of convergence of point spectra is germane. Indeed, there are two issues to consider. The first is the question of how well the eigenvalues of $\mathcal{L}$ acting on $L^2[-L, L]$ approximate those eigenvalues of $\mathcal{L}$ on $L^2(\mathbb{R})$ that reside in the natural domain Ω of the Evans function (the set to the right of the essential spectrum). The operator $\mathcal{L}$ has only a finite number of eigenvalues in Ω, while acting on $L^2([-L, L])$ the operator has a countably infinite number of eigenvalues in $\mathbb{C}$. The second question is, do the remainder of the eigenvalues on the bounded domain accumulate near the essential spectrum associated with the unbounded domain problem, to the absolute spectrum, or to something else? We address the first question, and sketch the ideas behind the results of the second one, referring the interested reader to [252, 254] for details.

Consider the eigenvalue problem for the exponentially asymptotic operator $\mathcal{L}: H^2(\mathbb{R}) \subset L^2(\mathbb{R}) \to L^2(\mathbb{R})$ given by

$$\mathcal{L}p := \partial_x^2 p + a_1(x)\partial_x p + a_0(x)p = \lambda p, \tag{9.6.1}$$

where the real-valued coefficients satisfy

$$|a_0(x) - a_0^0| + |a_1(x) - a_1^0| \le Ce^{-\alpha|x|}. \tag{9.6.2}$$

We compare this to the eigenvalue problem on $H^2([-L, L])$ for the same operator subject to the separated boundary conditions

$$b_1^\pm p(\pm L) + b_2^\pm \partial_x p(\pm L) = 0. \tag{9.6.3}$$

The Riemann surface for the unbounded domain is,

$$\lambda = \gamma^2 + a_0^0 - (a_1^0)^2/4, \tag{9.6.4}$$

with the point spectrum lying in the natural domain of the Evans function,

$$\Omega = \{\gamma \in \mathbb{C} : \operatorname{Re}\gamma > |a_1^0|/2\}.$$

For the unbounded domain problem the essential spectrum is $\partial\Omega$, and the absolute spectrum lies on the imaginary axis $\{\gamma : \operatorname{Re}\gamma = 0\}$ (see Fig. 9.2).

In order to compare the spectra we first vectorize the eigenvalue problems. Setting $\mathbf{Y} = (p, \partial_x p)^{\mathrm{T}}$, we obtain

$$\partial_x \mathbf{Y} = \mathbf{A}(x,\gamma)\mathbf{y}, \quad \mathbf{A}(x,\gamma) = \mathbf{A}_0(\gamma) + \mathbf{R}(x), \tag{9.6.5}$$

where

$$\mathbf{A}_0(\gamma) = \begin{pmatrix} 0 & 1 \\ \gamma^2 - (a_1^0)^2/4 & -a_1^0 \end{pmatrix}, \quad \mathbf{R}(x) = \begin{pmatrix} 0 & 0 \\ a_0^0 - a_0(x) & a_1^0 - a_1(x) \end{pmatrix}.$$

For the bounded domain problem the Jost solutions $\mathbf{J}_L^\pm(x,\gamma)$ are entire in γ for fixed x and solve (9.6.5) subject to the boundary conditions

$$\mathbf{J}_L^\pm(\pm L, \gamma) = \mathbf{b}^\pm, \quad \mathbf{b}^\pm = \begin{pmatrix} -b_2^\pm \\ b_1^\pm \end{pmatrix} \tag{9.6.6}$$

[see (8.1.5)]. For the unbounded domain problem the matrix eigenvalues and associated eigenvectors are given by

$$\mu_1(\gamma) = -\frac{a_1^0}{2} + \gamma,\ \mathbf{v}_1(\gamma) = \begin{pmatrix} 1 \\ \mu_1(\gamma) \end{pmatrix}; \quad \mu_2(\gamma) = -\frac{a_1^0}{2} - \gamma,\ \mathbf{v}_2(\gamma) = \begin{pmatrix} 1 \\ \mu_2(\gamma) \end{pmatrix}, \tag{9.6.7}$$

and the corresponding Jost solutions, $J_\infty^\pm(x,\gamma)$, satisfy (9.6.5) subject to

$$\lim_{x\to-\infty} e^{-\mu_1(\gamma)x} J_\infty^-(x,\gamma) = v_1(\gamma), \quad \lim_{x\to+\infty} e^{-\mu_2(\gamma)x} J_\infty^+(x,\gamma) = v_2(\gamma).$$

For fixed x, the $J_\infty^\pm$ are analytic for $\{\gamma : \mathrm{Re}\,\gamma > -\alpha/2\}$. From these Jost solutions we construct the two Evans functions,

$$E_L(\gamma) = \det\left(J_L^-, J_L^+\right)(0,\gamma) = e^{-\int_0^x \mathrm{tr}A(s,\gamma)\,ds} \det\left(J_L^-, J_L^+\right)(x,\gamma)$$
$$E_\infty(\gamma) = \det\left(J_\infty^-, J_\infty^+\right)(0,\gamma) = e^{-\int_0^x \mathrm{tr}A(s,\gamma)\,ds} \det\left(J_\infty^-, J_\infty^+\right)(x,\gamma).$$

The second equalities, which shift the point of evaluation of the Jost determinant, follow from Liouville's formula. The spectral convergence questions can now be phrased in terms of the zeros of $E_L(\gamma)$ and $E_\infty(\gamma)$.

From the Jost Theorem 9.2.3, and the discussion of Chapter 9.3.2, for $L \gg 1$, the Jost solutions for the unbounded domain problem satisfy the asymptotic relations

$$e^{-\mu_1(\gamma)x} J_\infty^-(x,\gamma) = \begin{cases} v_1(\gamma) + \mathcal{O}(e^{-\alpha L}), & x \le -L \\ t(\gamma)v_1(\gamma) + r(\gamma)e^{(\mu_2-\mu_1)(\gamma)x} v_2(\gamma) + \mathcal{O}(e^{-\alpha L}), & x \ge L \end{cases}$$

$$e^{-\mu_2(\gamma)x} J_\infty^+(x,\gamma) = \begin{cases} r(\gamma)e^{(\mu_1-\mu_2)(\gamma)x} v_1(\gamma) + t(\gamma)v_2(\gamma) + \mathcal{O}(e^{-\alpha L}), & x \le -L \\ v_2(\gamma) + \mathcal{O}(e^{-\alpha L}), & x \ge L. \end{cases}$$

Here $t(\gamma)$ is the transmission coefficient and $r(\gamma)$ is the reflection coefficient. In particular, fixing $\delta \in (0,\alpha)$, the spectral set $\{\gamma : \mathrm{Re}\,\gamma > \delta\} \supset \Omega$, is strictly within the physical sheet of the Riemann surface. On this set the asymptotic relations reduce to

$$e^{-\mu_1(\gamma)x} J_\infty^-(x,\gamma) \sim \begin{cases} v_1(\gamma), & x \le -L \\ t(\gamma)v_1(\gamma), & x \ge L \end{cases}, \quad e^{-\mu_2(\gamma)x} J_\infty^+(x,\gamma) \sim \begin{cases} t(\gamma)v_2(\gamma), & x \le -L \\ v_2(\gamma), & x \ge L, \end{cases} \tag{9.6.8}$$

where the errors in these approximations are $\mathcal{O}(e^{-\delta L})$. The Evans function for the unbounded domain problem thus satisfies

$$E_\infty(\gamma) = \lim_{L\to+\infty} e^{-\int_0^L \mathrm{tr}A(s,\gamma)\,ds} \det(J_\infty^-, J_\infty^+)(L,\gamma) = t(\gamma)\det(v_1, v_2)(\gamma) = -2\gamma t(\gamma), \tag{9.6.9}$$

and we (again) see that the Evans function for the unbounded domain problem is proportional to the transmission coefficient. Since $\delta > 0$ was arbitrary we deduce that on its natural domain $E_\infty(\gamma)$ and $t(\gamma)$ have zeros of identical multiplicity.

Turning to the bounded domain problem, we pick γ satisfying $\mathrm{Re}\,\gamma \ge \delta$ and $E_\infty(\gamma) \ne 0$, so that the Jost solutions $J_\infty^\pm(x,\gamma)$ are linearly independent for each value of x. In particular there exist constants $c_\pm$ such that

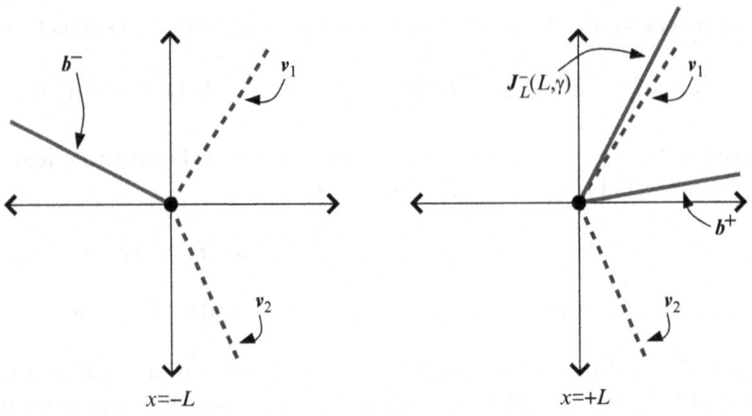

Fig. 9.9 The *left figure*, for $x = -L$, depicts the vectors $v_{1,2}(\gamma)$ (dashed *blue* line) and the vector b^- (solid *red* line). The *right figure*, for $x = +L$, depicts the vectors $v_{1,2}(\gamma)$ (dashed *blue* lines) and the vector b^+ (dashed *red* line). The forward image of b^- under the flow, given by the Jost solution evaluated at L, $J_L^-(L,\gamma)$ (solid *red* line), approaches v_1. The bounded and unbounded domain Evans functions are related as long as the sets $\{v_1, b^+\}$ and $\{v_2, b^-\}$ are linearly independent. (Color figure online.)

$$J_L^-(-L,\gamma) = b^- = c_- J_\infty^-(-L,\gamma) + c_+ J_\infty^+(-L,\gamma),$$

and since each of J_L^- and $J_\infty^\pm$ solve (9.6.5), it follows that

$$J_L^-(+L,\gamma) = c_- J_\infty^-(+L,\gamma) + c_+ J_\infty^+(+L,\gamma).$$

The relations (9.6.8) yield the asymptotic behavior

$$\begin{aligned} e^{-\mu_1(\gamma)L} J_L^-(+L,\gamma) &= c_- e^{-\mu_1(\gamma)L} J_\infty^-(+L,\gamma) + c_+ e^{(\mu_2-\mu_1)(\gamma)L} e^{-\mu_2(\gamma)L} J_\infty^+(+L,\gamma) \\ &\sim c_- t(\gamma) v_1(\gamma) + c_+ e^{(\mu_2-\mu_1)(\gamma)L} v_2(\gamma). \end{aligned}$$

However, from (9.6.7), $\mathrm{Re}(\mu_2(\gamma) - \mu_1(\gamma)) = -2\,\mathrm{Re}\,\gamma \le -2\delta$, and so, as long as $c_- \neq 0$, or equivalently $\det(v_2(\gamma), b^-) \neq 0$, we have the leading order behavior

$$e^{-\mu_1(\gamma)L} J_L^-(+L,\gamma) = c_- t(\gamma) v_1(\gamma) + \mathcal{O}(e^{-\delta L})$$

(see Fig. 9.9). The other Jost solution satisfies $J_L^+(+L,\gamma) = b^+$, and we may approximate the bounded-domain Evans function as

$$\begin{aligned} E_L(\gamma) &= e^{-\int_0^L \mathrm{tr} A(s,\gamma)\,ds} \det(J_L^-, J_L^+)(+L,\gamma) \\ &= e^{\mu_1(\gamma)L - \int_0^L \mathrm{tr} A(s,\gamma)\,ds} \det(e^{-\mu_1(\gamma)L} J_L^-(+L,\gamma), b^+) \\ &= c_- t(\gamma) e^{\mu_1(\gamma)L - \int_0^L \mathrm{tr} A(s,\gamma)\,ds} [\det(v_1(\gamma), b^+) + \mathcal{O}(e^{-\delta L})]. \end{aligned}$$

Normalizing the bounded domain Evans function

$$\tilde{E}_L(\gamma) = \mathrm{e}^{-\mu_1(\gamma)L + \int_0^L \operatorname{tr} A(s,\gamma)\,\mathrm{d}s} E_L(\gamma),$$

and using the relationship between the transmission coefficient and the Evans function given in (9.6.9), we obtain

$$\tilde{E}_L(\gamma) = -c_- \frac{\det(\boldsymbol{v}_1(\gamma), \boldsymbol{b}^+)}{2\gamma} E_\infty(\gamma) + \mathcal{O}(\mathrm{e}^{-\delta L}). \tag{9.6.10}$$

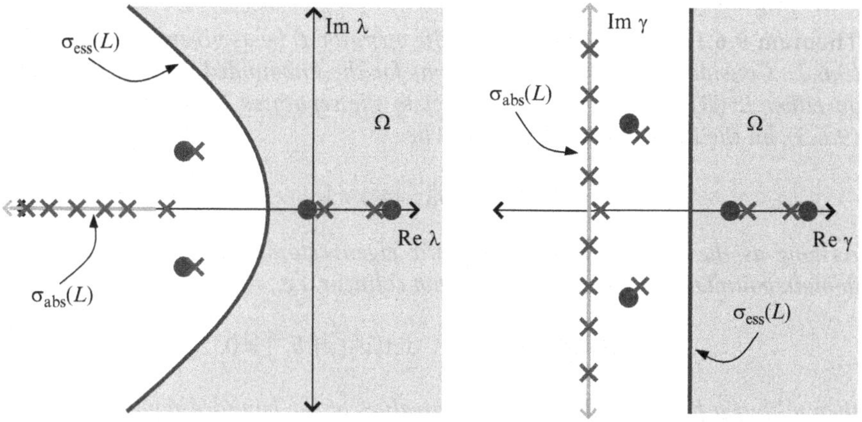

Fig. 9.10 A comparison in the λ-plane (*left*) and γ-plane (*right*) of the spectra for the bounded and unbounded domain problems. The point spectra for the unbounded domain problem (*blue* circles) are exponentially shadowed, up to multiplicity, by point spectra of the bounded domain problem (*red* crosses) while the remainder of the bounded-domain spectra accumulates around the absolute spectrum of the unbounded domain problem, and not about the essential spectrum. (Color figure online.)

For $L \gg 1$ the relation (9.6.10) shows that the two Evans functions are exponentially close so long as

$$\det(\boldsymbol{v}_2(\gamma), \boldsymbol{b}^-) \neq 0, \quad \det(\boldsymbol{v}_1(\gamma), \boldsymbol{b}^+) \neq 0, \tag{9.6.11}$$

and $\operatorname{Re}\gamma$ is uniformly bounded away from zero. Since both Evans functions are analytic on the physical sheet, $\{\gamma : \operatorname{Re}\gamma > 0\}$, it follows from Rouché's theorem that if $E_\infty(\gamma_0) = 0$ for some $\operatorname{Re}\gamma_0 > 0$, then within an $\mathcal{O}(\mathrm{e}^{-(\operatorname{Re}\gamma_0)L})$ neighborhood of γ_0, the bounded domain Evans function, $E_L(\gamma)$, has zeros with multiplicity that sums to $E_\infty(\gamma_0)$. This is significant: the zeros of the unbounded domain Evans function, which are strictly on the physical sheet of the Riemann surface, but which do not correspond to eigenvalues, are exponentially close to eigenvalues of the bounded domain problem.

The essential spectrum also typically lies strictly within the physical sheet, i.e., within the set $\{\gamma : \operatorname{Re}\gamma > \delta\}$ for $\delta > 0$ sufficiently small; see Fig. 9.2. Thus, if $E_\infty(\gamma) \neq 0$ on the essential spectrum, then $E_L(\gamma) \neq 0$ for L sufficiently large; that is, the essential spectrum of the unbounded domain problem is not seen by the bounded domain problem. However, E_∞, and hence E_L, have only a finite number of zeros for $\mathrm{Re}\,\gamma > \delta$, while E_L has a countably infinite number of zeros on the physical sheet $\{\gamma : \mathrm{Re}\,\gamma \geq 0\}$. It follows that all but a finite number of the eigenvalues for the unbounded domain problem lie within $\mathcal{O}(\delta)$ of the absolute spectrum (see Fig. 9.10). While we will not prove it here, it is shown in [252, 254] that the branch point of the absolute spectrum is an accumulation point for the point spectrum of the bounded domain problem as $L \to \infty$. We summarize the results of this section in the following theorem.

Theorem 9.6.1. *Let the Sturm–Liouville operator $\mathcal{L}$ be as given in (9.6.1) and (9.6.2). Consider the eigenvalue problems for the unbounded operator $\mathcal{L}$ acting on either $L^2(\mathbb{R})$, or on $L^2[-L,L]$ subject to the separated boundary conditions (9.6.3), on the Riemann surface defined by*

$$\lambda = \gamma^2 + a_0^0 - (a_1^0)^2/4.$$

As long as the left and right asymptotic eigenvectors, (9.6.7), and bounded-domain boundary-vectors, (9.6.6), are not colinear, i.e.,

$$\det\left(\boldsymbol{v}_1(\gamma), \boldsymbol{b}^+\right) \neq 0 \quad \det\left(\boldsymbol{v}_2(\gamma), \boldsymbol{b}^-\right) \neq 0, \tag{9.6.12}$$

then all but a finite number of the eigenvalues of the bounded domain problem converge to the absolute spectrum of the unbounded domain problem as $L \to \infty$, with the remainder converging exponentially and up to multiplicity to the zeros of the Evans function for the unbounded domain problem. In particular, the point spectra on the natural domain of the unbounded domain problem is exponentially close to point spectra of the bounded domain problem.

Remark 9.6.2. If the essential spectrum of the unbounded domain problem crosses into the set $\{\lambda : \mathrm{Re}\,\lambda > 0\}$, but the absolute spectrum does not, then the bounded domain problem need not be unstable. In this case, one needs to verify that all the zeros of E_∞, not just those on the natural domain, lie in $\{\lambda : \mathrm{Re}\,\lambda < 0\}$. That is, a zero of the extended Evans function that does not correspond to an eigenvalue can signify instability for the bounded domain problem.

Remark 9.6.3. As we show in the example below, if condition (9.6.12) does not hold, then it is possible that the there are eigenvalues of the bounded domain problem that tend to a limit within the physical sheet as $L \to \infty$ and that remain an $\mathcal{O}(1)$ distance from any eigenvalues of the unbounded domain problem. The interested reader should consult [252, Section 4] for further details.

Example 9.6.4. Recall the constant coefficient problem first discussed in Chapter 3.2,

$$\partial_x^2 p + 2\partial_x p = \lambda p.$$

On the Riemann surface $\lambda = \gamma^2 - 1$ the matrix eigenvalues and associated eigenvectors are given by

$$\mu_1(\gamma) = -1 + \gamma,\ v_1(\gamma) = \begin{pmatrix} 1 \\ \mu_1(\gamma) \end{pmatrix}; \quad \mu_2(\gamma) = -1 - \gamma,\ v_2(\gamma) = \begin{pmatrix} 1 \\ \mu_2(\gamma) \end{pmatrix}.$$

The Jost solutions take the form

$$J_\infty^-(x,\gamma) = \mathrm{e}^{\mu_1(\gamma)x} v_1(\gamma), \quad J_\infty^+(x,\gamma) = \mathrm{e}^{\mu_2(\gamma)x} v_2(\gamma),$$

and we easily calculate that

$$E_\infty(\gamma) = \det\left(J_\infty^-, J_\infty^+\right)(0,\gamma) = -2\gamma.$$

To examine the bounded domain Evans function, $E_L(\gamma)$, we start with the fundamental matrix solution

$$\boldsymbol{\Phi}(x,\gamma) = \begin{pmatrix} \mathrm{e}^{\mu_1(x+L)} & \mathrm{e}^{\mu_2(x+L)} \\ \mu_1 \mathrm{e}^{\mu_1(x+L)} & \mu_2 \mathrm{e}^{\mu_2(x+L)} \end{pmatrix} \quad \Rightarrow \quad \boldsymbol{\Phi}^{-1}(-L,\gamma) = \frac{1}{2\gamma} \begin{pmatrix} -\mu_2 & 1 \\ \mu_1 & -1 \end{pmatrix},$$

from which we construct the Jost solution

$$\begin{aligned} J_L^-(x,\gamma) &:= \boldsymbol{\Phi}(x,\gamma)\boldsymbol{\Phi}(-L,\gamma)^{-1}\boldsymbol{b}^- \\ &= \frac{\mathrm{e}^{-2L}}{\gamma} \begin{pmatrix} \gamma\cosh(2\gamma L) - \sinh(2\gamma L) & \sinh(2\gamma L) \\ (\gamma^2-1)\sinh(2\gamma L) & \gamma\cosh(2\gamma L) - \sinh(2\gamma L) \end{pmatrix} \boldsymbol{b}^-. \end{aligned}$$

Since $\operatorname{tr}\boldsymbol{A}(x,\gamma) = -2$, the Evans function for the bounded domain problem takes the form

$$\begin{aligned} E_L(\gamma) &= \mathrm{e}^{\int_0^L 2\,\mathrm{d}s} \det\left(J_L^-(L,\gamma), \boldsymbol{b}^+\right) \\ &= \frac{1}{\gamma}\det\left(\begin{pmatrix} \gamma\cosh(2\gamma L) - \sinh(2\gamma L) & \sinh(2\gamma L) \\ (\gamma^2-1)\sinh(2\gamma L) & \gamma\cosh(2\gamma L) - \sinh(2\gamma L) \end{pmatrix}\boldsymbol{b}^-, \boldsymbol{b}^+\right). \end{aligned}$$

We have $E_\infty(\gamma) \neq 0$ on the physical sheet of the Riemann surface. Assuming that (9.6.12) holds, Theorem 9.6.1 implies that $E_L(\gamma) \neq 0$ for $\operatorname{Re}\gamma \geq \delta > 0$ where $\delta > 0$ and $L > 0$ is sufficiently large. What happens if (9.6.12) does not hold? Specifically, fix a real $\gamma_* > 0$ outside the spectrum of the unbounded domain problem and choose $\boldsymbol{b}^- = v_2(\gamma^*)$. In this case the expression for bounded domain Evans function satisfies

$$E_L(\gamma_*) = \det\left(\mathrm{e}^{-2\gamma_* L} v_2(\gamma_*), \boldsymbol{b}^+\right).$$

Upon setting $\boldsymbol{b}^+ = v_2(\gamma^*)$ we see that $E_L(\gamma^*) = 0$. However, we do not require both nondegeneracy conditions in (9.6.12) to fail, for if we perturb $\boldsymbol{b}^+(\epsilon) = v_2(\gamma^*) + \epsilon \boldsymbol{b}_1^+$, and apply the implicit function theorem , we can conclude that there is a smooth curve $\gamma(\epsilon)$ with $\gamma(0) = \gamma_*$ such that the bounded domain problem with boundary vectors $\{\boldsymbol{b}_-, \boldsymbol{b}_+(\epsilon)\}$ satisfies $E_L(\gamma(\epsilon), \epsilon) = 0$ for $\epsilon > 0$ sufficiently small. Thus, the violation of one of the conditions (9.6.12) can induce an eigenvalue of the bounded domain problem at an arbitrary point in $\mathbb{R}^+$ that does not correspond to a zero of the unbounded domain Evans function.

Exercises

Exercise 9.6.1. In the preceding example suppose that $\boldsymbol{b}^+ = v_1(\gamma_*)$ for some real $\gamma_* > 0$. Show that $\boldsymbol{b}^-$ can be chosen so that $E_L(\gamma) = 0$ for γ close to γ^*.

Exercise 9.6.2. Consider the eigenvalue problem for the square-well operator introduced in (9.3.1)

$$\mathcal{L}p := \partial_x^2 p + a_1^0 \partial_x p + a_0(x)p = \lambda p,$$

where a_0 is the piecewise constant square-well potential. The Evans function for $\mathcal{L}$ for $a_1^0 = 0$ is given in (9.3.8).

(a) Determine $\sigma_{\text{ess}}(\mathcal{L})$ and $\sigma_{\text{abs}}(\mathcal{L})$, and construct the Evans function for $\mathcal{L}$ to find the point spectrum.
(b) Determine the Evans function for the eigenvalue problem associated with $\mathcal{L}$ subject to the separated boundary conditions

$$b_1^- p(-L) + b_2^- \partial_x p(-L) = 0, \quad b_1^+ p(+L) + b_2^+ \partial_x p(+L) = 0.$$

Plot the zeros of the Evans function for various values of L and compare to the whole-line point spectrum and the absolute spectrum. Identify values of $b_1^\pm$ and $b_2^\pm$ for which spurious eigenvalues are created, or for which known eigenvalues are not approximated.

Exercise 9.6.3. Consider the eigenvalue problem on the real line given by

$$\partial_x^2 p + 2\partial_x p + \ell(\ell+1)\operatorname{sech}^2(x)p = \lambda p, \quad \ell \in \mathbb{N}.$$

(a) Compute $\sigma_{\text{ess}}(\mathcal{L})$ and $\sigma_{\text{abs}}(\mathcal{L})$.
(b) Verify that the Evans function $E_\infty(\gamma)$ is analytic on the Riemann surface $\lambda = \gamma^2 - 1$ for $\operatorname{Re}\gamma > -1$.
(c) Show that

$$E_\infty(\sqrt{j}) = 0, \quad j = 1, \ldots, \ell.$$

Which of these zeros correspond to point eigenvalues?
(d) Now consider the eigenvalue problem with the separated boundary conditions

$$b_1^- p(-L) + b_2^- \partial_x p(-L) = 0, \quad b_1^+ p(+L) + b_2^+ \partial_x p(+L) = 0.$$

Using a finite-difference scheme to approximate the differential operators, numerically compute the spectrum for various values of L.

How large does L have to be in order to accurately locate the known eigenvalues? Are there specific values of $b_1^\pm$ and $b_2^\pm$ for which spurious eigenvalues are created, or for which known eigenvalues are not approximated?

9.7 Application: Eigenvalue Problems for Periodic Problems with Large Spatial Period

Nonseparated boundary conditions couple the values of the unknown function at both boundaries; of these, periodic boundary conditions are the classic example. We consider the eigenvalue problem,

$$\mathcal{L}p := \partial_x^2 p + a_{1,0}(x)\partial_x p + a_{0,0}(x)p = \lambda p, \quad x \in \mathbb{R}, \tag{9.7.1}$$

where the unbounded operator $\mathcal{L}$ acts on $L^2(\mathbb{R})$ and the coefficients satisfy the asymptotic decay estimates as in (9.6.2),

$$|a_{0,0}(x) - a_0^0| + |a_{1,0}(x) - a_1^0| \le Ce^{-\alpha|x|}.$$

In this section we compare the spectrum of $\mathcal{L}$ to that of $\mathcal{L}_\eta$, given by

$$\mathcal{L}_\eta p := \partial_x^2 p + a_{1,\eta}(x)\partial_x p + a_{0,\eta}(x)p = \lambda p, \quad x \in [-L_\eta, L_\eta], \tag{9.7.2}$$

subject to $2L_\eta$-periodic boundary conditions. Here for $j = 0,1$ the coefficients $a_{j,\eta}$ are $2L_\eta$-periodic,

$$a_{j,\eta}(x + 2L_\eta) = a_{j,\eta}(x), \quad \lim_{\eta \to 0^+} L_\eta = +\infty, \tag{9.7.3}$$

with

$$|a_{j,\eta}(x) - a_{j,0}(x)| = \mathcal{O}(\eta e^{-\alpha|x|}), \quad -L_\eta \le x \le L_\eta. \tag{9.7.4}$$

The L_η-periodic problem of (9.7.2) has periodic coefficients with large period, and over one period the coefficients of the L_η-problem converge uniformly to the coefficients of the unbounded problem. A typical example of such a coefficient is

$$a_{0,\eta}(x) = \sum_{n=-\infty}^{\infty} \operatorname{sech}(x - nL_\eta),$$

with $a_{0,0}(x) = \operatorname{sech}(x)$. Potentials of this form arise naturally when periodic orbits bifurcate from a limiting homoclinic orbit with hyperbolic end-states, as the cnoidal waves of KdV bifurcate from the solitary wave; see Fig. 5.2. It is interesting to ask in what sense the spectra of the linearization about the periodic wave, $\mathcal{L}_\eta$ approaches that of the limiting homoclinic, $\mathcal{L}$.

We compare the point spectrum for the Bloch-wave decomposition of $\mathcal{L}$, defined in Chapter 8.4, to the point, essential, and absolute spectrum of the unbounded domain problem. The Bloch-wave decomposition writes the spectra of $\mathcal{L}$ as the union of the point spectra of the operators

$$\mathcal{L}_{\mu,\eta}q := (\partial_x + \mathrm{i}\mu)^2 q + a_{1,\eta}(x)(\partial_x + \mathrm{i}\mu)q + a_{0,\eta}(x)q = \lambda q, \quad -1/L_\eta < \mu \leq 1/L_\eta. \tag{9.7.5}$$

Recall from Theorem 8.4.3 that in the case of simple eigenvalues for fixed μ the spectra for (9.7.5) consists of continuous μ-dependent curves; hence, in this situation we are asking in what sense the curves of spectra for the Bloch-wave problem approximate the various types of spectra for the unbounded domain problem.

First consider the unbounded domain problem. The Riemann surface for (9.7.1) is defined by

$$\lambda = \gamma^2 + a_{0,0}^0 - (a_{1,0}^0)^2/4,$$

and the point spectrum for the unbounded domain problem lies within the natural domain

$$\Omega = \{\gamma \in \mathbb{C} : \mathrm{Re}\,\gamma > |a_{1,0}^0|/2\},$$

the boundary of which forms the essential spectrum. The Evans function for the unbounded domain problem,

$$E_\infty(\gamma) = \det\left(\boldsymbol{J}^-, \boldsymbol{J}^+\right)(0,\gamma), \tag{9.7.6}$$

is determined through the Jost solutions which satisfy

$$\lim_{x\to-\infty} \mathrm{e}^{-\mu_1(\gamma)x} \boldsymbol{J}_\infty^-(x,\gamma) = \boldsymbol{v}_1(\gamma), \quad \lim_{x\to+\infty} \mathrm{e}^{-\mu_2(\gamma)x} \boldsymbol{J}_\infty^+(x,\gamma) = \boldsymbol{v}_2(\gamma),$$

where the matrix eigenvalues and associated eigenvectors are,

$$\mu_1(\gamma) = -\frac{a_{1,0}^0}{2} + \gamma,\ \boldsymbol{v}_1(\gamma) = \begin{pmatrix} 1 \\ \mu_1(\gamma) \end{pmatrix}; \quad \mu_2(\gamma) = -\frac{a_{1,0}^0}{2} - \gamma,\ \boldsymbol{v}_2(\gamma) = \begin{pmatrix} 1 \\ \mu_2(\gamma) \end{pmatrix}.$$

Now consider the Bloch-wave problem (9.7.5). Let $\boldsymbol{\Phi}_L(x,\gamma,\mu) \in \mathbb{C}^{2\times 2}$ represent the fundamental matrix solution to the vectorized form of the eigenvalue problem. From (8.4.11) we know that the associated Evans function takes the form

$$E_L(\gamma,\mu) = \det\left(\boldsymbol{\Phi}_\eta(L_\eta,\gamma,\mu)\boldsymbol{\Phi}_\eta^{-1}(-L_\eta,\gamma,\mu) - \boldsymbol{I}_2\right).$$

Rewriting the matrix in the determinant as

$$\boldsymbol{\Phi}_L(L_\eta)\boldsymbol{\Phi}_L^{-1}(-L_\eta) - \boldsymbol{I}_2 = \boldsymbol{\Phi}_L(L_\eta)\boldsymbol{\Phi}_L^{-1}(0)\underbrace{\left[\boldsymbol{\Phi}_L(0)\boldsymbol{\Phi}_L^{-1}(-L_\eta) - \boldsymbol{\Phi}_L(0)\boldsymbol{\Phi}_L^{-1}(L_\eta)\right]}_{\boldsymbol{E}_L(\gamma,\mu)},$$

and normalizing the Evans function by the nonzero determinant of the prefactor matrix, we obtain

$$E_L(\gamma,\mu) = \det \boldsymbol{E}_L(\gamma,\mu). \tag{9.7.7}$$

We wish to show for $0<\eta\ll 1$, i.e., $L_\eta \gg 1$, that for each $|\mu|\le L_\eta^{-1}$ the zeros of $E_L(\gamma,\mu)$ on the natural domain are asymptotically close, up to multiplicity, to those of $E_\infty(\gamma)$. This was first shown by Gardner [**92**] using topological ideas, and by Sandstede and Scheel [**252, 254**] via an analytic approach. We sketch the ideas of this latter approach. Without loss of generality we suppose that $\mu=0$ as the case for $|\mu|\le L_\eta^{-1}$ follows from a perturbative argument; henceforth, we suppress the μ-dependence of solutions to the Bloch-wave problem. The matrices $\boldsymbol{\Phi}_L(0,\gamma)\boldsymbol{\Phi}_L^{-1}(\pm L_\eta,\gamma)$ are the solution map for the vectorized form of (9.7.5) from $x=\pm L_\eta$ to $x=0$. Under the assumption (9.7.4) the indefinite integrals of $(a_{j,\eta}-a_{j,0})(x)$ are *uniformly* $\mathcal{O}(\eta)$ over $[-L,L]$. As a consequence, the flow generated by the bounded domain problem is also *uniformly* $\mathcal{O}(\eta)$-close over $[-L,L]$ to the corresponding map generated by the vectorized form of the unbounded domain problem (9.7.2). Thus, for $0<\eta\ll 1$, the matrix

$$\boldsymbol{E}_\infty(\gamma) := \boldsymbol{\Phi}_\infty(0,\gamma)\boldsymbol{\Phi}_\infty^{-1}(-L_\eta,\gamma) - \boldsymbol{\Phi}_\infty(0,\gamma)\boldsymbol{\Phi}_\infty^{-1}(L_\eta,\gamma),$$

where $\boldsymbol{\Phi}_\infty(x,\gamma)$ is a fundamental matrix for the unbounded domain problem, is close to the matrix $\boldsymbol{E}_L(\gamma,\mu)$.

Theorem 9.7.1. *Consider the eigenvalue problem for the Bloch-wave decomposition, (9.7.5), under the approximation assumption, (9.7.4). Suppose that for the unbounded domain problem, (9.7.1), $\gamma=\gamma_0\subset\Omega$ is an isolated eigenvalue with multiplicity m. Let $C\subset\mathbb{C}$ be a positively oriented simple closed curve of fixed radius that contains γ_0 within its interior and is an $\mathcal{O}(1)$ distance contains γ_0, but is an $\mathcal{O}(1)$ distance from all other spectra of the unbounded domain problem. Then for η sufficiently small, for each $-1/L_\eta<\mu\le 1/L_\eta$ the Block-wave eigenvalue problem has precisely m eigenvalues in the interior of C.*

Proof. Assume that $\gamma\in\Omega$ lies to the right of the essential spectrum and is not an eigenvalue for the unbounded domain problem. We show that, up to multiplication by an analytic nonzero constant, $\det\boldsymbol{E}_\infty(\gamma)=E_\infty(\gamma)+O(\eta)$. Since the matrix $\boldsymbol{\Phi}_\infty(x_1,\gamma)\boldsymbol{\Phi}_\infty(x_2,\gamma)^{-1}\boldsymbol{v}$ represents the flow of the vector $\boldsymbol{v}$ from $x=x_2$ to $x=x_1$, we have

$$J_\infty^-(0,\gamma) = \boldsymbol{\Phi}_\infty(0,\gamma)\boldsymbol{\Phi}_\infty(-L_\eta,\gamma)^{-1}J_\infty^-(-L_\eta,\gamma)$$
$$J_\infty^+(0,\gamma) = \boldsymbol{\Phi}_\infty(0,\gamma)\boldsymbol{\Phi}_\infty(L_\eta,\gamma)^{-1}J_\infty^+(L_\eta,\gamma).$$

Now, from the construction of the Jost solutions (see Theorem 9.2.3) we know that

$$|J_\infty^-(-L_\eta,\gamma)| = \mathcal{O}(\mathrm{e}^{-\kappa_{\mathrm{u}}(\gamma)L_\eta}), \quad |J_\infty^+(L_\eta,\gamma)| = \mathcal{O}(\mathrm{e}^{\kappa_{\mathrm{s}}(\gamma)L_\eta}),$$

where for γ in the natural domain,

$$\kappa_{\rm u}(\gamma) = \operatorname{Re}\mu_1(\gamma) > 0, \quad \kappa_{\rm s}(\gamma) = \operatorname{Re}\mu_2(\gamma) < 0.$$

Since up to an exponentially small error

$$J_\infty^-(-L_\eta,\gamma) \sim e^{-\mu_1(\gamma)L_\eta}\boldsymbol{v}_1(\gamma), \quad J_\infty^+(L_\eta,\gamma) \sim e^{\mu_2(\gamma)L_\eta}\boldsymbol{v}_2(\gamma), \tag{9.7.8}$$

it follows from the exponential dichotomy of the solution operator that

$$\begin{aligned}|\boldsymbol{\Phi}_\infty(0,\gamma)\boldsymbol{\Phi}_\infty(L_\eta,\gamma)^{-1}J_\infty^-(-L_\eta,\gamma)| &= \mathcal{O}(e^{-\kappa_{\rm u}(\gamma)L_\eta})\\ |\boldsymbol{\Phi}_\infty(0,\gamma)\boldsymbol{\Phi}_\infty(-L_\eta,\gamma)^{-1}J_\infty^+(L_\eta,\gamma)| &= \mathcal{O}(e^{\kappa_{\rm s}(\gamma)L_\eta}).\end{aligned}$$

In conclusion, we deduce the asymptotic matrix relation

$$\boldsymbol{E}_\infty(\gamma)\Big(J_\infty^-(-L_\eta,\gamma), -J_\infty^+(L_\eta,\gamma)\Big) = \Big(J_\infty^-, J_\infty^+\Big)(0,\gamma) + \mathcal{O}(e^{-\kappa L_\eta}), \tag{9.7.9}$$

where $\kappa := \min\{\kappa_{\rm u}, -\kappa_{\rm s}\} > 0$.

The matrix

$$\boldsymbol{C}(\gamma) := \Big(J_\infty^-(-L_\eta,\gamma), -J_\infty^+(L_\eta,\gamma)\Big)$$

is nonsingular and analytic for all γ in the natural domain: this follows from the asymptotics (9.7.8) and the fact that the eigenvectors for the asymptotic matrix $\boldsymbol{A}_{0,\infty}(\gamma)$ associated with the vectorized version of the unbounded domain problem form a basis for $\mathbb{C}^2$. Since $\det \boldsymbol{C}(\gamma)$ is analytic and nonzero, we now see from (9.7.9) that for γ in the natural domain of the Evans function, $\det \boldsymbol{E}_\infty(\gamma)$ and $E_\infty(\gamma)$ will be exponentially close as $L_\eta \to +\infty$ ($\eta \to 0^+$). The closeness of the matrix $\boldsymbol{E}_L(\gamma)$ to $\boldsymbol{E}_\infty(\gamma)$ and a standard winding-number argument yields the result. □

Remark 9.7.2. Unlike the case of separated boundary conditions, for periodic boundary conditions no nondegeneracy condition is required to prevent spurious eigenvalues. Moreover, it can be shown that the periodic point spectrum, which does not converge to the point spectrum of the unbounded domain problem, instead converges to the *essential* spectrum of the unbounded domain problem. This is somewhat intuitive, as the Jost solutions for $\gamma \in \sigma_{\rm ess}(\mathcal{L})$ are asymptotically periodic for $|x| \ge L_\eta$ and $\eta \ll 1$ (see Fig. 9.11).

Exercises

Exercise 9.7.1. Suppose that γ_0 is a simple eigenvalue for the unbounded domain problem (9.7.1). Using Theorem 8.4.3, show for the appropriate approximating periodic problem (9.7.2) there is a nearby continuous simple closed curve of spectra.

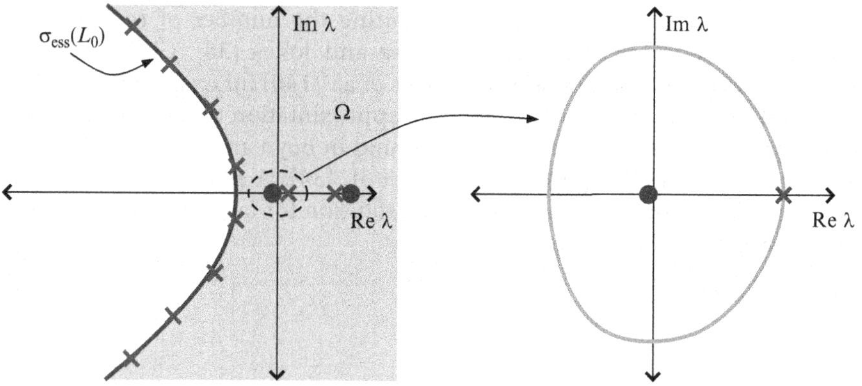

Fig. 9.11 (*left*) A comparison of the spectra of the unbounded domain problem and of the Bloch decomposition problem for $\mu = 0$. The point eigenvalues (*blue* circles) for the unbounded domain problem are approximated, up to multiplicity, within $O(\eta)$ by eigenvalues of the periodic problem (*red* crosses). The point eigenvalues for the periodic problem converge onto the essential spectrum of the unbounded domain problem, not its absolute spectrum. (*right*) A blow-up of the spectrum near $\lambda = 0$, under the assumption that the point eigenvalue of the unbounded domain problem is simple. The (solid *green*) curve is the essential spectrum for the periodic problem traced out by varying the Floquet exponent, μ. (Color figure online.)

Exercise 9.7.2. Consider the square-wave potential operator $\mathcal{L}$ given in Exercise 9.6.2. Compare the operator's spectrum on the unbounded domain with the spectrum determined below.

(a) Determine the spectrum of $\mathcal{L}$ on $[-L, L]$ subject to periodic boundary conditions for Floquet multiplier $\mu = 0$.
(b) Plot the $\mu = 0$ spectrum of $\mathcal{L}$ for a range of values of L and compare to the unbounded domain spectrum.

Exercise 9.7.3. It is well known that the Jacobi elliptic function $\mathrm{dn}(x,k)$ approximates $\mathrm{sech}(x)$ as $k \to 1^-$. This Jacobi elliptic function has period $2K(k)$, where $K(k)$ is the elliptic integral of the first kind with the property that $K(k) \to +\infty$ as $k \to 1^-$. Consider the eigenvalue problem on the real line given by

$$\partial_x^2 p + 2\partial_x p + \ell(\ell+1)\mathrm{dn}(x,k)^2 p = \gamma p, \quad \ell \in \mathbb{N}.$$

Numerically compute the spectrum for $0 < 1-k \ll 1$ using a finite difference scheme in order to approximate the differential operators. How small must $0 < 1-k$ be in order to achieve a good approximation of the point spectrum?

9.8 Additional Reading

An approach to the analysis of point spectra that combines the Levinson theorem with ingredients of the Jost functions and the orientation index can be found in Eastham [75] and Ma [195]. The Morse index and the Maslov index have also proven useful in computing the number of real positive eigenvalues of linear operators; see Bose and Jones [**38**], Chardard [**48**], Chardard et al. [**49**, **50**, **51**, **52**], and Jones et al. [**140**] for examples.

An alternate approach to proving the approximation theorems discussed in Chapter 9.6 and Chapter 9.7 can be found in Beyn and Lorenz [**32**], Beyn and Rottmann-Matthes [**33**], and Beyn et al. [**34**]. A relationship between the Jost solution on the whole line and a solution for large domains is given in Latushkin and Sukhtayev [**184**].

Chapter 10
The Evans Function for nth-Order Operators on the Real Line

The primary goal of this chapter is the construction of the Evans function for eigenvalue problems associated with nth-order, exponentially asymptotic linear operators acting on $L^2(\mathbb{R})$. The construction, through the Jost solutions, is distinguished from the construction for second-order operators by the fact that the matrix eigenvalues and associated eigenvectors for the nth-order problem may not be analytic in the natural domain of the Evans function. Moreover, while it is relatively easy to determine the essential spectrum for these problems, the matrix eigenvalues and the absolute spectrum do not generally have an explicit representation. We sidestep these issues via an analytic extension of the stable and unstable spaces of the asymptotic matrix which leads to the construction of Jost matrices. Once constructed, we apply the Evans function to develop an orientation index to study edge bifurcations and to investigate the relation between the spectra of whole-line and large-domain eigenvalue problems. Since many elements of the analysis of the nth-order systems are generalizations of that for the Sturm–Liouville problems of Chapter 9, we will only sketch the proofs of the main results, focusing on the areas of distinction between the two approaches.

We consider the eigenvalue problem for the nth-order operator

$$\mathcal{L}p := \partial_x^n p + a_{n-1}(x)\partial_x^{n-1} p + \cdots + a_1(x)\partial_x p + a_0(x)p = \lambda p, \tag{10.0.1}$$

where $\mathcal{L} : H^n(\mathbb{R}) \subset L^2(\mathbb{R}) \to L^2(\mathbb{R})$ is exponentially asymptotic as in (3.1.1). To minimize notation we will only consider the case that the asymptotic limits at $x = \pm\infty$ are equal, that is, $a_j^\pm = a_j^0$ for $j = 0,\dots,n-1$. However, the construction has a natural extension to the case of distinct asymptotic limits at $x = \pm\infty$, as well as to systems, for $\boldsymbol{p} \in \mathbb{C}^k$, with exponentially asymptotic coefficients $\boldsymbol{a}_j(x) \in \mathbb{R}^{k\times k}$. We vectorize the eigenvalue problem, introducing $\boldsymbol{Y} = (p, \partial_x p, \dots, \partial_x^{n-1} p)^{\mathrm{T}}$, and write (10.0.1) as the first-order system

T. Kapitula and K. Promislow, *Spectral and Dynamical Stability of Nonlinear Waves*, Applied Mathematical Sciences 185, DOI 10.1007/978-1-4614-6995-7_10,

$$\partial_x Y = \underbrace{[A_0(\lambda) + R(x)]}_{A(x,\lambda)} Y, \tag{10.0.2}$$

in terms of the asymptotic matrix

$$A_0(\lambda) := \begin{pmatrix} 0 & 1 & \dots & 0 & 0 \\ 0 & 0 & \dots & 0 & 0 \\ \vdots & \vdots & \vdots\vdots\vdots & \vdots & \vdots \\ 0 & 0 & \dots & 0 & 1 \\ \lambda - a_0^0 & -a_1^0 & \dots & -a_{n-2}^0 & -a_{n-1}^0 \end{pmatrix}, \tag{10.0.3}$$

and the exponentially localized matrix

$$R(x) = \begin{pmatrix} 0 & 0 & \dots & 0 & 0 \\ 0 & 0 & \dots & 0 & 0 \\ \vdots & \vdots & \vdots\vdots\vdots & \vdots & \vdots \\ 0 & 0 & \dots & 0 & 0 \\ a_0^0 - a_0(x) & a_1^0 - a_1(x) & \dots & a_{n-2}^0 - a_{n-2}(x) & a_{n-1}^0 - a_{n-1}(x) \end{pmatrix}. \tag{10.0.4}$$

10.1 The Jost Matrices

The fundamental distinction between the second-order operators studied in Chapter 9 and the nth-order operator (10.0.1) lies in the possible behavior of the matrix eigenvalues, that is, the eigenvalues of the asymptotic matrix $A_0(\lambda)$. For the second-order operator, there are two matrix eigenvalues, $\{\mu_j(\lambda)\}$, which are analytic in λ except where they coincide, which corresponds to the branch point of the absolute spectrum of the corresponding operator. For the nth-order operator the matrix eigenvalues may have many points at which they lose analyticity due to collision. Many of these points have no significance for the invertibility of the operator $\mathcal{L} - \lambda$. Avoiding this issue requires a different formulation of the Jost solutions.

From Theorem 3.1.13 we know that we may divide the complex plane into a finite collection of disjoint sets, $\{S_i\}_{i=1}^N$, each of which is comprised either entirely of an essential spectrum, or is part of the resolvent set except for a finite collection of point spectra. Following the convention of Hypothesis 3.2.1, we assume that $\mathcal{L}$ is well-posed and define the *natural domain*, Ω, of the Evans functions to be the element S_1 of the disjoint collection that contains $\operatorname{Re}\lambda \gg 1$. Fixing $\lambda \in \Omega$, then for each $\mu \in \sigma(A_0(\lambda))$ we define the spectral set $\mathbb{E}_{\mu(\lambda)} := \operatorname{gker}(A_0(\lambda) - \mu)$ and recall the stable and unstable subspaces of A_0,

$$\mathbb{E}^{\mathrm{s}}(\lambda) = \bigcup_{\operatorname{Re}\mu<0} \mathbb{E}_{\mu(\lambda)}, \quad \mathbb{E}^{\mathrm{u}}(\lambda) = \bigcup_{\operatorname{Re}\mu>0} \mathbb{E}_{\mu(\lambda)}, \tag{10.1.1}$$

with the associated collections of eigenvalues $\sigma_{u,s}(\lambda) := \sigma_{u,s}(A_0(\lambda))$. For $\lambda \in \Omega$ we may enumerate the eigenvalues (according to multiplicity) so that

$$\operatorname{Re}\mu_1(\lambda) \leq \cdots \leq \operatorname{Re}\mu_k(\lambda) < 0 < \operatorname{Re}\mu_{k+1}(\lambda) \leq \cdots \leq \operatorname{Re}\mu_n(\lambda). \tag{10.1.2}$$

The Fredholm border, which contains $\partial\Omega$, is precisely the set of $\lambda \in \mathbb{C}$ for which either $\operatorname{Re}\mu_{k+1}(\lambda) = 0$ or $\operatorname{Re}\mu_k(\lambda) = 0$. This fact confirms that $k := \dim(\mathbb{E}^s(\lambda))$ is in fact independent of $\lambda \in \Omega$.

We would like to extend the definition of these sets beyond the natural domain in such a way that we maximize analyticity, which can be lost when the matrix eigenvalues coincide. However, we ignore collisions of the matrix eigenvalues within σ_u or within σ_s and concern ourselves instead with the collision of a matrix eigenvalue from σ_u with one from σ_s. This is consistent with an analytic extension of $\sigma_{u,s}$. More specifically, for λ_0 outside of Ω we define the set $\sigma_u(\lambda_0)$ by choosing an analytic path connecting λ_0 to $\lambda \in \Omega$ such for all $\tilde{\lambda}$ on this path no matrix eigenvalue from $\sigma_u(\tilde{\lambda})$ coincides with a matrix eigenvalue from $\sigma_s(\tilde{\lambda})$. We then define the set $\sigma_u(\lambda_0)$ to be its analytic limit obtained along this path. Within these sets we enumerate the matrix eigenvalues so as to preserve the ordering

$$\begin{aligned} &\operatorname{Re}\mu_1(\lambda) \leq \cdots \leq \operatorname{Re}\mu_k(\lambda), \\ &\operatorname{Re}\mu_{k+1}(\lambda) \leq \cdots \leq \operatorname{Re}\mu_n(\lambda). \end{aligned} \tag{10.1.3}$$

However, it may be that $\operatorname{Re}\mu_k(\lambda) > \operatorname{Re}\mu_{k+1}(\lambda)$ for λ outside of Ω; see Fig. 10.1. To track the upper and lower bounds over these sets we introduce

$$\eta_s(\lambda) := \operatorname{Re}\mu_k(\lambda), \quad \eta_u(\lambda) := \operatorname{Re}\mu_{k+1}(\lambda).$$

The analytic extension of $\sigma_{u,s}$ fails precisely for λ at which $\mu_k(\lambda) = \mu_{k+1}(\lambda)$: such λ are called branch points of the operator $\mathcal{L}$. It is also clear that the definition of $\sigma_{u,s}(\lambda_0)$ may depend upon the choice of path that connects λ_0 to Ω. Sorting this out requires the introduction of branch cuts, which restrict the paths of analytic continuation, or of a Riemann surface that segregates the multiple possibilities onto distinct sheets. The absolute spectrum, which is associated with a particular choice of branch cut, is defined as the curve for which $\eta_s(\lambda) = \eta_u(\lambda)$.

The Riemann surface unfolds the multiplicity of the zeros, that is. the matrix eigenvalues, of the dispersion relation

$$d(\lambda, \mu) := \det(A_0(\lambda) - \mu I_n), \tag{10.1.4}$$

[compare to (3.1.18)] via the Riemann map $\lambda = g(\gamma)$. An explicit construction of the Riemann surface is not possible for the general nth-order problem. However a local construction in a neighborhood of a branch point, $\lambda = \lambda_{br}$ is possible. At the branch point we have $\mu_k(\lambda_{br}) = \mu_{k+1}(\lambda_{br}) = \mu_{br}$, and for simplicity we assume that $\dim[\operatorname{gker}(A_0(\lambda_{br}) - \mu_{br})] = 2$: equivalently, no other matrix eigenvalues equal μ_{br} when $\lambda = \lambda_{\mathrm{br}}$. The dispersion relation then satisfies

$$d(\lambda_{\mathrm{br}}, \mu_{\mathrm{br}}) = \partial_\mu d(\lambda_{\mathrm{br}}, \mu_{\mathrm{br}}) = 0, \quad \text{and} \quad \partial_\mu^2 d(\lambda_{\mathrm{br}}, \mu_{\mathrm{br}}) \neq 0.$$

In addition, assuming that $\partial_\lambda d(\lambda_{\mathrm{br}}, \mu_{\mathrm{br}}) \neq 0$, so that the matrix eigenvalues $\mu_k(\lambda)$ and $\mu_{k+1}(\lambda)$ are distinct for $\lambda \neq \lambda_{\mathrm{br}}$, then Taylor expanding $d(\lambda, \mu)$ near $(\lambda_{\mathrm{br}}, \mu_{\mathrm{br}})$ yields the dominant terms

$$d(\lambda, \mu) \sim \partial_\lambda d(\lambda_{\mathrm{br}}, \mu_{\mathrm{br}})(\lambda - \lambda_{\mathrm{br}}) + \frac{1}{2}\partial_\mu^2 d(\lambda_{\mathrm{br}}, \mu_{\mathrm{br}})(\mu - \mu_{\mathrm{br}})^2. \tag{10.1.5}$$

To leading order the Riemann surface, g, is defined as

$$\lambda = g(\gamma) \sim \lambda_{\mathrm{br}} + \gamma^2,$$

and near the branch point the colliding matrix eigenvalues have the expansions

$$\mu_k(\gamma) = \mu_{\mathrm{br}} + d_{\mathrm{br}}\gamma + \mathcal{O}(\gamma^2), \quad \mu_{k+1}(\gamma) = \mu_{\mathrm{br}} - d_{\mathrm{br}}\gamma + \mathcal{O}(\gamma^2),$$

where $d_{\mathrm{br}} := \sqrt{-2\partial_\lambda d / \partial_\mu^2 d}$. The matrix eigenvalues are analytic in γ; see [**162**, Chapter II.1] for details and treatment of degenerate cases.

The motivation for our particular extension of the sets $\sigma_{\mathrm{u,s}}$ beyond the natural domain is that it generates analytic spectral projections. Let $C_{\mathrm{u,s}} \subset \mathbb{C}$ be simple, positively oriented, closed curves that do not intersect; furthermore, let $\sigma_{\mathrm{u,s}}$ lie in the interior of $C_{\mathrm{u,s}}$ respectively; see the left panel in Fig. 10.1. The spectral projections onto the stable and unstable subspaces of the asymptotic system, $P^{\mathrm{s}}(\lambda) : \mathbb{C} \mapsto \mathbb{E}^{\mathrm{s}}(\lambda)$ and $P^{\mathrm{u}}(\lambda) : \mathbb{C} \mapsto \mathbb{E}^{\mathrm{u}}(\lambda)$, are

$$P^{\mathrm{s}}(\lambda)\boldsymbol{v} = \frac{1}{2\pi\mathrm{i}} \oint_{C_{\mathrm{s}}} (\mu \boldsymbol{I}_n - \boldsymbol{A}_0(\lambda))^{-1} \boldsymbol{v} \, \mathrm{d}\mu,$$

$$P^{\mathrm{u}}(\lambda)\boldsymbol{v} = \frac{1}{2\pi\mathrm{i}} \oint_{C_{\mathrm{u}}} (\mu \boldsymbol{I}_n - \boldsymbol{A}_0(\lambda))^{-1} \boldsymbol{v} \, \mathrm{d}\mu.$$

The projections are analytic away from the branch points, furthermore,

(a) the projections are mutually orthogonal, i.e., $P^{\mathrm{s}}P^{\mathrm{u}} = P^{\mathrm{u}}P^{\mathrm{s}} = 0$,
(b) $P^{\mathrm{s}}\boldsymbol{v} + P^{\mathrm{u}}\boldsymbol{v} = \boldsymbol{v}$ for λ not a branch point,
(c) $P^{\mathrm{s,u}}\boldsymbol{A}_0(\lambda) = \boldsymbol{A}_0(\lambda)P^{\mathrm{s,u}} \quad \Rightarrow \quad P^{\mathrm{s,u}}\mathrm{e}^{\boldsymbol{A}_0(\lambda)x} = \mathrm{e}^{\boldsymbol{A}_0(\lambda)x}P^{\mathrm{s,u}}$.

The spectral projections generate the following exponential dichotomies for the FMS of $\boldsymbol{A}_0(\lambda)$,

$$|P^{\mathrm{s}}(\lambda)\mathrm{e}^{\boldsymbol{A}_0(\lambda)(x-\tau)}\boldsymbol{v}| = |\mathrm{e}^{\boldsymbol{A}_0(\lambda)(x-\tau)}P^{\mathrm{s}}(\lambda)\boldsymbol{v}| \leq C\mathrm{e}^{\eta_{\mathrm{s}}(\lambda)(x-\tau)}|\boldsymbol{v}|, \quad x \geq \tau, \tag{10.1.6}$$

and

$$|P^{\mathrm{u}}(\lambda)e^{A_0(\lambda)(x-\tau)}v| = |e^{A_0(\lambda)(x-\tau)}P^{\mathrm{u}}(\lambda)v| \le Ce^{\eta_{\mathrm{u}}(\lambda)(x-\tau)}|v|, \quad x \le \tau. \tag{10.1.7}$$

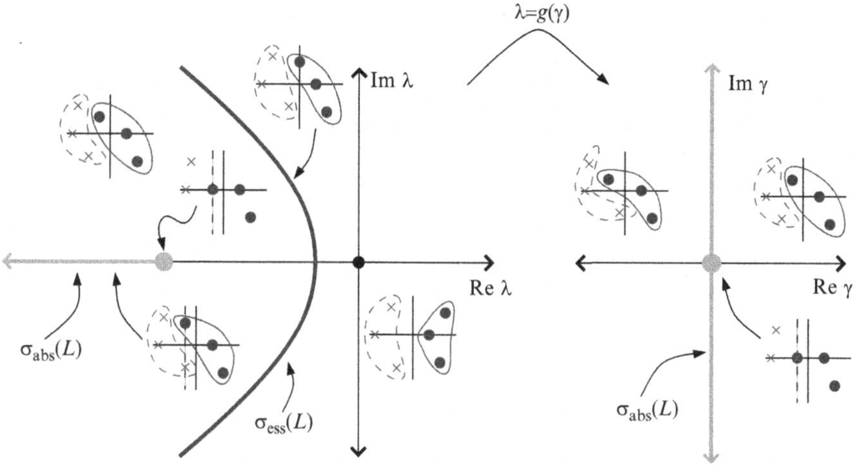

Fig. 10.1 The *left panel* depicts the sets σ_{s} (*red* crosses) and σ_{u} (*blue* circles) as well as the essential and absolute spectrum in the λ-plane. The *right panel* depicts $\sigma_{\mathrm{u,s}}$ and the absolute spectrum in the γ-plane that unfolds the Riemann surface. The branch point is marked by a (*green*) circle, and is assumed to be mapped to $\gamma = 0$. The insets depict $\sigma_{\mathrm{u,s}}$ for $n = 6$ and $k = 3$ with the contour C_{u} denoted by a (*blue*) solid curve, and the contour C_{s} by a (*red*) dashed curve. At the branch point, $\mu_3(\lambda) = \mu_4(\lambda)$, and the curves $C_{\mathrm{u,s}}$ cannot be defined. On the resonance sheet of the Riemann surface, $\eta_{\mathrm{u}} > \eta_{\mathrm{s}}$. (Color figure online.)

Since the matrix $\boldsymbol{R}$ is exponentially localized, we anticipate that this exponential dichotomy extends to the full problem, (10.0.2). That is, for $\lambda \in \Omega$ we expect that it has k linearly independent solutions which decay exponentially fast as $x \to -\infty$, and $n-k$ linearly independent solutions that decay exponentially fast as $x \to +\infty$. The individual solutions may not be analytic in λ, even on the natural domain Ω, but we can define *matrix-valued* solutions that are analytic on an extended domain containing Ω. Indeed, away from branch points, where $\eta_{\mathrm{u}} = \eta_{\mathrm{s}}$, and the associated branch cuts, the spectral projections P^{s} and P^{u} are analytic, even if the projections onto the constituent subspaces $\mathbb{E}_{\mu(\lambda)}$ are not. Consequently we may choose a basis $\{v_1(\lambda),\dots,v_k(\lambda)\}$ for $\mathbb{E}^{\mathrm{s}}(\lambda)$ and a basis $\{v_{k+1}(\lambda),\dots,v_n(\lambda)\}$ for $\mathbb{E}^{\mathrm{u}}(\lambda)$, for which the elements depend analytically in λ. We introduce the corresponding matrices

$$V_{\mathrm{s}} = \left(v_1,\dots,v_k\right) \in \mathbb{C}^{n\times k} \quad \text{and} \quad V_{\mathrm{u}} = \left(v_{k+1},\dots,v_n\right) \in \mathbb{C}^{n\times(n-k)}, \tag{10.1.8}$$

with ranges that coincide with $\mathbb{E}^{\mathrm{s}}$ and $\mathbb{E}^{\mathrm{u}}$, respectively.

Remark 10.1.1. An equivalent approach is to consider the problem on Grassmanian manifolds. The interested reader should consult [**11**] for the details.

The matrices $e^{A_0 x}V_{s,u}$ are solutions of the asymptotic ODE $\partial_x Y = A_0(\lambda)Y$. To each of these we wish to associate a Jost matrix that solves (10.0.2) and is asymptotically equal to $e^{A_0 x}V_{s,u}$ at either $\pm\infty$. This is possible on the *extended domain*

$$\Omega_e := \{\lambda \in \mathbb{C} : \eta_u - \eta_s > -\alpha\}, \tag{10.1.9}$$

where α is the decay rate in the exponential localization of R, see (10.0.4). Moreover, on the open subset of Ω_e on which the matrix eigenvalues are distinct, denoted Ω_e°, we may refine the structure of the Jost matrices.

Theorem 10.1.2. *Consider the system (10.0.2) with a perturbation matrix R that is exponentially localized with rate $\alpha > 0$. For each $\lambda \in \Omega_e$ there exist Jost matrices $J^\pm(x,\lambda)$ that are analytic in λ for fixed x and satisfy the asymptotic relations*

$$\begin{aligned} |J^+(x,\lambda) - e^{A_0(\lambda)x}V_s| &\le Ce^{-\alpha L/2}|V_s|, \quad x \ge +L \gg 0 \\ |J^-(x,\lambda) - e^{A_0(\lambda)x}V_u| &\le Ce^{-\alpha L/2}|V_u|, \quad x \le -L \ll 0, \end{aligned} \tag{10.1.10}$$

where $V_s(\lambda) \in \mathbb{C}^{n\times(n-k)}$ and $V_u(\lambda) \in \mathbb{C}^{n\times k}$ are each analytic matrices of full rank with ranges that span $\mathbb{E}^{s,u}(\lambda)$, respectively. The Jost matrices are analytic in λ on Ω_e except at the absolute spectrum (branch cuts), and are analytic in γ on the regions of the Riemann surface within $\Omega_e(\gamma)$. Furthermore, for $\lambda \in \Omega_e^\circ$ the Jost matrices take the form

$$J^+(x,\lambda) = \left(J_1^+,\dots,J_k^+\right)(x,\lambda), \quad J^-(x,\lambda) = \left(J_1^-,\dots,J_{n-k}^-\right)(x,\lambda),$$

where each Jost solution $J_j^\pm(x,\lambda)$ is analytic in λ for fixed x and satisfies the refined estimate (10.1.16).

Remark 10.1.3. If the matrix $A = A(x,\lambda,\epsilon)$ depends upon a parameter ϵ, then the Jost matrices are as smooth in ϵ as is the matrix A.

Proof. To construct the Jost matrix, we follow (9.2.8), writing

$$J^+(x) = e^{A_0 x}V_s + K(x), \tag{10.1.11}$$

where $J^+, K \in \mathbb{C}^{n\times(n-k)}$. The Jost function satisfies (10.0.2) if K satisfies

$$\partial_x K = A_0 K + R(x)\left[e^{A_0 x}V_s + K(x)\right],$$

which, from variation of parameters, has the solution

$$K(x) = \int_{x_0}^{x} e^{A_0(x-\tau)}R(\tau)\left[e^{A_0\tau}V_s + K(\tau)\right] d\tau.$$

for any $x_0 \in \mathbb{R}$ [compare with (9.2.9)]. Using the decomposition $P^{\mathrm{u}} + P^{\mathrm{s}} = \mathcal{I}$ we split the integrand into two parts, and change the limits of each integral, which amounts to a shift of initial data. We obtain

$$\begin{aligned} \boldsymbol{K}(x) = & \int_{+\infty}^{x} \mathrm{e}^{A_0(x-\tau)} P^{\mathrm{u}} \boldsymbol{R}(\tau) \left[\mathrm{e}^{A_0 \tau} \boldsymbol{V}_{\mathrm{s}} + \boldsymbol{K}(\tau)\right] \mathrm{d}\tau \\ & + \int_{L}^{x} \mathrm{e}^{A_0(x-\tau)} P^{\mathrm{s}} \boldsymbol{R}(\tau) \left[\mathrm{e}^{A_0 \tau} \boldsymbol{V}_{\mathrm{s}} + \boldsymbol{K}(\tau)\right] \mathrm{d}\tau, \end{aligned} \tag{10.1.12}$$

and following the argument from (9.2.10)–(9.2.13) we conclude that as long as the gap condition

$$\eta_{\mathrm{u}} - \eta_{\mathrm{s}} > -\alpha \tag{10.1.13}$$

holds, equivalently as long as $\gamma \in \Omega_{\mathrm{e}}$, then $\boldsymbol{K}(x)$ satisfies the estimate

$$|\boldsymbol{K}(x)| \leq C \mathrm{e}^{-\alpha L/2} \mathrm{e}^{\eta_{\mathrm{s}} x} |\boldsymbol{V}_{\mathrm{s}}|, \quad x \geq L,$$

which establishes (10.1.10) for $\boldsymbol{J}^{+}$. The analyticity of the Jost matrices in γ and their analytic extension across the branch cut, defined by $\operatorname{Re} \mu_k(\gamma) = \operatorname{Re} \mu_{k+1}(\gamma)$, follows from the analyticity of the matrix eigenvalues and spectral projections in γ, away from the branch point $\mu_k(\gamma) = \mu_{k+1}(\gamma)$. The analytic extension of the spectral projection follows naturally; see the inset of the right γ-panel Fig. 10.1 which depicts the contours $C_{\mathrm{u,s}}$ on the resonance sheet of the Reimann surface, where we have $\eta_{\mathrm{s}} < \eta_{\mathrm{u}}$. However, as long as (10.1.13) holds, the Jost matrix still satisfies (10.1.10) and is analytic in γ for fixed x away from the branch points. Analyticity in γ at the branch points follows from a standard continuity argument.

On the set $\Omega_{\mathrm{e}}^{\circ}$, for which the matrix eigenvalues are simple, the Jost matrices can be written in terms of linearly independent Jost solutions. In this case the matrix eigenvalues are analytic in λ, and we may take the basis vectors, $\{\boldsymbol{v}_j(\lambda)\}$, for $\boldsymbol{V}^{\mathrm{u}}$ and $\boldsymbol{V}^{\mathrm{s}}$ to be eigenvectors of $\boldsymbol{A}_0(\lambda)$. We consider $\boldsymbol{J}^{+}$, and for each $\ell = 1, \ldots, k$, we introduce $C_{1,\ell} \subset \mathbb{C}$, a positively oriented closed curve that encloses $\mu_1(\lambda), \ldots, \mu_\ell(\lambda)$ and no other matrix eigenvalues, and let $C_{1,\ell}^{\mathrm{c}}$ be a positively oriented closed curve that encloses the remainder of the matrix eigenvalues (see the left panel of Fig. 10.2). The spectral projection $P_{1,\ell}^{\mathrm{s}}(\lambda)$, defined as the integral of the resolvent over $C_{1,\ell}$, has an ℓ-dimensional range, while the range of the complementary projection $P_{1,\ell}^{\mathrm{u}}(\lambda)$ is $(n-\ell)$-dimensional. Introducing the spectral cutoffs

$$\eta_\ell^{\mathrm{s}}(\lambda) := \max_{j \leq \ell} \operatorname{Re} \mu_j(\lambda), \quad \eta_\ell^{\mathrm{u}}(\lambda) := \min_{j > \ell} \operatorname{Re} \mu_j(\lambda),$$

we remark that there is the spectral gap

$$\eta_\ell^{\mathrm{u}} - \eta_\ell^{\mathrm{s}} > -\alpha, \tag{10.1.14}$$

since on the portion of the resonant sheet of the Riemann surface within the extended domain, Ω_e, the real values of the "stable" and "unstable" matrix eigenvalues can overlap, as in (10.1.3), but not by more than the asymptotic decay rate $\alpha > 0$. This leads to the estimates

$$|P^s_{1,\ell}(\lambda)e^{A_0(\lambda)(x-\tau)}v| = |e^{A_0(\lambda)(x-\tau)}P^s_{1,\ell}(\lambda)v| \le Ce^{\eta^s_\ell(\lambda)(x-\tau)}|v|, \quad x \ge \tau,$$

and

$$|P^u_{1,\ell}(\lambda)e^{A_0(\lambda)(x-\tau)}v| = |e^{A_0(\lambda)(x-\tau)}P^u_{1,\ell}(\lambda)v| \le Ce^{\eta^u_\ell(\lambda)(x-\tau)}|v|, \quad x \le \tau.$$

The partial Jost matrix is constructed in the form

$$J^+_{1,\ell}(x) = e^{A_0 x}V_{1,\ell} + K_{1,\ell}(x),$$

where the columns of the matrix

$$V_{1,\ell} := \left(v_\ell, \ldots, v_\ell\right),$$

lie within the range of $P^s_{1,\ell}$. Using the decomposition $P^u_{1,\ell} + P^s_{1,\ell} = \mathcal{I}$, we find that the correction term satisfies the integral equation

$$\begin{aligned} K_{1,\ell}(x) = &\int_{+\infty}^{x} e^{A_0(x-\tau)}P^u_{1,\ell}R(\tau)\left[e^{A_0\tau}V_{1,\ell} + K_{1,\ell}(\tau)\right]d\tau \\ &+ \int_{L}^{x} e^{A_0(x-\tau)}P^s_{1,\ell}R(\tau)\left[e^{A_0\tau}V_{1,\ell} + K_{1,\ell}(\tau)\right]d\tau. \end{aligned}$$

In light of the gap condition, (10.1.14), the arguments of Theorem 9.2.3 show that the remainder $K_{1,\ell}(x)$ satisfies

$$|K_{1,\ell}(x)| \le Ce^{-\alpha L}e^{\eta^n_s x}|V_{1,\ell}|, \quad x \ge L.$$

We deduce that the partial Jost matrix solution satisfies

$$|J^+_{1,\ell}(x) - e^{A_0 x}V_{1,\ell}| \le Ce^{-\alpha L/2}e^{\eta^s_\ell x}|V_{1,\ell}|, \quad x \ge L. \tag{10.1.15}$$

Recalling that the matrix eigenvalues are simple for $\lambda \in \Omega^\circ_e$, for $\ell = 1,\ldots,k$ we introduce P^s_ℓ, the spectral projection associated with the curve C_ℓ, which surrounds only the matrix eigenvalue μ_ℓ. We have $P^s_{1,\ell} = P^s_1 \oplus P^s_2 \oplus \cdots \oplus P^s_\ell$ (see the right panel of Fig. 10.2). Given the estimates (10.1.15) we deduce that for each ℓ there is a Jost solution that satisfies

$$|J^+_\ell(x) - e^{A_0 x}P^s_\ell v| \le Ce^{-\alpha L/2}e^{\eta^s_\ell x}|P^s_\ell v|, \quad x \ge L. \tag{10.1.16}$$

In other words, we have $J^+_{1,\ell} = \left(J^+_1, \ldots, J^+_\ell\right)$. □

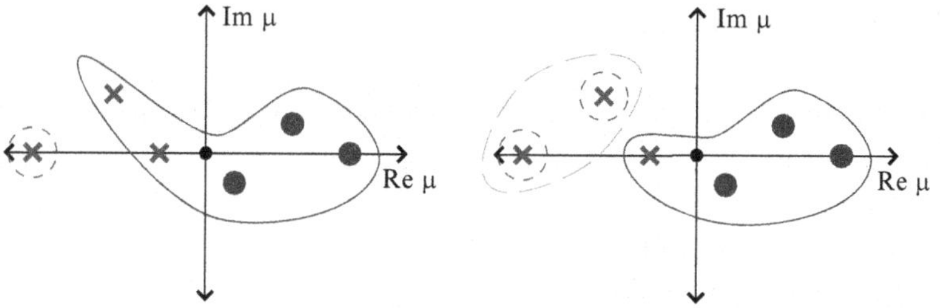

Fig. 10.2 The set $\sigma_s = \{\mu_1, \mu_2, \mu_3\}$ is denoted with (*red*) crosses, while $\sigma_u = \{\mu_4, \mu_5, \mu_6\}$ is denoted with (*blue*) circles. The *left panel* shows the contour $C_{1,1}$, (*red*) dashed circle, and the contour $C^c_{1,1}$ that surrounds the other five matrix eigenvalues, (*blue*) solid curve. The *right panel* depicts the contour $C_{1,2}$ as a (*green*) dashed curve and $C^c_{1,2}$ which surrounds the remaining four matrix eigenvalues as a (*blue*) solid curve. Also depicted are the (*red*) dashed contours, C_1 and C_2 which enclose μ_1 and μ_2. (Color figure online.)

Exercises

Exercise 10.1.1. This exercise provides an alternate proof of (10.1.16). Suppose that the matrix eigenvalues satisfy

$$\operatorname{Re}\mu_1(\lambda) < \operatorname{Re}\mu_2(\lambda) < \cdots < \operatorname{Re}\mu_k(\lambda).$$

Further suppose that $\operatorname{Re}\mu_{k+1}(\lambda) \le \operatorname{Re}\mu_j(\lambda)$ for $j = k+1, \dots, n$. Fix a value of λ and $\ell = 1, \dots, k$.

(a) Set $\tilde{\boldsymbol{Y}} = \boldsymbol{Y}\mathrm{e}^{-\operatorname{Re}\mu_\ell(\lambda)x}$. Show that for the transformed eigenvalue problem the asymptotic matrix $\tilde{\boldsymbol{A}}_0(\lambda)$ has an $(\ell-1)$-dimensional stable space, a one-dimensional center space, and an $(n-\ell)$-dimensional unstable space.

(b) Dropping the tildes in the transformed system, introduce three positively-oriented closed curves, $C_\ell^{\mathrm{s,c,u}} \subset \mathbb{C}$, which enclose the matrix eigenvalues $\{\mu_1, \dots, \mu_{\ell-1}\}$, $\mu_\ell(\lambda)$, and $\{\mu_{\ell+1}, \dots, \mu_n\}$, respectively. Let $P_\ell^{\mathrm{s,c,u}}$ denote the associated spectral projections, which satisfy $P_\ell^{\mathrm{u}} + P_\ell^{\mathrm{c}} + P_\ell^{\mathrm{s}} = \mathcal{I}$. Show that there is a Jost solution

$$\boldsymbol{J}_\ell^+ = \mathrm{e}^{\boldsymbol{A}_0 x} P_\ell^{\mathrm{c}} \boldsymbol{v} + \boldsymbol{K}_\ell,$$

for which

$$|\boldsymbol{K}_\ell(x)| \le C\mathrm{e}^{-\alpha L/2}\mathrm{e}^{\operatorname{Re}\mu_{\ell+1}(\lambda)x}|P_\ell^{\mathrm{u}}\boldsymbol{v}|, \quad x \le -L,$$
$$|\boldsymbol{K}_\ell(x)| \le C\mathrm{e}^{-\alpha L/2}\mathrm{e}^{\operatorname{Re}\mu_{\ell-1}(\lambda)x}|P_\ell^{\mathrm{s}}\boldsymbol{v}|, \quad x \ge L.$$

(c) Derive upper and lower bounds for the solution $\boldsymbol{J}_\ell^+$ of the original system. *Hint*: Write $\boldsymbol{K}$ in the form (10.1.12) with P_ℓ^{c} grouped with P_ℓ^{s} and estimate $\boldsymbol{K}$.

10.2 The Evans Function

The Jost matrices provide a natural generalization of the Evans function.

Definition 10.2.1. Let $\mathcal{L}$ be an exponentially asymptotic nth-order operator as in (10.0.1) On its extended domain, $\Omega_e \subset \mathbb{C}$, given in (10.1.9), the Evans matrix is the concatenation of the two Jost matrices, defined in Theorem 10.1.2, evaluated at $x = 0$,

$$\boldsymbol{E}(\lambda) := \left(\boldsymbol{J}^{-}, \boldsymbol{J}^{+}\right)(0\lambda),$$

and the Evans function is its determinant

$$E(\lambda) := \det \boldsymbol{E}(\lambda).$$

The Evans function for the exponentially asymptotic nth-order operator $\mathcal{L}$ enjoys many of the same properties as for the exponentially asymptotic Sturm–Liouville operators of Chapter 9. The structure of the proof is quite similar to that of Theorem 8.2.9, and we present an outline for the special case in which the matrix eigenvalues of the asymptotic matrix $\boldsymbol{A}_0(\lambda)$ are simple. A general proof can be found in, e.g., Alexander et al. [**11**], Pego and Weinstein [**220**].

Theorem 10.2.2. *Let $\mathcal{L}$ be an exponentially asymptotic operator of the form (10.0.1). The associated Evans function is analytic in λ on its natural domain, and has an analytic extension onto Ω_e. For $\lambda \in \Omega$ the operator $\mathcal{L}$ has an eigenvalue of multiplicity d if and only if the Evans function has a zero of order d. More specifically, if $\lambda = \lambda_0 \in \Omega$ is an eigenvalue of $\mathcal{L}$ with $\mathrm{m_g}(\lambda_0) = m$ and $\mathrm{m_a}(\lambda_0) = d \geq m$, then the first $d-1$ derivatives of E at λ_0 are zero, while*

$$E^{(d)}(\lambda_0) = d! \det \boldsymbol{F}(\lambda_0) \det \boldsymbol{\Phi}(0, \lambda_0) \neq 0, \tag{10.2.1}$$

where $\boldsymbol{\Phi}$ is the FMS given in (10.2.2) and $\boldsymbol{F}(\lambda_0) \in \mathbb{C}^{m \times m}$ defined in (10.2.3) is nonsingular.

Proof. (sketch) We only address the derivative formula. Consider the eigenvalue problem (10.0.1) and suppose that $\lambda_0 \in \sigma_{\mathrm{pt}}(\mathcal{L})$ satisfies $\mathrm{m_g}(\lambda_0) = m \leq \min(k, n-k)$ and $\mathrm{m_a}(\lambda_0) = d \geq m$. Then there exist generalized eigenfunctions $p_{j,i}$ for $i = 0, \dots, m-1$ and $j = 0, \dots, d_i - 1$ with $\sum d_i = d$ with the Jordan block structure

$$(\mathcal{L} - \lambda_0) p_{j,i} = p_{j-1,i}, \quad p_{-1,i} = 0.$$

Assuming that the matrix eigenvalues are simple, the Jost matrices can be written in terms of the individual Jost solutions. In the case at hand we may normalize the Jost solutions so that

$$J_{i+1}^{-}(x,\lambda_0) = J_{i+1}^{+}(x,\lambda_0) = \begin{pmatrix} p_{0,i} \\ \partial_x p_{0,i} \\ \vdots \\ \partial_x^{n-1} p_{0,i} \end{pmatrix},$$

for $i = 0,\ldots,m-1$. Since $\mathcal{L} - \lambda_0$ is Fredholm of index zero we know that $\overline{\lambda_0} \in \sigma_{\text{pt}}(\mathcal{L}^{\text{a}})$, with the same multiplicities, and hence $\ker(\mathcal{L}^{\text{a}} - \overline{\lambda_0}) = \text{span}\{q_0,\ldots,q_{m-1}\}$. Each solution q_ℓ induces a vector solution of the vectorized adjoint system,

$$J_\ell^{\text{a}}(x) = \begin{pmatrix} j_{1,\ell}^{\text{a}}(x) \\ j_{2,\ell}^{\text{a}}(x) \\ \vdots \\ j_{n-1,\ell}^{\text{a}}(x) \\ q_\ell(x) \end{pmatrix}.$$

Moreover, as in Lemma 8.1.3, for any solution $\boldsymbol{Y}$ of the vector system and $\boldsymbol{Z}$ of the adjoint vector system the quantity $\boldsymbol{Y}(x,\lambda)\cdot\boldsymbol{Z}(x,\lambda)$ is independent of $x \in \mathbb{R}$. Since the asymptotic matrix $\boldsymbol{A}_0^{\text{a}}$ of the adjoint problem is the adjoint of the asymptotic matrix $\boldsymbol{A}_0$, the two sets of matrix eigenvectors form a bi-orthonormal set; in particular, $J_j^{\pm}\cdot J_k^{\text{a}\pm} = 0$ unless $j = k$ and both are + or both are −. This permits us choose the initial condition for the fundamental matrix solution to (10.0.2) in the form

$$\boldsymbol{\Phi}(0,\lambda_0) = \left(J_0^{\text{a}}, J_1^{\text{a}},\ldots,J_{m-1}^{\text{a}}, J_m^{-},\ldots,J_{n-k}^{-}, J_1^{+},\ldots,J_{m-1}^{+}, J_m^{+},\ldots,J_k^{+}\right)(0,\lambda_0), \quad (10.2.2)$$

where each adjoint solution $J_j^{\text{a}}(x)$ has length one when $x = 0$. Following the argument of Theorem 8.2.9 we arrive at the factored form (10.2.1) for the dth derivative of E, in terms of the nonsingular matrix

$$\boldsymbol{F}(\lambda_0) = \begin{pmatrix} \langle p_{d_0,0}, q_0\rangle & \langle p_{d_0,0}, q_1\rangle & \cdots & \langle p_{d_0,0}, q_{m-1}\rangle \\ \vdots & \vdots & & \vdots \\ \langle p_{d_{m-1},m-1}, q_0\rangle & \langle p_{d_{m-1},m-1}, q_1\rangle & \cdots & \langle p_{d_{m-1},m-1}, q_{m-1}\rangle \end{pmatrix} \quad (10.2.3)$$

formed from the inner product of the *top* of the Jordan chain of $\mathcal{L}^{\text{a}}$ and the *bottom* of the Jordan chain of $\mathcal{L}$. □

Exercises

Exercise 10.2.1. For $\lambda_0 \in \sigma_{\text{pt}}(\mathcal{L})$, prove that

$$\text{m}_{\text{g}}(\lambda_0) \le \min(k, n-k).$$

Exercise 10.2.2. Assume that the matrix eigenvalues are simple. Show that there is a normalization of the Jost solutions and the adjoint eigenvectors such that the matrix $\mathbf{\Phi}(0,\lambda_0)$ of (10.2.2) is nonsingular.

Exercise 10.2.3. Assume that the matrix eigenvalues are simple. Show that the Evans function of Definition 10.2.1 enjoys the following properties:

(a) It is analytic in $\lambda \in \Omega_e \backslash \sigma_{abs}(\mathcal{L})$.
(b) For $\lambda \in \Omega$, $E(\lambda) = 0$ if and only if $\lambda \in \sigma_{pt}(\mathcal{L})$.
(c) There is a $\theta > 0$ such that if $|\arg \lambda| < \theta$ and $|\lambda|$ is sufficiently large, then $E(\lambda) \neq 0$.

Exercise 10.2.4. Consider the eigenvalue problem

$$\mathcal{L}\boldsymbol{p} := \boldsymbol{d}\partial_x^2\boldsymbol{p} + \boldsymbol{a}_1(x)\partial_x\boldsymbol{p} + \boldsymbol{a}_0(x)\boldsymbol{p} = \lambda\boldsymbol{p},$$

where $\boldsymbol{a}_1(x), \boldsymbol{a}_0(x) \in \mathbb{R}^{\ell\times\ell}$ are smooth and exponentially asymptotic to constant matrices $\boldsymbol{a}_0^0, \boldsymbol{a}_1^0$, and $\boldsymbol{d}$ is a diagonal, positive-definite matrix. Introducing $\boldsymbol{Y} = (\boldsymbol{p}, \partial_x\boldsymbol{p})^{\mathrm{T}} \in \mathbb{C}^{2\ell}$, the eigenvalue problem can be written as

$$\partial_x\boldsymbol{Y} = [\boldsymbol{A}_0(\lambda) + \boldsymbol{R}(x)]\boldsymbol{Y},$$

where

$$\boldsymbol{A}_0(\lambda) = \begin{pmatrix} 0 & \boldsymbol{I}_\ell \\ \boldsymbol{d}^{-1}(\lambda\boldsymbol{I}_\ell - \boldsymbol{a}_0^0) & -\boldsymbol{d}^{-1}\boldsymbol{a}_1^0 \end{pmatrix}, \quad \boldsymbol{R}(x) = \begin{pmatrix} 0 & 0 \\ \boldsymbol{d}^{-1}(\boldsymbol{a}_0^0 - \boldsymbol{a}_0(x)) & \boldsymbol{d}^{-1}(\boldsymbol{a}_1^0 - \boldsymbol{a}_1(x)) \end{pmatrix}.$$

(a) Show that there exists a $\pi/2 < \theta < \pi$ such that if $|\arg \lambda| < \theta$ and $|\lambda|$ is sufficiently large, then there are ℓ matrix eigenvalues with a positive real part and ℓ matrix eigenvalues with a negative real part.

Let $\Omega \subset \mathbb{C}$ be the unbounded domain for which there are an equal numbers of matrix eigenvalues with positive and negative real part. Show that:

(b) There are Jost matrices $\boldsymbol{J}^\pm(x,\lambda) \in \mathbb{C}^{2\ell\times\ell}$ that satisfy asymptotic relations of the form (10.1.10).
(c) The Evans function defined by $E(\lambda) = \det(\boldsymbol{J}^-, \boldsymbol{J}^+)(0,\lambda)$ is analytic except at the absolute spectrum.
(d) For $\lambda \in \Omega$, $E(\lambda) = 0$ if and only if $\lambda \in \sigma_{pt}(\mathcal{L})$.
(e) If $D \subset \Omega$ is a bounded domain, then there are at most a finite number of eigenvalues in D [*Hint*: Show that if $|\lambda|$ is sufficiently large, then $E(\lambda) \neq 0$].

10.3 Application: The Orientation Index

The orientation index, introduced in Chapter 9.4, detects real, positive eigenvalues by comparing the sign of the Evans function near zero with its sign at $\lambda = +\infty$. In many applications the operator $\mathcal{L}$ has a kernel and hence $E(0) = 0$. However from Theorem 10.2.2 we may attempt to deduce the sign of the first nonzero derivative, and hence determine the sign of E near zero. It then remains to determine the sign of the Evans function for large positive λ. We construct the orientation index for an operator $\mathcal{L}$ of the form (10.0.1), which is Fredholm of index zero, with a simple eigenvalue at $\lambda = 0$. Extensions to nonsimple eigenvalues at $\lambda = 0$ are considered in Exercise 10.3.1.

Theorem 10.3.1 (Orientation index). *Let $\mathcal{L}$ be an exponentially localized operator of the form (10.0.1). Suppose that $\lambda = 0 \in \Omega$ is a simple eigenvalue of $\mathcal{L}$ with eigenfunction $p_0(x)$ and adjoint eigenfunction $q_0(x)$ and that the matrix eigenvalues $\{\mu_j^\pm\}$ at $\lambda = 0$ are simple. Denote the vector solutions induced by p_0 and q_0 by $\boldsymbol{J}_0$ and $\boldsymbol{J}_0^{\mathrm{a}}$. Then the Jost solutions can be normalized so that $\boldsymbol{J}_1^-(x,0) = \boldsymbol{J}_1^+(x,0) = \boldsymbol{J}_0(x)$, and*

$$\lim_{x\to-\infty} \mathrm{e}^{\mathrm{Re}\,\mu_j^- x}\boldsymbol{J}_j^-(x,0) = \boldsymbol{v}_j^-, \quad \lim_{x\to+\infty} \mathrm{e}^{\mathrm{Re}\,\mu_j^+ x}\boldsymbol{J}_i^+(x,0) = \boldsymbol{v}_j^+ \tag{10.3.1}$$

where $\boldsymbol{v}_j^\pm$ are the basis for the stable and unstable eigenspaces of $\boldsymbol{A}_0(0)$. The orientation index, defined as

$$\begin{aligned}\mathbb{O} := \operatorname{sign}\Big[&\langle p_0, q_0\rangle \det\big(\boldsymbol{v}_1^-,\dots,\boldsymbol{v}_k^-,\boldsymbol{v}_1^+,\dots,\boldsymbol{v}_{n-k}^+\big)\\ &\times\det\big(\boldsymbol{J}_0^{\mathrm{a}},\boldsymbol{J}_2^-,\dots,\boldsymbol{J}_k^-,\boldsymbol{J}_0,\boldsymbol{J}_2^+,\dots,\boldsymbol{J}_{n-k}^+\big)(0,0)\Big],\end{aligned}$$

has the property that if $\mathbb{O} = +1$ then $\mathcal{L}$ has an even number of real positive eigenvalues, whereas if $\mathbb{O} = -1$ there $\mathcal{L}$ has an odd (nonzero) number of real positive eigenvalues.

Remark 10.3.2. An equivalent formulation is to define

$$\mathbb{O} = \mathbb{O}_0 \operatorname{sign}[E'(0)],$$

with $E'(0)$ given in (10.3.2) and the basis orientation, $\mathbb{O}_0$, of $\mathbb{R}^n$ given in (10.3.3).

Proof. (sketch)Since $\lambda = 0$ is in the natural domain and the matrix eigenvalues are simple, the matrix eigenvectors form a basis, indeed $\mathbb{C}^n = \mathbb{E}^{\mathrm{u}}(0) \oplus \mathbb{E}^{\mathrm{s}}(0)$. Denoting the basis of $\mathbb{E}^{\mathrm{u}}(0)$ by $\{\boldsymbol{v}_j^-\}$ and of $\mathbb{E}^{\mathrm{s}}(0)$ by $\{\boldsymbol{v}_j^+\}$ then from (10.1.16) we may choose the Jost solutions so that (10.3.1) holds, and moreover, since $E(0) = 0$, we may order and normalize them so that

$$\boldsymbol{J}_1^-(x,0) = \boldsymbol{J}_1^+(x,0) \equiv \boldsymbol{J}_0(x) := \begin{pmatrix} p_0 \\ \partial_x p_0 \\ \vdots \\ \partial_x^{n-1} p_0 \end{pmatrix}.$$

From Theorem 10.2.2 the derivative of the Evans function at $\lambda = 0$ is

$$E'(0) = \langle p_0, q_0 \rangle \det\left(J_0^{\mathrm{a}}, J_2^-, \dots, J_k^-, J_0, J_2^+, \dots, J_{n-k}^+\right)(0,0), \tag{10.3.2}$$

where the adjoint eigenvector has been scaled so that $|J_0^{\mathrm{a}}(0)| = 1$.

The Jost solutions orient a basis for $\mathbb{R}^n$ via their asymptotic behavior at $x = \pm\infty$, see (9.4.8). From (10.3.1) the basis orientation at $\lambda = 0$ is given by

$$\mathbb{O}_0 := \operatorname{sign}\left[\det\left(v_1^-, \dots, v_k^-, v_1^+, \dots, v_{n-k}^+\right)\right]. \tag{10.3.3}$$

Moreover, following the arguments of Corollary 9.2.6, we can show that for $\lambda \in \Omega \cap \mathbb{R}$ with $\lambda \gg 1$ the Jost matrices satisfy the asymptotic relation

$$J^-(x,\lambda) = \mathrm{e}^{A_0(\lambda)x} P^{\mathrm{u}}(\lambda) V + \mathcal{O}(|\lambda|^{-1/n}), \quad x \le 0$$
$$J^+(x,\lambda) = \mathrm{e}^{A_0(\lambda)x} P^{\mathrm{s}}(\lambda) V + \mathcal{O}(|\lambda|^{-1/n}), \quad x \ge 0$$

From the argument of Lemma 9.3.9 we find that $E(\lambda) \neq 0$ for $\lambda \gg 1$. Moreover, the arguments of Theorem 9.4.2 show that

$$\operatorname{sign}[E(\lambda)] = \mathbb{O}_0, \quad \lambda \gg 1.$$

□

10.3.1 Example: Generalized Korteweg–de Vries Equation

In Chapter 5.2.1 we considered the orbital stability of the equilibria ϕ_c of the generalized Korteweg-de Vries equation (gKdV) written in the traveling coordinates,

$$u_t + \partial_x\left(\partial_x^2 u - cu + \frac{1}{r+1} u^{r+1}\right) = 0, \tag{10.3.4}$$

where the parameter $c > 0$ corresponds to the wave speed. In particular, it was shown that the equilibria is orbitally stable for $1 \le r < 4$. Via the orientation index we show that if $r > 4$, then the eigenvalue problem for the linearization $\mathbb{L} := \partial_x \mathcal{L}$,

$$\partial_x \mathcal{L} p = \lambda p, \quad \mathcal{L} = -\partial_x^2 + c - \phi_c^r(x),$$

has an odd number of real positive eigenvalues, and the equilibria is unstable. This application of the orientation index was first presented in Pego and Weinstein [220].

In Chapter 3.1.1.1 we established that the essential spectrum of $\mathbb{L}$ is composed of the imaginary axis, $\sigma_{\mathrm{ess}}(\mathbb{L}) = \mathrm{i}\mathbb{R}$. The natural domain $\Omega \subset \mathbb{C}$ is comprised of the open right-half of the complex plane. Moreover, the matrix eigenvalues at $\lambda = 0$,

$$\mu_3(0) = -\sqrt{c}, \quad \mu_2(0) = 0, \quad \mu_1(0) = \sqrt{c},$$

are simple, and for $\lambda \in \Omega$ they satisfy

$$\operatorname{Re}\mu_1(\lambda) < 0 < \operatorname{Re}\mu_2(\lambda) \le \operatorname{Re}\mu_3(\lambda).$$

Consequently the Jost matrices have dimension $J^+(x,\lambda) \in \mathbb{C}^{3\times 1}$ and $J^-(x,\lambda) \in \mathbb{C}^{3\times 2}$. The absolute spectrum of $\mathbb{L}$ is uniformly bounded away from the imaginary axis (see Chapter 3.2.1.1), so that the extended domain Ω_{e} of the Jost matrices and the Evans function

$$E(\lambda) = \det(J^-, J^+)(0,\lambda),$$

contains the half-space $\operatorname{Re}\lambda \ge -\delta$ for some $\delta > 0$.

To determine the orientation index, we recall the eigenvalue picture at $\lambda = 0$. From (5.2.61) and (5.2.62), we know that as long as $\partial_c\langle \phi_c, \phi_c\rangle \neq 0$, then $\mathrm{m}_{\mathrm{g}}(0) = 1$ and $\mathrm{m}_{\mathrm{a}}(0) = 2$, with

$$\partial_x\mathcal{L}(\partial_x\phi_c) = 0, \quad \partial_x\mathcal{L}(-\partial_c\phi_c) = \partial_x\phi_c.$$

In particular, the kernel of $\mathbb{L}$ is spanned by the translational eigenfunction $p_0 = p_{0,0} := \partial_x\phi_c$, while $p_{1,0} = -\partial_c\phi_c$ and the kernel of the adjoint operator $\mathbb{L}^{\mathrm{a}} := -\mathcal{L}\partial_x$ is spanned by $q_0 := \phi_c$. For λ near the origin we write the Jost matrix as the concatenation of two Jost solutions

$$J^-(x,\lambda) = \left(J_1^-, J_2^-\right)(x,\lambda),$$

with the normalization

$$J_1^-(x,0) = \begin{pmatrix} \partial_x\phi_c \\ \partial_x^2\phi_c \\ \partial_x^3\phi_c \end{pmatrix}, \quad J_2^-(x,0) = \begin{pmatrix} \mathcal{L}^{-1}(1) \\ \partial_x\mathcal{L}^{-1}(1) \\ \partial_x^2\mathcal{L}^{-1}(1) \end{pmatrix}.$$

The formula for the second Jost solution is derived from the observation

$$\mathbb{L}\mathcal{L}^{-1}(1) = \partial_x\mathcal{L}\mathcal{L}^{-1}(1) = \partial_x(1) = 0,$$

where $\mathcal{L}^{-1}(1)$ is asymptotically equal to the inverse wave speed, $1/c$, at $\pm\infty$. The third Jost solution, $J_1^+(x,\lambda)$, is normalized at $\lambda = 0$ so that $J_1^+(x,0) = J_1^-(x,0) \equiv J_0(x)$. The adjoint ODE Jost eigenfunction at $\lambda = 0$ induced by $q_0 = \phi_c$ is

$$J_0^{\mathrm{a}}(x) = \begin{pmatrix} j_1^{\mathrm{a}}(x) \\ j_2^{\mathrm{a}}(x) \\ \phi_c(x) \end{pmatrix}.$$

The operator $\mathbb{L}$ arises from a Hamiltonian system, and has the generic Jordan chain of length two. We deduce that $E(0) = E'(0) = 0$. From Theorem 10.2.2, with $p_{1,0}$ and q_0 given above, we calculate that

$$E''(0) = 2\langle -\partial_c \phi_c, \phi_c \rangle \det \boldsymbol{\Phi}(0,0) = -\partial_c Q(u) \det \boldsymbol{\Phi}(0,0), \tag{10.3.5}$$

where the charge $Q(u) := \langle u, u \rangle / 2$, and

$$\det \boldsymbol{\Phi}(0,0) = \det\left(J_0^{\mathrm{a}}, J_2^-, J_0\right)(0,0)$$

(we ignore the scaling associated with $|J_0^{\mathrm{a}}(0)|$, since this will not affect the sign of $E''(0)$). Indeed, the variation of the charge, which played a defining role in the orbital stability analysis of Chapter 5.2.1, also plays a central role in the orientation index analysis.

To apply the results of Exercise 10.3.1, we must determine the orientation $\mathbb{O}_0$, which determines the sign of E for $\lambda \gg 1$. This requires an understanding the relation between the Jost solutions and the choice of basis vectors. At $\lambda = 0$ we have

$$\lim_{x \to -\infty} \mathrm{e}^{-\mu_1(0)x} J_1^-(x,0) = d_- \begin{pmatrix} c^{1/2} \\ c \\ c^{3/2} \end{pmatrix}, \quad \lim_{x \to -\infty} \mathrm{e}^{-\mu_2(0)x} J_2^-(x,0) = \begin{pmatrix} c^{-1} \\ 0 \\ 0 \end{pmatrix},$$

and

$$\lim_{x \to +\infty} \mathrm{e}^{-\mu_3(0)x} J_1^+(x,0) = d_+ \begin{pmatrix} c^{1/2} \\ -c \\ c^{3/2} \end{pmatrix}.$$

Here $d_\pm \neq 0$, and since $\partial_x \phi_c$ is odd, it follows that $d_+ = -d_-$. Consequently,

$$\mathbb{O}_0 = \operatorname{sign}\left[\det \begin{pmatrix} d_- c^{1/2} & c^{-1} & -d_- c^{1/2} \\ d_- c & 0 & d_- c \\ d_- c^{3/2} & 0 & -d_- c^{3/2} \end{pmatrix}\right] = \operatorname{sign}\left[2 d_-^2 c^{3/2}\right] = +1,$$

and the orientation index simplifies to

$$\mathbb{O} = -\operatorname{sign}\left[\partial_c Q(\phi_c)\right] \operatorname{sign}\left[\det \boldsymbol{\Phi}(0,0)\right].$$

The equilibrium ϕ_c is orbitally stable for $1 \le r < 4$, and there are no eigenvalues with positive real part. It follows that $\mathbb{O} = +1$ for these values of r. Moreover, from the simple scaling argument of (5.2.12), $\operatorname{sign}\left[\partial_c \langle \phi_c, \phi_c \rangle\right] = \operatorname{sign}[4 - r]$. In particular,

$$\operatorname{sign}\left[\det \boldsymbol{\Phi}(0,0)\right] < 0, \quad 1 \le r < 4,$$

but the fundamental matrix $\boldsymbol{\Phi}(x,0)$ is always nonsingular, and is smooth in r, hence its determinant cannot change sign for any $r \ge 1$. We deduce that

$$\mathbb{O} = \text{sign}\,[\partial_c Q(\phi_c)] = \text{sign}[4-r].$$

The failure of the orbital stability proof at $r = 4$ coincides with the existence of unstable positive eigenvalues of the linearization, $\mathbb{L}$, of the gKdV about the equilibrium ϕ_c.

10.3.2 Example: Parametrically Forced Ginzburg–Landau Equation

We continue the discussion of Chapter 6.1.1 concerning the parametrically forced Ginzburg–Landau equation

$$\partial_t u = \partial_x^2 u - u + |u|^2 u + \mathrm{i}\left(\mu u - \mathrm{e}^{-\mathrm{i}\theta}\overline{u}\right), \tag{10.3.6}$$

where $(x,t) \in \mathbb{R}\times\mathbb{R}^+$, $u \in \mathbb{C}$, $0 < \mu < 1$, $-\pi/2 < \theta < \pi/2$, and $\mu = \cos\theta$.

The system possesses a real-valued equilibria

$$U_\theta(x) = \sqrt{2}A\,\text{sech}(Ax), \quad A^2 := 1 + \sin\theta. \tag{10.3.7}$$

The value $\theta = 0$ is a bifurcation point for the system, and we use the orientation index to distinguish the eigenvalue structure for $\theta > 0$ from $\theta < 0$. Decomposing the solution as $u = U_\theta + u_\mathrm{r} + \mathrm{i}u_\mathrm{i}$ and linearizing, we obtain the eigenvalue problem for $\mathcal{L}$,

$$\mathcal{L}\begin{pmatrix} u_\mathrm{r} \\ u_\mathrm{i} \end{pmatrix} = \lambda \begin{pmatrix} u_\mathrm{r} \\ u_\mathrm{i} \end{pmatrix}, \quad \mathcal{L} := \begin{pmatrix} \mathcal{L}_+ & -2\mu \\ 0 & \mathcal{L}_- \end{pmatrix}, \tag{10.3.8}$$

where the suboperators are

$$\mathcal{L}_+ = \partial_x^2 - A^2 + 3U_\theta^2, \quad \mathcal{L}_- = \partial_x^2 - (A^2 - 2\sin\theta) + U_\theta^2.$$

Rescaling the independent variable $y = Ax$, the operators become

$$\mathcal{L}_+ = A^2\left(\partial_y^2 - 1 + 6\,\text{sech}^2(y)\right), \quad \mathcal{L}_- = A^2\left(\partial_y^2 - \frac{A^2 - 2\sin\theta}{A^2} + 2\,\text{sech}^2(y)\right),$$

which is of the form studied in (9.4.1), with $b_0 = 2$ (for $\mathcal{L}_-$) or $b_0 = 6$ (for $\mathcal{L}_+$). Using (9.4.2), and noting that $\gamma = 0$ corresponds to the edge of the essential spectrum, after simplification we find

$$\sigma_\mathrm{pt}(\mathcal{L}_-) = \{2\sin\theta\}, \quad \sigma_\mathrm{pt}(\mathcal{L}_+) = \{0, 3A^2\}. \tag{10.3.9}$$

In particular, $\mathcal{L}_-$ is invertible for $\theta \neq 0$ while $\ker(\mathcal{L}_+) = \text{span}\{\partial_x U_\theta\}$, and $\lambda = 0$ is an eigenvalue of $\mathcal{L}$ with eigenfunction $\boldsymbol{p}_0(x) = (\partial_x U_\theta(x), 0)^\mathrm{T}$.

The adjoint operator is

$$\mathcal{L}^{\mathrm{a}} = \begin{pmatrix} \mathcal{L}_+ & 0 \\ -2\mu & \mathcal{L}_- \end{pmatrix},$$

and for $\theta \neq 0$, its kernel is spanned by

$$q_0 = \begin{pmatrix} \partial_x U_\theta \\ 2\mu\mathcal{L}_-^{-1}(\partial_x U_\theta) \end{pmatrix}. \tag{10.3.10}$$

From [**74**, Eq. (3.21)], the solution $u = u_-$ to $\mathcal{L}_- u = 0$ and that decays as $x \to -\infty$ is given by

$$u_-(x) = \left(\frac{\operatorname{sech}(Ax)}{2}\right)^B F(B+2, B-1, B+1, z), \quad z := \frac{1}{2}(1 + \tanh(Ax)),$$

where F is the hypergeometric function and

$$B^2 = \frac{A^2 - 2\sin\theta}{A^2} = \frac{1-\sin\theta}{1+\sin\theta}. \tag{10.3.11}$$

Moreover, from [**189**, Chapter 9] we have the compact expression

$$F(B+2, B-1, B+1, z) = (1-z)^{-B}\left(1 - \frac{2}{B+1}z\right);$$

so that

$$u_-(x) = \left(\frac{\operatorname{sech}(Ax)}{2}\right)^B (1-z)^{-B}\left(1 - \frac{2}{B+1}z\right). \tag{10.3.12}$$

In particular, we can readily verify that

$$\lim_{x\to-\infty} \mathrm{e}^{-ABx} u_-(x) = 1, \quad u_-(0) = \frac{B}{B+1}, \quad \partial_x u_-(0) = A(B-1). \tag{10.3.13}$$

Since $\mathcal{L}_-$ has preserves parity, the solution

$$u_+(x) = u_-(-x), \tag{10.3.14}$$

decays as $x \to +\infty$ and satisfies

$$\lim_{x\to+\infty} \mathrm{e}^{ABx} u_+(x) = 1, \quad u_+(0) = \frac{B}{B+1}, \quad \partial_x u_+(0) = -A(B-1). \tag{10.3.15}$$

Introducing the vector notation $\boldsymbol{u} = (u_\mathrm{r}, u_\mathrm{i})^\mathrm{T}$, the system $\mathcal{L}\boldsymbol{u} = 0$ has two linearly independent solutions that decay as $x \to -\infty$,

$$\boldsymbol{u}_1^- = \begin{pmatrix} \partial_x U_\theta \\ 0 \end{pmatrix}, \quad \boldsymbol{u}_2^- = \begin{pmatrix} 2\mu\mathcal{L}_+^{-1} u_- \\ u_- \end{pmatrix}, \tag{10.3.16}$$

and two which decay as $x \to +\infty$,

$$u_1^+ = \begin{pmatrix} \partial_x U_\theta \\ 0 \end{pmatrix}, \quad u_2^+ = \begin{pmatrix} -2\mu \mathcal{L}_+^{-1} u_+ \\ -u_+ \end{pmatrix}. \tag{10.3.17}$$

The Jost solutions at $\lambda = 0$ of the vectored eigenvalue problem associated to (10.3.8) are

$$J_i^\pm(x,0) = \begin{pmatrix} u_i^\pm \\ \partial_x u_i^\pm \end{pmatrix}, \tag{10.3.18}$$

for $i = 1,2$, and the decaying eigenfunction solution of the vector problem is

$$J_0(x) = J_1^-(x) = \begin{pmatrix} \partial_x U_\theta \\ 0 \\ \partial_x^2 U_\theta \\ 0 \end{pmatrix}.$$

To determine the orientation of $\mathbb{R}^4$ from these Jost solutions, we observe

$$\lim_{x\to-\infty} e^{-Ax} J_1^-(x,0) = v_1^- = \sqrt{2}A^2 \begin{pmatrix} 1 \\ 0 \\ A \\ 0 \end{pmatrix}; \quad \lim_{x\to-\infty} e^{-ABx} J_2^-(x,0) = v_2^- = \begin{pmatrix} -\cot\theta \\ 1 \\ -AB\cot\theta \\ AB \end{pmatrix},$$

and

$$\lim_{x\to+\infty} e^{Ax} J_1^+(x,0) = v_1^+ = \sqrt{2}A^2 \begin{pmatrix} -1 \\ 0 \\ A \\ 0 \end{pmatrix}; \quad \lim_{x\to+\infty} e^{ABx} J_2^+(x,0) = v_2^+ = \begin{pmatrix} \cot\theta \\ -1 \\ -AB\cot\theta \\ AB \end{pmatrix}.$$

Consequently, the orientation, (10.3.3), of $\mathbb{R}^4$ satisfies

$$\mathbb{O}_0 = \operatorname{sign}\left[\det(v_1^-, v_2^-, v_1^+, v_2^+)\right] = \operatorname{sign}[4A^6 B] = +1.$$

To evaluate the orientation index it remains to determine the sign of $E'(0)$. The adjoint kernel q_0 from (10.3.10) induces the adjoint ODE solution

$$J_0^{\mathrm{a}}(x) = \begin{pmatrix} -\partial_x q_0(x) \\ q_0(x) \end{pmatrix}.$$

Since

$$\langle p_0, q_0 \rangle = \langle \partial_x U_\theta, \partial_x U_\theta' \rangle > 0,$$

from Theorem 10.2.2, the derivative of the Evans function is given by

$$E'(0) = \langle \partial_x U_\Theta, \partial_x U_\Theta \rangle \det\left(J_0^{\mathrm{a}}, J_2^-, J_0, J_2^+ \right)(0,0)$$

(again, we ignore the scaling associated with $|J_0^{\mathrm{a}}(0)|$, since this will not affect the sign of $E'(0)$). Since $\partial_x U_\Theta(x)$ is odd, and $\mathcal{L}_\pm$ preserve parity, we have

$$J_0(0) = \begin{pmatrix} 0 \\ 0 \\ \partial_x^2 U_\Theta(0) \\ 0 \end{pmatrix}, \quad J_0^{\mathrm{a}}(0) = -\begin{pmatrix} \partial_x^2 U_\Theta(0) \\ 2\mu \partial_x \mathcal{L}_-^{-1}(\partial_x U_\Theta)(0) \\ 0 \\ 0 \end{pmatrix},$$

and, moreover, via (10.3.14),

$$[\mathcal{L}_+^{-1} u_+](0) = [\mathcal{L}_+^{-1} u_-](0), \quad \partial_x [\mathcal{L}_+^{-1} u_+](0) = -\partial_x [\mathcal{L}_+^{-1} u_-](0).$$

From this we deduce that

$$J_2^-(0) = \begin{pmatrix} 2\mu \mathcal{L}_+^{-1} u_- \\ u_- \\ 2\mu \partial_x \mathcal{L}_+^{-1} u_- \\ \partial_x u_- \end{pmatrix}(0), \quad J_2^+(0) = \begin{pmatrix} -2\mu \mathcal{L}_+^{-1} u_- \\ -u_- \\ 2\mu \partial_x \mathcal{L}_+^{-1} u_- \\ \partial_x u_- \end{pmatrix}(0).$$

Substituting these formulas for the Jost solutions into $E'(0)$ and simplifying, we obtain

$$E'(0) = -2\langle \partial_x U_\Theta, \partial_x U_\Theta \rangle \partial_x^2 U_\Theta \partial_x u_- \left[\partial_x^2 U_\Theta u_- - 4\mu^2 (\mathcal{L}_+^{-1} u_-)(\partial_x \mathcal{L}_-^{-1}(\partial_x U_\Theta)) \right]. \tag{10.3.19}$$

where the expression on the right is evaluated at $x = 0$.

It remains to evaluate the quantities in brackets. We first consider $\mathcal{L}_+^{-1} u_-$. One solution to $\mathcal{L}_+ u = 0$ is $u = \partial_x U_\Theta$: let $\tilde{u}$ represent the other solution with initial condition

$$\tilde{u}(0) = -\frac{1}{\partial_x^2 U_\Theta(0)}, \quad \partial_x \tilde{u}(0) = 0.$$

From variation of parameters we obtain the expression

$$\left[\mathcal{L}_+^{-1} u_- \right](x) = \tilde{u}(x) \int_{-\infty}^{x} \partial_t U_\Theta(t) u_-(t)\,\mathrm{d}t - \partial_x U_\Theta(x) \int_0^x \tilde{u}(t) u_-(t)\,\mathrm{d}t + c\,\partial_x U_\Theta(x).$$

Since $\partial_x U_\Theta$ is odd, evaluating at $x = 0$ yields

$$\left[\mathcal{L}_+^{-1} u_- \right](0) = C_- \tilde{u}(0) = -\frac{C_-}{\partial_x^2 U_\Theta(0)}; \quad C_- := \int_{-\infty}^{0} \partial_t U_\Theta(t) u_-(t)\,\mathrm{d}t. \tag{10.3.20}$$

Now consider $\partial_x[\mathcal{L}_-^{-1}(\partial_x U_\theta)]$. Using variation of parameters to invert $\mathcal{L}_-$ yields

$$\left[\mathcal{L}_-^{-1}(\partial_x U_\theta)\right](x) = \frac{1}{\Delta}\left(u_-(x)\int_x^{+\infty} u_+(t)\partial_x U_\theta(t)\,\mathrm{d}t + u_+(x)\int_{-\infty}^{x} u_+(t)\partial_x U_\theta(t)\,\mathrm{d}t\right),$$

where from (10.3.14) and (10.3.15) we have

$$\Delta := (u_-\partial_x u_+ - \partial_x u_- u_+)(0) = -2u_-(0)\partial_x u_-(0).$$

Since U_θ is even, taking the derivative and evaluating at $x = 0$ yields

$$\partial_x\left[\mathcal{L}_-^{-1}\partial_x U_\theta\right](0) = \frac{1}{\Delta}\left(u_-'(0)\int_0^{+\infty} u_+(t)\partial_x U_\theta(t)\,\mathrm{d}t + u_+'(0)\int_{-\infty}^{0} u_-(t)\partial_x U_\theta(t)\,\mathrm{d}t\right). \tag{10.3.21}$$

Moreover, from (10.3.14) we have

$$\int_0^{+\infty} u_+(t)\partial_t U_\theta(t)\,\mathrm{d}t = \int_0^{+\infty} u_-(-t)\partial_t U_\theta(t)\,\mathrm{d}t = -\int_{-\infty}^{0} u_+(t)\partial_x U_\theta(t)\,\mathrm{d}t.$$

However, $\partial_x u_+(0) = -\partial_x u_-(0)$, the expression (10.3.21) reduces to

$$\partial_x\left[\mathcal{L}_-^{-1}(\partial_x U_\theta)\right](0) = -\frac{2\partial_x u_-(0)}{\Delta}\int_{-\infty}^{0} u_-(t)\partial_t U_\theta(t)\,\mathrm{d}t = \frac{C_-}{u_-(0)}, \tag{10.3.22}$$

where the second equality follows from the definitions of C_- and Δ.

Combining (10.3.20) and (10.3.22), the derivative of the Evans function given in (10.3.19) can be simplified to

$$E'(0) = -2\langle\partial_x U_\theta, \partial_x U_\theta\rangle(\partial_x^2 U_\theta)\partial_x u_-\left[\partial_x U_\theta u_- + 4\mu^2\frac{C_-^2}{\partial_x U_\theta U_\theta u_-}\right] = -C_E\partial_x u_-,$$

where $C_E > 0$ is given by

$$C_E := \frac{2\langle\partial_x U_\theta, \partial_x U_\theta\rangle}{u_-}\left[((\partial_x U_\theta)u_-)^2 + 4\mu^2 C_-^2\right]\Big|_{x=0},$$

and $u_-(0) > 0$ by (10.3.13). Finally, from (10.3.13)

$$\partial_x u_-(0) = -A(1-B),$$

we determine from (10.3.11) that

$$\operatorname{sign}[(E'(0)] = \operatorname{sign}[1-B] = \operatorname{sign}\left[\frac{\sqrt{1+\sin\theta} - \sqrt{1-\sin\theta}}{A}\right] = \operatorname{sign}[\theta].$$

Consequently,

$$\mathbb{O} = \mathbb{O}_0 \operatorname{sign}[(E'(0)] = \operatorname{sign}[(E'(0)] = \operatorname{sign}[\theta], \tag{10.3.23}$$

which demonstrates the bifurcation at $\theta = 0$. For $\theta > 0$ the linearization has an even number of real, positive eigenvalues, while for $\theta < 0$ it has an odd number. Indeed, when $\theta = 0$, i.e., $\mu = 1$, the waves merge in a saddle-node bifurcation, so that $E(0) = E'(0) = 0$ and $\mathcal{L}$ has a double eigenvalue at $\lambda = 0$.

Exercises

Exercise 10.3.1. Suppose that $\lambda = 0$ is an eigenvalue with multiplicity $\mathrm{m_a}(0) = d$. Define the orientation index as

$$\mathbb{O} = \mathbb{O}_0 \operatorname{sign}[E^{(d)}(0)].$$

Show that if $\mathbb{O} = +1$, then the Evans function has an even number of real, positive zeros, and if $\mathbb{O} = -1$ it has an odd number of such zeros.

Exercise 10.3.2. State and prove an orientation theorem, similar to Theorem 10.3.1, for exponentially asymptotic systems of the form

$$\mathcal{L}\boldsymbol{p} := \boldsymbol{d}\partial_x^2\boldsymbol{p} + \boldsymbol{a}_1(x)\partial_x\boldsymbol{p} + \boldsymbol{a}_0(x)\boldsymbol{p} = \lambda\boldsymbol{p},$$

(see Exercise 10.2.4).

Exercise 10.3.3. Reconsider the problem discussed in Chapter 10.3.1. Calculate $\det \boldsymbol{\Phi}(0,0)$ explicitly in order to show that it is negative for all values of r.

Exercise 10.3.4. Compute $E''(0)$ for $\mathcal{L}$ given in (10.3.8) when $\mu = 1$. Verify that $E''(0) < 0$ and deduce the sign of E for real λ near 0.

Exercise 10.3.5. Use the information presented in (10.3.9) to show that $\mathcal{L}$ given in (10.3.8) has $\sigma_{\mathrm{pt}}(\mathcal{L}) = \{2\sin\theta, 3A^2\}$, and compare this with the orientation result.

10.4 Application: Edge Bifurcations

As in Chapter 9.5, we consider a family of smoothly parameterized, nth-order, exponentially localized operators, $\mathcal{L} = \mathcal{L}(\epsilon)$ with a common domain $H^n(\mathbb{R})$. A key application of the Evans function is to track zeros that bifurcate from the branch point of the absolute spectrum of $\mathcal{L}(\epsilon)$, possibly entering the physical sheet of the Riemann surface and becoming point spectra. In terms of the Riemann parameter, γ, the implicit function theorem guarantees that simple zeros, $\gamma = \gamma(\epsilon)$, of the Evans function, $E = E(\gamma, \epsilon)$, which bifurcate from the branch point, admit an expansion of the form

$$\gamma = \gamma_\epsilon := -\frac{\partial_\epsilon E(0,0)}{\partial_\gamma E(0,0)}\epsilon + \mathcal{O}(\epsilon^2). \tag{10.4.1}$$

If the zero moves onto the physical sheet of the Riemann surface, then in the λ-plane its position relative to a quadratic branch point, λ_{br}, takes the form

$$\lambda = \lambda_{\mathrm{br}} - c\gamma_\epsilon^2\epsilon^2 + \mathcal{O}(\epsilon^3),$$

for some $c \in \mathbb{C}$. The motion of the zero, and its relation to both the essential spectrum and the absolute spectrum, depends upon both the motion of the branch point and the motion of the zero on the Riemann surface. In the following theorem we evaluate the key components of this analysis.

Theorem 10.4.1. *Let $\mathcal{L}(\epsilon)$ be a family of exponentially asymptotic operators of the form (10.0.1), with coefficients that depend analytically upon ϵ near $\epsilon = 0$. If $\gamma = 0$ is a quadratic branch point for the Riemann surface associated with the Evans function, $E = E(\gamma,\epsilon)$, and $E(0,0) = 0$, then there exists $A \neq 0$ such that*

$$\partial_\gamma E(0,0) = A \lim_{L\to+\infty}\left(\boldsymbol{J}_0^{\mathrm{a}}(-L)\cdot\partial_\gamma \boldsymbol{v}_1^-(0,0) - \boldsymbol{J}_0^{\mathrm{a}}(+L)\cdot\partial_\gamma \boldsymbol{v}_1^+(0,0)\right), \tag{10.4.2}$$

and

$$\begin{aligned}\partial_\epsilon E(0,0) = \; & A\Bigg[\lim_{L\to+\infty}(\boldsymbol{J}_0^{\mathrm{a}}(-L)\cdot\partial_\epsilon \boldsymbol{v}_1^-(0,0) - \boldsymbol{J}_0^{\mathrm{a}}(+L)\cdot\partial_\epsilon \boldsymbol{v}_1^+(0,0)) \\ & + \int_{+\infty}^{+\infty} \boldsymbol{J}_0^{\mathrm{a}}(s)\cdot[\boldsymbol{A}_1 + \boldsymbol{R}_1(s)]\boldsymbol{J}_0(s)\,\mathrm{d}s\Bigg],\end{aligned} \tag{10.4.3}$$

where $\boldsymbol{J}_0$ is the Jost function common to the extended stable and unstable matrices, $\boldsymbol{J}_0^{\mathrm{a}}$ is the associated adjoint solution scaled so that $|\boldsymbol{J}_0^{\mathrm{a}}(0)| = 1$, and $\boldsymbol{v}_1^\pm$ are the asymptotic vectors defined in (10.4.8). Moreover, if both of these quantities are nonzero, then the zero is simple and its location on the Riemann surface is given by (10.4.1).

Remark 10.4.2. If the adjoint solution is not normalized, then the right-hand side of both (10.4.2) and (10.4.3) should be multiplied by $|\boldsymbol{J}_0^{\mathrm{a}}(0)|^2$. Since, to leading order, the perturbed zero of the Evans function satisfies

$$\gamma(\epsilon) = -\frac{\partial_\epsilon E(0,0)}{\partial_\gamma E(0,0)}\epsilon + \mathcal{O}(\epsilon^2).$$

[see (9.5.4)], the exact value of $|\boldsymbol{J}_0^{\mathrm{a}}(0)|$ is irrelevant in locating the zero.

Proof. (outline) The ϵ-dependent eigenvalue problem for $\mathcal{L}$ takes the vectorized form

$$\partial_x \boldsymbol{Y} = [\boldsymbol{A}_0(\lambda,\epsilon) + \boldsymbol{R}(x,\epsilon)]\boldsymbol{Y}. \tag{10.4.4}$$

Since the branch point at λ_{br} is assumed quadratic, for small ϵ the limiting matrix $A_0(\lambda,\epsilon)$ has a an eigenvalue $\mu_{br} = \mu_{br}(\epsilon)$ with a Jordan block of length two for $\lambda = \lambda_{br}(\epsilon)$, see the discussion surrounding (10.1.5). In particular, for λ near λ_{br} and μ near μ_{br} the dispersion relation, (10.1.4) admits the expansion

$$\det(A_0(\lambda,\epsilon) - \mu I_n) \sim a(\lambda - \lambda_{br}) + b(\mu - \mu_{br})^2, \tag{10.4.5}$$

and the Riemann surface is defined locally, at leading-order, by map

$$\gamma^2 = -\frac{a}{b}(\lambda - \lambda_{br}). \tag{10.4.6}$$

The Jost matrices, as well as the Evans function, are analytic on the Riemann surface in a neighborhood of $\gamma = 0$.

Rewriting the vector system in the Riemann variables it has the form

$$\partial_x Y = [A_0(\gamma,\epsilon) + R(x,\epsilon)]\,Y, \tag{10.4.7}$$

where the matrices admit the expansions

$$A_0(\gamma,\epsilon) = A_0(\gamma) + \epsilon A_1 + \mathcal{O}(\epsilon^2), \quad R(x,\epsilon) = R_0(x) + \epsilon R_1(x) + \mathcal{O}(\epsilon^2),$$

in particular, A_1 is independent of γ. By assumption, the Evans function has a zero at the branch point and hence the Jost solutions may be ordered so that $J_1^{\pm}(x,\gamma,\epsilon)$ satisfy $J_1^-(x,0,0) = J_1^+(x,0,0) \equiv J_0(x)$. Furthermore, these Jost solutions satisfy the asymptotic relations

$$\lim_{x\to-\infty} e^{-\gamma x} J_1^-(x,\gamma,\epsilon) = v_1^-(\gamma,\epsilon), \quad \lim_{x\to+\infty} e^{\gamma x} J_1^+(x,\gamma,\epsilon) = v_1^+(\gamma,\epsilon). \tag{10.4.8}$$

[compare with the transformations leading to (9.5.33)]. We denote by $J_0^{a}(x)$ the solution of the adjoint ODE with $\gamma = \epsilon = 0$, which

(a) grows at most linearly in x as $x \to \pm\infty$;
(b) satisfies $J_0^{a}(0) \cdot J_i^{\pm}(0,0,0) = 0$ for each Jost function.

Following the arguments of Lemmas 9.5.4 and 9.5.7 yields (10.4.2) and (10.4.3); the details are omitted. □

10.4.1 Example: The Nonlinear Schrödinger Equation

We derive a closed-form expression for the Evans function associated with the pulse solution of the nonlinear Schrödinger equation (NLS). This construction is relevant in its own right, but is also necessary to perform the edge-bifurcation calculations for the Manakov equation in Chapter 10.4.2. In Exercise 5.2.10 it is shown that the pulse solution of the NLS is orbitally stable, which implies that for the associated linearization there is no spectra with a positive real part. The derivation of the Evans function will identify the points in the absolute spectrum from which eigenvalues can bifurcate under a perturbation.

The NLS equation takes the form

$$\mathrm{i}\partial_t u + \partial_x^2 u - u + |u|^2 u = 0, \tag{10.4.9}$$

where $(x,t) \in \mathbb{R} \times \mathbb{R}^+$ and $u \in \mathbb{C}$. It supports a real-valued pulse solution,

$$U(x) = \sqrt{2}\,\mathrm{sech}(x).$$

We rewrite the NLS equation in terms of u and its complex-conjugate $v := \overline{u}$,

$$\mathrm{i}\partial_t u + \partial_x^2 u - u + u^2 v = 0, \quad -\mathrm{i}\partial_t v + \partial_x^2 v - v + uv^2 = 0. \tag{10.4.10}$$

This form is common in the physics literature, and yields an easier route to an explicit construction of the Evans via the *squared eigenfunctions* [**145**, **153**]. In these variables the equilibria becomes $(u,v) = (U,U)$, and introducing

$$u = U + p_1, \quad v = U + p_2,$$

the linearized problem for $\boldsymbol{p} = (p_1, p_2)^T$ is

$$\mathrm{i}\partial_t \boldsymbol{p} = \underbrace{[\left(\partial_x^2 - 1\right)\sigma_3 + 2\,\mathrm{sech}^2(x)(2\sigma_3 + \mathrm{i}\sigma_2)]}_{\mathcal{L}}\boldsymbol{p}, \tag{10.4.11}$$

where we have introduced the Pauli spin matrices

$$\sigma_1 = \begin{pmatrix} 0 & 1 \\ 1 & 0 \end{pmatrix}, \quad \sigma_2 = \mathrm{i}\begin{pmatrix} 0 & -1 \\ 1 & 0 \end{pmatrix}, \quad \sigma_3 = \begin{pmatrix} 1 & 0 \\ 0 & -1 \end{pmatrix}.$$

The separated variables ansatz, $\boldsymbol{p}(x,t) \mapsto \mathrm{e}^{\mathrm{i}\lambda t}\boldsymbol{p}(x)$, yields the eigenvalue problem in standard form,

$$\mathcal{L}\boldsymbol{p} = \lambda \boldsymbol{p}. \tag{10.4.12}$$

This formulation of $\mathcal{L}$ rotates is spectra by $\pi/2$ from its typical formulation. The problem (10.4.12) is of the form studied in Exercise 10.2.4, with

$$\boldsymbol{d} = \sigma_3, \quad \boldsymbol{a}_1(x) \equiv \boldsymbol{0}_2, \quad \boldsymbol{a}_0(x) = -\sigma_3 + 2\,\mathrm{sech}^2(x)\begin{pmatrix} 2 & 1 \\ -1 & -2 \end{pmatrix}.$$

Moreover, in that exercise it was not necessary, except for part (a), that $\boldsymbol{d}$ be positive definite.

The operator $\mathcal{L}$ has real-valued coefficients and satisfies $\sigma_1\mathcal{L}\sigma_1 = -\mathcal{L}$. These two facts imply, as expected for a Hamiltonian system, that eigenvalues come in quartets $\{\pm\lambda, \pm\overline{\lambda}\}$ with associated eigenfunctions $\{\boldsymbol{p}, \overline{\boldsymbol{p}}, \sigma_1\boldsymbol{p}, \sigma_1\overline{\boldsymbol{p}}\}$. This symmetry implies we need only consider $\operatorname{Re}\lambda \geq 0$ when constructing the Evans function.

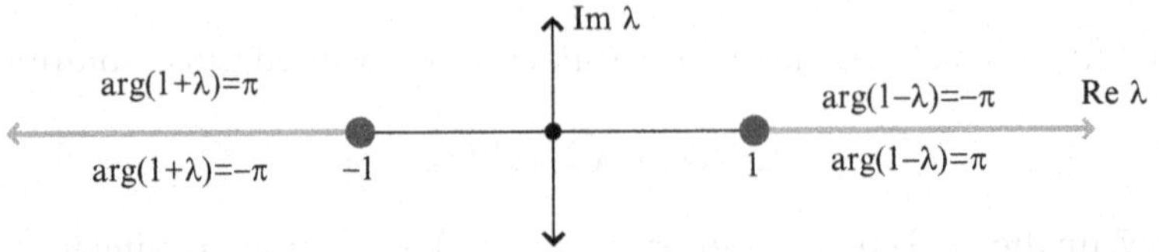

Fig. 10.3 The branch cuts for $\mu_\ell(\lambda)$ and $\mu_r(\lambda)$, which correspond to the absolute spectrum of $\mathcal{L}$, are denoted by a solid (*green*) curve. The branch points, which are at the edge of the absolute spectrum, are denoted by (*red*) circles. Note that the definition of $\mathcal{L}$ effectively rotates its spectrum by $\pi/2$ radians as compared to the usual framework. (Color figure online.)

The matrix eigenvalues for the vectorized version of (10.4.12) are

$$\mu_1(\lambda) = \mu_\ell(\lambda), \quad \mu_2(\lambda) = \mu_r(\lambda), \quad \mu_3(\lambda) = -\mu_\ell(\lambda), \quad \mu_4(\lambda) = -\mu_r(\lambda)$$

where we have defined the branch cuts

$$\mu_\ell(\lambda) := \sqrt{1-\lambda},\ \arg(1-\lambda) \in [-\pi, \pi); \quad \mu_r(\lambda) := \sqrt{1+\lambda},\ \arg(1+\lambda) \in [-\pi, \pi).$$

The function μ_ℓ is analytic in the (left) half-plane $\{\lambda : \operatorname{Re}\lambda < 1\}$, while μ_r is analytic in the (right) half-plane $\{\lambda : \operatorname{Re}\lambda > -1\}$ (see Fig. 10.3), and they satisfy $\mu_\ell(\lambda) = \mu_r(\lambda)$ if and only if $\lambda = 0$. The matrix eigenvectors become

$$\boldsymbol{v}_1(\lambda) = \begin{pmatrix} 0 \\ 1 \\ 0 \\ \mu_1(\lambda) \end{pmatrix}, \quad \boldsymbol{v}_2(\lambda) = \begin{pmatrix} 1 \\ 0 \\ \mu_2(\lambda) \\ 0 \end{pmatrix}, \quad \boldsymbol{v}_3(\lambda) = \begin{pmatrix} 0 \\ 1 \\ 0 \\ \mu_3(\lambda) \end{pmatrix}, \quad \boldsymbol{v}_4(\lambda) = \begin{pmatrix} 1 \\ 0 \\ \mu_4(\lambda) \\ 0 \end{pmatrix}.$$

The eigenvectors are linearly independent for all λ away from the branch cuts, and the Jost solutions are determined via the asymptotic relations

$$\lim_{x\to-\infty} \mathrm{e}^{-\mu_i(\lambda)x}\boldsymbol{J}_i^-(x,\lambda) = \boldsymbol{v}_i(\lambda), \quad \lim_{x\to+\infty} \mathrm{e}^{-\mu_{i+2}(\lambda)x}\boldsymbol{J}_i^+(x,\lambda) = \boldsymbol{v}_{i+2}(\lambda),$$

for $i = 1, 2$. Surprisingly, the Jost solutions can be explicitly computed. From Kaup [**164**], two solutions to the eigenvalue problem (10.4.12) are

$$\boldsymbol{p}_\ell(x,\lambda) := \frac{\mathrm{e}^{-\mu_\ell(\lambda)x}}{(\mu_\ell(\lambda)+1)^2}\left\{[(\mu_\ell(\lambda)+1)^2 - 2\mu_\ell(\lambda)\mathrm{e}^{-x}\operatorname{sech}(x)]\begin{pmatrix} 0 \\ 1 \end{pmatrix} - \operatorname{sech}^2(x)\begin{pmatrix} 1 \\ 1 \end{pmatrix}\right\}$$

and

$$\boldsymbol{p}_{\mathrm{r}}(x,\lambda) := \frac{\mathrm{e}^{-\mu_{\mathrm{r}}(\lambda)x}}{(\mu_{\mathrm{r}}(\lambda)+1)^2}\left\{[(\mu_{\mathrm{r}}(\lambda)+1)^2 - 2\mu_{\mathrm{r}}(\lambda)\mathrm{e}^{-x}\operatorname{sech}(x)]\begin{pmatrix}1\\0\end{pmatrix} - \operatorname{sech}^2(x)\begin{pmatrix}1\\1\end{pmatrix}\right\}.$$

Since $\mathcal{L}$ is even in x, we obtain two other solutions via the transformations $\boldsymbol{p}_\ell(-x,\lambda)$ and $\boldsymbol{p}_{\mathrm{r}}(-x,\lambda)$. The normalized Jost solutions are given by

$$J_1^-(x,\lambda) = \begin{pmatrix}\boldsymbol{p}_\ell\\ -\partial_x\boldsymbol{p}_\ell\end{pmatrix}(-x,\lambda),\quad J_2^-(x,\lambda) = \begin{pmatrix}\boldsymbol{p}_{\mathrm{r}}\\ -\partial_x\boldsymbol{p}_{\mathrm{r}}\end{pmatrix}(-x,\lambda)$$

$$J_1^+(x,\lambda) = \begin{pmatrix}\boldsymbol{p}_\ell\\ \partial_x\boldsymbol{p}_\ell\end{pmatrix}(x,\lambda),\quad J_2^+(x,\lambda) = \begin{pmatrix}\boldsymbol{p}_{\mathrm{r}}\\ \partial_x\boldsymbol{p}_{\mathrm{r}}\end{pmatrix}(x,\lambda).$$

From the definition of the Evans function we determine that

$$\begin{aligned} E(\lambda) &= \det\left(J_1^-, J_2^-, J_1^+, J_2^+\right)(0,\lambda), \\ &= 4\mu_\ell(\lambda)\mu_{\mathrm{r}}(\lambda)\left(\frac{\mu_\ell(\lambda)-1}{\mu_\ell(\lambda)+1}\right)^2\left(\frac{\mu_{\mathrm{r}}(\lambda)-1}{\mu_{\mathrm{r}}(\lambda)+1}\right)^2. \end{aligned} \tag{10.4.13}$$

What does Evans function tells us about the spectrum? Since $\mu_\ell(0) = \mu_{\mathrm{r}}(0) = 1$, we see that $\lambda = 0$ is a zero, and an eigenvalue, of multiplicity four. This is consistent with (5.2.61), as the system is Hamiltonian with two symmetries, and we expect a two-dimensional kernel and a four-dimensional generalized kernel. Apart from the origin, the Evans function is nonzero in the λ-plane off of the branch cuts. However, along its branch cuts, we have $E(\pm 1) = 0$. Significantly, the Evans function is zero at the two branch points. Focusing on $\operatorname{Re}\lambda > 0$, the matrix eigenvalues are analytic on the Riemann surface

$$\lambda = 1 - \gamma^2,$$

for which the Evans function becomes

$$E(\gamma) = 4\gamma\sqrt{2-\gamma^2}\left(\frac{\gamma-1}{\gamma+1}\right)^2\left(\frac{\sqrt{2-\gamma^2}-1}{\sqrt{2-\gamma^2}+1}\right)^2. \tag{10.4.14}$$

The Evans function is analytic at the branch point $\gamma = 0$, and since $E'(0) = \sqrt{2}$, the branch point is a simple zero.

Remark 10.4.3. The Evans function has also been explicitly constructed in [**149**] for the defocusing NLS

$$\mathrm{i}\partial_t u + \partial_x^2 u + u - |u|^2 u = 0.$$

Here it was seen that the branch point coincided with the null eigenvalue. A perturbative Evans function calculation in this setting was done by Pelinovsky and Kevrekidis [**225**].

To set up an edge bifurcation calculation, we first note that

$$\boldsymbol{p}_\ell(x,0) = \begin{pmatrix} 0 \\ 1 \end{pmatrix} - \operatorname{sech}^2(x) \begin{pmatrix} 1 \\ 1 \end{pmatrix},$$

so that $\boldsymbol{J}_1^-(x,0) = \boldsymbol{J}_1^+(x,0) \equiv \boldsymbol{J}_0(x)$. To evaluate (10.4.2) and (10.4.3), we need an expression for the adjoint solution $\boldsymbol{J}_0^{\mathrm{a}}(x)$ when $\gamma = 0$. The adjoint, $\mathcal{L}^{\mathrm{a}}$, conjugates via $\boldsymbol{\sigma}_3$ into $\mathcal{L}$, i.e.,

$$\sigma_3 \mathcal{L}^{\mathrm{a}} \sigma_3 = \mathcal{L},$$

and since $\boldsymbol{\sigma}_3^{-1} = \boldsymbol{\sigma}_3$, this implies that

$$\mathcal{L}^{\mathrm{a}}(\boldsymbol{\sigma}_3 \boldsymbol{p}) = \lambda \boldsymbol{\sigma}_3 \boldsymbol{p}.$$

For real λ the adjoint eigenfunction is given by $\boldsymbol{q} = \boldsymbol{\sigma}_3 \boldsymbol{p}$; in particular, when $\gamma = 0$, $\boldsymbol{q}_0 = \boldsymbol{\sigma}_3 \boldsymbol{p}_\ell$. The adjoint ODE solution takes the form

$$\boldsymbol{J}_0^{\mathrm{a}}(x) = \begin{pmatrix} -\boldsymbol{\sigma}_3^{\mathrm{T}} \partial_x \boldsymbol{q}_0 \\ \boldsymbol{\sigma}_3^{\mathrm{T}} \boldsymbol{q}_0 \end{pmatrix} = \begin{pmatrix} -\partial_x \boldsymbol{p}_\ell(x,0) \\ \boldsymbol{p}_\ell(x,0) \end{pmatrix}.$$

From Remark 10.4.2 we know that the scaling of the adjoint Jost solution is not important for the edge bifurcation problem.

We are now prepared to apply Theorem 10.4.1 to a perturbation $\mathcal{L}_\epsilon = \mathcal{L} + \epsilon \mathcal{L}_1$ of the NLS operator. This will be the topic of the next example.

10.4.2 Example: A Perturbed Manakov Equation

Under the assumption of negligible group velocity birefringence the governing equations of pulse propagation in linearly birefringent, lossless fibers is described by the perturbed Manakov equation

$$\begin{aligned} \mathrm{i}\partial_t u + \partial_x^2 u - (\omega + \kappa)u + \left(|u|^2 + (1-\epsilon)|v|^2\right)u + \epsilon \overline{u} v^2 &= 0 \\ \mathrm{i}\partial_t v + \partial_x^2 v - (\omega - \kappa)v + \left((1-\epsilon)|u|^2 + |v|^2\right)v + \epsilon u^2 \overline{v} &= 0. \end{aligned} \qquad (10.4.15)$$

Here, $u, v \in \mathbb{C}$, $\kappa > 0$ is the phase velocity differential, and $0 < \epsilon < 1$ is the four-wave-mixing coefficient. When $\epsilon = 0$, the system (10.4.15) is the Manakov equation, which, like the NLS, is an integrable Hamiltonian system. Edge bifurcations for the Manakov systems were first considered in [**192**, **193**].

The system (10.4.15) has a Hamiltonian structure, see (7.2.26)

$$\partial_t u = -\mathrm{i}\frac{\delta \mathcal{H}}{\delta \overline{u}}, \quad \partial_t v = -\mathrm{i}\frac{\delta \mathcal{H}}{\delta \overline{v}},$$

where the Hamiltonian is given by $\mathcal{H}(u,v) = \mathcal{H}_0(u,v) + \epsilon\mathcal{H}_1(u,v)$, with

$$\mathcal{H}_0(u,v) = \int_{\mathbb{R}} \left[|\partial_x u|^2 + |\partial_x v|^2 + \omega(|u|^2 + |v|^2) + \kappa(|u|^2 - |v|^2) - \frac{1}{2}(|u|^2 + |v|^2)^2 \right] \mathrm{d}x$$
$$\mathcal{H}_1(u,v) = -\frac{1}{2}\int_{\mathbb{R}} (\overline{u}v - u\overline{v})^2 \,\mathrm{d}x.$$

When $\epsilon = 0$, the system is invariant under three commuting symmetries: that of spatial translation,

$$T_1(\gamma_1)\begin{pmatrix} u \\ v \end{pmatrix}(x,t) = \begin{pmatrix} u \\ v \end{pmatrix}(x+\gamma_1,t),$$

and a two-parameter rotation,

$$T_2(\gamma_2)\begin{pmatrix} u \\ v \end{pmatrix}(x,t) = \begin{pmatrix} \mathrm{e}^{\mathrm{i}\gamma_2}u \\ \mathrm{e}^{\mathrm{i}\gamma_2}v \end{pmatrix}(x,t), \quad T_3(\gamma_3)\begin{pmatrix} u \\ v \end{pmatrix}(x,t) = \begin{pmatrix} \mathrm{e}^{-\mathrm{i}\gamma_3}u \\ \mathrm{e}^{\mathrm{i}\gamma_3}v \end{pmatrix}(x,t).$$

For $\epsilon > 0$ the spatial translation and co-rotating symmetry remains; however, the counter-rotating symmetry is broken. Recalling the discussion of perturbed Hamiltonian systems leading to Lemma 7.2.8, for $\epsilon = 0$ the operator obtained by linearizing about an equilibrium will (generically) have a six-dimensional generalized eigenspace at the origin, while for $\epsilon > 0$, but sufficiently small, there will (generically) be four eigenvalues at the origin, and the other two will be of $\mathcal{O}(\epsilon^{1/2})$. Since the system is Hamiltonian for all $\epsilon > 0$, the two small eigenvalues cannot form a quartet, and must form either a purely real or a purely imaginary pair.

For the unperturbed system there is a large class of solitary wave solutions: we focus solely upon the "up" wave, which is given by

$$\begin{pmatrix} u \\ v \end{pmatrix} = \sqrt{2\eta}\,\mathrm{sech}(\sqrt{\eta}\,x)\begin{pmatrix} 1 \\ 0 \end{pmatrix}, \quad \eta = \omega + \kappa.$$

Since this equilibria has no v-component, there will be only *four* eigenvalues at the origin when $\epsilon = 0$, and since the perturbed system retains both symmetries, no eigenvalues will bifurcate from the origin when $\epsilon > 0$. We will consider the linearized problem in real-valued coordinates. We rescale the spatial variable, setting $\tilde{x} = \sqrt{\eta}\,x$, and then drop the tilde. Decomposing $u = \phi + u_{\mathrm{r}} + \mathrm{i}u_{\mathrm{i}}$ and $v = v_{\mathrm{r}} + \mathrm{i}v_{\mathrm{i}}$, where $\phi = \sqrt{2}\,\mathrm{sech}(x)$ is the rescaled pulse, and setting $\boldsymbol{u} = (u_{\mathrm{r}}, u_{\mathrm{i}})^{\mathrm{T}}$ and $\boldsymbol{v} = (v_{\mathrm{r}}, v_{\mathrm{i}})^{\mathrm{T}}$, the linearized eigenvalue problem associated with (10.4.15) becomes the uncoupled system

$$\mathcal{J}\mathcal{L}_1\boldsymbol{u} = \lambda\boldsymbol{u}, \quad \mathcal{J}\mathcal{L}_2\boldsymbol{v} = \lambda\boldsymbol{v}; \quad \mathcal{J} = \begin{pmatrix} 0 & 1 \\ -1 & 0 \end{pmatrix}, \tag{10.4.16}$$

where the self-adjoint operators $\mathcal{L}_{1,2}$ are

$$\mathcal{L}_1 = \begin{pmatrix} -\partial_x^2 + 1 - 3\phi^2 & 0 \\ 0 & -\partial_x^2 + 1 - \phi^2 \end{pmatrix},$$
$$\mathcal{L}_2 = \begin{pmatrix} -\partial_x^2 + \rho - \phi^2 & 0 \\ 0 & -\partial_x^2 + \rho - (1-2\epsilon)\phi^2 \end{pmatrix},$$

where the nonnegative parameter

$$\rho = \frac{\omega - \kappa}{\omega + \kappa} \in (0,1),$$

has its range enforced by the assumption that $\omega > \kappa$.

The eigenvalue problem for $\mathcal{JL}_1$ is precisely that for the NLS investigated in Chapter 10.4.1; see also Exercise 5.2.10. The operator has a kernel of algebraic multiplicity 4, with the remainder of its spectrum comprised of essential spectrum on the imaginary axis. Any instability will therefore arise from the operator $\mathcal{JL}_2$. The essential (and absolute) spectrum of $\mathcal{JL}_2$ is purely imaginary,

$$\sigma_{\mathrm{ess}}(\mathcal{JL}_2) = \{\lambda \in \mathbb{C} : |\mathrm{Im}\,\lambda| \ge \rho,\ \mathrm{Re}\,\lambda = 0\},$$

and it has no kernel. For $\epsilon = 0$, the eigenvalue problem is equivalent to

$$(\mathcal{L}_+ + \mathrm{i}\lambda)(\mathcal{L}_+ - \mathrm{i}\lambda)v_{\mathrm{r}} = 0, \quad \mathcal{L}_+ := -\partial_x^2 + \rho - \phi^2. \tag{10.4.17}$$

Indeed, $\lambda_0 \in \sigma_{\mathrm{pt}}(\mathcal{L}_+)$ if and only if $\pm\mathrm{i}\lambda_0 \in \sigma_{\mathrm{pt}}(\mathcal{JL}_2)$. The Evans function calculation presented in Chapter 9.3.2 allows us to deduce that

$$\sigma_{\mathrm{pt}}(\mathcal{L}_+) = \{\rho - 1\},$$

and consequently, for $\epsilon = 0$,

$$\sigma_{\mathrm{pt}}(\mathcal{JL}_2) = \{\pm\mathrm{i}\lambda_0\}, \quad \lambda_0 = 1 - \rho.$$

We introduce the critical value

$$\rho_{\mathrm{c}} = \frac{1}{2},$$

and remark that if $\rho_{\mathrm{c}} < \rho < 1$, then the simple eigenvalues $\pm\mathrm{i}\lambda_0$ are isolated from the essential spectrum; otherwise, they are embedded in the essential spectrum.

First consider the case $\rho \in (\rho_{\mathrm{c}}, 1)$. When $\epsilon = 0$ the operator $\mathcal{JL}_2 - \mathrm{i}\lambda_0$ is Fredholm with index zero, and the perturbation theory of Chapter 6.1 for Hamiltonian systems implies that for ϵ sufficiently small, the two simple eigenvalues must remain on the imaginary axis. This implies spectral stability; however, we cannot use Theorem 5.2.11 to conclude that the equilibria is orbitally stable. Indeed, for $\epsilon = 0$, we have

$$n(\mathcal{L}_+) = 1 \quad \Rightarrow \quad n(\mathcal{L}_2) = 2,$$

while $n(\boldsymbol{D}) = 0$. From Theorem 7.1.5, and the fact that $\mathcal{L}_2$ is nonsingular, we deduce that the two imaginary eigenvalues $\{\pm i\lambda_0\}$ have a negative Krein signature.

For $\rho \in (0, \rho_c)$, the eigenvalues $\{\pm i\lambda_0\}$ are embedded in the essential spectrum when $\epsilon = 0$. We cannot directly apply the perturbation theory of Chapter 6.1 because the operator $\mathcal{J}\mathcal{L}_2 - i\lambda_0$ is not Fredholm. In Li and Promislow [**193**] the Evans function is used to track the evolution of its zeros as ϵ increases. For $\operatorname{Im}\lambda > 0$ the branch point is at $i\rho$ and the Riemann surface is defined by

$$\gamma^2 = \rho + i\lambda \quad \Rightarrow \quad \lambda = i(\rho - \gamma^2).$$

If $\rho < \rho_c$, then the embedded eigenvalue is not at the branch point, and it has been shown that for small $\epsilon > 0$, the two embedded zeros leave the

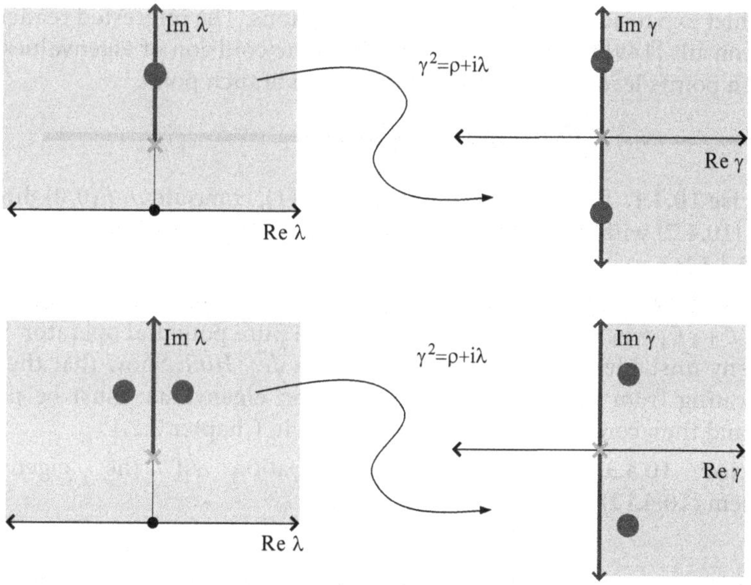

Fig. 10.4 The spectrum of $\mathcal{J}\mathcal{L}_2$ for $\epsilon = 0$ (*top two panels*) and $\epsilon > 0$ (*bottom two panels*). The *left panels* depict the upper-half of the λ-plane while the *right panels* show the corresponding region in the γ-plane, under the Riemann map $\gamma^2 = \rho + i\lambda$, with the physical sheet of the Riemann surface depicted as a shaded (*gray*) region. For $0 < \rho < \rho_c$ and $\epsilon = 0$ there is an eigenvalue, (*red*) circle, embedded in the essential spectrum, (*blue*) line. For $\epsilon > 0$ the eigenvalue splits into two eigenvalues that exit the essential spectrum. In the *right panel* there are two zeros of the Evans function, (*red*) circle, which move onto the physical sheet for $\epsilon > 0$. The branch point is denoted by a (*green*) cross. (Color figure online.)

essential spectrum and produce a complex quartet of eigenvalues, so that the equilibria is spectrally unstable (see Fig. 10.4). However, the calculation is nongeneric since the eigenvalues travel along the imaginary axis at first order in ϵ, and only leave the essential spectrum at second order. Indeed, the second *nonzero* derivative of the Evans function is evaluated in terms of explicit integrals, yielding an expansion of the form

$$E(\gamma,\epsilon) \sim A(\gamma-\gamma_0) + \mathrm{i}B\epsilon - C\epsilon^2 \quad \gamma_0^2 = \rho - \lambda_0 < 0,$$

where the real constants A, B, C are all positive. Finally, in the case $\rho = \rho_c$, the eigenvalue resides at the branch point when $\epsilon = 0$; moreover, the Evans function has a third-order zero at the branch point. A detailed calculation shows that for $\epsilon > 0$ one of the zeros moves onto the physical sheet of the Riemann surface, becoming a purely imaginary eigenvalue, while the other two move onto the resonance sheet.

In summary, the equilibria is spectrally stable for small $\epsilon > 0$ if $\rho_c \le \rho < 1$, and spectrally unstable otherwise. The discussion in [**193**, Section 5] presents a richly structured bifurcation diagram for larger values of ϵ obtained via a numerical evaluation of the Evans function based upon a Dirichlet expansion to compute the Jost solutions. The interested reader can also consult [**149**] for an example in which the collision of eigenvalues and branch points leads to higher-order zeros at a branch point.

Exercises

Exercise 10.4.1. For the operator $\mathcal{L}$ of (10.4.11), compute $\partial_\gamma E(0,0)$ directly from (10.4.2) without using the explicit form of the Evans function derived in (10.4.14).

Exercise 10.4.2. For the operator $\mathcal{L}$ of (10.4.11), consider the perturbation $\mathcal{L}_\epsilon := \mathcal{L} + \epsilon\mathcal{L}_1$ where $\boldsymbol{\sigma}_1\mathcal{L}_1\boldsymbol{\sigma}_1 = \mathcal{L}_1$, and $\mathcal{L}_1$ is a pure potential operator. Show that any unstable eigenvalues must be of $\mathcal{O}(\sqrt{\epsilon})$. *Hint*: Show that the zero bifurcating from the branch point, if it is an eigenvalue, must be purely real, and then consider the results presented in Chapter 7.2.1.

Exercise 10.4.3. Consider the perturbation of the eigenvalue problem (10.4.12),

$$\boldsymbol{a}_0(x) \mapsto \boldsymbol{a}_0(x) + \epsilon\,\mathrm{sech}^4(x)\begin{pmatrix} a & b \\ -b & -a \end{pmatrix},$$

where $a, b \in \mathbb{R}$.

(a) Verify that the perturbed problem has the eigenvalue-quartet symmetry.
(b) Compute $\partial_\epsilon E(0,0)$, and give a condition on the parameters a, b that ensures that the perturbed zero is an eigenvalue.

(c) Explain why the perturbed eigenvalue must be purely real to all orders in ϵ.

Exercise 10.4.4. Consider the perturbation of the eigenvalue problem (10.4.12),

$$\boldsymbol{d} \mapsto \boldsymbol{d} + \mathrm{i}\epsilon d\boldsymbol{I}_2, \quad \boldsymbol{a}_0(x) \mapsto \boldsymbol{a}_0(x) + \mathrm{i}\epsilon^2 b\boldsymbol{I}_2,$$

where $d < 0$ and $b > 0$.

(a) Verify that the perturbed problem no longer satisfies the reflection symmetry with respect to the real axis.
(b) Compute $\partial_\epsilon E(0,0)$, and give a condition on the parameters b, d which ensure that the perturbed zero is an eigenvalue.
(c) Give a condition on the parameters which ensures that the perturbed eigenvalue satisfies $\operatorname{Im}\lambda \geq 0$, so that it does not contribute to a spectral instability.

Exercise 10.4.5. Reconsider the spectral problem discussed in Chapter 10.4.2. It is possible to analytically determine part of the bifurcation diagram for $\mathcal{L}_2$ for $\epsilon > 0$ not necessarily small. Consider the suboperator of $\mathcal{L}_2$,

$$\mathcal{L}_- := -\partial_x^2 + \rho - (1-2\epsilon)\phi^2$$

[compare with $\mathcal{L}_+$ defined in (10.4.17)].

(a) If $0 \leq \epsilon < 1/2$, show that

$$\sigma_{\mathrm{pt}}(-\partial_x^2 - (1-2\epsilon)\phi^2) \cap \mathbb{R}^- = \{-\gamma_\epsilon^2\}, \quad \gamma_\epsilon = \frac{1}{2}\left(-1 + \sqrt{9-16\epsilon}\right),$$

while if $\epsilon > 1/2$, then

$$\sigma_{\mathrm{pt}}(-\partial_x^2 - (1-2\epsilon)\phi^2) \cap \mathbb{R}^- = \emptyset.$$

(b) For $0 \leq \epsilon < 1/2$ find a function $\rho = \rho(\epsilon)$ for $\epsilon > 0$ such that

$$\mathrm{n}(\mathcal{L}_-) = \begin{cases} 1, & \rho < \rho(\epsilon) \\ 0, & \rho > \rho(\epsilon). \end{cases}$$

(c) Show that if $\rho > \rho(\epsilon)$ for $0 \leq \epsilon < 1/2$, or if $\epsilon > 1/2$, then $\mathcal{J}\mathcal{L}_2$ has precisely one positive real eigenvalue, and that there are no other eigenvalues with a positive real part.

10.5 Eigenvalue Problems on Large Intervals: Separated Boundary Conditions

Extending the ideas of Chapter 9.6, we consider exponentially asymptotic nth-order operators $\mathcal{L}$, of the form (10.0.1) and compare the spectrum they generate when acting on $L^2(\mathbb{R})$ to the spectrum associated with the boundary-value problem with separated boundary conditions on large intervals, i.e., $-L \le x \le L$, where $L \gg 1$. We present the main results and provide a sketch of their proofs. A more detailed presentation can be found in [252, 254].

Consider the exponentially asymptotic nth-order operator $\mathcal{L}$ introduced in (10.0.1). We compare the spectrum of $\mathcal{L} : H^n(\mathbb{R}) \subset L^2(\mathbb{R}) \mapsto L^2(\mathbb{R})$ with that of the same operator acting on $H^n[-L,L] \subset L^2[-L,L]$ subject to separated boundary conditions of the form

$$\boldsymbol{B}_{-}\begin{pmatrix} p \\ \partial_x p \\ \vdots \\ \partial_x^{n-1} p \end{pmatrix}(-L) = \boldsymbol{0}, \quad \boldsymbol{B}_{+}\begin{pmatrix} p \\ \partial_x p \\ \vdots \\ \partial_x^{n-1} p \end{pmatrix}(+L) = \boldsymbol{0}, \tag{10.5.1}$$

where $\boldsymbol{B}_{-L} \in \mathbb{R}^{(n-k)\times n}$ and $\boldsymbol{B}_{+L} \in \mathbb{R}^{k\times n}$; see (8.2.1) and (8.2.2). The Jost matrices for the unbounded domain problem are denoted $\boldsymbol{J}_{\infty}^{\pm}(x,\lambda)$, with $\boldsymbol{J}_{\infty}^{-} \in \mathbb{C}^{n\times(n-k)}$ and $\boldsymbol{J}_{\infty}^{+} \in \mathbb{C}^{n\times k}$. For the bounded domain problem we denote

$$\mathbb{B}_{-L} := \ker(\boldsymbol{B}_{-}) = \operatorname{span}\{\boldsymbol{b}_1^-,\dots,\boldsymbol{b}_{n-k}^-\}, \quad \mathbb{B}_{+L} \ker(\boldsymbol{B}_{+}) = \operatorname{span}\{\boldsymbol{b}_1^+,\dots,\boldsymbol{b}_k^+\},$$

and define Jost solutions for $1 \le i \le (n-k)$ and $1 \le j \le k$ by

$$\boldsymbol{J}_{i,L}^-(x,\lambda) = \boldsymbol{\Phi}(x,\lambda)\boldsymbol{\Phi}(-L,\lambda)^{-1}\boldsymbol{b}_i^-, \quad \boldsymbol{J}_{j,L}^+(x,\lambda) = \boldsymbol{\Phi}(x,\lambda)\boldsymbol{\Phi}(+L,\lambda)^{-1}\boldsymbol{b}_j^+.$$

Here $\boldsymbol{\Phi}(x,\lambda)$ is a fundamental matrix solution to the vectorized version of the $\mathcal{L}$ eigenvalue problem. The Jost matrices for the bounded domain problem are formed by the concatenation of these Jost solutions

$$\boldsymbol{J}_L^-(x,\lambda) = \left(\boldsymbol{J}_{1,L}^-,\dots,\boldsymbol{J}_{k,L}^-\right)(x,\lambda), \quad \boldsymbol{J}_L^+(x,\lambda) = \left(\boldsymbol{J}_{1,L}^+,\dots,\boldsymbol{J}_{n-k,L}^+\right)(x,\lambda).$$

From these Jost matrices we construct the two Evans functions,

$$E_L(\lambda) = \det\left(\boldsymbol{J}_L^-,\boldsymbol{J}_L^+\right)(0,\lambda) = \mathrm{e}^{-\int_0^x \operatorname{tr}\boldsymbol{A}(s,\lambda)\,\mathrm{d}s} \det\left(\boldsymbol{J}_L^-,\boldsymbol{J}_L^+\right)(x,\lambda),$$
$$E_\infty(\lambda) = \det\left(\boldsymbol{J}_\infty^-,\boldsymbol{J}_\infty^+\right)(0,\lambda) = \mathrm{e}^{-\int_0^x \operatorname{tr}\boldsymbol{A}(s,\lambda)\,\mathrm{d}s} \det\left(\boldsymbol{J}_\infty^-,\boldsymbol{J}_\infty^+\right)(x,\lambda).$$

The bounded domain Evans function $E_L(\lambda)$ is entire, while the whole-line Evans function $E_\infty(\lambda)$ is analytic except at the absolute spectrum.

We recall the matrices $V^{\mathrm{s}}(\lambda)$ and $V^{\mathrm{u}}(\lambda)$ introduced in (10.1.8) which have columns that span the stable and unstable subspaces of $A_0(\lambda)$ for $\lambda \in \Omega$, and their analytic extensions for $\lambda \in \Omega_{\mathrm{e}}$. The Jost matrices for the unbounded domain problem satisfy the asymptotic relations

$$\lim_{x\to-\infty} \mathrm{e}^{-A_0(\lambda)x} J_\infty^-(x,\lambda) = V^{\mathrm{u}}(\lambda), \quad \lim_{x\to+\infty} \mathrm{e}^{-A_0(\lambda)x} J_\infty^+(x,\lambda) = V^{\mathrm{s}}(\lambda).$$

Generalizing the arguments leading to Lemma 9.3.1 and the subsequent definition of the transmission and reflection coefficients, the Jost matrices satisfy the "opposite-end" asymptotic relations

$$\lim_{x\to+\infty} \mathrm{e}^{-A_0(\lambda)x} J_\infty^-(x,\lambda) = V^{\mathrm{u}}(\lambda)t(\lambda) + V^{\mathrm{s}}(\lambda)r(\lambda)$$
$$\lim_{x\to-\infty} \mathrm{e}^{-A_0(\lambda)x} J_\infty^+(x,\lambda) = V^{\mathrm{s}}(\lambda)r(\lambda) + V^{\mathrm{u}}(\lambda)t(\lambda).$$

Here $t(\lambda) \in \mathbb{C}^{(n-k)\times(n-k)}$ and $r(\lambda) \in \mathbb{C}^{k\times k}$ are called the *transmission matrix* and the *reflection matrix*, respectively. As in (9.6.9), we find that the Evans function for the unbounded domain problem is related to the transmission matrix

$$E_\infty(\lambda) = \lim_{x\to+\infty} \mathrm{e}^{-\int_0^x \operatorname{tr} A(s,\lambda)\,\mathrm{d}s} \det\left(J_\infty^-, J_\infty^+\right)(x,\lambda) = \det t(\lambda) \det\left(V^{\mathrm{u}}, V^{\mathrm{s}}\right)(\lambda).$$

Moreover, away from the absolute spectrum, $\det\left(V^{\mathrm{u}}, V^{\mathrm{s}}\right)(\lambda) \neq 0$, and hence $E_\infty(\lambda)$ and $\det t(\lambda)$, share the same zeros up to multiplicity.

To evaluate the Evans function for the bounded domain problem, we assume that the left boundary space coincides with the unstable manifold of $A_0(\lambda)$

$$\dim[\mathbb{B}_{-L} \cap \mathbb{E}^{\mathrm{u}}(\lambda)] = n-k \quad \Leftrightarrow \quad \det\left(V^{\mathrm{s}}(\lambda), b_1^-, \ldots, b_{n-k}^-\right) \neq 0.$$

With this assumption the Jost matrix $J_L^-(x,\lambda)$ satisfies

$$J_L^-(+L,\lambda) \sim V^{\mathrm{u}}(\lambda)c_-(\lambda)t(\lambda) + \mathcal{O}(\mathrm{e}^{-\delta L})V^{\mathrm{s}}(\lambda),$$

where $c_-(\lambda) \in \mathbb{C}^{(n-k)\times(n-k)}$ is nonsingular [compare with (9.6.8)]. This yields

$$\begin{aligned} E_L(\lambda) &= \mathrm{e}^{-\int_0^L \operatorname{tr} A(s,\lambda)\,\mathrm{d}s} \det\left(J_L^-, J_L^+\right)(L,\lambda) \\ &\propto \det c_-(\lambda) \det t(\lambda) \det\left(b_1^+, \ldots, b_k^+, V^{\mathrm{u}}(\lambda)\right) + \mathcal{O}(\mathrm{e}^{-\delta L}) \\ &\propto E_\infty(\lambda) \det\left(b_1^+, \ldots, b_k^+, V^{\mathrm{u}}(\lambda)\right) + \mathcal{O}(\mathrm{e}^{-\delta L}). \end{aligned}$$

Under the additional assumption

$$\dim[\mathbb{B}_{+L} \cap \mathbb{E}^{\mathrm{s}}(\lambda)] = k, \quad \Leftrightarrow \quad \det\left(b_1^+, \ldots, b_k^+, V^{\mathrm{u}}(\lambda)\right) \neq 0,$$

we conclude that, up to an exponentially small error, $E_L(\lambda) \propto E_\infty(\lambda)$. These arguments lead us to the following generalization of Theorem 9.6.1.

Lemma 10.5.1. *Let $\mathcal{L}$ be an exponentially localized nth-order operator of the form (10.0.1). Let $\mathrm{E}_\mathrm{L}(\lambda)$ be the Evans function for the bounded domain problem subject to separated boundary conditions (10.5.1), and let $\mathrm{E}_\infty(\lambda)$ denote the Evans function for the unbounded domain problem. Let the stable and unstable matrices $\mathbf{V}^{\mathrm{s,u}}$ associated with the stable and unstable subspaces $\mathbb{E}^{\mathrm{s,u}}$ of the asymptotic system be defined as in (10.1.8). Assume the two boundary conditions are compatible in the sense that*

$$\det\left(\mathbf{V}^{\mathrm{s}}(\lambda), \boldsymbol{b}_1^-, \ldots, \boldsymbol{b}_{n-k}^-\right) \neq 0, \quad \det\left(\boldsymbol{b}_1^+, \ldots, \boldsymbol{b}_k^+, \mathbf{V}^{\mathrm{u}}(\lambda)\right) \neq 0, \tag{10.5.2}$$

for all $\lambda \in \Omega_0 \subset \Omega_\mathrm{e} \backslash \sigma_{\mathrm{abs}}(\mathcal{L})$. Then there exists $\delta > 0$ such that $\lambda_0 \in \Omega_0$ is a zero of order m for the Evans function $\mathrm{E}_\infty(\lambda)$ if and only if the Evans function $\mathrm{E}_\mathrm{L}(\lambda)$ has m zeros (counting multiplicity) within $\mathcal{O}(\mathrm{e}^{-\delta L})$ of λ_0, for L sufficiently large.

Remark 10.5.2. As in Theorem 9.6.1, if $\Omega_0 = \Omega_\mathrm{e} \backslash \sigma_{\mathrm{abs}}(\mathcal{L})$, we may conclude that the countably infinite remainder of zeros of $E_L(\lambda)$ that do not converge to the zeros of $E_\infty(\lambda)$ must approach the absolute spectrum of $\mathcal{L}$ as $L \to \infty$.

Remark 10.5.3. If the conditions of (10.5.2) do not hold, then it is possible for there to be zeros of $E_L(\lambda)$ that are not close to either the absolute spectrum or any zeros of $E_\infty(\lambda)$.

Exercises

Exercise 10.5.1. The unbounded domain eigenvalue problem associated with the linearization of the gKdV equation, (10.3.4), with speed $c = 1$, about its equilibria ϕ_1, takes the form

$$\partial_x\left(-\partial_x^2 + 1 - \phi(x)\right)p = \lambda p, \quad \phi(x) = \frac{1}{2}(r+1)(r+2)\operatorname{sech}^2(rx/2).$$

If $1 \leq r < 4$, then from the orbital stability result of Chapter 5.2.1 we know the system has no eigenvalues with positive real part. On the other hand, if $r > 4$, we know from the orientation index (see Chapter 10.3.1), that there is at least one positive real eigenvalue. Consider the bounded domain problem with $\dim\ker(\boldsymbol{B}_-) = 1$ and $\dim\ker(\boldsymbol{B}_+) = 2$. Compute the spectrum numerically for $r = 3$ and $r = 5$ using a finite difference approximation for the differential operators. Find choices of the matrix $\boldsymbol{B}_-$ that generate spurious eigenvalues for $r = 3$ and that suppress the known positive eigenvalue for $r = 5$.

10.6 Eigenvalue Problems: Periodic Coefficients with a Large Spatial Period

Periodic boundary conditions are the canonical example of nonseparated boundary conditions. We follow [252, 254], generalizing the results of Chapter 9.7, relating the spectrum of an unbounded domain problem to that of a large bounded domain subject to periodic boundary conditions. Extending (9.7.1), the unbounded domain eigenvalue problem takes the form

$$\mathcal{L}p := \partial_x^n p + a_{n-1,0}(x)\partial_x^{n-1} p + \cdots + a_{1,0}(x)\partial_x p + a_{0,0}(x)p = \lambda p, \tag{10.6.1}$$

where the coefficients satisfy

$$\sum_{j=0}^{n-1} |a_{j,0}(x) - a_{j,0}^0| \le C\mathrm{e}^{-\alpha|x|}. \tag{10.6.2}$$

After a Bloch-wave decomposition the bounded domain eigenvalue problem is given by

$$\begin{aligned}\mathcal{L}_{\mu,\eta} q :=& (\partial_x + \mathrm{i}\mu)^n q + a_{n-1,\eta}(x)(\partial_x + \mathrm{i}\mu)^{n-1} q \\ &+ \cdots + a_{1,\eta}(x)(\partial_x + \mathrm{i}\mu)q + a_{0,\eta}(x)q = \lambda q,\end{aligned} \tag{10.6.3}$$

subject to periodic boundary conditions, where the parameter μ satisfies

$$-1/L_\eta < \mu \le 1/L_\eta.$$

The coefficients are all $2L_\eta$-periodic, where $L_\eta \to +\infty$ as $\eta \to 0^+$, and uniformly approximate the unbounded domain problem coefficients over one spatial period, that is,

$$|a_{j,\eta}(x) - a_{j,0}(x)| = \mathcal{O}(\eta\mathrm{e}^{-\alpha|x|}), \quad -L_\eta \le x \le L_\eta, \tag{10.6.4}$$

for $j = 0, \dots, n-1$.

As in Chapter 10.5 the Jost matrices for the unbounded domain satisfy the asymptotic relations

$$\lim_{x\to-\infty} \mathrm{e}^{-A_0(\lambda)x} J_\infty^-(x,\lambda) = \boldsymbol{V}^{\mathrm{u}}(\lambda), \quad \lim_{x\to+\infty} \mathrm{e}^{-A_0(\lambda)x} J_\infty^+(x,\lambda) = \boldsymbol{V}^{\mathrm{s}}(\lambda),$$

and the Evans function is given by $E_\infty(\lambda) = \det\left(J^-, J^+\right)(0,\lambda)$. For the Bloch-wave problem the Evans function is

$$E_L(\lambda,\mu) = \det \boldsymbol{E}_L(\lambda,\mu),$$

where the Evans matrix is defined by

$$\boldsymbol{E}_L(\lambda,\mu) = \boldsymbol{\Phi}_L(0,\lambda,\mu)\boldsymbol{\Phi}_L(-L_\eta,\lambda,\mu)^{-1} - \boldsymbol{\Phi}_L(0,\lambda,\mu)\boldsymbol{\Phi}_L(L_\eta,\lambda,\mu)^{-1},$$

[see (9.7.7)]. Here $\boldsymbol{\Phi}_L(x,\lambda,\mu)$ is a fundamental matrix solution to the vectorized Bloch-wave problem. As in Chapter 9.7, we compare the bounded domain Evans matrix with the unbounded domain Evans matrix defined by

$$\boldsymbol{E}_\infty(\lambda) := \boldsymbol{\Phi}_\infty(0,\lambda)\boldsymbol{\Phi}_\infty(-L_\eta,\lambda)^{-1} - \boldsymbol{\Phi}_\infty(0,\lambda)\boldsymbol{\Phi}_\infty(L_\eta,\lambda)^{-1},$$

where $\boldsymbol{\Phi}_\infty(x,\lambda)$ is a fundamental matrix generated by the vectorized version of the unbounded domain problem. From the assumptions on the coefficients $a_{j,\eta}$, it follows that $\boldsymbol{E}_\infty(\lambda)$ is $\mathcal{O}(\eta, L_\eta^{-1})$-close to the matrix $\boldsymbol{E}_L(\lambda,\mu)$ as $\eta \to 0$.

As in Chapter 9.7, we show that $E_\infty(\lambda)$ will be exponentially close to $\det \boldsymbol{E}_\infty(\lambda)$ as $L_\eta \to +\infty\,(\eta \to 0^+)$. This implies that on the natural domain Ω the zeros of the unbounded domain problem Evans function are close to those of the Bloch-wave problem Evans function. As a consequence, the eigenvalues for the unbounded domain problem are well-approximated by the Bloch-wave problem for each admissible value of μ.

Similar to the estimates leading to (9.7.9), for $\lambda \in \Omega_{\mathrm{e}} \backslash \sigma_{\mathrm{abs}}(\mathcal{L})$ we have

$$\begin{aligned}
\boldsymbol{\Phi}_\infty(0,\lambda)\boldsymbol{\Phi}_\infty(-L_\eta,\lambda)^{-1}\boldsymbol{J}_\infty^-(-L_\eta,\lambda) &= \boldsymbol{J}_\infty^-(0,\lambda)\\
\boldsymbol{\Phi}_\infty(0,\lambda)\boldsymbol{\Phi}_\infty(-L_\eta,\lambda)^{-1}\boldsymbol{J}_\infty^+(L_\eta,\lambda) &= \mathcal{O}(\mathrm{e}^{-\delta L_\eta}),
\end{aligned}$$

and

$$\begin{aligned}
\boldsymbol{\Phi}_\infty(0,\lambda)\boldsymbol{\Phi}_\infty(L_\eta,\lambda)^{-1}\boldsymbol{J}_\infty^-(-L_\eta,\lambda) &= \mathcal{O}(\mathrm{e}^{-\delta L_\eta})\\
\boldsymbol{\Phi}_\infty(0,\lambda)\boldsymbol{\Phi}_\infty(L_\eta,\lambda)^{-1}\boldsymbol{J}_\infty^+(L_\eta,\lambda) &= \boldsymbol{J}_\infty^+(0,\lambda).
\end{aligned}$$

Here $\delta > 0$ depends upon the stable–unstable gap, $\eta_{\mathrm{u}} - \eta_{\mathrm{s}} > 0$, which is strictly positive for $\lambda \in \Omega$. Thus, as in (9.7.9) we have

$$\boldsymbol{E}_\infty(\lambda)\big(\boldsymbol{J}_\infty^-(-L_\eta,\lambda), -\boldsymbol{J}_\infty^+(L_\eta,\lambda)\big)(\lambda) = \big(\boldsymbol{J}^-,\boldsymbol{J}^+\big)(0,\lambda) + \mathcal{O}(\mathrm{e}^{-\delta L_\eta}).$$

Since the Jost matrices are well-approximated by the asymptotic subspaces $V^{\mathrm{s,u}}(\lambda)$, and the sum of these subspaces span $\mathbb{C}^n$ for $\lambda \in \Omega$, we obtain the following result.

Lemma 10.6.1. *Consider the eigenvalue problem for the Bloch-wave operator $\mathcal{L}_{\mu,\eta}$ introduced in (10.6.3) subject to the conditions (10.6.4). Suppose that for the unbounded domain problem, (10.6.1), there is a $\lambda = \lambda_0 \in \Omega$, which is an isolated eigenvalue with multiplicity m. Let $C \subset \mathbb{C}$ be a positively oriented simple closed curve of radius at least $\mathcal{O}(\eta)$, which contains λ_0 but no other spectra of the unbounded domain problem in its interior. Then, for η sufficiently small, and for each $-1/L_\eta < \mu \le 1/L_\eta$, the associated Bloch-wave problem has precisely m eigenvalues in the interior of C.*

Remark 10.6.2. As in Chapter 9.7, the remainder of the spectrum of the Bloch-wave problem that does not converge to the point spectrum of the unbounded domain problem will accumulate on the essential spectrum of the unbounded domain problem.

Example 10.6.3. Consider the NLS equation

$$\mathrm{i}\partial_t u + \partial_x^2 u - u + |u|^2 u = 0.$$

In Chapter 10.4.1, an explicit construction of the Evans function revealed that the operator

$$\mathcal{L} = \left(\partial_x^2 - 1\right)\sigma_3 + U^2(x)(2\sigma_3 + \mathrm{i}\sigma_2),$$

obtained by linearizing about the equilibria $U(x) = \sqrt{2}\,\mathrm{sech}(x)$, has the spectrum

$$\sigma_{\mathrm{pt}}(\mathcal{L}) = \{0\}, \quad \sigma_{\mathrm{ess}}(\mathcal{L}) = \{\lambda \in \mathbb{C} : |\mathrm{Re}\,\lambda| \geq 1,\ \mathrm{Im}\,\lambda = 0\}.$$

Furthermore, from the Hamiltonian structure of the NLS, the rotational and translational symmetries imply that $\mathrm{m_g}(0) = 2$, $\mathrm{m_a}(0) = 4$.

The NLS equation also supports a family of spatially periodic equilibria given by

$$U_k(x) = \sqrt{2(2-k^2)}\,\mathrm{dn}(x/\sqrt{2-k^2}, k), \quad 0 \leq k < 1,$$

where $\mathrm{dn}(z,k)$ is the Jacobi elliptic function that approaches a sum of periodic translations of $\mathrm{sech}(z)$ as $k \to 1^-$. Linearizing the NLS equation about U_k leads to the operator

$$\mathcal{L}_k = \left(\partial_x^2 - 1\right)\sigma_3 + U_k^2(x)(2\sigma_3 + \mathrm{i}\sigma_2).$$

The equilibria $U_k(x)$ has period $2L_k$, with $L_k = \sqrt{2-k^2}\,K(k)$, where $K(k)$ is the elliptic integral of the first kind with $K(k) \to +\infty$ as $k \to 1^-$. Moreover as $k \to 1^-$, $U_k(x)$ uniformly approximates $U(x)$ on the interval $-L_k \leq x \leq L_k$. Thus, we may invoke Lemma 10.6.1 to conclude that for $0 < 1 - k \ll 1$ the Bloch-decomposition of the operator $\mathcal{L}_k$ will have four eigenvalues near the origin for each value of μ. The remainder of the spectrum will accumulate on the essential spectrum for the unbounded domain problem.

Remark 10.6.4. A formal perturbative calculation for the location of these eigenvalues near the origin was presented in Arnold [**19**]. The spectrum of $\mathcal{L}_k$ for any value of k was determined by Ivey and Lafortune [**129**] via an explicit formulation of the Evans function.

10.7 Additional Reading

The Evans function has been extensively applied to the analysis of spectral stability for waves in systems of conservation laws; see, e.g., Benzoni-Gavage et al. [**31**], Humpherys and Zumbrun [**119**], Plaza and Zumbrun [**235**] and Zumbrun [**294**].

In systems for which an analysis of the Evans function is too complicated, numerical investigations, typically through a winding number calculation, have been employed. See Humpherys and Zumbrun [**120**, **121**], Humpherys et al. [**122**], and Zumbrun [**291**] for examples arising from the linearization about equilibria of systems conservation laws. The numerical evaluation of the Evans function has been performed in other contexts; see Aparicio et al. [**16**], Ledoux et al. [**190**, **191**], and Malham and Niesen [**201**] for examples.

The Evans function can be constructed using a functional analytic formulation via Fredholm determinants. To the best of our knowledge, the first connections between the dynamical systems approach and the functional analytic formulation appeared in Kapitula and Sandstede [**154**]. Refinements and substantial extensions of this connection are presented in Gesztesy et al. [**101**], Latushkin and Sukhtavey [**183**], Latushkin and Sukhtayev [**184**], Latushkin and Tomilov [**185**], and Latushkin et al. [**187**]. In particular, this approach leads to a generalization of Theorem 10.2.2; see Gesztesy et al. [**100**].

The Evans function has been constructed for nonlocal eigenvalue problems; see Kapitula et al. [**158**], Rubin [**246**], Zhang [**287**, **288**, **289**] for examples. The Evans function has also been constructed to analyze transverse stability of wave structures arising in more than one space dimension; see Deconinck et al. [**69**], Oh and Sandstede [**212**], Oh and Zumbrun [**213**], and Terman [**271**].

The spectral stability of equilibria arising in singularly perturbed systems, which have structure arising on widely separated spatial scales, has not been discussed in this book. The Evans function has been extensively used for problems of this type; relevant references include, Alexander and Jones [**9**, **10**], Doelman et al. [**71**, **72**, **73**], and Kapitula [**142**]. In the construction of multipulse equilibria within singularly perturbed problems it is typical to find eigenvalues that are exponentially small functions of the pulse spacing. These can be tracked using Lin's method (also known as the homoclinic Lin–Sandstede method, or the homoclinic Lyapunov–Schmidt method): For examples, see Alexander et al. [**12**], Sandstede [**248**], Sandstede and Scheel [**253**, **256**], Sandstede et al. [**258**], Yew [**285**], and Yew et al. [**286**].

References

[1] In P. Kevrekidis, D. Frantzeskakis, and R. Carretero-González, editors, *Emergent Nonlinear Phenomena in Bose–Einstein Condensates*, volume 45 of *Springer Series in Atomic, Molecular and Optical Physics*. Springer-Verlag, New York, 2008.

[2] F. Abdullaev. *Theory of Solitons in Inhomogeneous Media*. John Wiley, New York, 1994.

[3] M. Ablowitz and P. Clarkson. *Solitons, Nonlinear Evolution Equations, and Inverse Scattering*, volume 149 of *London Math. Soc. Lecture Note Series*. Cambridge University Press, Cambridge, 1991.

[4] M. Ablowitz, D. Kaup, A. Newell, and H. Segur. The inverse scattering transform: Fourier analysis for nonlinear problems. *Stud. Appl. Math.*, 53(4):249–315, 1974.

[5] M. Ablowitz, B. Prinari, and A. Trubatch. *Discrete and Continuous Nonlinear Schrödinger Systems*, volume 302 of *London Math. Soc. Lecture Note Series*. Cambridge University Press, Cambridge, 2004.

[6] R. Adams and J. Fournier. *Sobolev Spaces*. Academic Press, New York, second edition, 2003.

[7] J. Albert, J. Bona, and D. Henry. Sufficient conditions for stability of solitary-wave solutions of model equations for long waves. *Phys. D*, 24(13):343–366, 1987.

[8] S. Albeverio and R. Høegh-Krohn. Perturbation of resonances in quantam mechanics. *J. Math. Anal. Appl.*, 101:491–513, 1984.

[9] J. Alexander and C.K.R.T. Jones. Existence and stability of asymptotically oscillatory triple pulses. *Z. Angew. Math. Phys.*, 44:189–200, 1993.

[10] J. Alexander and C.K.R.T. Jones. Existence and stability of asymptotically oscillatory double pulses. *J. Reine Angew. Math.*, 446:49–79, 1994.

[11] J. Alexander, R. Gardner, and C.K.R.T. Jones. A topological invariant arising in the stability of travelling waves. *J. Reine Angew. Math.*, 410:167–212, 1990.

[12] J. Alexander, M. Grillakis, C.K.R.T. Jones, and B. Sandstede. Stability of pulses on optical fibers with phase-sensitive amplifiers. *Z. Angew. Math. Phys.*, 48(2): 175–192, 1997.

[13] C. De Angelis. Self-trapped propagation in the nonlinear cubic-quintic Schrödinger equation: a variational approach. *IEEE J. Quantum Elect.*, 30(3): 818–821, 1994.

[14] J. Angulo. Nonlinear stability of periodic travelling wave solutions to the Schrödinger and the modified Korteweg–de Vries equations. *J. Diff. Eq.*, 235:1–30, 2007.

[15] J. Angulo and J. Quintero. Existence and orbital stability of cnoidal waves for a 1D Boussinesq equation. *Int. J. Math. Math. Sci.*, 2007:52020, 2007.

T. Kapitula and K. Promislow, *Spectral and Dynamical Stability of Nonlinear Waves*, Applied Mathematical Sciences 185, DOI 10.1007/978-1-4614-6995-7,

[16] N. Aparicio, S. Malham, and M. Oliver. Numerical evaluation of the Evans function by Magnus integration. *BIT*, 45:219–258, 2005.

[17] I. Aranson and L. Kramer. The world of the complex Ginzburg–Landau equation. *Rev. Mod. Phys.*, 74(1):99–143, 2002.

[18] W. Arendt, C. Batty, M. Hieber, and F. Neubrander. *Vector-Valued Laplace Transforms and Cauchy Problems*. Springer-Verlag, New York, 2011.

[19] J. Arnold. Stability theory for periodic pulse train solutions of the nonlinear Schrödinger equation. *IMA J. Appl. Math.*, 52:123–140, 1994.

[20] V. Arnold. *Mathematical Methods of Classical Mechanics*. Springer-Verlag, New York, second edition, 1989.

[21] D. Bambusi. Long-time stability of some small-amplitude solutions in nonlinear Schrödinger equations. *Comm. Math. Phys.*, 189(1):205–226, 1997.

[22] D. Bambusi. On long-time stability in Hamiltonian perturbations of nonresonant linear PDEs. *Nonlinearity*, 12(4):823–850, 1999.

[23] D. Bambusi, A. Carati, and A. Ponno. The nonlinear Schrödinger equation as a resonant normal form. *Disc. Cont. Dyn. Sys.*, 2(1):109–128, 2002.

[24] B. Barker, P. Noble, L. Rodrigues, and K. Zumbrun. Stability of periodic Kuramoto–Sivashinsky waves. *Appl. Math. Lett.*, 25(5):824–829, 2012.

[25] P. Bates and C.K.R.T. Jones. Invariant manifolds for semilinear partial differential equations. *Dynamics Reported*, 2:1–38, 1989.

[26] R. Beals and R. Coifman. Inverse scattering and evolution equations. *Comm. Pure Appl. Math.*, 38:29–42, 1985.

[27] R. Beals, P. Deift, and C. Tomei. *Direct and Inverse Scattering on the Line*, volume 28 of *Math. Sur. Mon.* American Mathematical Society, Providence, RI, 1988.

[28] M. Beck, H. Hupkes, B. Sandstede, and K. Zumbrun. Nonlinear stability of semidiscrete shocks for two-sided schemes. *SIAM J. Math. Anal.*, 42:857–903, 2010.

[29] T. Bellsky, A. Doelman, T. Kaper, and K. Promislow. Adiabatic stability under semi-strong interactions: the weakly damped regime. to appear in *Indiana U. Math. J.*, 2013. (in press).

[30] A. Ben-Artzi and I. Gohberg. Dichotomy of systems and invertibility of linear ordinary differential operators. *Operator Theory Adv. Appl.*, 56:91–119, 1992.

[31] S. Benzoni-Gavage, D. Serre, and K. Zumbrun. Alternate Evans functions and viscous shock waves. *SIAM J. Math. Anal.*, 32(5):929–962, 2001.

[32] W.-J. Beyn and J. Lorenz. Stability of traveling waves: dichotomies and eigenvalue conditions on finite intervals. *Num. Funct. Anal. Opt.*, 20:201–244, 1999.

[33] W.-J. Beyn and J. Rottmann-Matthes. Resolvent estimates for boundry-value problems on large intervals via the theory of discrete approximations. *Num. Funct. Anal. Opt.*, 28:603–629, 2007.

[34] W.-J. Beyn, Y. Latushkin, and J. Rottmann-Matthes. Finding eigenvalues of holomorphic Fredholm operator pencils using boundary value problems and contour integrals. arXiv:1210.3952, 2013.

[35] J. Bona and R. Sachs. Global existence of smooth solutions and stability of solitary waves for a generalized Boussinesq equation. *Comm. Math. Phys.*, 118:15–29, 1988.

[36] J. Bona, P. Souganidis, and W. Strauss. Stability and instability of solitary waves of Korteweg–de Vries type. *Proc. R. Soc. London A*, 411:395–412, 1987.

[37] J. Bona, K. Promislow, and C. Wayne. On the asymptotic behavior of solutions to nonlinear, dispersive, dissipative wave equations. *Math. Comp. Sim.*, 37:265–277, 1994.

[38] A. Bose and C.K.R.T. Jones. Stability of the in-phase travelling wave solution in a pair of coupled nerve fibres. *Indiana U. Math. J.*, 44(1):189–220, 1995.

[39] J. Boussinesq. Théorie des ondes et des remous qui se propagent le long d'un canal rectangulaire horizontal, en communiquant au liquide contenu dans ce canal des vitesses sensiblement pareilles de la surface au fond. *J. Math. Pures et Appl.*, 17: 55–108, 1872.

[40] P. Bressloff and S. Folias. Front bifurcations in an excitatory neural network. *SIAM J. Appl. Math.*, 65:131–151, 2004.

[41] J. Bronski and M. Johnson. The modulational instability for a generalized KdV equation. *Arch. Rat. Mech. Anal.*, 197(2):357–400, 2010.

[42] J. Bronski and Z. Rapti. Modulational instability for nonlinear Schrödinger equations with a periodic potential. *Dyn. Part. Diff. Eq.*, 2(4):335–355, 2005.

[43] J. Bronski, M. Johnson, and T. Kapitula. An index theorem for the stability of periodic traveling waves of KdV type. *Proc. Roy. Soc. Edinburgh: Section A*, 141(6): 1141–1173, 2011.

[44] J. Bronski, M. Johnson, and T. Kapitula. An instability index theory for quadratic pencils and applications. preprint, 2012.

[45] R. Carmona and J. Lacroix. *Spectral Theory of Random Schrödinger Operators: Probability and Its Applications*. Birkhäuser, Boston, 1990.

[46] T. Cazenave and P. Lions. Orbital stabilty of standing waves for some nonlinear Scrödinger equations. *Comm. Math. Phys.*, 85(4):549–561, 1982.

[47] K. Chadan and P. Sabatier. *Inverse Problems in Quantum Scattering Theory*. Springer-Verlag, New York, second edition, 1989.

[48] F. Chardard. Maslov index for solitary waves obtained as a limit of the Maslov index for periodic waves. *C.R. Acad. Sci. Paris, Ser. I*, 345:689–694, 2007.

[49] F. Chardard, F. Dias, and T. Bridges. Fast computation of the Maslov index for hyperbolic periodic orbits. *J. Phys. A: Math. Gen.*, 39:14545–14557, 2006.

[50] F. Chardard, F. Dias, and T. Bridges. On the Maslov index of multipulse homoclinic orbits. *Proc. Royal. Soc. London A*, 465:2897–2910, 2009.

[51] F. Chardard, F. Dias, and T. Bridges. Computing the Maslov index of solitary waves. Part 1: Hamiltonian systems on a 4-dimensional phase space. *Physica D*, 238:1841–1867, 2010.

[52] F. Chardard, F. Dias, and T. Bridges. Computing the Maslov index of solitary waves. Part 2: Phase space with dimension greater than four. *Physica D*, 240:1334–1344, 2011.

[53] C. Chicone and Y. Latushkin. *Evolution Semigroups in Dynamical Systems and Differential Equations*, volume 70 of *Math. Surv. Monogr.* American Mathematical Society, Providence, RI, 1999.

[54] P. Chossat and R. Lauterbach. *Methods in Equivariant Bifurcations and Dynamical Systems*, volume 15 of *Advanced Series in Nonlinear Dynamics*. World Scientific, Singapore, 2000.

[55] M. Chugunova and D. Pelinovsky. On quadratic eigenvalue problems arising in stability of discrete vortices. *Lin. Alg. Appl.*, 431:962–973, 2009.

[56] M. Chugunova and D. Pelinovsky. Count of eigenvalues in the generalized eigenvalue problem. *J. Math. Phys.*, 51(5):052901, 2010.

[57] A. Comech and D. Pelinovsky. Purely nonlinear instability of standing waves with minimal energy. *Comm. Pure Appl. Math.*, 56:1565–1607, 2003.

[58] A. Comech, S. Cuccagna, and D. Pelinovsky. Nonlinear instability of a critical traveling wave in the generalized Korteweg–de Vries equation. *SIAM J. Math. Anal.*, 39: 1–33, 2007.

[59] W.A. Coppel. Dichotomies in stability theory. In *Lecture Notes in Mathematics 629*. Springer-Verlag, New York, 1978.

[60] R. Courant and D. Hilbert. *Methods of Mathematical Physics*, volume 1. Interscience Publishers, New York, 1953.

[61] W. Craig and C. Wayne. Newton's method and periodic solutions of nonlinear wave equations. *Comm. Pure Appl. Math.*, 46:1409–1498, 1993.

[62] W. Craig and W. Wayne. Periodic solutions of nonlinear Schrödinger equations and the Nash–Moser method. In J. Seimenis, editor, *Hamiltonian Mechanics, Integrability, and Chaotic Behavior*. Plenum, New York, 1995.

[63] S. Cuccagna, D. Pelinovsky, and V. Vougalter. Spectra of positive and negative energies in the linearized NLS problem. *Comm. Pure Appl. Math.*, 58:1–29, 2005.
[64] E. Cytrynbaum and J. Keener. Stability conditions for the traveling pulse: Modifying the restitution hypothesis. *Chaos*, 12(3):788–789, 2002.
[65] J. Daleckii and M. Krein. *Stability of Solutions of Differential Equations in Banach Space*, volume 43 of *Trans. Math. Monogr.* American Mathematical Society, Providence, RI, 1974.
[66] B. Deconinck and T. Kapitula. On the spectral and orbital stability of spatially periodic stationary solutions of generalized Korteweg–de Vries equations. submitted.
[67] B. Deconinck and T. Kapitula. The orbital stability of the cnoidal waves of the Korteweg–de Vries equation. *Phys. Letters A*, 374:4018–4022, 2010.
[68] B. Deconinck and M. Nivala. The stability analysis of the periodic traveling wave solutions of the mkdv equation. *Stud. Appl. Math.*, 126:17–48, 2010.
[69] B. Deconinck, D. Pelinovsky, and J. Carter. Transverse instabilities of deep-water solitary waves. *Proc. Royal Soc. A*, 462:2039–2061, 2006.
[70] A. Doelman and T. Kaper. Semi-strong pulse interactions in a class of coupled reaction–diffusion equations. *SIAM J. Appl. Dyn. Sys.*, 2(1):53–96, 2003.
[71] A. Doelman, R. Gardner, and T. Kaper. Stability analysis of singular patterns in the 1-D Gray–Scott model I: a matched asymptotics approach. *Physica D*, 122(1–4): 1–36, 1998.
[72] A. Doelman, R. Gardner, and T. Kaper. A stability index analysis of the 1-D Gray–Scott model. *Memoirs AMS*, 155(737), 2002.
[73] A. Doelman, T. Kaper, and K. Promislow. Nonlinear asymptotic stability of the semi-strong pulse dynamics in a regularized Gierer–Meinhardt model. *SIAM J. Math. Anal.*, 38(6):1760–1787, 2007.
[74] P. Drazin and R. Johnson. *Solitons: An Introduction*. Cambridge University Press, Cambridge, 1989.
[75] M. Eastham. *The Asymptotic Solution of Linear Differential Systems: Applications of the Levinson Theorem*. Clarendon Press, Oxford, 1989.
[76] K. Engel and R. Nagel. *One-Parameter Semigroups for Linear Evolution Equations*. Springer-Verlag, New York, 2000.
[77] J. Evans. Nerve axon equations, I: Linear approximations. *Indiana U. Math. J.*, 21: 877–955, 1972.
[78] J. Evans. Nerve axon equations, II: Stability at rest. *Indiana U. Math. J.*, 22:75–90, 1972.
[79] J. Evans. Nerve axon equations, III: Stability of the nerve impulse. *Indiana U. Math. J.*, 22:577–594, 1972.
[80] J. Evans. Nerve axon equations, IV: The stable and unstable impulse. *Indiana U. Math. J.*, 24:1169–1190, 1975.
[81] L. Evans. *Partial Differential Equations*, volume 19 of *Graduate Studies in Mathematics*. American Mathematical Society, Providence, RI, 1998.
[82] B. Fiedler and A. Scheel. *Spatio-Temporal Dynamics of Reaction–Diffusion Patterns*, Trends in Nonlinear Analysis. Springer-Verlag, Berlin, 2003.
[83] F. Finkel, A. González-López, and M. Rodríguez. A new algebraization of the Lamé equation. *J. Phys. A: Math. Gen.*, 33:1519–1542, 2000.
[84] S. Folias and P. Bressloff. Breathing pulses in an excitatory neural network. *SIAM J. Appl. Dyn. Sys.*, 3:378–407, 2004.
[85] S. Folias and P. Bressloff. Stimulus-locked waves and breathers in an excitatory neural network. *SIAM J. Appl. Math.*, 65:2067–2092, 2005.
[86] R. Froese. Asymptotic distribution of resonances in one dimension. *J. Diff. Eqs.*, 137 (2):251–272, 1997.
[87] R. Froese. Upper bounds for the resonance-counting function of Schrödinger operators in odd dimensions. *Canad. J. Math.*, 50(3):538–546, 1998.

[88] T. Gallay and M. Hărăguş. Orbital stability of periodic waves for the nonlinear Schrödinger equation. *J. Dyn. Diff. Eqns.*, 19:825–865, 2007.
[89] T. Gallay and M. Hărăguş. Stability of small periodic waves for the nonlinear Schrödinger equation. *J. Diff. Eq.*, 234:544–581, 2007.
[90] C. Gardner, J. Greene, M. Kruskal, and R. Miura. Method for solving the Korteweg–de Vries equation. *Phys. Rev. Lett.*, 19(19):1095–1097, 1967.
[91] R. Gardner. On the structure of the spectra of periodic travelling waves. *J. Math. Pures Appl.*, 72:415–439, 1993.
[92] R. Gardner. Spectral analysis of long-wavelength periodic waves and applications. *J. Reine Angew. Math.*, 491:149–181, 1997.
[93] R. Gardner and C.K.R.T. Jones. Travelling waves of a perturbed diffusion equation arising in a phase field model. *Indiana U. Math. J.*, 38(4):1197–1222, 1989.
[94] R. Gardner and C.K.R.T. Jones. A stability index for steady-state solutions of boundary-value problems for parabolic systems. *J. Diff. Eq.*, 91:181–203, 1991.
[95] R. Gardner and C.K.R.T. Jones. Stability of travelling wave solutions of diffusive predator–prey systems. *Trans. AMS*, 327(2):465–524, 1991.
[96] R. Gardner and K. Zumbrun. The gap lemma and geometric criteria for instability of viscous shock profiles. *Comm. Pure Appl. Math.*, 51(7):797–855, 1998.
[97] S. Gatz and J. Herrmann. Soliton propagation in materials with saturable nonlinearity. *J. Opt. Soc. Am. B*, 8(11):2296–2302, 1991.
[98] F. Gesztesy and H. Holden. A unified approach to eigenvalues and resonances of Schrödinger operators using Fredholm determinants. *J. Math. Anal. Appl.*, 123: 181–198, 1987.
[99] F. Gesztesy, C.K.R.T. Jones, Y. Latushkin, and M. Stanislavova. A spectral mapping theorem and invariant manifolds for nonlinear Schrödinger equations. *Indiana U. Math. J.*, 49(1):221–243, 2000.
[100] F. Gesztesy, Y. Laushkin, and K. Makarov. Evans functions, Jost functions, and Fredholm determinants. *Arch. Rat. Mech. Anal.*, 186:361–421, 2007.
[101] F. Gesztesy, Y. Latushkin, and K. Zumbrun. Derivatives of (modified) Fredholm determinants and stability of standing and traveling waves. *J. Math. Pures Appl.*, 9 (2):160–200, 2008.
[102] A. Ghazaryan, Y. Latushkin, S. Schechter, and A. de Souza. Stability of gasless combustion fronts in one-dimensional solids. *Arch. Rat. Mech. Anal.*, 198:981–1030, 2010.
[103] A. Ghazaryan, Y. Latushkin, and S. Schecter. Stability of traveling waves for a class of reaction–diffusion systems that arise in chemical reaction models. *SIAM J. Math. Anal.*, 42:2434–2472, 2010.
[104] A. Ghazaryan, Y. Latushkin, and S. Schechter. Stability of traveling waves for degenerate systems of reaction –diffusion equations. *Indiana U. Math. J.*, 60:443–472, 2011.
[105] H. Goldstein. *Classical Mechanics*. Addison-Wesley, New York, second edition, 1980.
[106] J. Goldstein. *Semigroups of Linear Operators and Applications*. Oxford University Press, New York, 1985.
[107] M. Grillakis. Linearized instability for nonlinear Schrödinger and Klein–Gordon equations. *Comm. Pure Appl. Math.*, 46:747–774, 1988.
[108] M. Grillakis. Analysis of the linearization around a critical point of an infinite-dimensional Hamiltonian system. *Comm. Pure Appl. Math.*, 43:299–333, 1990.
[109] M. Grillakis, J. Shatah, and W. Strauss. Stability theory of solitary waves in the presence of symmetry, I. *J. Funct. Anal.*, 74:160–197, 1987.
[110] M. Grillakis, J. Shatah, and W. Strauss. Stability theory of solitary waves in the presence of symmetry, II. *J. Funct. Anal.*, 94:308–348, 1990.
[111] R. Haberman. *Applied Partial Differential Equations with Fourier Series and Bounday Value Problems*. Pearson, New York, fourth edition, 2004.

[112] J. Hale. *Ordinary Differential Equations*. Robert E. Krieger Publishing, Malabar, FL, second edition, 1980.

[113] J. Härterich, B. Sandstede, and A. Scheel. Exponential dichotomies for linear non-autonomous functional differential equations of mixed type. *Indiana U. Math. J.*, 51:1081–1109, 2002.

[114] P. Hartman. *Ordinary Differential Equations*. John Wiley, New York, 1964.

[115] B. Helffer and J. Sjöstrand. From resolvent bounds to semigroup bounds. arXiv1001.4171v1, 2013.

[116] R. Horn and C. Johnson. *Matrix Analysis*. Cambridge University Press, New York, 1985.

[117] P. Howard. *Pointwise estimates for the stability for scalar conservation laws*. PhD thesis, Indiana University, 1998.

[118] P. Howard and K. Zumbrun. Stability of undercompressive shock profiles. *J. Diff. Eq.*, 225(1):308–360, 2006.

[119] J. Humpherys and K. Zumbrun. Spectral stability of small-amplitude shock profiles for dissipative symmetric hyperbolic–parabolic systems. *Z. Angew. Math. Phys.*, 53(1):20–34, 2002.

[120] J. Humpherys and K. Zumbrun. An efficient shooting algorithm for Evans function calculations in large systems. *Physica D*, 220(2):116–126, 2006.

[121] J. Humpherys and K. Zumbrun. Efficient numerical stability analysis of detonation waves in ZND. *Quart. Appl. Math.*, 70(4):685–703, 2012.

[122] J. Humpherys, B. Sandstede, and K. Zumbrun. Efficient computation of analytic bases in Evans function analysis of large systems. *Numer. Math.*, 103(4):631–642, 2006.

[123] H. Hupkes and B. Sandstede. Stability of pulse solutions for the discrete Fitzhugh–Nagumo system. *Trans. Amer. Math. Soc.*, 365:251–301, 2013.

[124] M. Hărăguş. Stability of periodic waves for the generalized BBM equation. *Rev. Roumaine Maths. Pures Appl.*, 53:445–463, 2008.

[125] M. Hărăguş. Transverse spectral stability of small periodic traveling waves for the KP equation. *Stud. Appl. Math.*, 126:157–185, 2011.

[126] M. Hărăguş and G. Iooss. *Local Bifurcations, Center Manifolds, and Normal Forms in Infinite-Dimensional Dynamical Systems*. Springer, New York, 2011.

[127] M. Hărăguş and T. Kapitula. On the spectra of periodic waves for infinite-dimensional Hamiltonian systems. *Physica D*, 237(20):2649–2671, 2008.

[128] M. Hărăguş, E. Lombardi, and A. Scheel. Spectral stability of wave trains in the kawahara equation. *J. Math. Fluid Mech.*, 8:482–509, 2006.

[129] T. Ivey and S. Lafortune. Spectral stability analysis for periodic traveling wave solutions of NLS and CGL perturbations. *Physica D*, 237:1750–1772, 2008.

[130] R. Jackson and M. Weinstein. Geometric analysis of bifurcation and symmetry breaking in a Gross–Pitaevskii equation. *J. Stat. Phys.*, 116:881–905, 2004.

[131] A. Jensen and M. Melgaard. Perturbation of eigenvalues embedded at a threshold. *Proc. Roy. Soc. Edinburgh*, 132A:163–179, 2002.

[132] M. Johansson and Y. Kivshar. Discreteness-induced oscillatory instabilities of dark solitons. *Phys. Rev. Lett.*, 82(1):85–88, 1999.

[133] M. Johnson. Nonlinear stability of periodic traveling wave solutions of the generalized Korteweg–de Vries equation. *SIAM J. Math. Anal.*, 41(5):1921–1947, 2009.

[134] C.K.R.T. Jones. Stability of the travelling wave solutions of the Fitzhugh–Nagumo system. *Trans. AMS*, 286(2):431–469, 1984.

[135] C.K.R.T. Jones. Instability of standing waves for nonlinear Schrödinger-type equations. *Ergod. Th. & Dynam. Sys.*, 8:119–138, 1988.

[136] C.K.R.T. Jones and J. Moloney. Instability of standing waves in nonlinear optical waveguides. *Phys. Lett. A*, 117:175–180, 1986.

[137] C.K.R.T. Jones and J. Moloney. Stability and instability of nonlinear standing waves in planar optical waveguides. In H. Gibbs, P. Mandel, N. Peyghambarian, and S. Smith, editors, *Optical Bistability III*, volume 8 of *Proceedings in Physics*. Springer-Verlag, New York, 1986.

[138] C.K.R.T. Jones and M. Romeo. Stability of neuronal pulses composed of two concatenated unstable kinks. *Phys. Rev. E*, 63:011904, 2001.

[139] C.K.R.T. Jones, R. Gardner, and T. Kapitula. Stability of travelling waves for non-convex scalar conservation laws. *Comm. Pure Appl. Math.*, 46:505–526, 1993.

[140] C.K.R.T. Jones, Y. Latushkin, and R. Marangell. The Morse and Maslov indices for matrix Hills equations. *Proceedings of Symposia in Pure Mathematics*, preprint, 2013.

[141] T. Kapitula. On the stability of travelling waves in weighted L^∞ spaces. *J. Diff. Eq.*, 112(1):179–215, 1994.

[142] T. Kapitula. Existence and stability of singular heteroclinic orbits for the Ginzburg–Landau equation. *Nonlinearity*, 9(3):669–686, 1996.

[143] T. Kapitula. Stability criterion for bright solitary waves of the perturbed cubic–quintic Schrödinger equation. *Physica D*, 116(1–2):95–120, 1998.

[144] T. Kapitula. The Evans function and generalized Melnikov integrals. *SIAM J. Math. Anal.*, 30(2):273–297, 1999.

[145] T. Kapitula. On the stability of N-solitons in integrable systems. *Nonlinearity*, 20 (4):879–907, 2007.

[146] T. Kapitula. The Krein signature, Krein eigenvalues, and the Krein oscillation theorem. *Indiana U. Math. J.*, 59:1245–1276, 2010.

[147] T. Kapitula and P. Kevrekidis. Linear stability of perturbed Hamiltonian systems: theory and a case example. *J. Phys. A: Math. Gen.*, 37(30):7509–7526, 2004.

[148] T. Kapitula and K. Promislow. Stability indices for constrained self-adjoint operators. *Proc. Amer. Math. Soc.*, 140(3):865–880, 2012.

[149] T. Kapitula and J. Rubin. Existence and stability of standing hole solutions to complex Ginzburg–Landau equations. *Nonlinearity*, 13(1):77–112, 2000.

[150] T. Kapitula and B. Sandstede. A novel instability mechanism for bright solitary-wave solutions to the cubic–quintic Ginzburg–Landau equation. *J. Opt. Soc. Am. B*, 15:2757–2762, 1998.

[151] T. Kapitula and B. Sandstede. Instability mechanism for bright solitary wave solutions to the cubic–quintic Ginzburg–Landau equation. *J. Opt. Soc. Am. B*, 15(11): 2757–2762, 1998.

[152] T. Kapitula and B. Sandstede. Stability of bright solitary wave solutions to perturbed nonlinear Schrödinger equations. *Physica D*, 124(1–3):58–103, 1998.

[153] T. Kapitula and B. Sandstede. Edge bifurcations for near-integrable systems via Evans function techniques. *SIAM J. Math. Anal.*, 33(5):1117–1143, 2002.

[154] T. Kapitula and B. Sandstede. Eigenvalues and resonances using the Evans function. *Disc. Cont. Dyn. Sys.*, 10(4):857–869, 2004.

[155] T. Kapitula and A. Stefanov. A Hamiltonian–Krein (instability) index theory for KdV-like eigenvalue problems. preprint, 2013.

[156] T. Kapitula, P. Kevrekidis, and B. Malomed. Stability of multiple pulses in discrete systems. *Phys. Rev. E*, 63(036604), 2001.

[157] T. Kapitula, P. Kevrekidis, and B. Sandstede. Counting eigenvalues via the Krein signature in infinite-dimensional Hamiltonian systems. *Physica D*, 195(3&4): 263–282, 2004.

[158] T. Kapitula, J. N. Kutz, and B. Sandstede. The Evans function for nonlocal equations. *Indiana U. Math. J.*, 53(4):1095–1126, 2004.

[159] T. Kapitula, P. Kevrekidis, and B. Sandstede. Addendum: Counting eigenvalues via the Krein signature in infinite-dimensional Hamiltonian systems. *Physica D*, 201 (1&2):199–201, 2005.

[160] T. Kapitula, P. Kevrekidis, and Z. Chen. Three is a crowd: Solitary waves in photorefractive media with three potential wells. *SIAM J. Appl. Dyn. Sys.*, 5(4):598–633, 2006.
[161] T. Kapitula, E. Hibma, H.-P. Kim, and J. Timkovich. Instability indices for matrix pencils. preprint, 2013.
[162] T. Kato. *Perturbation Theory for Linear Operators*. Springer-Verlag, Berlin, 1980.
[163] D. Kaup. A perturbation expansion for the Zakharov–Shabat inverse scattering transform. *SIAM J. Appl. Math.*, 31(1):121–133, 1976.
[164] D. Kaup. Perturbation theory for solitons in optical fibers. *Phys. Rev. A*, 42(9):5689–5694, 1990.
[165] D. Kaup and T. Lakoba. The squared eigenfunctions of the massive Thirring model in laboratory coordinates. *J. Math. Phys.*, 37(1):308–323, 1996.
[166] D. Kaup and A. Newell. Evolution equations, singular dispersion relations, and moving eigenvalues. *Adv. Math.*, 31:67–100, 1979.
[167] D. Kaup and J. Yang. Stability and evolution of solitary waves in perturbed generalized nonlinear Schrödinger equations. *SIAM J. Appl. Math.*, 60(3):967–989, 2000.
[168] S. Kawashima and A. Matsumura. Asymptotic stability of travelling wave solutions of systems for one-dimensional gas motion. *Comm. Math. Phys.*, 101:97–127, 1985.
[169] P. Kevrekidis. *The Discrete Nonlinear Schrödinger Equation: Mathematical Analysis, Numerical Computation, and Physical Perspectives*, volume 232 of *Springer Tracts in Modern Physics*. Springer-Verlag, New York, 2009.
[170] P. Kevrekidis and D. Pelinovsky. Discrete vector on-site vortices. *Proc. Royal Soc. A*, 462:2671–2694, 2006.
[171] P. Kevrekidis and M. Weinstein. Dynamics of lattice kinks. *Physica D*, 142(1–2): 113–152, 2000.
[172] P. Kevrekidis, R. Carretero-González, G. Theocharis, D. Frantzeskakis, and B. Malomed. Stability of dark solitons in a Bose–Einstein condensate trapped in an optical lattice. *Phys. Rev. A*, 68(3):035602, 2003.
[173] P. Kevrekidis, D. Pelinovsky, and A. Stefanov. Asymptotic stability of small solitons in the discrete nonlinear Schrödinger equation in one dimension. *SIAM J. Math. Anal.*, 41:2010–2030, 2009.
[174] Y. Kivshar, D. Pelinovsky, T. Cretegny, and M. Peyrard. Internal modes of solitary waves. *Phys. Rev. Lett.*, 80(23):5032–5035, 1998.
[175] Y. Kodama, M. Romagnoli, and S. Wabnitz. Soliton stability and interactions in fibre lasers. *Elect. Lett.*, 28(21):1981–1983, 1992.
[176] R. Kollár. Homotopy method for nonlinear eigenvalue pencils with applications. *SIAM J. Math. Anal.*, 43(2):612–633, 2011.
[177] R. Kollár and P. Miller. Graphical Krein signature theory and Evans–Krein functions. preprint, 2013.
[178] M. Krein. *Topics in Differential and Integral Equations and Operator Theory*, volume 7 of *Operator Theory: Advances and Applications*, pp. 1–98. Birkhäuser, Basel, 1983.
[179] S. Lafortune and J. Lega. Instability of local deformations of an elastic rod. *Physica D*, 182(1–2):103–124, 2003.
[180] S. Lafortune and J. Lega. Spectral stability of local deformations of an elastic rod: Hamiltonian formalism. *SIAM J. Math. Anal.*, 36(6):1726–1741, 2005.
[181] T. Lakoba and D. Kaup. Perturbation theory for the Manakov soliton and its applications to pulse propagation in randomly birefringent fibers. *Phys. Rev. E*, 56(5): 6147–6165, 1997.
[182] Y. Latushkin and A. Pogan. The dichotomy theorem for evolution bi-families. *J. Diff. Eq.*, 245:2267–2306, 2008.
[183] Y. Latushkin and A. Sukhtavey. The algebraic multiplicity of eigenvalues and the Evans function revisited. *Math. Model. Nat. Phenom.*, 5:269–292, 2010.
[184] Y. Latushkin and A. Sukhtayev. The Evans function and the Weyl–Titchmarsh function. *Disc. Cont. Dyn. Sys. Ser. S*, 5:939–970, 2012.

[185] Y. Latushkin and Y. Tomilov. Fredholm differential operators with unbounded coefficients. *J. Diff. Eq.*, 208:388–429, 2005.
[186] Y. Latushkin and V. Yurov. Stability estimates for semigroups on Banach spaces. to appear in *Disc. Cont. Dyn. Sys. Ser. B*, 2013.
[187] Y. Latushkin, A. Pogan, and R. Schnaubelt. Dichotomy and Fredholm properties of evolution equations. *J. Operator Theory*, 58:387–414, 2007.
[188] P. Lax. *Linear Algebra*. John Wiley, New York, 1997.
[189] N. Lebedev. *Special Functions and Their Applications*. Dover, New York, 1972.
[190] V. Ledoux, S. Malham, J. Niesen, and V. Thummler. Computing stability of multidimensional travelling waves. *SIAM J. Appl. Dyn. Sys.*, 8(1):480–507, 2009.
[191] V. Ledoux, S. Malham, and V. Thümmler. Grassmannian spectral shooting. *Math. Comp.*, 79:1585–1619, 2010.
[192] Y. Li and K. Promislow. Structural stability of non–ground-state traveling waves of coupled nonlinear Schrödinger equations. *Physica D*, 124(1–3):137–165, 1998.
[193] Y. Li and K. Promislow. The mechanism of the polarization mode instability in birefringent fiber optics. *SIAM J. Math. Anal.*, 31(6):1351–1373, 2000.
[194] M. Lukas, D. Pelinovsky, and P. Kevrekidis. Lyapunov–Schmidt reduction algorithm for three-dimensional discrete vortices. *Physica D*, 212:339–350, 2008.
[195] Z.-Q. Ma. The Levinson theorem. *J. Phys. A: Math. Gen.*, 39:R625–R659, 2006.
[196] R. MacKay. Stability of equilibria of Hamiltonian systems. In R. MacKay and J. Meiss, editors, *Hamiltonian Dynamical Systems*, pp. 137–153. Adam Hilger, Briston, UK, 1987.
[197] R. MacKay. Movement of eigenvalues of Hamiltonian equilibria under non-Hamiltonian perturbation. *Phys. Lett. A*, 155:266–268, 1991.
[198] J. Maddocks. Restricted quadratic forms and their application to bifurcation and stability in constrained variational principles. *SIAM J. Math. Anal.*, 16(1):47–68, 1985.
[199] J. Maddocks and R. Sachs. On the stability of KdV multi-solitons. *Comm. Pure Appl. Math.*, 46:867–901, 1993.
[200] W. Magnus and S. Winkler. *Hill's Equation*, volume 20 of *Interscience Tracts in Pure and Applied Mathematics*. Interscience, New York, 1966.
[201] S. Malham and J. Niesen. Evaluating the Evans function: order reduction in numerical methods. *Math. Comp.*, 77:159–179, 2008.
[202] J. Marion. *Classical Dynamics of Particles and Systems*. Academic Press, New York, second edition, 1970.
[203] A. Markushevich. *Theory of Functions*. Chelsea Publishing, New York, 1985.
[204] J. Marzuola and M. Weinstein. Long-time dynamics near the symmetry-breaking bifurcation for nonlinear Schrödinger/Gross–Pitaevskii equations. *Disc. Cont. Dyn. Sys.*, 28(4):1505–1554, 2010.
[205] L. Meirovitch. *Analytical Methods in Vibrations*. Macmillan Series in Applied Mechanics, Macmillan, New York, 1967.
[206] J. Miller and M. Weinstein. Asymptotic stability of solitary waves for the regularized long wave equation. *Comm. Pure Appl. Math.*, 49:399–441, 1996.
[207] P. Miller. *Applied Asymptotic Analysis*, volume 75 of *Graduates Studies in Mathematics*, American Mathematical Society, Providence, RI, 2006.
[208] T. Mizumachi and D. Pelinovsky. On the asymptotic stability of localized modes in the discrete nonlinear Schrödinger equation. *Disc. Cont. Dyn. Sys. Series S*, 5: 971–987, 2012.
[209] R. Moore and K. Promislow. Renormalization group reduction of pulse dynamics in thermally loaded optical parametric oscillators. *Physica D*, 206:62–81, 2005.
[210] R. Moore, W. Kath, B. Sandstede, C.K.R.T. Jones, and J. Alexander. Stability of multiple pulses in optical fibers with phase-sensitive amplification and noise. *Optics Comm.*, 195:1–28, 2001.

[211] A. Newell. *Solitons in Mathematics and Physics*, volume 48 of *CBMS-NSF Regional Conference Series in Applied Mathematics*. SIAM, Philadelphia,1985.

[212] M. Oh and B. Sandstede. Evans function for periodic waves on infinite cylindrical domains. *J. Diff. Eq.*, 248(3):544–555, 2010.

[213] M. Oh and K. Zumbrun. Low-frequency stability analysis of periodic traveling-wave solutions of viscous conservation laws in several dimensions. *J. Anal. Appl.*, 25(1):1–21, 2006.

[214] E. Ostrovskaya, Y. Kivshar, D. Skryabin, and W. Firth. Stability of multihump optical solitons. *Phys. Rev. Lett.*, 83(2):296–299, 1999.

[215] K. Palmer. Exponential dichomoties and transversal homoclinic points. *J. Diff. Eq.*, 55(2):225–256, 1984.

[216] K. Palmer. Exponential dichotomies and Fredholm operators. *Proc. Amer. Math. Soc.*, 104(1):149–156, 1988.

[217] J. Pava. Nonlinear stability of periodic traveling wave solutions to the Schrödinger and the modified Korteweg–de Vries equations. *J. Diff. Eq.*, 235(1):1–30, 2007.

[218] J. Pava, J. Bona, and M. Scialom. Stability of cnoidal waves. *Adv. Diff. Eq.*, 11(12): 1321–1374, 2006.

[219] A. Pazy. *Semigroups of Linear Operators and Applications to Partial Differential Equations*. Springer-Verlag, New York, 1983.

[220] R. Pego and M. Weinstein. Eigenvalues, and instabilities of solitary waves. *Phil. Trans. R. Soc. Lond. A*, 340:47–94, 1992.

[221] R. Pego and M. Weinstein. Asymptotic stability of solitary waves. *Comm. Math. Phys.*, 164:305–349, 1994.

[222] D. Pelinovsky. Inertia law for spectral stability of solitary waves in coupled nonlinear Schrödinger equations. *Proc. Royal Soc. London A*, 461:783–812, 2005.

[223] D. Pelinovsky. *Localization in Periodic Potentials*. Cambridge University Press, New York, 2011.

[224] D. Pelinovsky. Spectral stability of nonlinear waves in KdV-type evolution equations. preprint, 2013.

[225] D. Pelinovsky and P. Kevrekidis. Dark solitons in external potentials. *Z. Angew. Math. Phys.*, 59:559–599, 2008.

[226] D. Pelinovsky and P. Kevrekidis. Stability of discrete dark solitons in nonlinear Schrödinger lattices. *J. Phys. A: Math. Gen.*, 41:185206, 2008.

[227] D. Pelinovsky and Y. Kivshar. Stability criterion for multicomponent solitary waves. *Phys. Rev. E*, 62(6):8668–8676, 2000.

[228] D. Pelinovsky and A. Sakovich. Internal modes of discrete solitons near the anti-continuum limit of the dNLS equation. *Physica D*, 240:265–281, 2011.

[229] D. Pelinovsky and A. Stefanov. On the spectral theory and dispersive estimates for a discrete Schrödinger equation in one dimension. *J. Math. Phys.*, 49:113501, 2008.

[230] D. Pelinovsky and A. Stefanov. Asymptotic stability of small gap solitons in the nonlinear Dirac equations. *J. Math. Phys.*, 53:073705, 2012.

[231] D. Pelinovsky and J. Yang. Instabilities of multihump vector solitons in coupled nonlinear Schrödinger equations. *Stud. Appl. Math.*, 115(1):109–137, 2005.

[232] D. Pelinovsky, P. Kevrekidis, and D. Frantzeskakis. Persistence and stability of discrete vortices in nonlinear Schrödinger lattices. *Physica D*, 212:20–53, 2005.

[233] D. Pelinovsky, P. Kevrekidis, and D. Frantzeskakis. Stability of discrete solitons in nonlinear Schrödinger lattices. *Physica D*, 212:1–19, 2005.

[234] L. Perko. *Differential Equations and Dynamical Systems*. Springer-Verlag, New York, second edition, 1996.

[235] R. Plaza and K. Zumbrun. An Evans function approach to spectral stability of small-amplitude shock profiles. *Disc. Cont. Dyn. Syst.*, 10(4):885–924, 2004.

[236] K. Promislow. A renormalization method for modulational stability of quasi-steady patterns in dispersive systems. *SIAM J. Math. Anal.*, 33(6):1455–1482, 2002.

[237] K. Promislow and J. Kutz. Bifurcation and asymptotic stability in the large detuning limit of the optical parametric oscillator. *Nonlinearity*, 13:675–698, 2000.

[238] J. Prüss. On the spectrum of C^0-semigroups. *Trans. Amer. Math. Soc.*, 284:847–857, 1984.

[239] J. Rademacher, B. Sandstede, and A. Scheel. Computing absolute and essential spectra using continuation. *Physica D*, 229(1&2):166–183, 2007.

[240] A. Ramm. Perturbation of resonances. *J. Math. Anal. Appl.*, 88:1–7, 1982.

[241] J. Rauch. Perturbation theory for eigenvalues and resonances of Schrödinger Hamiltonians. *J. Func. Anal.*, 35:304–315, 1980.

[242] M. Reed and B. Simon. *Methods of Modern Mathematical Physics IV: Analysis of Operators*. Academic Press, New York, 1978.

[243] M. Reed and B. Simon. *Methods of Modern Mathematical Physics III: Scattering Theory*. Academic Press, New York, 1979.

[244] M. Reed and B. Simon. *Methods of Modern Mathematical Physics I: Functional Analysis*. Academic Press, New York, 1980.

[245] J. Rubin. Stability, bifurcations and edge oscillations in standing pulse solutions to an inhomogeneous reaction–diffusion system. *Proc. Roy. Soc. Edinburgh A*, 129(5): 1033–1079, 1999.

[246] J. Rubin. A nonlocal eigenvalue problem for the stability of a traveling wave in a neuronal medium. *Disc. Cont. Dyn. Sys. A*, 4:925–940, 2004.

[247] G. Samaey and B. Sandstede. Determining stability of pulses for partial differential equations with time delays. *Dyn. Sys.*, 20:201–222, 2005.

[248] B. Sandstede. Stability of multiple-pulse solutions. *Trans. Amer. Math. Soc.*, 350: 429–472, 1998.

[249] B. Sandstede. Center manifolds for homoclinic solutions. *J. Dyn. Diff. Eq.*, 12: 449–510, 2000.

[250] B. Sandstede. Stability of travelling waves. In *Handbook of Dynamical Systems*, volume 2, chapter 18, pp. 983–1055. Elsevier, New York, 2002.

[251] B. Sandstede. Evans functions and nonlinear stability of travelling waves in neuronal network models. *Int. J. Bif. Chaos*, 17:2693–2704, 2007.

[252] B. Sandstede and A. Scheel. Absolute and convective instabilities of waves on unbounded and large bounded domains. *Physica D*, 145:233–277, 2000.

[253] B. Sandstede and A. Scheel. Gluing unstable fronts and backs together can produce stable pulses. *Nonlinearity*, 13:1465–1482, 2000.

[254] B. Sandstede and A. Scheel. On the stability of travelling waves with large spatial period. *J. Diff. Eq.*, 172:134–188, 2001.

[255] B. Sandstede and A. Scheel. On the structure of spectra of modulated travelling waves. *Math. Nachr.*, 232:39–93, 2001.

[256] B. Sandstede and A. Scheel. Evans function and blow-up methods in critical eigenvalue problems. *Disc. Cont. Dyn. Sys.*, 10:941–964, 2004.

[257] B. Sandstede and A. Scheel. Relative Morse indices, Fredholm indices, and group velocities. *Disc. Cont. Dyn. Syst.*, 20:139–158, 2008.

[258] B. Sandstede, J. Alexander, and C.K.R.T. Jones. Existence and stability of n-pulses on optical fibers with phase-sensitive amplifiers. *Physica D*, 106(1&2):167–206, 1997.

[259] D. Sattinger. On the stability of waves of nonlinear parabolic systems. *Advances in Math.*, 22:312–355, 1976.

[260] M. Schecter. *Fredholm operators and the essential spectrum*. Courant Inst. Math. Sciences, New York U., 1965. (see `http://archive.org/details/fredholmoperator00sche`).

[261] B. Simon. Notes on infinite determinants of Hilbert space operators. *Adv. Math.*, 24:244–273, 1977.

[262] B. Simon. On the absorption of eigenvalues by continuous spectrum in regular perturbation problems. *J. Func. Anal.*, 25:338–344, 1977.

[263] B. Simon. Resonances in one dimension and Fredholm determinants. *J. Func. Anal.*, 178:396–420, 2000.
[264] D. Skryabin. Energy of internal modes of nonlinear waves and complex frequencies due to symmetry breaking. *Phys. Rev. E*, 64(055601(R)), 2001.
[265] J. Smoller. *Shock Waves and Reaction Diffusion Equations*. Springer-Verlag, New York, 1983.
[266] A. Soffer and M. Weinstein. Resonances, radiation damping and instability in Hamiltonian nonlinear wave equations. *Invent. Math.*, 136:9–74, 1999.
[267] A. Soffer and M. Weinstein. Selection of the ground state for nonlinear Schrödinger equations. *Rev. Math. Phys.*, 16(8):977–1071, 2004.
[268] J. Soto-Crespo, N. Akhmediev, and V. Afanasjev. Stability of the pulselike solutions of the quintic complex Ginzburg–Landau equation. *J. Opt. Soc. Am. B*, 13(7): 1439–1449, 1996.
[269] T. Tao. Why are solitons stable? *Bull. Amer. Math. Soc.*, 46(1):1–33, 2009.
[270] R. Temam. *Navier–Stokes equations and nonlinear functional analysis*. CBMS-NSF Regional Conference Series in Applied Mathematics. SIAM, Philadelphia, second edition, 1995.
[271] D. Terman. Stability of planar wave solutions to a combustion model. *SIAM J. Math. Anal.*, 21(5):1139–1171, 1990.
[272] E. Titchmarsh. *Eigenfunction Expansions Associated with Second-Order Differential Equations. Part I*. Oxford University Press, Oxford, UK, edition, 1946.
[273] E. Titchmarsh. *Eigenfunction Expansions Associated with Second-Order Differential Equations. Part II*. Oxford University Press, Oxford, UK, first edition, 1958.
[274] E. Titchmarsh. *The Theory of Functions*. Oxford University Press, Oxford, UK, second edition, 1976.
[275] V. Vougalter and D. Pelinovsky. Eigenvalues of zero energy in the linearized NLS problem. *J. Math. Phys.*, 47:062701, 2006.
[276] C. Wayne. Periodic and quasi-periodic solutions of nonlinear wave equations via KAM theory. *Comm. Math. Phys.*, 127:479–528, 1990.
[277] J. Weidmann. *Spectral Theory of Ordinary Differential Operators*, volume 1258 of *Lecture Notes in Mathematics*. Springer-Verlag, New York, 1987.
[278] J. Weidmann. Spectral theory of Sturm–Liouville operators. Approximation by regular problems. In W. Amrein, A. Hinz, and D. Pearson, editors, *Sturm–Liouville Theory: Past and Present*, pp. 75–98. Birkhäuser, Boston, 2005.
[279] M. Weinstein. Modulational stability of ground states of nonlinear Schrödinger equations. *SIAM J. Math. Anal.*, 16(3):472–491, 1985.
[280] M. Weinstein. Lyapunov stability of ground states of nonlinear dispersive evolution equations. *Comm. Pure Appl. Math.*, 39:51–68, 1986.
[281] S. Wiggins. *Introduction to Applied Nonlinear Dynamical Systems and Chaos*, volume 2 of *Texts in Applied Mathematics*. Springer-Verlag, New York, second edition, 2003.
[282] J. Yang. Complete eigenfunctions of linearized integrable equations expanded around a soliton solution. *J. Math. Phys.*, 41(9):6614–6638, 2000.
[283] J. Yang. Structure of linearization operators of the Korteweg–de Vries heirarchy equations expanded around single-soliton solutions. *Phys. Lett. A*, 279:341–346, 2001.
[284] J. Yang. *Nonlinear Waves in Integrable and Nonintegrable Systems*. SIAM, 2010.
[285] A. Yew. Stability analysis of multipulses in nonlinearly coupled Schrödinger equations. *Indiana U. Math. J.*, 49(3):1079–1124, 2000.
[286] A. Yew, B. Sandstede, and C.K.R.T. Jones. Instability of multiple pulses in coupled nonlinear Schrödinger equations. *Phys. Rev. E*, 61(5):5886–5892, 2000.
[287] L. Zhang. On stability of traveling wave solutions in synaptically coupled neuronal networks. *Diff. Int. Eq.*, 16:513–536, 2003.

[288] L. Zhang. Existence, uniqueness and exponential stability of traveling wave solutions of some integral differential equations arising from neuronal networks. *J. Diff. Eq.*, 197:162–196, 2004.

[289] L. Zhang. Evans functions and bifurcations of standing wave solutions in delayed synaptically coupled neuronal networks. *J. Appl. Anal. Comp.*, 2:213–240, 2012.

[290] K. Zumbrun. Multidimensional stability of planar viscous shock waves. In *Advances in the theory of shock waves*, volume 47 of *Prog. Nonlinear Diff. Eq. Appl.*, pp. 307–516. Birkhäuser Boston, Boston, 2001.

[291] K. Zumbrun. Numerical error analysis for Evans function computations: a numerical gap lemma, centerd-coordinate methods, and the unreasonable effectiveness of continuous orthogonalization. arXiv:0904.0268v2, 2009.

[292] K. Zumbrun. Instantaneous shock location and one-dimensional nonlinear stability of viscous shock waves. *Quart. Appl. Math.*, 69(1):177–202, 2011.

[293] K. Zumbrun. Stability and dynamics of viscous shock waves. In *Nonlinear Conservation Laws and Applications*, volume 153 of *IMA Vol. Math. Appl.*, pp. 123–167. Springer, New York, 2011.

[294] K. Zumbrun. Stability of detonation profiles in the ZND limit. *Arch. Rat. Mech. Anal.*, 200(1):141–182, 2011.

[295] K. Zumbrun and P. Howard. Pointwise semigroup methods and stability of viscous shock waves. *Indiana U. Math. J.*, 47:741–781, 1998.

[296] M. Zworski. Resonances in physics and geometry. *Notices Amer. Math. Soc.*, 46(3): 319–328, 1999.

Index

T. Kapitula and K. Promislow, *Spectral and Dynamical Stability of Nonlinear Waves*, Applied Mathematical Sciences 185, DOI 10.1007/978-1-4614-6995-7,

Printed in the United States
By Bookmasters